Studies in Computational Intelligence

Volume 709

Series editor

Janusz Kacprzyk, Polish Academy of Sciences, Warsaw, Poland
e-mail: kacprzyk@ibspan.waw.pl

About this Series

The series "Studies in Computational Intelligence" (SCI) publishes new developments and advances in the various areas of computational intelligence—quickly and with a high quality. The intent is to cover the theory, applications, and design methods of computational intelligence, as embedded in the fields of engineering, computer science, physics and life sciences, as well as the methodologies behind them. The series contains monographs, lecture notes and edited volumes in computational intelligence spanning the areas of neural networks, connectionist systems, genetic algorithms, evolutionary computation, artificial intelligence, cellular automata, self-organizing systems, soft computing, fuzzy systems, and hybrid intelligent systems. Of particular value to both the contributors and the readership are the short publication timeframe and the worldwide distribution, which enable both wide and rapid dissemination of research output.

More information about this series at http://www.springer.com/series/7092

Sundarapandian Vaidyanathan
Chang-Hua Lien

Editors

Applications of Sliding Mode Control in Science and Engineering

 Springer

Editors
Sundarapandian Vaidyanathan
Research and Development Centre
Vel Tech University
Chennai, Tamil Nadu
India

Chang-Hua Lien
Maritime College and Department of Marine
  Engineering
National Kaohsiung Marine University
Kaohsiung
Taiwan

ISSN 1860-949X   ISSN 1860-9503   (electronic)
Studies in Computational Intelligence
ISBN 978-3-319-85704-6   ISBN 978-3-319-55598-0   (eBook)
DOI 10.1007/978-3-319-55598-0

Printed on acid-free paper

This Springer imprint is published by Springer Nature
The registered company is Springer International Publishing AG
The registered company address is: Gewerbestrasse 11, 6330 Cham, Switzerland

Preface

About the Subject

Sliding mode control (SMC) is a nonlinear control method. The sliding mode control method alters the dynamics of a given dynamical system (linear or nonlinear) by applying a discontinuous control signal that forces the system to "slide" along a cross-section (manifold) of the system's normal behaviour. SMC is a special class of the variable-structure systems (VSS). In sliding mode control method, the state feedback control law is not a continuous function of time. Instead, the state feedback control law can switch from one continuous structure to other structure based on the current position in the state space. For over 50 years, SMC has been extensively studied and widely used in many scientific and industrial applications due to its simplicity and robustness against parameter variations and disturbances. The sliding mode control scheme involves (1) the selecting a hyper-surface or a manifold (i.e. the sliding manifold) such that the system trajectory exhibits desirable behavior when confined to this sliding manifold, and (2) finding feedback gains so that the system trajectory intersects and stays on the sliding manifold. Important types of SMC are classical sliding mode control, integral sliding mode control, higher order sliding mode control, terminal sliding mode control, and super-twisting sliding mode control. The new SMC approaches such as super-twisting sliding mode control show promising dynamical properties such as finite time convergence and chattering alleviation. Sliding mode control has applications in several branches of science and engineering like control systems, chaos theory, mechanical engineering, robotics, electrical engineering, chemical engineering, and network engineering.

About the Book

The new Springer book, *Applications of Sliding Mode Control in Science and Engineering,* consists of 20 contributed chapters by subject experts who are specialized in the various topics addressed in this book. The special chapters have been brought out with a focus on applications of sliding mode control in the broad areas of chaos theory, robotics, electrical engineering, physics, chemical engineering, memristors, mechanical engineering, environmental engineering, finance, and biology. Importance has been given to the chapters offering practical solutions, design and modeling with new types of sliding mode control such as higher order sliding mode control, terminal sliding mode control, super-twisting sliding mode control, and integral sliding mode control.

Objectives of the Book

This volume presents a selected collection of contributions focused on recent advances and applications of sliding mode control in science and engineering. The book focuses on multi-disciplinary applications of SMC in chaos theory, robotics, unmanned aerial vehicles, electrical engineering, physics, chemical engineering, memristors, memristive devices, mechanical engineering, environmental engineering, finance, and biology. These are among those multi-disciplinary applications where computational intelligence has excellent potentials for use. Both novice and expert readers should find this book a useful reference for SMC.

Organization of the Book

This well-structured book consists of 20 full chapters

Book Features

- The book chapters deal with the recent research problems such as applications of SMC.
- The book has contributed chapters by subject experts in sliding mode control.
- The book chapters contain a good literature survey with a long list of references.
- The book chapters are well-written with a good exposition of the research problem, methodology, block diagrams, and simulations.
- The book chapters discuss details of engineering applications and future research areas.

Audience

The book is primarily meant for researchers from academia and industry, who are using SMC in the research areas—electrical engineering, control engineering, robotics, mechanical engineering, computer science, and information technology. The book can also be used at graduate or advanced undergraduate level as a textbook or a major reference for courses such as power systems, control systems, robotics, electrical devices, scientific modeling, and computational science.

Acknowledgements

As the editors, we hope that the chapters in this well-structured book will stimulate further research using SMC and utilize them in multi-disciplinary applications in both science and engineering.

We hope sincerely that this book, covering so many different topics, will be very useful for all readers.

We would like to thank all the reviewers for their diligence in reviewing the chapters.

Special thanks go to Springer, especially the book's Editorial team.

Chennai, India

Kaohsiung, Taiwan

Sundarapandian Vaidyanathan
Professor and Dean
Chang-Hua Lien
Professor and Dean

Contents

Sliding Mode Control Design for Some Classes of Chaotic Systems

Yi-You Hou, Cheng-Shun Fang and Chang-Hua Lien

Abstract In this chapter, the synchronous controls for some classes of chaotic systems (Horizontal Platform, Coronary Artery, Rikitake) are considered and investigated. Sliding mode control is used to solve the synchronization problem of some classes of chaotic systems. The proposed scheme guarantees the synchronization between the master and slave chaotic systems based on the use of Lyapunov stability theory. Moreover, the selection of switching surface and the existence of sliding mode is addressed. Finally, the experimental results validate the proposed chaotic synchronization approach.

Keywords Chaotic systems · Synchronous control · Sliding mode control

1 Introduction

Chaos theory is a branch of the theory of nonlinear system and has been intensively studied in the past four decades. In 1963, E. N. Lorenz presented the first well-known chaotic system, which was a third-order autonomous system with only two multiplication-type quadratic terms but displayed very complex dynamical behaviors [22]. Chaos phenomenon which is a deterministic nonlinear dynamical system has been generally developed over the past two decades. Based on its

Y.-Y. Hou (✉)
Department of Electronic Engineering, Southern Taiwan University
of Science and Technology, Tainan, Taiwan
e-mail: hou.yi_you@msa.hinet.net; yyhou@stust.edu.tw

C.-S. Fang
HAMASTAR Technology Co., Ltd. Products Division, Kaohsiung, Taiwan
e-mail: ezpstmb2013@gmail.com

C.-H. Lien
Department of Marine Engineering, National Kaohsiung Marine University,
Kaohsiung, Taiwan
e-mail: chlien@mail.nkmu.edu.tw

© Springer International Publishing AG 2017
S. Vaidyanathan and C.-H. Lien (eds.), *Applications of Sliding Mode Control in Science and Engineering*, Studies in Computational Intelligence 709,
DOI 10.1007/978-3-319-55598-0_1

particular properties, such as broadband noise-like waveform, and depending sensitively on the system's precise initial conditions, etc. [4, 10, 16, 17, 35]. These properties offer some advantages in secure communication systems [2, 16, 17, 31, 34]. Due to its powerful applications in engineering systems, both control and synchronization problems have extensively been studied in the past decades for chaotic/hyperchaotic systems such as Lorenz system, Chua's system, Rössler system, Chen's system, Lur'e system, Lü system, horizontal platform system, coronary artery system, Rikitake system, and chaotic neural networks [1, 3, 7–9, 11, 14, 15, 21, 24, 26, 27, 30, 32, 33]. Therefore, the synchronization of chaotic circuits for the secure communication has received much attention in the literature [1, 18, 19, 23]. Until now, many control methods for chaotic systems have extensively been studied extensively in the literature, such as linear state observer design, impulsive control, adaptive control, sampled driving signal via Takagi-Sugeno (T-S) fuzzy model, sliding mode control design, etc. [6, 20, 21, 24].

Sliding-mode control (SMC) is a characteristic kind of variable structure systems. In these two decades, sliding-mode control has been a useful and distinctive robust control strategy for many kinds of engineer systems. Depending on the proposed switching surface and discontinuous controller, the trajectories of dynamic systems can be guide to the fixed sliding manifold. The performance under consideration can be achieved. In general, there are two main advantages of SMC which are the reducing order of dynamics from the purposed switching functions and robustness of restraining system uncertainties. Many studies have been conducted on SMC [5, 13, 28]. By designing a switching surface and using a discontinuous control law, the trajectories of dynamic systems can be forced to slide along the fixed sliding manifold. The sliding mode control technique has successfully applied to synchronization of chaotic system [29]. Since SMC technique-based control law is effective and guarantees both the occurrence of sliding motion and synchronization of the master-slave unified chaotic systems [12].

To verify the above systems performance, in this paper a SMC based chaotic secure communication system, which master and slave (Horizontal Platform Systems, Coronary Artery systems, and Rikitake chaotic circuits), are derived to not only guarantee synchronization between the master and slave chaotic systems. The simulations results are presented to show the effectiveness of the proposed method.

The rest of the chapter is organized as follows. The next section provides a problem statement of three nonlinear differential equations/chaotic systems (Horizontal Platform Systems, Coronary Artery systems, and Rikitake systems), which consists the base of this work. In Sect. 3, the proposed SMC controller is designed to synchronize the three classes of master-slave systems. Some numerical examples are given Sect. 4. Finally, the conclusion and some thoughts for future work are presented in Sect. 5. Note that throughout the remainder of this chapter, the notation $|x(t)|$ denotes the modulus of $x(t)$ and $sign(y)$ is sign function of y, if $y > 0$, $sign(y) = 1$; if $y = 0$, $sign(y) = 0$; if $y < 0$, $sign(y) = -1$.

2 Problem Statement

The main goal of synchronization is to force the trajectory of the slave system to be identical to that of the master system. Generally speaking, chaotic systems are described by a set of nonlinear differential equations. Consider the following three nonlinear differential equations/chaotic systems.

(1) Horizontal Platform Systems; HPS [8]

The platform is free to rotate around a horizontal axis through the center itself. HPS has the accelerometer to detect the location. Accelerometer outputs a signal to the brake torque generated by the rotation of the inverse of its balance of HPS. If the platform deviates from the horizon, the torque generated by the reverse rotation of the balance of HPS as shown in Fig. 1.

HPS state equation can be described as follows:

$$A\ddot{x}(t) + D\dot{x}(t) + rg \sin x(t) - 3g/R(B - C) \cos x(t) \cdot \sin x(t) = F \cos \omega t \quad (1)$$

where A, B, C is the moment of inertia axis 1, 2, and 3 platform. D is Damping. R is Radius of the Earth, r is proportionality constant acceleration, g is Gravitational constant. X represents relative to the rotation of the platform on the planet, $F \cos \omega t$ is harmonic torque. Let $x(t) = x_1(t), \dot{x}(t) = x_2(t), A = 1, D = a, \ rg = b, F = h,$ and $3g/R(B - C) = l$. Then the state equations of horizontal platform systems (1) can be represented as follows:

Fig. 1 The horizontal platform system

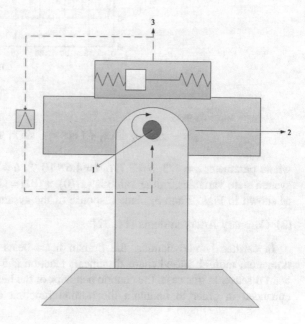

Fig. 2 The state x_1 response of horizontal platform systems (2)

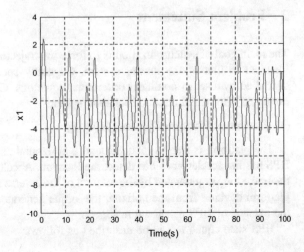

Fig. 3 The state x_2 response of horizontal platform systems (2)

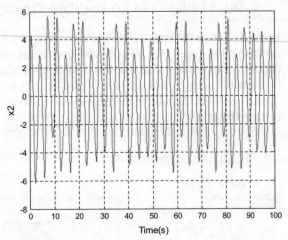

$$\dot{x}_1 = x_2$$
$$\dot{x}_2 = -ax_2 - b\,\sin x_1 + l\,\cos x_1 \cdot \sin x_1 + h\,\cos \omega t. \qquad (2)$$

where parameter $a = 4/3$, $b = 3.776$, $l = 4.6 * 10^{-6}$, $h = 34/3$, $\omega = 1.8$, x_1 and x_2 is system state variables, Initial value is $(x_1(0), x_2(0)) = (1, -1)$. And time response as shown in Figs. 2 and 3, plane response of the system shown in Fig. 4.

(2) Coronary Artery systems [14, 32]

In cardiac hemodynamics, the human heart beats has a certain chaos phenomenon, morbid caused chaos disappears reaction is a sign of some symptoms of heart disease. In this case, the chaotic behavior of the heart should be maintained or enhanced in order to maintain the normal function of the heart. In medicine,

Fig. 4 The state $x_1 - x_2$ responses of horizontal platform systems (2)

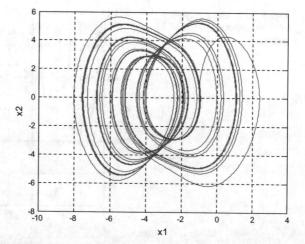

inhibition of disease through drug therapy heartbeat, heart and healthy heart disease synchronization will help medical research in the prevention of coronary artery disease and myocardial infarction and other heart disease reduction. Coronary heart disease, also known as ischemic heart disease, is the most common form of heart disease, because of coronary artery stenosis and insufficiency caused by myocardial dysfunction or organic disease, also called ischemic cardiomyopathy. Blood can't supply to the heart result in severe angina and heart failure will be the most serious can cause death. So in this exhibit chaotic behavior will propose how to coronary heart, with many of the previous studies Chaos in the heart of the dynamic behavior of sorting out the research papers presented to the coronary artery, and proposes synchronous controller, will be proposed for improve the pathological changes in the cardiovascular system and other research methods. According tracking control research paper proposed chaotic coronary system, the heart of the chaotic phenomena can be obtained, and the state equation is expressed as:

$$\dot{x} + bx + cy = 0$$
$$\dot{y} + (\lambda + b\lambda)x + (\lambda + c\lambda)y - \lambda x^3 - E \cos \omega \tau = 0 \tag{3}$$

where x is the amount of change in vessel diameter, y is intravascular pressure differential, τ is time, E is blood vessels suffered periodic external disturbance factor, b, c and λ is system parameters. The so-called chaotic state of blood vessels, means that the blood vessels spasm. Then the state equations of coronary artery systems (3) can be represented as follows:

$$\dot{x} = -0.153x + 1.74y$$
$$\dot{y} = 0.745x - 0.453y - 0.7865x^3 + 0.3 \cos(t) \tag{4}$$

Fig. 5 The state x response of Coronary Artery systems (4)

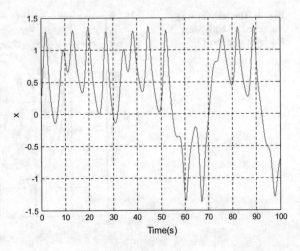

Fig. 6 The state y response of Coronary Artery systems (4)

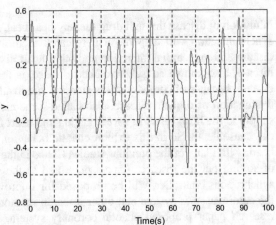

where x and y is system state variables, initial value is $(x(0), y(0)) = (0.2, 0.2)$. Time response as shown in Figs. 5 and 6, response plane view of the system shown in Fig. 7.

(3) Rikitake systems [25]

Many various effective methods have been proposed to explain the origin of the earth's main dipole field, the Rikitake system is a mathematical model obtained from a simple mechanical engineering system used by Rikitake to study the reversals of the Earth's magnetic field [25]. The Rikitake is a two-disk dynamo as shown in Fig. 8 [25], two conducting rotating disks, D_1 (Left) and D_2 (Right) are subjected to a common torque of magnitude G. Both disks D_1 and D_2 rotate in the same sense. Disk D_1 has an angular velocity ω_1 around the axis of rotation A_1 and D_2 has an angular velocity ω_2 around the axis of rotation A_2. A Current loop L_1 is coaxial with disk D_2 with axis A_2 and L_1 is below D_2, while current loop L_2 is

Fig. 7 The state $x -$ y responses of Coronary Artery systems (4)

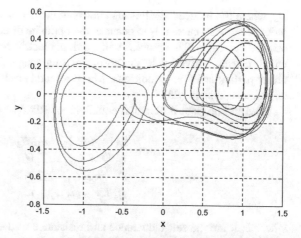

Fig. 8 Rikitake two-disk dynamo

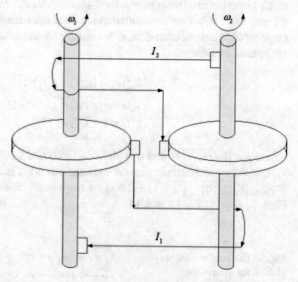

coaxial with disk D_1 with axis A_1 and L_2 is above D_1. Currents 1 I and I_2 both run upwards in the axial wires A_1 and A_2 respectively. From the configuration of Fig. 8 we see that current I_1 runs radially outward in Disk D_1 while current I_2 runs radially inward in Disk D_2. Loop L_1 is connected to Disk D_1 and its axis of rotation A_1 by conducting brushes, similarly loop L_2 is connected to Disk D_2 and its axis of rotation A_2 by conducting brushes. Current loop L_1 causes a magnetic field B_2 to pierce through the rotating disk D_2, and current loop L_2 causes a magnetic field B_1 to pierce through the rotating disk D_1. According to Faraday's law and Lenz's law magnetic field B_1 crossing disk D_1 induces an EMF (electromagnetic field) between the center of D_1 and its rim causing an induced inward current I_1 to occur in the opposite direction to I_1 which reduces the original current I_1. A similar situation happens in disk D_2 and magnetic field B_2, but of opposite polarity to that of B_1 and

D_1, where the induced EMF is also between the center of D_2 and its rim causing an induced outward current I_2 to occur in the opposite direction to I_2 which reduces the original current I_2. This process will result in electric current reversals in both loops which causes a reversal in the corresponding magnetic fields. Chaos shall result under particular initial conditions. Many researchers have discussed the dynamics of Rikitake system [25].

Then the original differential equations of Rikitake system described by

$$L_1 \frac{dI_1}{dt} + R_1 I_1 = \omega_1 M I_2$$
$$L_2 \frac{dI_2}{dt} + R_2 I_2 = \omega_2 M I_1$$
$$C_1 \frac{d\omega_1}{dt} = G_1 - M I_1 I_2 \tag{5}$$
$$C_2 \frac{d\omega_2}{dt} = G_2 - N I_1 I_2$$

where L, R are the self-inductance and resistance of the coil, the electric currents I, ω, C, G are the electric currents, the angular velocity, momentum of inertia, and the driving force, M, N are the mutual inductance between the coils and the disks. If the moment of inertia of each disk, C is considered, the system mathematical model can be written as follows:

$$\dot{x}(t) = -bx(t) + z(t)y(t)$$
$$\dot{y}(t) = -by(t) + (z - a)x(t) \tag{6}$$
$$\dot{z}(t) = 1 - x(t)y(t)$$

where $u = R\sqrt{LC/GM}$ and $a = (\omega_1 - \omega_2)\sqrt{CM/GL}$.

Consider the Rikitake chaotic systems (6) with parameters $b = 1.2$, $a = 3$, and $(x(0), y(0), z(0)) = (1, 1, 1)$. The state responses' simulation results are shown in Figs. 9, 10, 11, 12, 13 and 14.

Fig. 9 The state x response of Rikitake system (6)

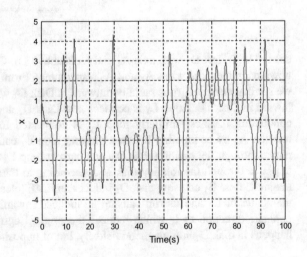

Time(s)

Fig. 10 The state y response of Rikitake system (6)

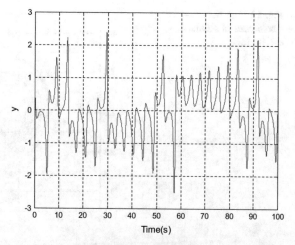

Fig. 11 The state z response of Rikitake system (6)

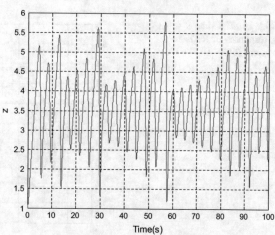

Fig. 12 The state $x - y$ responses of Rikitake system (6)

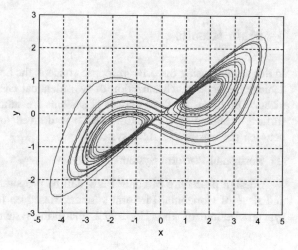

Fig. 13 The state $y -$ z responses of Rikitake system (6)

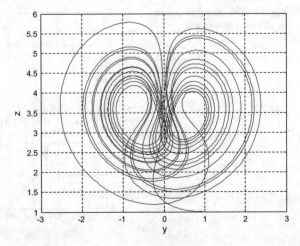

Fig. 14 The state $x -$ z responses of Rikitake system (6)

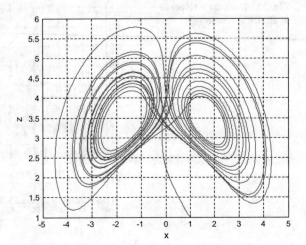

3   Main Results

In this section, using the drive-response concept, the SMC control law is derived to achieve the state synchronization of two identical chaotic systems. The synchronization problems for three classes of systems are studied. Based on the switching surface, a sliding mode control (SMC) is derived to guarantee synchronization between the master and slave systems.

(1) Horizontal Platform Systems; HPS

A design procedure of control law will be proposed to synchronize the master and slave of Horizontal Platform systems. Based on the master-slave concept for synchronization of the Horizontal Platform systems, the master and slave

Horizontal Platform systems are described by the following differential Eqs. (7) and (8), respectively.

Master system:

$$\dot{x}_{m1} = x_{m2}$$
$$\dot{x}_{m2} = -\frac{4}{3}x_{m2} - 3.776 \sin x_{m1}$$
$$+4.6 \times 10^{-6} \cos x_{m1} \cdot \sin x_{m1} + \frac{34}{3}\cos(1.8t) \qquad (7)$$

Slave system:

$$\dot{x}_{s1} = x_{s2}$$
$$\dot{x}_{s2} = -\frac{4}{3}x_{s2} - 3.776 \sin x_{s1}$$
$$+4.6 \times 10^{-6} \cos x_{s1} \cdot \sin x_{s1} + \frac{34}{3}\cos(1.8t) \qquad (8)$$
$$+u+d(t)$$

The initial value is set to $(x_{m1}(0), x_{m2}(0)) = (0.2, 0.2)$, $(x_{s1}(0), x_{s2}(0)) = (-0.3, -0.3)$ and u is set to 0, u is expressed as synchronous controller, $d(t)$ is external interference, $d(t) = 0.1\sin(t)$, and define $e_1(t) = x_{s1}(t) - x_{m1}(t)$, $e_2(t) = x_{s2}(t) - x_{m2}(t)$. The following Figs. 15, 16, 17 and 18 is time response of the master-slave status with controller $u = 0$.

This section aims at proposing a sliding mode controller to synchronize the master and slave systems (7)–(8). To achieve this design goal, the dynamics of synchronization error $e(t) = x_s(t) - x_m(t)$ between the master-slave systems given in (7) and (8) can be described by the following:

Fig. 15 Time responses of state x_{m1} and x_{s1}

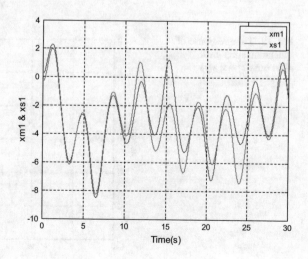

Fig. 16 Time responses of state x_{m2} and x_{s2}

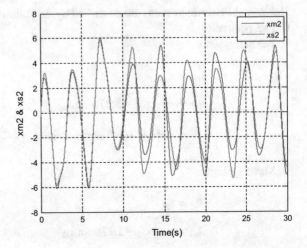

Fig. 17 Time response of synchronization error $e_1(t) = x_{s1}(t) - x_{m1}(t)$

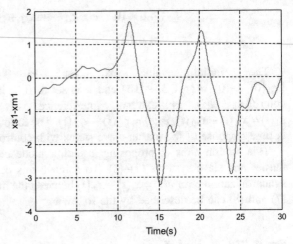

Fig. 18 Time response of synchronization error $e_2(t) = x_{s2}(t) - x_{m2}(t)$

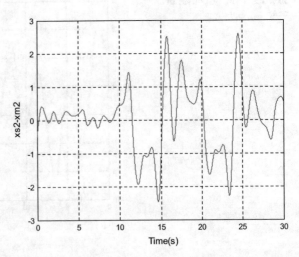

$$e_1 = x_{s1} - x_{m1}$$
$$e_2 = x_{s2} - x_{m2}$$

(9)

and

$$\dot{e}_1 = \dot{x}_{s1} - \dot{x}_{m1} = x_{s2} - x_{m2} = e_2$$
$$\dot{e}_2 = \dot{x}_{s2} - \dot{x}_{m2}$$
$$= -\frac{4}{3}e_2 - 3.776(\sin x_{s1} - \sin x_{m1})$$
$$+ 4.6 \times 10^{-6}(\cos x_{s1} \cdot \sin x_{s1} - \cos x_{m1} \cdot \sin x_{m1})$$
$$+ u + d(t)$$

(10)

Now, the controller u is defined as

$$u = u_{eq} + u_{sw}$$

(11)

and switching surface is define as

$$s = c_1 e_1 + e_2$$

(12)

where c_1 is constant value and the equivalent control u_{eq} in the sliding manifold is obtained by $\dot{s} = 0$. However, the equivalent control u_{eq} cannot obtain the sliding motion if the initial state is not on the switching surface. The control input u_{sw} design for the error dynamics satisfying the reaching condition $(s(t)\dot{s}(t) < 0)$ of sliding motion onto the sliding surface $s(t) = 0$ is derived below.

When the system operates in the sliding mode, it satisfies the following conditions [29]:

$$s(t) = 0 \text{ and } \dot{s} = 0.$$

(13)

Differentiating (13) with respect to time, and then substituting (10) yields

$$\dot{s} = c_1 \dot{e}_1 + \dot{e}_2$$
$$= c_1 e_2 - \frac{4}{3}e_2 - 3.776(\sin x_{s1} - \sin x_{m1})$$
$$+ 4.6 \times 10^{-6}(\cos x_{s1} \cdot \sin x_{s1} - \cos x_{m1} \cdot \sin x_{m1})$$
$$+ u_{eq} + d(t) = 0$$

(14)

Consequently, the equivalent control $u_{eq}(t)$ in the sliding manifold is obtained by $\dot{s} = 0$. The equivalent control $u_{eq}(t)$ in the sliding mode is given by

$$u_{eq} = -c_1 e_2 + \frac{4}{3} e_2 + 3.776(\sin x_{s1} - \sin x_{m1})$$
$$-4.6 \times 10^{-6}(\cos x_{s1} \sin x_{s1} - \cos x_{m1} \sin x_{m1}) \qquad (15)$$
$$-d(t)$$

Next, design hitting control input u_{sw}:

$$u_{sw} = -w \cdot sign(s) \qquad (16)$$

where w is constant value.

Due to the interference signal $d(t)$ is bounded and unknown, so the controller u overall design follows equation:

$$
\begin{aligned}
u &= u_{eq} + u_{sw} \\
&= -c_1 e_2 + \frac{4}{3} e_2 + 3.776(\sin x_{s1} - \sin x_{m1}) \\
&\quad -4.6 \times 10^{-6}(\cos x_{s1} \cdot \sin x_{s1} \\
&\quad - \cos x_{m1} \cdot \sin x_{m1}) - w \cdot sign(s)
\end{aligned}
\qquad (17)
$$

Next, we solve the switching surface such that equivalent sliding mode dynamics (13) is asymptotically stable.

Define the Lyapunov function:

$$V = \frac{1}{2} s^2 \qquad (18)$$

Evaluating the time derivative of $V(t)$ along the trajectory given in Eq. (10) gives:

$$
\begin{aligned}
\dot{V} &= s\dot{s} = s[c_1 \dot{e}_1 + \dot{e}_2] \\
&= s[c_1 e_2 - \frac{4}{3} e_2 - 3.776(\sin x_{s1} - \sin x_{m1}) \\
&\quad + 4.6 \times 10^{-6}(\cos x_{s1} \cdot \sin x_{s1} - \cos x_{m1} \cdot \sin x_{m1}) \\
&\quad + u + d(t)] \\
&= s\{[c_1 e_2 - \frac{4}{3} e_2 - 3.776(\sin x_{s1} - \sin x_{m1}) \\
&\quad + 4.6 \times 10^{-6}(\cos x_{s1} \cdot \sin x_{s1} - \cos x_{m1} \cdot \sin x_{m1})] \\
&\quad + [-c_1 e_2 + \frac{4}{3} e_2 + 3.776(\sin x_{s1} - \sin x_{m1}) \\
&\quad - 4.6 \times 10^{-6}(\cos x_{s1} \cdot \sin x_{s1} - \cos x_{m1} \cdot \sin x_{m1}) \\
&\quad - w \cdot sign(s)] + d(t)\} \\
&= s[d(t) - w \cdot sign(s)] \\
&= s[d(t)] - w \cdot |s|
\end{aligned}
\qquad (19)
$$

Let $d(t)$ is a bounded and satisfy $|d(t)| \leq \gamma$.

$$\begin{aligned} \dot{V} &\leq s[d(t)] - w \cdot |s| \\ &\leq |s| \cdot [|d(t)|] - w \cdot |s| \\ &\leq |s| \cdot (\gamma - w) \end{aligned} \quad (20)$$

Then, there exist a constant $w > \gamma$ such that $\dot{V} < 0$. Based on Lyapunov stability theory, the error dynamics system (9) under the controller $u(t)$ in (17) is asymptotically stable.

(2) Coronary Artery systems

A design procedure of control law will be proposed to synchronize the master and slave of Coronary Artery systems. Based on the master-slave concept for synchronization of the Coronary Artery systems, the master and slave Coronary Artery systems are described by the following differential Eqs. (21) and (22), respectively.

Master system:

$$\begin{aligned} \dot{x}_m &= -0.153 x_m + 1.74 y_m \\ \dot{y}_m &= 0.745 x_m - 0.453 y_m - 0.7865 x_m^3 \\ &\quad + 0.3 \cos(t) \end{aligned} \quad (21)$$

Slave system:

$$\begin{aligned} \dot{x}_s &= -0.153 x_s + 1.74 y_s \\ \dot{y}_s &= 0.745 x_s - 0.453 y_s - 0.7865 x_s^3 \\ &\quad + 0.3 \cos(t) + u + d(t) \end{aligned} \quad (22)$$

The initial value is set to $(x_m(0), y_m(0)) = (0.3, 0.3)$, $(x_s(0), y_s(0)) = (-0.2, -0.2)$ and u is set to 0, u is expressed as synchronous controller, $d(t)$ is external interference, $d(t) = 0.1 \sin(t)$, and define $e_1(t) = x_s(t) - x_m(t)$, $e_2(t) = y_s(t) - y_m(t)$. The time responses of the master and slave systems with controller $u = 0$ are shown in Figs. 19, 20, 21 and 22.

This section aims at proposing a sliding mode controller to synchronize the master and slave systems (21)–(22). To achieve this design goal, the dynamics of synchronization error between the master-slave systems given in (21) and (22) can be described by the following:

Fig. 19 Time response of
state x_m and x_s

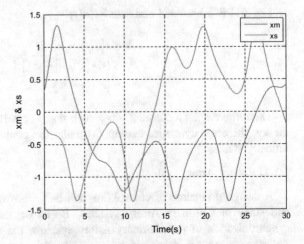

Fig. 20 Time response of
state y_m and y_s

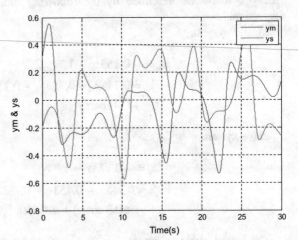

Fig. 21 Time response of
synchronization error
$e_1(t) = x_s(t) - x_m(t)$

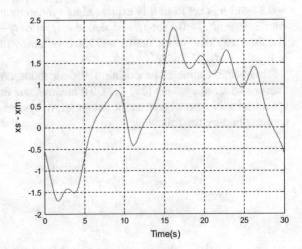

Fig. 22 Time response of
synchronization error
$e_2(t) = y_s(t) - y_m(t)$

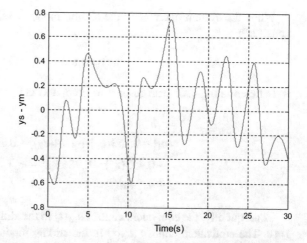

$$e_1 = x_s - x_m$$
$$e_2 = y_s - y_m \qquad (23)$$

and

$$\dot{e}_1 = \dot{x}_s - \dot{x}_m = -0.153(e_1) + 1.74(e_2)$$
$$\dot{e}_2 = \dot{y}_s - \dot{y}_m$$
$$= 0.745(e_1) - 0.453(e_2) - 0.7865(x_s^3 - x_m^3) \qquad (24)$$
$$+ u + d(t)$$

Now, the controller u is defined as

$$u = u_{eq} + u_{sw} \qquad (25)$$

and switching surface is define as

$$s = c_1 e_1 + e_2 \qquad (26)$$

where c_1 is constant value and the equivalent control u_{eq} in the sliding manifold is
obtained by $\dot{s} = 0$. However, the equivalent control u_{eq} cannot obtain the sliding
motion if the initial state is not on the switching surface. The control input u_{sw}
design for the error dynamics satisfying the reaching condition $(s(t)\dot{s}(t) < 0)$ of
sliding motion onto the sliding surface $s(t) = 0$ is derived below.

When the system operates in the sliding mode, it satisfies the following conditions [29]:

$$s(t) = 0 \text{ and } \dot{s} = 0. \tag{27}$$

Differentiating (26) with respect to time, and then substituting (24) yields

$$
\begin{aligned}
\dot{s} &= c_1 \dot{e}_1 + \dot{e}_2 \\
&= c_1[-0.153(e_1) + 1.74(e_2)] + 0.745(e_1) \\
&\quad - 0.453(e_2) - 0.7865(x_s^3 - x_m^3) \\
&\quad + u_{eq} + d(t) = 0
\end{aligned} \tag{28}
$$

Consequently, the equivalent control $u_{eq}(t)$ in the sliding manifold is obtained by $\dot{s} = 0$. The equivalent control $u_{eq}(t)$ in the sliding mode is given by

$$
\begin{aligned}
u_{eq} &= -c_1[-1.53(e_1) + 1.74(e_2)] - 0.745(e_1) \\
&\quad + 0.453(e_2) + 0.7865(x_s^3 - x_m^3) - d(t)
\end{aligned} \tag{29}
$$

Next, design hitting control input u_{sw}:

$$u_{sw} = -w \cdot sign(s) \tag{30}$$

where w is constant value.

Due to the interference signal $d(t)$ is bounded and unknown, so the controller u overall design follows equation:

$$
\begin{aligned}
u &= u_{eq} + u_{sw} \\
&= -c_1[-0.153(e_1) + 1.74(e_2)] - 0.745(e_1) \\
&\quad + 0.453(e_2) + 0.7865(x_s^3 - x_m^3) - w \cdot sign(s)
\end{aligned} \tag{31}
$$

Next, we solve the switching surface such that equivalent sliding mode dynamics (26) is asymptotically stable.

Define the Lyapunov function:

$$V = \frac{1}{2}s^2 \tag{32}$$

Evaluating the time derivative of $V(t)$ along the trajectory given in Eq. (24) gives:

$$
\begin{aligned}
\dot{V} &= s\dot{s} \\
&= s[c_1\dot{e}_1 + \dot{e}_2] \\
&= s\{c_1[-0.153(e_1) + 1.74(e_2)] + 0.745(e_1) \\
&\quad - 0.453(e_2) - 0.7865(x_s^3 - x_m^3) \\
&\quad + u + d(t)\} \\
&= s\{[c_1[-0.153(e_1) + 1.74(e_2)] + 0.745(e_1) \\
&\quad - 0.453(e_2) - 0.7865(x_s^3 - x_m^3) \\
&\quad - c_1[-0.153(e_1) + 1.74(e_2)] - 0.745(e_1) \\
&\quad + 0.453(e_2) + 0.7865(x_s^3 - x_m^3) \\
&\quad - w \cdot sign(s)] + d(t)\} \\
&= s[d(t) - w^*sign(s)] \\
&= s[d(t)] - w \cdot |s|
\end{aligned}
\tag{33}
$$

Let $d(t)$ is a bounded and satisfy $|d(t)| \leq \gamma$.

$$
\begin{aligned}
\dot{V} &\leq s[d(t)] - w \cdot |s| \\
&\leq |s| \cdot [|d(t)|] - w \cdot |s| \\
&\leq |s| \cdot (\gamma - w)
\end{aligned}
\tag{34}
$$

Then, there exist a constant $w > \gamma$ such that $\dot{V} < 0$. Based on Lyapunov stability theory, the error dynamics system (23) under the controller $u(t)$ in (26) is asymptotically stable.

(3) Rikitake systems

A design procedure of control law will be proposed to synchronize the master and slave of Rikitake systems. Based on the master-slave concept for synchronization of the Rikitake systems, the master and slave Rikitake systems are described by the following differential Eqs. (35) and (36), respectively.

Master system:

$$
\begin{aligned}
\dot{x}_m &= -1.2x_m + z_m y_m \\
\dot{y}_m &= -1.2y_m + (z_m - 3)x_m \\
\dot{z}_m &= 1 - x_m y_m
\end{aligned}
\tag{35}
$$

Slave system:

$$\dot{x}_s = -1.2x_s + z_s y_s + u + d(t)$$
$$\dot{y}_s = -1.2y_s + (z_s - 3)x_s \qquad (36)$$
$$\dot{z}_s = 1 - x_s y_s$$

The initial value is set to $x_m(0) = 1, y_m(0) = 1, z_m(0)) = 1$, $x_s(0) = 0.5, y_s(0) = 0.5, z_s(0) = 0.5$ and u is set to 0, u is expressed as synchronous controller, $d(t)$ is external interference, $d(t) = 0.1\sin(t)$, and define $e_1(t) = x_s(t) - x_m(t)$, $e_2(t) = y_s(t) - y_m(t)$, $e_3(t) = z_s(t) - z_m(t)$. The following Figs. 23, 24, 25, 26, 27 and 28 is time response of the master-slave status with controller $u = 0$.

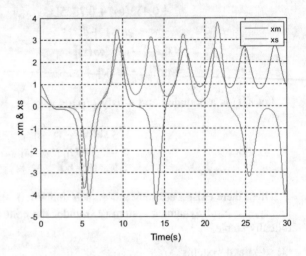

Fig. 23 Time responses of state x_m and x_s

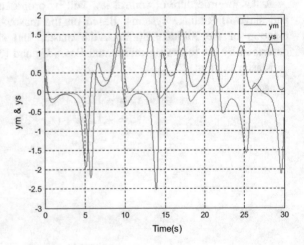

Fig. 24 Time responses of state y_m and y_s

Fig. 25 Time responses of state z_m and z_s

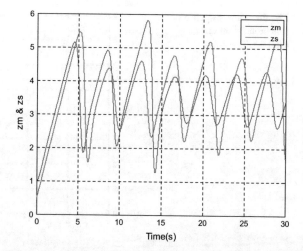

Fig. 26 Time response of synchronization error $e_1(t) = x_s(t) - x_m(t)$

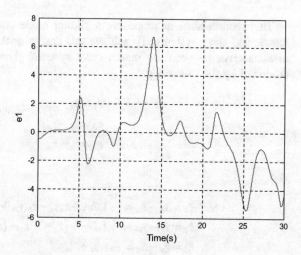

Fig. 27 Time response of synchronization error $e_2(t) = y_s(t) - y_m(t)$

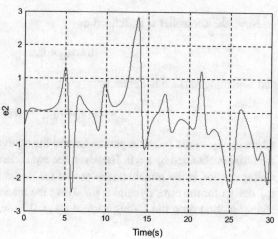

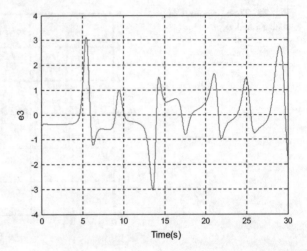

Fig. 28 Time response of synchronization error $e_3(t) = z_s(t) - z_m(t)$

This section aims at proposing a sliding mode controller to synchronize the master and slave systems. To achieve this design goal, the dynamics of synchronization error between the master-slave systems given in (37) and (38) can be described by the following:

$$
\begin{aligned}
e_1 &= x_s - x_m \\
e_2 &= y_s - y_m \\
e_3 &= z_s - z_m
\end{aligned}
\tag{37}
$$

and

$$
\begin{aligned}
\dot{e}_1 &= \dot{x}_s - \dot{x}_m = -1.2e_1 + z_s y_s - z_m y_m + u + d(t) \\
\dot{e}_2 &= \dot{y}_s - \dot{y}_m = -1.2e_2 + (z_s - 3)x_s - (z_m - 3)x_m \\
\dot{e}_3 &= \dot{z}_s - \dot{z}_m = -x_s y_s + x_m y_m
\end{aligned}
\tag{38}
$$

Now, the controller u is defined as

$$
u = u_{eq} + u_{sw}
\tag{39}
$$

and switching surface is define as

$$
s = c_1 e_1 + c_2 e_2 + c_3 e_3
\tag{40}
$$

where c_1, c_2, c_3 are constant parameter and the equivalent control u_{eq} in the sliding manifold is obtained by $\dot{s} = 0$. However, the equivalent control u_{eq} cannot obtain the sliding motion if the initial state is not on the switching surface. The control input u_{sw} design for the error dynamics satisfying the reaching condition $(s(t)\dot{s}(t) < 0)$ of sliding motion onto the sliding surface $s(t) = 0$ is derived below.

When the system operates in the sliding mode, it satisfies the following conditions [29]:

$$s(t) = 0 \text{ and } \dot{s} = 0. \tag{41}$$

Differentiating (40) with respect to time, and then substituting (38) yields

$$\begin{aligned}
\dot{s} &= c_1 \dot{e}_1 + c_2 \dot{e}_2 + c_3 \dot{e}_3 \\
&= c_1(-1.2e_1 + z_s y_s - z_m y_m + u_{eq} + d(t)) \\
&\quad + c_2(-1.2e_2 + (z_s - 3)x_s - (z_m - 3)x_m) \\
&\quad + c_3(-x_s y_s + x_m y_m) = 0
\end{aligned} \tag{42}$$

Consequently, the equivalent control $u_{eq}(t)$ in the sliding manifold is obtained by $\dot{s} = 0$. The equivalent control $u_{eq}(t)$ in the sliding mode is given by

$$\begin{aligned}
u_{eq} &= [-c_1(-1.2e_1 + z_s y_s - z_m y_m + d(t)) \\
&\quad - c_2(-1.2e_2 + (z_s - 3)x_s - (z_m - 3)x_m) \\
&\quad - c_3(-x_s y_s + x_m y_m)]/c_1
\end{aligned} \tag{43}$$

Next, design hitting control input u_{sw}:

$$u_{sw} = -w \cdot sign(s) \tag{44}$$

where w is constant value.

Due to the interference signal $d(t)$ is bounded and unknown, so the controller u overall design follows equation:

$$\begin{aligned}
u &= u_{eq} + u_{sw} \\
&= [-c_1(-1.2e_1 + z_s y_s - z_m y_m) \\
&\quad - c_2(-1.2e_2 + (z_s - 3)x_s \\
&\quad - (z_m - 3)x_m) - c_3(-x_s y_s + x_m y_m)]/c_1 \\
&\quad - w \cdot sign(s)
\end{aligned} \tag{45}$$

Next, we solve the switching surface such that equivalent sliding mode dynamics (37) is asymptotically stable. Define the Lyapunov function:

$$V = \frac{1}{2}s^2 \tag{46}$$

Evaluating the time derivative of $V(t)$ along the trajectory given in Eq. (38) gives:

$$
\begin{aligned}
\dot{V} &= s\dot{s} \\
&= s[c_1\dot{e}_1 + c_2\dot{e}_2 + c_3\dot{e}_3] \\
&= s[c_1(-1.2e_1 + z_s y_s - z_m y_m + u + d(t)) \\
&\quad + c_2(-1.2e_2 + (z_s - 3)x_s - (z_m - 3)x_m) \\
&\quad + c_3(-x_s y_s + x_m y_m)] \\
&= s\{c_1[-1.2e_1 + z_s y_s - z_m y_m \\
&\quad + (-c_1(-1.2e_1 + z_s y_s - z_m y_m) \\
&\quad - c_2(-1.2e_2 + (z_s - 3)x_s - (z_m - 3)x_m) \\
&\quad - c_3(-x_s y_s + x_m y_m))/c_1 - w \cdot sign(s) + d(t)] \\
&\quad + c_2[-1.2e_2 + ((z_s - 3)x_s - (z_m - 3)x_m)] \\
&\quad + c_3[-x_s y_s + x_m y_m]\} \\
&= s[c_1 d(t) - c_1 w \cdot sign(s)] \\
&= s[d(t) - w \cdot sign(s)] \\
&= s[d(t)] - w \cdot |s|
\end{aligned}
\tag{47}
$$

Let $d(t)$ is a bounded and satisfy $|d(t)| \leq \gamma$.

$$
\begin{aligned}
\dot{V} &\leq s[d(t)] - w \cdot |s| \\
&\leq |s| \cdot [|d(t)|] - w \cdot |s| \\
&\leq |s| \cdot (\gamma - w)
\end{aligned}
\tag{48}
$$

Then, there exist a constant $w > \gamma$ such that $\dot{V} < 0$. Based on Lyapunov stability theory, the error dynamics system (37) under the controller $u(t)$ in (45) is asymptotically stable.

4   Illustrative Examples and Simulation Results

To demonstrate the synchronization performance of the proposed control scheme, simulations results of three classes of chaotic systems are given in this section.

(1) Horizontal Platform Systems; HPS

To demonstrate the proposed synchronization approach, we consider the Horizontal Platform system with the following system parameters $c_1 = 8$, $w = 8$. By applying the controller $u(t)$ to (17), the states responses with initial conditions

Fig. 29 State responses of master x_{m1} and slave x_{s1} systems

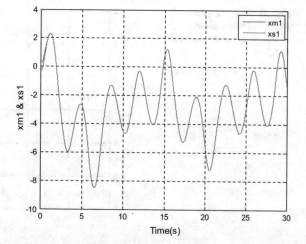

Fig. 30 State responses of master x_{m2} and slave x_{s2} systems

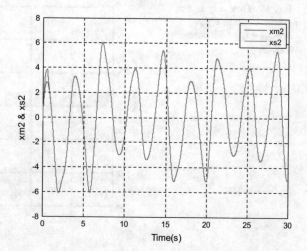

$(x_{m1}(0), x_{m2}(0)) = (0.2, 0.2)$ and $(x_{s1}(0), x_{s2}(0)) = (-0.3, -0.3)$, respectively, is shown in Figs. 29 and 30, The synchronization error between the master system and slave system, respectively, is shown in Figs. 31 and 32. It shows that the synchronization error converges to zero. Figure 33 displays that the sliding mode surface $s(t)$ converge to zero under the proposed control.

(2) Coronary Artery systems

To demonstrate the proposed synchronization approach, we consider the Coronary Artery system with the following system parameters $c_1 = 5$, $w = 5$. By applying the controller $u(t)$ to (31), the state responses with initial conditions $(x(0), y(0)) = (0.2, 0.2)$, respectively, is shown in Figs. 34 and 35, The synchronization error between the master system and slave system, respectively, is shown

Fig. 31 Time response of
synchronization error
$e_1 = x_{s1} - x_{m1}$ with controller
SMC

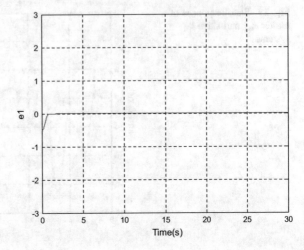

Fig. 32 Time response of
synchronization error
$e_2 = x_{s2} - x_{m2}$ with controller
SMC

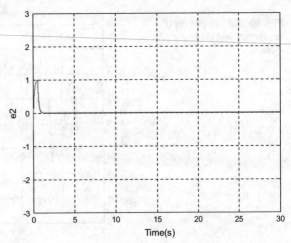

in Figs. 36 and 37. It shows that the synchronization error converges to zero. Figure 38 displays that the sliding mode surface $s(t)$ converge to zero under the proposed control.

(3) Rikitake systems

To demonstrate the proposed synchronization approach, we consider the Rikitake system with the following system parameters $c_1 = 2$, $c_2 = 1$, $c_3 = 1$, $w = 0.3$. By applying the controller $u(t)$ to (45), the states responses with initial conditions $x_m(0) = 1, y_m(0) = 1, z_m(0)) = 1, x_s(0) = 0.5, y_s(0) = 0.5, z_s(0) = 0.5$, respectively, is shown in Figs. 39, 40 and 41, The synchronization error between the master system and slave system, respectively, is shown in Figs. 42, 43 and 44. It shows that the synchronization error converges to zero. Figure 45 displays that the sliding mode surface $s(t)$ converge to zero under the proposed control.

Fig. 33 Simulation results of switch surface $s(t)$

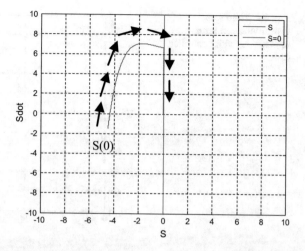

Fig. 34 State responses of master x_m and slave x_s systems

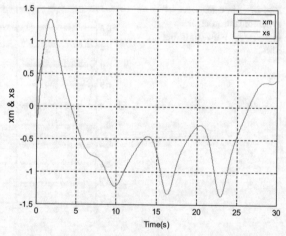

Fig. 35 State responses of master y_m and slave y_s systems

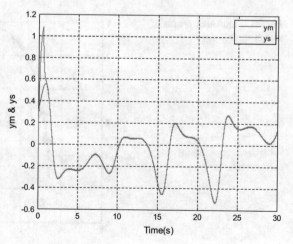

Fig. 36 Time response of
synchronization error
$e_1 = x_s - x_m$ with controller
SMC

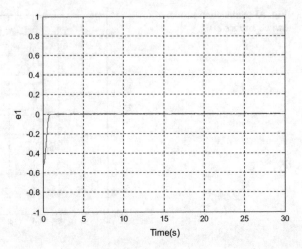

Fig. 37 Time response of
synchronization error
$e_2 = y_s - y_m$ with controller
SMC

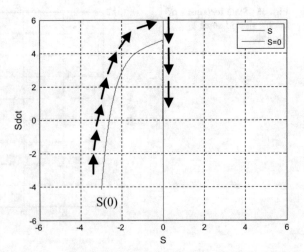

Fig. 38 Simulation results of
switch surface $s(t)$

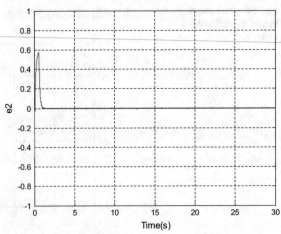

Fig. 39 State responses of master x_m and slave x_s systems

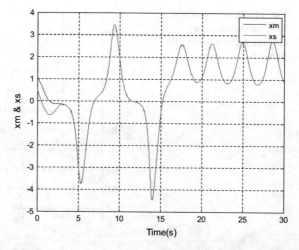

Fig. 40 State responses of master y_m and slave y_s systems

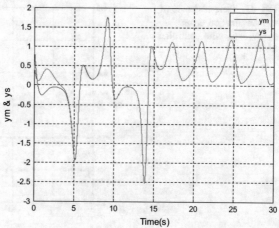

Fig. 41 State responses of master z_m and slave z_s systems

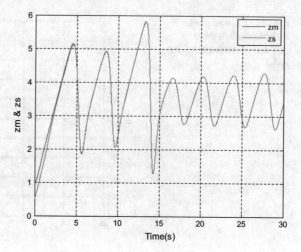

Fig. 42 Time response of synchronization error $e_1 = x_s - x_m$ with controller SMC

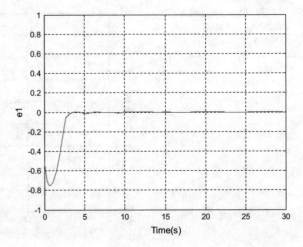

Fig. 43 Time response of synchronization error $e_2 = y_s - y_m$ with controller SMC

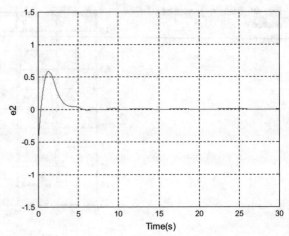

Fig. 44 Time response of synchronization error $e_3 = z_s - z_m$ with controller SMC

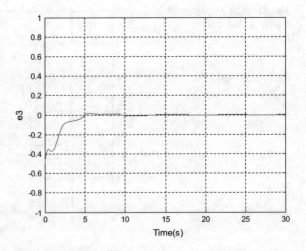

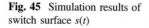

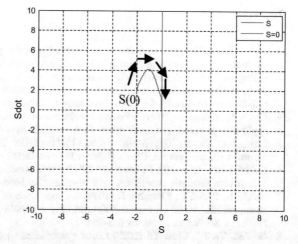

Fig. 45 Simulation results of switch surface $s(t)$

5 Conclusion

In this work, the case for three classes of coupled identical nonlinear chaotic systems (Horizontal Platform Systems, Coronary Artery systems, and Rikitake chaotic circuits) via the proposed SMC was presented. By applying sliding mode control and Lyapunov stability theory, the problem of synchronization problem for three classes of chaotic systems has been investigated. The switching surface is first designed for the considered error dynamics, and then, based on it, a sliding mode controller is derived to guarantee synchronization of both the master and slave chaotic systems. Furthermore, some simulation examples have demonstrated the validity of the proposed theoretical results.

So, some research directions are worthy to be studied in the further. The possible future works are suggested below: (1) Development of other effective synchronization and control schemes such as impulsive control scheme and stochastic synchronization should be further studied to deal with the synchronization of coupled nonlinear chaotic systems. (2) Based on the main synchronization results of this study, future research can also be aimed at lots of applications such as image processing, secure communication, and so on. (3) With some extension and modification, the results in this dissertation may be applied to practical nonlinear dynamical systems. For practical systems, some limitations such as saturation, backlash, and nonlinearities on control input should be further investigated. (4) Some important issues in control system design for the stochastic model may be considered, e.g., guaranteed cost, H_∞ control, tracking control, LQR, and synchronization, etc.

References

1. Agiza HN (2004) Chaos synchronization of *Lü* dynamical system. Nonlinear Anal 58:11–20
2. Almeida DIR, Alvarez J, Barajas JG (2006) Robust synchronization of Sprott circuits using sliding mode control. Chaos, Solitons Fractals 30:11–18
3. Arena P, Caponetto R, Fortuna L, Porto D (1998) Bifurcation and chaos in non-integer order cellular neural networks. Int J Bifurcat Chaos 8:1527–1539
4. Chadli M, Zelinka I (2014) Chaos synchronization of unknown inputs Takagi-Sugeno fuzzy: application to secure communications. Comput Math Appl 68:2142–2147
5. Chang WD, Yan JJ (2005) Adaptive robust PID controller design based on a sliding mode for uncertain chaotic systems. Chaos, Solitons Fractals 26:167–175
6. Elabbasy EM, Agiza HN, El-Dessoky MM (2004) Controlling and synchronization of Rossler system with uncertain parameters. Int J Nonlinear Sci Numer Simul 5:171–181
7. Fallahi K, Raoufi R, Khoshbin H (2008) An application of Chen system for secure chaotic communication based on extended Kalman filter and multi-shift cipher algorithm. Commun Nonlinear Sci Numer Simul 13:763–781
8. Ge ZM, Yu TC, Chen YS (2003) Chaos synchronization of a horizontal platform system. J Sound Vib 268:731–749
9. Gholipour Y, Mola M (2014) Investigation stability of Rikitake system. J Control Eng Technol 4:82–85
10. Gui A, Ge W (2006) Periodic solution and chaotic strange attractor for shunting inhibitory cellular neural networks with impulses. CHAOS 16:033116
11. Hartley TT, Lorenzo CF, Qammer HK (1995) Chaos on a fractional Chua's system. IEEE Trans Circuits Syst I Fundam Theory Appl 42:485–490
12. Hou YY, Liau BY, Chen HC (2012) Synchronization of unified chaotic systems using sliding mode controller. Math Probl Eng 632712:1–10
13. Jang MJ, Chen CC, Chen CO (2002) Sliding mode control of chaos in the cubic Chua's circuit system. Int J Bifurcat Chaos 12:1437–1449
14. Li W (2012) Tracking control of chaotic coronary artery system. Int J Syst Sci 43:21–30
15. Li X, Lu YJ, Zhang QM, Zhang ZY (2016) Circuit implementation and antisynchronization of an improved Lorenz chaotic system. Shock Vib 1–12
16. Li Z, Li K, Wen C, Soh YC (2003) A new chaotic secure communication system. IEEE Trans Commun 51:1306–1312
17. Li Z, Xu D (2004) A secure communication scheme using projective chaos synchronization. Chaos, Solitons Fractal 22:477–481
18. Lian KY, Chiang TS, Chiu CS, Liu P (2001) Synthesis of fuzzy mode-based designs to synchronization and secure communications for chaotic systems. IEEE Trans Syst Man, Cybern Part B 31:66–83
19. Lian KY, Chiu CS, Chiang TS, Liu P (2001) Secure communications of chaotic systems with robust performance via fuzzy observer-based design. IEEE Trans Fuzzy Syst 9:212–220
20. Liao TL, Huang NS (1999) An observer-based approach for chaotic synchronization with applications to secure communications. IEEE Trans. Circuits Syst I 46:1144–1150
21. Liao X, Yu P (2006) Chaos control for the family of Rössler systems using feedback controllers. Chaos, Solitons Fractals 29:91–107
22. Lorenz EN (1963) Deterministic non-periodic flows. J Atmos Sci 20:130–141
23. Park JH (2005) Adaptive synchronization of a unified chaotic system with an uncertain parameter. Int J Nonlinear Sci Numer Simul 6:201–206
24. Rafikov M, Balthazar JM (2004) On an optimal control design for Rössler system. Phys Lett A 333:241–245
25. Rikitake T (1958) Oscillation of a system of disk dynamos. Proc Camb Philos Soc 54:89–105
26. Satnesh S (2015) Single input sliding mode control for hyperchaotic Lu system with parameter uncertainty. Int J Dyn Control 1–11

27. Shi X, Zhu Q (2007) An exact linearization feedback control of CHEN equation. Int J Nonlinear Sci 3:58–62
28. Sun J, Zhang Y (2004) Impulsive control and synchronization of Chua's oscillators. Math Comput Simul 66:499–508
29. Utkin VI (1978) Sliding mode and their applications in variable structure systems. IEEE Trans Autom Control 22:212–222
30. Wang X, Tian L, Yu L (2006) Linear feedback controlling and synchronization of the Chen's chaotic system. Int J Nonlinear Sci 2:43–49
31. Wen G, Wang QG, Lin C, Han X, Li G (2006) Synthesis for robust synchronization of chaotic systems under output feedback control with multiple random delays. Chaos, Solitons Fractals 29:1142–1146
32. Yeh CY, Shiu J, Yau HT (2012) Circuit implementation of coronary artery chaos phenomenon and optimal PID synchronization controller design. Math Probl Eng 1–13
33. Zhang C (2016) Mirror symmetry multi-wing attractors generated from a novel four-dimensional hyperchaotic system. Opt—Int J Light Electron Opt 127:2924–2930
34. Zhou J, Huang HB, Qi GX, Yang P, Xie X (2005) Communication with spatial periodic chaos synchronization. Phys Lett A 335:191–196
35. Zhou L, Wang C, He H, Lin Y (2015) Time-controllable combinatorial inner synchronization and outer synchronization of anti-star networks and its application in secure communication. Commun Nonlinear Sci Numer Simul 22:623–640

Sliding Mode Based Control and Synchronization of Chaotic Systems in Presence of Parametric Uncertainties

Moez Feki

Abstract This chapter deals with the control and synchronization of chaos where both of them are regarded as a case of a control problem. The proposed approach consists in using the sliding mode control theory. We first compare the pioneering OGY control method to the sliding mode control method. We next present a sliding surface design based on the Lyapunov theory. We show that for the class of chaotic systems that can be stabilized using a smooth feedback controller, a sliding manifold can be easily constructed using the Lyapunov function. Besides, we prove that the designed sliding surface is a stable manifold for the originally chaotic system. Thus, should the state behavior be confined to it, then the trajectory will slide towards the equilibrium. We also prove that the proposed controller is robust to mismatched parametric uncertainties. To diminish the effect of the unwanted chattering phenomenon resulting from high sliding gains, an adaptive sliding controller is finally designed to present a robust model independent controller that achieves stabilization of the equilibrium points as well as synchronization of two systems. All these results will be confirmed through numerical simulations on Rossler's system.

Keywords Sliding mode control · Controlling chaos · Parametric uncertainties · Robust control · Adaptive gain · Rossler's system

1 Introduction

Chaotic behavior has been observed in several natural systems such as chaos in the brain [28], cardiac chaos [14], as well as in engineering systems and mainly in power electrical systems [5, 21, 26]. Due to its complexity and unpredictability, designers have more often than not attempted to develop methods to avoid it. During the last three decades, newly developed mathematical and simulation tools incited

M. Feki (✉)
École Supérieure des Sciences et de la Technologie de Hammam Sousse,
University of Sousse, Sousse, Tunisia
e-mail: moez.feki@enig.rnu.tn; wwfekimo@gmail.com

© Springer International Publishing AG 2017
S. Vaidyanathan and C.-H. Lien (eds.), *Applications of Sliding Mode Control in Science and Engineering*, Studies in Computational Intelligence 709,
DOI 10.1007/978-3-319-55598-0_2

35

researchers to develop methods to harness the very peculiar behavior of chaos. While the first developed methods assume total knowledge of the system to be controlled, recently researchers focused on uncertain chaotic systems. Our present work falls within this new stream, where we present a control method that can achieve control and synchronization of chaotic systems with minimal knowledge of the system.

The interest in controlling chaotic systems has known a boost after the pioneering work of Ott, Grebogi and York (OGY) [25]. Since then several strategies to control chaos have been developed, see [2, 4] and the references therein. A widely considered controlling method consists in adding an input control signal to attempt to stabilize an unstable equilibrium point or an unstable periodic orbit. This input control signal can be constructed using linear state feedback [15, 37] or nonlinear state feedback [32, 37]. PI and PD regulators have also been used for chaos control of Chua's system [18, 38]. Recently, a new method to stabilize the unstable periodic orbit has been developed [24]. This method concerns switched chaotic systems and it is based on perturbation of the switching instances. In most of these cited works, full knowledge of the system model is required.

To circumvent such restrictive requirement, fuzzy controller has been used to control uncertain chaotic systems in [3, 34]. In [35] authors presented an extension of the OGY method to be one based on the sliding mode control concept. In [20, 23, 29] authors used sliding mode controllers to suppress chaos in Lorenz and Liu system with parametric uncertainties. Authors in [12, 30, 31] designed adaptive control laws to stabilize and synchronize a new chaotic system with unknown parameters based on adaptive control theory and Lyapunov stability theory. In [7] we suggested a model independent adaptive controller to efficiently control and synchronize Chua's system with cubic nonlinearity. The synchronization of uncertain chaotic systems has also been considered using robust observer design [8, 10, 13].

In this work we address chaos control using sliding mode theory. First, the OGY method is compared to the sliding mode method where the design of the sliding surface rely on the eigenvectors corresponding to the stable eigenvalues of the system Jacobian. Next, we propose a simple method based on Lyapunov theory to construct the sliding surface. We show that if the states are confined to the sliding surface the originally chaotic system becomes asymptotically stable and the trajectory will slide along the stable manifold until it reaches an equilibrium point. Robustness of the proposed controller with respect to modeling uncertainties is also proven. To decrease the chattering behavior of the sliding control, the sliding gain is adjusted adaptively. We finally confirm our results with numerical simulations on Rossler's system.

This chapter is outlined as follows. In Sect. 2, we present the chaotic Rossler's system and a brief analysis of its behavior. In Sect. 3, we compare the OGY control method to the sliding control strategy. In Sect. 4, we present a sliding surface construction method based on the Lyapunov function, we show the controller design method and we prove its robustness and how to apply it for synchronization. In Sect. 5 we illustrate the efficiency of our method using numerical simulations. Section 6 will be devoted to the adaptive sliding controller design. Finally, conclusions and remarks will be brought in the last section.

2 Rossler's System

In this chapter, we will consider Rossler's system as a workhorse although most of the presented results can be applied to several other well known chaotic systems such as Lorenz system, Chen System or Chua system [11].

The defining equations of the Rossler's system are [27]:

$$\dot{x}_1 = -x_2 - x_3 \tag{1}$$
$$\dot{x}_2 = x_1 + ax_2 \tag{2}$$
$$\dot{x}_3 = b + x_3(x_1 - c) \tag{3}$$

where $x = (x_1, x_2, x_3)^T$ is the state vector and $(a, b, c)^T$ is a parameter vector, here the superscript T denotes the transpose of the vector. Rossler studied the system with $a = 0.2$, $b = 0.2$ and $c = 5.7$. In our simulations, we have used $c = 7$. One of the advantages of Rossler's equations is the linear aspect of the first two equations which can be easily studied on the $x_3 = 0$ plane. Indeed, the second order subsystem

$$\dot{x}_1 = -x_2 \tag{4}$$
$$\dot{x}_2 = x_1 + ax_2 \tag{5}$$

has complex eigenvalues with positive real part for $0 < a < 2$, thus the system trajectory will spiral outwards on the (x_1, x_2) plane. Now, if we consider back the x_3 state, then as far as x_1 is small compared to c, the system behavior keeps close to $x_3 = 0$ plane. Once the system orbit approches $x_1 \geq c$, the x_3 variable grows exponentially and reverse the dynamics of x_1 that is the states x_1 and x_2 start to decrease.

To make a further analysis, we may easily determine the system fixed points which are:

$$x_{eq}^1 = \left(\frac{c - \sqrt{c^2 - 4ab}}{2}, \frac{-c + \sqrt{c^2 - 4ab}}{2a}, \frac{c - \sqrt{c^2 - 4ab}}{2a} \right) \tag{6}$$
$$= (0.5719, -2.8595, 2.8595) \times 10^{-2}$$

$$x_{eq}^2 = \left(\frac{c + \sqrt{c^2 - 4ab}}{2}, \frac{-c - \sqrt{c^2 - 4ab}}{2a}, \frac{c + \sqrt{c^2 - 4ab}}{2a} \right) \tag{7}$$
$$= (6.9943, -34.9714, 34.9714)$$

We note that the first fixed point is very close to the origin and according to Fig. 1 it resides in the center of the attractor. Whereas the second fixed point does not belong to the attractor and will not be studied in this work.

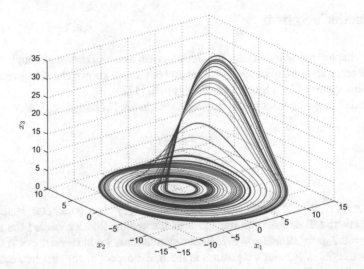

Fig. 1 Rossler's attractor

The behavior of the system in the vicinity of the fixed point can be analyzed through the determination of the eigenvalues and eigenvectors of the Jacobian matrix J.

$$J = \begin{bmatrix} 0 & -1 & -1 \\ 1 & a & 0 \\ x_3^* & 0 & x_1^* - c \end{bmatrix}$$

where x_i^* is the ith component of the fixed point. In our simulations, the eigenvalues are:

$$\lambda_1 = 0.0980 + 0.9951j, \ \lambda_2 = 0.0980 - 0.9951j, \ \lambda_3 = -6.9903$$

and the eigenvectors are:

$$V_1 = \begin{pmatrix} 0.7072 \\ -0.0721 - 0.7033j \\ 0.0028 - 0.0004j \end{pmatrix}, V_2 = \begin{pmatrix} 0.7072 \\ -0.0721 + 0.7033j \\ 0.0028 + 0.0004j \end{pmatrix}, V_3 = \begin{pmatrix} 0.1389 \\ -0.0193 \\ 0.9901 \end{pmatrix}$$

Clearly, the plane defined in the state space by V_1 and V_2 is an unstable manifold and the system orbit will be diverging in its direction. However, the line defined in the state space by V_3 is a stable manifold and the system orbit will be converging in its direction.

The overall behavior thus can be described as follows. An orbit within the Rossler attractor starting near the $x_3 = 0$ plane, spirals outwards around the unstable fixed point and close to the plane. As soon as the spiral is large enough, the orbit leaves the $x_3 = 0$ plane in the positive direction while getting closer to the stable manifold.

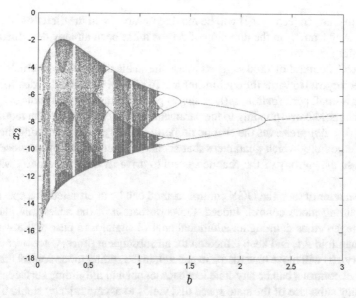

Fig. 2 Bifurcation diagram with varying b

Once the x_1 variable is small enough, the orbit plunges again to the $x_3 = 0$ plane. The behavior is thus dominated by a homoclinic orbit.

In Fig. 2, we show the bifurcation diagram with varying b. It is clear that the system undergoes a series of period doubling bifurcations when the parameter b decreases. We also note the existence of windows of period three behavior within the chaotic region confirming the premise of period three implies chaos [22].

3 The OGY Control Method and Sliding Mode

Consider a continuous time chaotic system defined by:

$$\dot{x} = f(x, p_0),\tag{8}$$

where x is the state vector and p_0 is an accessible parameter vector. We assume that system (8) behaves chaotically for a range of parameter vector p such that $|p - p_0| \leq \Delta p_0$. Let x_0 be an equilibrium point of (8), due to the chaotic nature of the system, x_0 is rather unstable equilibrium and for many well known systems, such as Rossler's system and Chua's system, x_0 is a saddle. Thus, the Jacobian $\frac{\partial f}{\partial x}|_{(x,p)=(x_0,p_0)}$ has stable and unstable eigenvalues respectively denoted λ_s and λ_u. In the case of Rossler system, $\lambda_s = \lambda_3$ and $\lambda_u = (\lambda_1, \lambda_2)$. To these eigenvalues correspond eigenvectors that divide the state space into stable and unstable manifolds $\mathcal{M}_s$ and $\mathcal{M}_u$. It is clear that any trajectory $x(t)$ will be moving towards x_0 in the direction of $\mathcal{M}_s$ and getting away from x_0 in the direction of $\mathcal{M}_u$ as it has been already described earlier for Rossler's system.

The OGY control method suggests to use the stable manifold as a vehicle to drive the trajectory $x(t)$ towards the equilibrium x_0. To achieve this aim, the idea consists in defining a small perturbation $\Delta p(t) < \Delta p_0$ to push the trajectory $x(t)$ onto $\mathcal{M}_s(t)$ (the stable manifold corresponding to the parameter vector $p_0 + \Delta p(t)$). The requirement that $\Delta p(t) < \Delta p_0$ preserves the chaotic property of the system; permitting thereby to use the ergodicity which guarantees that sooner or later $x(t)$ will pass close to $\mathcal{M}_s$ and to use the entropy of the chaotic system to drive its orbit towards x_0 with little efforts.

As a matter of fact, the OGY control method can be interpreted as a special case of the sliding mode control. Indeed, the perturbation of the accessible parameter can be regarded as defining an additional control signal u to push $x(t)$ towards the stable manifold $\mathcal{M}_s$ and keep it thereon for all subsequent times. Once over $\mathcal{M}_s$, the trajectory $x(t)$ will move towards x_0. In general, in the sliding mode control theory the additional control u can be any state feedback signal and the sliding surface $\sigma \equiv \mathcal{M}_s$ can be any subspace of the state space that yields to asymptotically stable behavior if the system is confined to it.

In this work, we attempt to stabilize the equilibrium point of Rossler's system x_{eq}^1. We can easily verify that this equilibrium is saddle type. More precisely, the equilibrium x_{eq}^1 has a stable line and an unstable plane. In $\mathbb{R}^3$ the line is defined by the intersection of planes that we will denote here σ_1 and σ_2.

To stabilize the origin, we will consider a perturbation to the parameter b which will be calculated using the sliding mode theory. However, stabilizing x_{eq}^1 using a scalar control signal is not a straight forward task. In fact, to reach our aim, we need to push the trajectory $x(t)$ towards the planes σ_1 and σ_2 simultaneously. To do so, we suggested to push the trajectory $x(t)$ alternately towards the furthest plane; the trajectory $x(t)$ will finally end up on the intersection, that is on the stable line. Eventually, the trajectory $x(t)$ will slide along that line until it reaches x_{eq}^1. Illustrations of the suggested controller will be presented in Sect. 5.

4   Sliding Mode Controller Design

In this section we present a new method to construct the stable manifold based on the Lyapunov stability theory. In fact, we attempt to avoid the dependence on the eigenvalues and eigenvectors of the Jacobian. The sliding manifold will be kept invariant using a state feedback controller. We show that such construction has several advantages such as robustness to mismatched parametric uncertainties.

4.1 Constructing New Sliding Surface

We consider the following autonomous continuous-time chaotic system:

$$\dot{x} = f(x) , \tag{9}$$

where $x \in \mathbb{R}^n$. An input control signal $u \in \mathbb{R}$ will be applied to stabilize the unstable equilibrium points or to synchronize the system with a master chaotic system. The input signal is injected using a vector field $g(x)$ as follows

$$\dot{x} = f(x) + g(x)u . \tag{10}$$

We suppose that there exists a stabilizing state feedback $\gamma(x)$ such that the closed-loop system

$$\dot{x} = f(x) + g(x) . \gamma(x) , \tag{11}$$

is uniformly asymptotically stable. According to the converse theorem of Lyapunov stability theory [33], there exists a C^∞ Lyapunov function $V(x)$ and class-$\mathcal{K}$ functions $\alpha_i \in \mathcal{K}$ $(i = 1, 2, 3)$ such that

(i) $\alpha_1(\|x\|) \leq V(x) \leq \alpha_2(\|x\|)$
(ii) $\dot{V}(x) \leq -\alpha_3(\|x\|)$

Now if we choose a nonlinear sliding surface of the form,

$$\sigma(x) = \frac{dV(x)}{dx} g(x) = 0 , \tag{12}$$

then on the sliding surface $\sigma(x) = 0$ the following relations hold for system (11).

$$\dot{V}(x) = \frac{dV(x)}{dx} f(x) + \frac{dV(x)}{dx} g(x) . \gamma(x) \tag{13a}$$

$$\dot{V}(x) = \frac{dV(x)}{dx} f(x) \tag{13b}$$

$$\dot{V}(x) \leq -\alpha_3(\|x\|) \tag{13c}$$

Remark 1 Using equalities (13), we can say that should the system orbit be confined to the specifically chosen sliding surface $\sigma(x) = 0$ defined by (12) the controlled system behaves similarly to the uncontrolled one. In addition, the originally autonomous chaotic system (9) becomes asymptotically stable, and admits $V(x)$ as a Lyapunov function as far as the system trajectory is restricted to $\sigma(x) = 0$.

4.2   Constructing Robust Sliding Controller

In order to restrict the states to the sliding surface we should choose a feedback control that ensures the attractivity of the surface and that guarantees the sliding behavior. To accomplish this aim the necessary condition $\sigma\dot{\sigma} < 0$ should be satisfied. One way, is to choose

$$\dot{\sigma} = -W_1\sigma - W_2\text{sign}(\sigma), \quad W_1 > 0, \ W_2 > 0, \tag{14}$$

which yields to

$$\sigma\dot{\sigma} = -W_1\sigma^2 - W_2|\sigma| < 0. \tag{15}$$

Knowing that

$$\dot{\sigma} = \frac{d\sigma}{dx}f(x) + \frac{d\sigma}{dx}g(x).u, \tag{16}$$

and combining with (14) we get

$$u(x) = \frac{-\frac{d\sigma}{dx}f(x) - W_1\sigma(x) - W_2\text{sign}(\sigma(x))}{\frac{d\sigma}{dx}g(x)}. \tag{17}$$

If we apply this feedback control, then the states will be attracted to the surface $\sigma(x) = 0$ and will be confined to it for all subsequent time. The trajectories will then slide along $\sigma(x) = 0$ to the equilibrium point. This inherently means that the equilibrium point should belong to the sliding surface.

From (13) we can deduce that since the effect of the controller vanishes on the sliding surface, then any uncertainties in the system parameters that are used to calculate the feedback law (17) will not affect the behavior of the closed loop system if the states are restricted to the sliding surface.

Fact 1 *The proposed controller (17) is robust to matched and mismatched variations of parameters that are not used to define $\sigma(x) = 0$.*

To prove this fact we consider that the controlled chaotic system is in reality

$$\dot{x} = f(x) + \Delta f(x) + g(x)u, \tag{18}$$

where $f(x)$ is the nominal part of the system and $\Delta f(x)$ represents the uncertainty part. We need to show that the attractivity and sliding condition $\sigma\dot{\sigma} < 0$ is still satisfied.

$$\dot{\sigma} = \frac{d\sigma}{dx}f(x) + \frac{d\sigma}{dx}\Delta f(x) + \frac{d\sigma}{dx}g(x).u \tag{19}$$

applying (17) yields

$$\dot{\sigma} = -W_1\sigma - W_2\text{sign}(\sigma) + \frac{d\sigma}{dx}\Delta f(x) \qquad (20)$$

Let $W_2 > \delta > \|\frac{d\sigma}{dx}\Delta f(x)\|$

$$\sigma\dot{\sigma} < -W_1\sigma^2 - W_2|\sigma| + \delta|\sigma| \qquad (21)$$
$$\sigma\dot{\sigma} < -W_1\sigma^2 - (W_2 - \delta)|\sigma| < 0 \qquad (22)$$

Fact 2 *The controller described by*

$$u(x) = \frac{-W_1\sigma(x) - W_2\text{sign}(\sigma(x))}{\frac{d\sigma}{dx}g(x)} . \qquad (23)$$

is independent of $f(x)$ and stabilizes the equilibrium points of (9).

The proof of this second fact follows directly from the proof of the first fact by choosing $W_2 > \delta > \|\frac{d\sigma}{dx}f(x)\|$.

4.3 Synchronization Using Sliding Mode Controller

The chaotic synchronization problem can be regarded as an output tracking problem [6]. We consider the slave system (10) with the following output

$$y = h(x) , \qquad (24)$$

We suppose that if the output y becomes equal to the output y_m of a master chaotic system then states of the master and slave system will be synchronized. This hypothesis is not restrictive as it can be satisfied by many chaotic systems in literature, such as Rossler and Lorenz system [9] Chua's system [7], and jerk systems [6] with $h(x) = x_1$.

Thus, to obtain synchronization, we want $y(t)$ to be equal to $y_m(t)$. Now let's assume that system (10) with the output (24) has relative degree $\rho = n$ (see [16]). Therefore we can define a nonlinear transformation $z = \phi(x)$

$$z_1 = h(x) , \quad z_{i+1} = L_f^i h(x) , \quad i = 1, \ldots, n-1 .$$

Using z as the new state variable, system (10) becomes

$$\dot{z}_i = z_{i+1} , \qquad i = 1, \ldots, n-1 , \qquad (25a)$$
$$\dot{z}_n = F(z) + G(z)u , \qquad (25b)$$

with $y = z_1$, $F(z) = L_f^n h(\Phi^{-1}(z))$ and $G(z) = L_g L_f^{n-1} h(\Phi^{-1}(z))$. We want to design a feedback control that forces the output y to track y_m. To accomplish this aim we define a tracking error $e = Y_m - z$ where $Y_m = (y_m, \dot{y}_m, \ldots, y_m^{(n-1)})^T$. The error dynamics are described by

$$\dot{e}_i = e_{i+1}, \qquad i = 1, \ldots, n-1, \tag{26a}$$

$$\dot{e}_n = y_m^{(n)} - F(Y_m - e) - G(Y_m - e)u. \tag{26b}$$

It is well known that the smooth controller

$$\gamma(e) = \frac{1}{G(z)}\left(-F(z) + y_m^{(n)} + K^T e\right), \tag{27}$$

with $K = (k_1, k_2, \ldots, k_n)^T$, leads to a linear closed loop system that can be written as $\dot{e} = Ae$

$$\begin{bmatrix} \dot{e}_1 \\ \dot{e}_2 \\ \vdots \\ \dot{e}_{n-1} \\ \dot{e}_n \end{bmatrix} = \begin{bmatrix} 0 & 1 & 0 & \ldots & 0 \\ 0 & 0 & 1 & \ldots & 0 \\ \vdots & \vdots & \ddots & \ddots & \vdots \\ \vdots & \vdots & \vdots & \ddots & 1 \\ -k_1 & -k_2 & -k_3 & \ldots & -k_n \end{bmatrix} \begin{bmatrix} e_1 \\ e_2 \\ \vdots \\ e_{n-1} \\ e_n \end{bmatrix}. \tag{28}$$

Eventually, the problem is solved if K is chosen such that all the roots of the polynomial

$$p(s) = s^n + k_n s^{n-1} + \cdots + k_3 s^2 + k_2 s + k_1$$

have negative real parts. Indeed we have

$$\lim_{t \to \infty} e(t) = 0 \Leftrightarrow \lim_{t \to \infty} z(t) = Y_m,$$

Taking the inverse transformation $x = \phi^{-1}(z)$ and $x_m = \phi^{-1}(Y_m)$, we easily deduce that synchronization is achieved, i.e.

$$\lim_{t \to \infty} x(t) = x_m(t).$$

We now have shown that a smooth stabilizing control law exists for the error dynamic system. Since (28) is a linear system, it is easy to find a Lyapunov function $V(e)$ in the form $V(e) = e^T P e$ where $P > 0$ is a positive definite matrix such that $\dot{V}(e) = -e^T Q e$, $Q > 0$. Next we construct a sliding surface

$$\sigma_e = -\frac{dV(e)}{de_n} G(Y_m - e) = 0.$$

Hence using (17) we obtain an expression for the feedback sliding control law

$$u(e) = \frac{-\frac{d\sigma_e}{de}\mathcal{F}(e) - W_1\sigma_e - W_2\text{sign}(\sigma_e)}{\frac{d\sigma_e}{de}\mathcal{G}(e)} . \tag{29}$$

where

$$\mathcal{F}(e) = \begin{pmatrix} e_2 \\ \vdots \\ e_n \\ y_m^{(n)} - F(Y_m - e) \end{pmatrix}, \qquad \mathcal{G}(e) = \begin{pmatrix} 0 \\ \vdots \\ 0 \\ -G(Y_m - e) \end{pmatrix}$$

$u(e)$ will force trajectories of (26) to slide on the surface σ_e until they reach $e = 0$ and consequently synchronization will be achieved.

5 Illustrative Example

The controlled Rossler system that we will consider in this section is merely similar to the one presented in Sect. 2 with a control signal u added to the dynamics of x_3:

$$\dot{x}_1 = -x_2 - x_3 \tag{30a}$$
$$\dot{x}_2 = x_1 + ax_2 \tag{30b}$$
$$\dot{x}_3 = b + x_3(x_1 - c) + u \tag{30c}$$

System (30) is in the form of (10) and the addition of signal u makes $g(x) = (0, 0, 1)^T$ besides it can be regarded as a perturbation to the parameter b.

5.1 Sliding Mode Control Issued from the OGY Method

Eigenvectors corresponding to x_{eq}^1 have been calculated in Sect. 2 and V_3 defines the stable manifold $\mathcal{M}_s$ which is a line passing through x_{eq}^1. To define σ_1 and σ_2, e just need to choose two vectors V_{31} and V_{32} that may or not be orthogonal to V_3 and to each other. In fact, they just need to be linearly independent. Then V_{31} and V_{32} will be considered as the normals to σ_1 and σ_2 respectively. Therefore we have:

$$\sigma_1(x) = V_{31}^T(x - x_{eq}^1) \tag{31}$$
$$\sigma_2(x) = V_{32}^T(x - x_{eq}^1) \tag{32}$$

and

$$\frac{\sigma_1(x)}{dx}f(x) = V_{31}^T f(x) \qquad\qquad \frac{\sigma_2(x)}{dx}f(x) = V_{32}^T f(x) \qquad (33)$$

$$\frac{\sigma_1(x)}{dx}g(x) = V_{31}(3) \qquad\qquad \frac{\sigma_2(x)}{dx}g(x) = V_{32}(3) \qquad (34)$$

where $V(i)$ is the ith component of vector V. It follows that from the expression of the controller $u(x)$ 17, we need to have a non zero third components in V_{31} and V_{32}. Next, we present two choices for V_{31} and V_{32}.

5.1.1 Choice Leading to First Order Sliding Mode

In this case, we have chosen:

$$V_{31} = \left(\frac{1}{V_3(1)}, \ -\frac{1}{2V_3(2)}, \ -\frac{1}{2V_3(3)} \right)^T \qquad (35)$$

and

$$V_{32} = \left(\frac{\frac{1}{2V_3(2)} - \frac{V_3(2)}{2V_3(3)^2}}{\frac{1}{V_3(1)} + \frac{V_3(1)}{2V_3(3)^2}}, \ 1, \ -\frac{V_3(1)}{V_3(3)}\frac{\frac{1}{2V_3(2)} - \frac{V_3(2)}{2V_3(3)^2}}{\frac{1}{V_3(1)} + \frac{V_3(1)}{2V_3(3)^2}} - \frac{V_3(2)}{V_3(3)} \right)^T \qquad (36)$$

We can easily verify that V_3, V_{31} and V_{32} are orthogonal to each other thus to surfaces σ_1 and σ_2 will also be orthogonal. Applying the control signal $u(x)$ 17 such that the system orbit will be attracted to the furthest surface leads to the results presented in Figs. 3 and 4. First, we need to mention here that we have saturated the controller at $U_{max} = 10$ to avoid large controller values in transients. We have also used $W_1 = 0$ and $W_2 = 5$ as controller gains and the controller has been switched on at time $t = 100$ s.

Clearly, the stabilization has been achieved although the controller signal amplitude is quite large that it cannot be really considered simply as a parameter perturbation. Besides, the chattering phenomenon is dominating the controller signal once the equilibrium is reached.

5.1.2 Choice Leading to Second Order Sliding Mode

In this case, we have chosen:

$$V_{31} = \left(-\frac{1}{V_3(1)}, \ \frac{1}{V_3(2)}, \ 0 \right)^T \qquad (37)$$

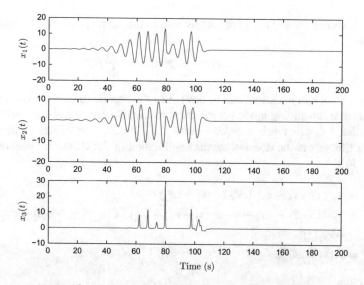

Fig. 3 Stabilization of x_{eq}^1 using first order sliding mode issued from OGY idea

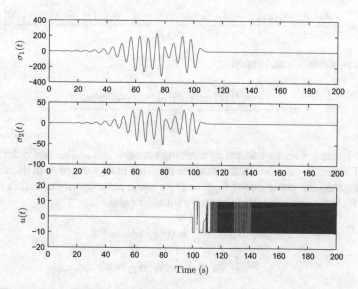

Fig. 4 Time evolution of σ_1 and σ_2 and the sliding mode controller

and

$$V_{32} = \left(\frac{1}{V_3(2)}, \frac{1}{V_3(1)}, -\frac{V_3(1)^2 + V_3(2)^2}{V_3(1)V_3(2)V_3(3)} \right)^T \qquad (38)$$

With this choice 16 will not immediately lead to the expression of $u(x)$ when using σ_1 since

$$\dot{\sigma}_1 = \frac{d\sigma_1}{dx}f(x) + \frac{d\sigma_1}{dx}g(x) \cdot u = \frac{d\sigma_1}{dx}f(x) \tag{39}$$

therefore, to obtain $u(x)$, we need to determine the second time derivative of $\sigma_1(x)$ and thus to reach sliding mode we will need to have $\sigma_1(x) = 0$ and $\dot{\sigma}_1(x) = 0$ which means that we will reach a second order sling mode. One of the main advantages of this fact is the decrease or the annihilation of the chattering phenomenon [1, 17, 36].

$$\begin{aligned}
\ddot{\sigma}_1 &= V_{31}(1)(-\dot{x}_2 - \dot{x}_3) + V_{31}(2)(\dot{x}_1 + a\dot{x}_2) \\
&= V_{31}(1)(-x_1 - ax_2 - b - x_3(x_1 - c)) + V_{31}(2)(-x_2 - x_3 + a(x_1 + ax_2)) \\
&\quad - V_{31}(1)u \\
&= A(x) + B(x)u
\end{aligned}$$

by applying

$$u(x) = \frac{-A(x) - W_{11}\sigma_1(x) - W_{21}\text{sign}(\sigma_1(x)) - W_{12}\dot{\sigma}_1(x) - W_{22}\text{sign}(\dot{\sigma}_1(x))}{B(x)} \tag{40}$$

to make σ_1 attractive and applying

$$u(x) = \frac{-\frac{d\sigma_2}{dx}f(x) - W_1\sigma_2(x) - W_2\text{sign}(\sigma_2(x))}{\frac{d\sigma_2}{dx}g(x)} . \tag{41}$$

to make σ_2 attractive, we achieve intermitting between first and second order sliding mode. The results obtained in this case are depicted in Figs. 5 and 6. In this case, we have saturated the controller at $U_{\max} = 15$ to avoid large controller values in transients. We have also used the following controller gains

$$W_1 = 0.1, \ W_{11} = 0.01, \ W_{12} = 0.2$$

$$W_2 = 40, \ W_{21} = 40, \ W_{22} = 40$$

again the controller has been switched on at time $t = 100\,\text{s}$. On Fig. 6, we clearly notice the discontinuities in the controller signal at the transient time due to the switchings from one controller to another. In addition, we notice the absence of the chattering phenomenon due to the second order sliding mode.

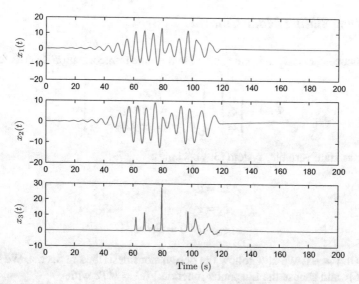

Fig. 5 Stabilization of x_{eq}^1 using second order sliding mode issued from OGY idea

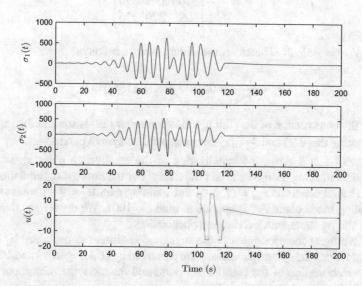

Fig. 6 Time evolution of σ_1 and σ_2 and the sliding mode controller

5.2 New Sliding Mode Control

If we choose $h(x) = x_2$ then we can define the state transformation $\Phi(x)$ as follows:

$$\Phi(x) = \begin{bmatrix} z_1 \\ z_2 \\ z_3 \end{bmatrix} = \begin{bmatrix} x_2 \\ x_1 + ax_2 \\ ax_1 + (a^2 - 1)x_2 - x_3 \end{bmatrix}.$$

Using z as state variable, system (30) becomes:

$$\dot{z}_1 = z_2 , \tag{42a}$$

$$\dot{z}_2 = z_3 , \tag{42b}$$

$$\dot{z}_3 = F(z) + G(z)u . \tag{42c}$$

where $G(z) = -1$. We now follow the design steps described in Sect. 4. We let $K = (8, 12, 6)^T$ and choose the Lyapunov function $V(e) = e^T Pe$ with

$$P = \frac{1}{512} \begin{bmatrix} 4880 & 3160 & 160 \\ 3160 & 5360 & 370 \\ 160 & 370 & 275 \end{bmatrix}$$

so that $\dot{V}(e) = -5\|e\|^2$. Therefore, the sliding surface becomes:

$$\sigma_e = \frac{1}{256}(160e_1 + 370e_2 + 275e_3) = 0$$

Finally the construction of the sliding mode controller (29) is straight forward. It is worth noting that $\sigma(e)$ and $\frac{d\sigma_e}{de}G(e)$ do not depend on system parameters.

Figures 7 and 8 show, the stabilization of the equilibrium point x_{eq}^1 which is regarded as tracking a time-invariant trajectory. For the numerical simulations, we used an initial condition $x_{ini} = (7, 2, 0)$ and a sliding gain $W_2 = 4$ whereas $W_1 = 2$. The sliding mode controller is applied at time $t = 100$ s. We clearly see that when $\sigma_e = 0$, sliding starts with a chattering phenomenon.

To investigate the robustness of the suggested controller as explained by Fact 1, we consider that the system parameters are perturbed by 10%. Figures 9 and 10 delineate the stabilization of the equilibrium point and therefore the robustness of the controller to mismatched perturbation.

Figure 11 shows that the trajectories diverge from each other due to sensitivity to the initial condition $\left(x_{ini}^m = (7, 2, 0.1)\right.$ and $\left. x_{ini}^s = (7, 2, 0)\right)$ and due to mismatched parameters. However, as soon as the controller is applied synchronization is achieved.

In the design of the sliding mode controller (29), the main encountered difficulty is to construct the nonlinear part $\frac{d\sigma_e}{de}F(e)$, due to the nonlinear expressions it could contain and the need for the system parameters. Actually, we have shown that the

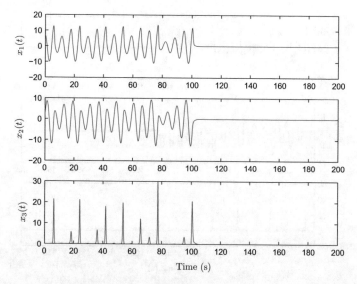

Fig. 7 Stabilization of the unstable equilibrium point x_{eq}^1

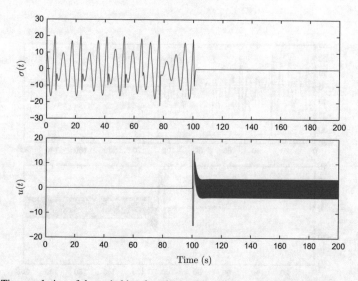

Fig. 8 Time evolution of the switching function and the sliding mode controller

controller is robust to parameters variation. Furthermore, should we choose the sliding gain W_2 such that $\|\frac{d\sigma_e}{de}\mathcal{F}(e)\| < W_2$, then we can use the following sliding controller (as explained by Fact 2):

$$u(e) = \frac{-W_1\sigma_e - W_2\text{sign}(\sigma_e)}{\frac{d\sigma_e}{de}\mathcal{G}(e)} . \tag{43}$$

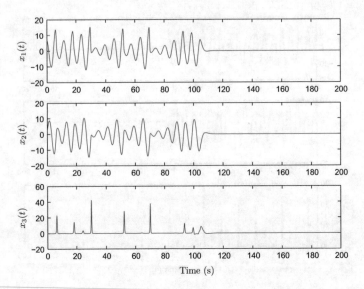

Fig. 9 Stabilization of the unstable equilibrium point x_{eq}^1 in presence of 10% perturbation

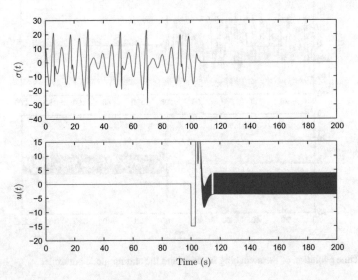

Fig. 10 Time evolution of the switching function and the sliding mode controller in presence of 10% perturbation

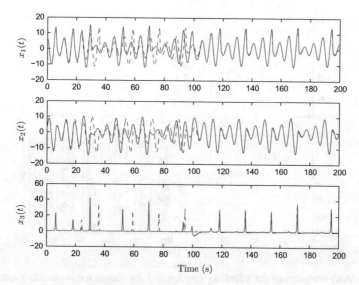

Fig. 11 Synchronization of two Rossler's systems, ($W_2 = 50$ and $W_1 = 10$)

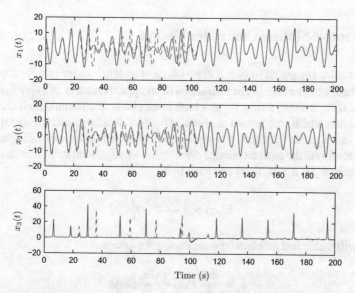

Fig. 12 Synchronization of two Rossler's systems in presence of 10% perturbation, ($W_2 = 150$ and $W_1 = 0$)

Figure 12 depicts the evolution of the state variables and Fig. 13 depicts the sliding controller used to synchronize two Rossler's systems using the robust controller (43) with $W_2 = 150$ and $W_1 = 0$.

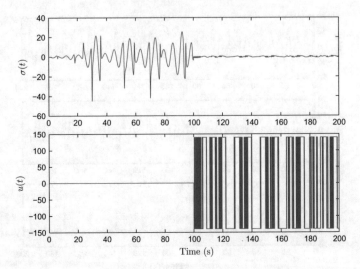

Fig. 13 Time evolution of the switching function and the sliding mode controller with model independent expression

6 Adaptive Sliding Controller Design

The controller suggested in (43) though efficient to synchronize two Rossler's systems, it has two drawbacks: The estimation of an upper boundary of $\|\frac{d\sigma_e}{de}\mathcal{F}(e)\|$ is not a simple task, besides, fixing W_2 at a high level yields to unwanted high chattering phenomenon which was indeed at about 150 which is a large control signal due to the fact that it needs to control a large signal $x_3 \approx 30$ with relatively fast dynamics.

To circumvent the aforementioned drawbacks, we suggest to substitute an adapted sliding gain $W_2(t)$ for the fixed one such that

$$W_2(t) > \left\| \frac{d\sigma_e}{de}\mathcal{F}(e) \right\|. \tag{44}$$

We will tackle our problem by considering the system

$$\dot{\sigma}_e = \frac{d\sigma_e}{de}\mathcal{F}(e) + \frac{d\sigma_e}{de}\mathcal{G}(e)u \tag{45}$$

and design the signal u such that $\sigma_e = 0$ is attained. This can be obtained by applying an adaptive linear controller

$$u_{ad}(\sigma_e) = \tilde{\kappa}(t)\sigma_e(t). \tag{46}$$

The adaptive linear controller is guaranteed to yield $\sigma_e = 0$ if it mimics the following linearizing controller [6]

$$u_{lin}(\sigma_e) = \frac{-\frac{d\sigma_e}{de}\mathcal{F}(e) - \kappa\sigma_e}{\frac{d\sigma_e}{de}\mathcal{G}(e)} \qquad \kappa > 0. \tag{47}$$

That is in the limit we have

$$u_{ad}^*(\sigma_e) = \tilde{\kappa}^*(t)\sigma_e(t) = u_{lin}(\sigma_e)$$

where $\tilde{\kappa}^*(t)$ is the constant limit of $\tilde{\kappa}(t)$

Fact 3 *The adaptive linear controller described in (46) where*

$$\tilde{\kappa}(t) = -\int_0^t \left(\gamma\frac{d\sigma_e}{de}\mathcal{G}(e)\sigma_e^2\right)dt, \qquad \gamma > 0 \tag{48}$$

yields to $\sigma_e = 0$.

To prove this fact, we proceed as follows:

$$
\begin{aligned}
\dot{\sigma}_e &= \frac{d\sigma_e}{de}\mathcal{F}(e) + \frac{d\sigma_e}{de}\mathcal{G}(e)u_{ad} \\
&= \frac{d\sigma_e}{de}\mathcal{F}(e) + \frac{d\sigma_e}{de}\mathcal{G}(e)u_{lin} - \frac{d\sigma_e}{de}\mathcal{G}(e)(u_{lin} - u_{ad}) \\
&= -\kappa\sigma_e - \frac{d\sigma_e}{de}\mathcal{G}(e)(u_{lin} - u_{ad}) \\
&= -\kappa\sigma_e - \frac{d\sigma_e}{de}\mathcal{G}(e)\left(\tilde{\kappa}^*(t) - \tilde{\kappa}(t)\right)\sigma_e
\end{aligned}
$$

Now we consider a Lyapunov function candidate

$$V(\sigma_e, \tilde{\kappa}) = \frac{1}{2}\sigma_e^2 + \frac{1}{2\gamma}\left(\tilde{\kappa}^* - \tilde{\kappa}\right)^2$$

$$
\begin{aligned}
\dot{V}(\sigma_e, \tilde{\kappa}) &= \sigma_e\dot{\sigma}_e - \frac{1}{\gamma}\left(\tilde{\kappa}^* - \tilde{\kappa}\right)\dot{\tilde{\kappa}} \\
&= -\kappa\sigma_e^2 - \left(\tilde{\kappa}^* - \tilde{\kappa}\right)\left(\frac{d\sigma_e}{de}\mathcal{G}(e)\sigma_e^2 + \frac{1}{\gamma}\dot{\tilde{\kappa}}\right)
\end{aligned}
$$

Therefore, if we choose $\tilde{\kappa}$ as in (48), the time derivative of the Lyapunov function becomes

$$\dot{V}(\sigma_e, \tilde{\kappa}) = -\kappa\sigma_e^2 < 0, \forall \sigma_e \neq 0$$

We can deduce that

$$\dot{\sigma}_e = -\kappa\sigma_e - \frac{d\sigma_e}{de}\mathcal{G}(e)\big(\tilde{\kappa}^*(t) - \tilde{\kappa}(t)\big)\sigma_e \tag{49}$$

is stable and hence $\sigma_e \in L_\infty$. Moreover, we have

$$\int_0^t \sigma_e^2\, dt = \frac{V(0) - V(t)}{\kappa}.$$

Since $V(t) \in L_\infty$ and $V(0)$ is finite, this implies that $\sigma_e \in L_2$. Also from (49) we obviously have $\dot{\sigma}_e \in L_\infty$ in addition to $\sigma_e \in L_\infty$ and $\sigma_e \in L_2$. Eventually by Barbalat's lemma [19] $\lim\limits_{t\to\infty}\sigma_e(t) = 0$. This ends the proof of Fact 3.

Now since by construction of u_{ad}, when $\tilde{\kappa}(t)$ reaches its limit we have

$$\tilde{\kappa}(t)\sigma_e(t) = u_{lin}(\sigma_e)$$

then using (46) and (47) we get

$$\frac{d\sigma_e}{de}\mathcal{F}(e) = -\Big(\frac{d\sigma_e}{de}\mathcal{G}(e)\tilde{\kappa}(t) + \kappa\Big)\sigma_e \tag{50}$$

Finally, to satisfy the sliding condition we need to choose the adaptive sliding gain as follows

$$W_2(t) = \Big|\frac{d\sigma_e}{de}\mathcal{G}(e)\tilde{\kappa}(t) + \kappa\Big|.|\sigma_e| + \eta, \quad \eta > 0. \tag{51}$$

Eventually, the adaptive sliding controller is given by the following expression

$$u(e) = \frac{-W_1\sigma_e - W_2(t)\text{sign}(\sigma_e)}{\frac{d\sigma_e}{de}\mathcal{G}(e)}. \tag{52}$$

where $W_2(t)$ is given in (51). We can verify that this controller does not depend on the system parameter since the controller was injected with a constant function $g(x) = (0, 0, 1)^T$.

Figures 14 and 15 depict the obtained results when applying the adaptive sliding controller on Rossler's system. The design parameters are

$$\kappa = 1, \quad \eta = 10, \quad \gamma = 5 \times 10^{-8}$$

Figure 14 shows that synchronization is achieved with an adaptive sliding controller that does not induce high chattering phenomenon. Figure 15 depicts the evolution of the sliding function and of the sliding gain $W_2(t)$. We notice that this gain is large when the controller action starts, as soon as the synchronization is attained, the

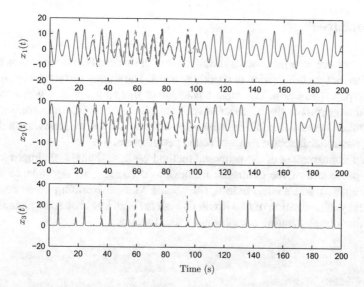

Fig. 14 Synchronization of two Rossler's systems using sliding mode controller with adaptive gain $W_2(t)$

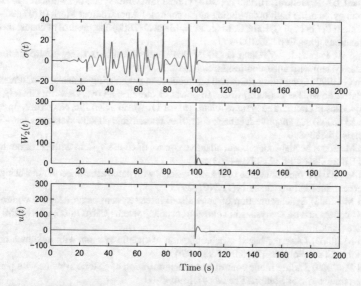

Fig. 15 Time evolution of the switching function, the adaptive sliding gain $W_2(t)$ and the adaptive sliding mode controller with model independent expression

value of $W_2(t)$ decreases and it is less than 40 at $t = 102$ s and less than 5 at $t = 105$ s which is relatively low compared to $W_2 = 150$ set for a non-adaptive robust controller. Therefore, a low chattering phenomenon is obtained.

7 Conclusion

In this work, we have proposed a sliding controller that can be used for chaos control as well as for synchronization of chaotic systems. The sliding surface design is based on a Lyapunov function. The main advantage of this choice is that when confined to the sliding surface the behavior of the controlled system becomes similar to the uncontrolled system. In addition, with this particular choice, we have shown that the sliding controller is robust to mismatched perturbation.

As an improvement of the proposed method we have thought of simplifying the design procedure of the controller meanwhile decreasing its dependence to system parameters and model. Indeed, the obtained adaptive sliding mode controller achieves synchronization of two Rossler's systems with low chattering phenomenon and without implementing complex nonlinear functions.

References

1. Behjameh MR, Delavari H, Vali A (2015) Global finite time synchronization of two nonlinear chaotic gyros using high order sliding mode control. J Appl Comput Mech 01(01):26–34
2. Boccaletti S, Grebogi C, Lai YC, Mancini H, Maza D (2000) The control of chaos: theory and applications. Phys Rep 329:103–197
3. Chang W, Park J, Joo Y, Chen G (2003) Static output-feedback fuzzy controller for chen's chaotic system with uncertainties. Inf Sci 151:227–244
4. Chen G (1999) Controlling chaos and bifurcations in engineering systems. CRC-Press
5. Di Bernardo M, Tse C (2002) Chaos in power electronics: an overview. In: Chen G, Ueta T (eds) Chaos in circuits and systems, chap. 16, vol 11. World Scientific, New York, pp 317–340
6. Feki M (2003) An adaptive feedback control of linearizable chaotic systems. Chaos Solitons Fractals 15:883–890
7. Feki M (2004) Model-independent adaptive control of chua's system with cubic nonlinearity. Int J Bifurc Chaos 14(12):4249–4263
8. Feki M (2004) Synchronization of chaotic systems with parametric uncertainties using sliding observers. Int J Bifurc Chaos 14(7):2467–2475
9. Feki M (2006) Synchronization of generalized lorenz system using adaptive controller. In: IFAC conference on analysis and control of chaotic systems CHAOS'06, CD–ROM. Reims-France (2006)
10. Feki M (2009) Observer-based synchronization of chaotic systems with unknown nonlinear function. Chaos Solitons Fractals 39:981–990
11. Feki M (2009) Sliding mode control and synchronization of chaotic systems with parametric uncertainties. Chaos Solitons Fractals 41:1390–1400
12. Fourati A, Feki M, Derbel N (2010) Stabilizing the unstable periodic orbits of a chaotic system using model independent adaptive time-delayed controller. Nonlinear Dyn 62(3):687–704
13. Gammoudi IE, Feki M (2013) Synchronization of integer order and fractional order chua's systems using robust observer. Commun Nonlinear Sci Numer Simul 18(3):625–638
14. Garfinkel A, Spano ML, Ditto WL, Weiss JN (1992) Controlling cardiac chaos. Science 257:1230–1235
15. Hwang CC, Hsieh JY, Lin RS (1997) A linear continuous feedback control of chua's circuit. Chaos Solitons Fractals 8:1507–1515
16. Isidori A (1995) Nonlinear control systems, 3rd edn. Springer, United Kingdom

17. Jemaâ-Boujelben SB, Feki M (2016) Integral higher order sliding mode control for mimo uncertain systems: application to chaotic pmsm. In: Proceedings of engineering & technology (PET), pp. 632–637. Hammamet-Tunisia
18. Jiang GP, Chen G, Tang WKS (2002) Stabilizing unstable equilibrium points of a class of chaotic systems using a state pi regulator. IEEE Trans Circuits Syst I 49(12):1820–1826
19. Khalil HK (1992) Nonlinear systems. Macmillan, New York
20. Konishi K, Hirai M, Kokame H (1998) Sliding mode control for a class of chaotic systems. Phys Lett A 245:511–517
21. Koubaâ K, Feki M (2014) Quasi-periodicity, chaos and coexistence in the time delay controlled two-cell dc-dc buck converter. Int J Bifurc Chaos 24(10): 1450124
22. Li TY, Yorke JA (1975) Period three implies chaos. Am Math Mon 82(10):985–992
23. Lin JS, Yan JJ, Liao TL (2006) Robust control of chaos in lorenz systems subject to mismatch uncertainties. Chaos Solitons Fractals 27:501–510
24. Miladi Y, Feki M, Derbel N (2015) Stabilizing the unstable periodic orbits of a hybrid chaotic system using optimal control. Commun Nonlinear Sci Numer Simul 20:1043–1056
25. Ott E, Grebogi C, Yorke JA (1990) Controlling chaos. Phys Rev Lett 64:1196–1199
26. Robert B, Feki M, Iu H (2006) Control of a pwm inverter using proportional plus extended time-delayed feedback. Int J Bifurc Chaos 16(1):113–128
27. Rössler O (1976) An equation for continuous chaos. Phys Lett A 57(5):397–398
28. Schiff S, Jerger K, Duong D, Chang T, Spano M, Ditto W (1994) Controlling chaos in the brain. Nature 370:615–620
29. Sundarapandian V (2012) Global chaos control of hyperchaotic liu system via sliding control method. Int J Control Theory Appl 5(2):117–123
30. Sundarapandian V (2013) Analysis and adaptive synchronization of two novel chaotic systems with hyperbolic sinusoidal and cosinusoidal nonlinearity and unknown parameters. J Eng Sci Technol Rev 6(4):53–65
31. Sundarapandian V, Pehlivan I (2012) Analysis, control, synchronization, and circuit design of a novel chaotic system. Math Comput Model 55(7–8):1904–1915
32. Tian YC, Tadé MO, Levy D (2002) Constrained control of chaos. Phys Lett A 296:87–90
33. Vidyasagar M (1993) Nonlinear systems analysis, 2nd edn. Prentice Hall, New Jersey
34. Yang CH, Wu CL, Chen YJ, Shiao SH (2015) Reduced fuzzy controllers for lorenz-stenflo system control and synchronization. Int J Fuzzy Syst 17(2):158–169
35. Yu X, Chen G, Xia Y, Song Y, Cao Z (2001) An invariant-manifold-based method for chaos control. IEEE Trans Circuits Syst I 48(8):930–937
36. Zhan-Shan Z, Jing Z, Gang D, Da-Kun Z (2015) Chaos synchronization of coronary artery system based on higher order sliding mode adaptive control. Acta Phys Sin 64(21): 210–508
37. Zhang H, Ma X-K, Li M, Zou J-L (2005) Controlling and tracking hyperchaotic rössler system via active backstepping design. Chaos Solitons Fractals 26: 353–361
38. Zou YL, Luo XS, Jiang P-Q, Wang BH, Chen G, Fang JQ, Quan HJ (2003) Controlling the chaotic n-scroll chua's circuit. Int J Bifurc Chaos 13(9): 2709–2714

Chattering Free Sliding Mode Controller Design for a Quadrotor Unmanned Aerial Vehicle

Nour Ben Ammar, Soufiene Bouallègue, Joseph Haggège
and Sundarapandian Vaidyanathan

Abstract In this paper, a nonlinear model for a Quadrotor Vertical Take-Off and Landing (VTOL) type of Unmanned Aerial Vehicles (UAVs) is firstly established for the attitude and position control. All aerodynamic forces and moments of the studied Quadrotor UAV are described within an inertial frame. Such a dynamical model is obtained using the Newton-Euler formalism. Secondly, an improved nonlinear Sliding Mode Control (SMC) approach is designed for this aircraft in order to stabilize its vertical flight dynamics, while avoiding the classical chattering problem. Since chattering phenomena is the most problematic issue in the sliding mode control applications, a Quasi Sliding Mode Control (QSMC) technique is used as a solution for the chattering avoidance in Quadrotor dynamics control. Demonstrative simulations are carried out in order to show the effectiveness of the proposed QSMC approach for the stabilization and tracking of various desired trajectories.

Keywords VTOL aircraft · Quadrotor UAV · Modeling · Sliding mode control (SMC) · Chattering avoidance · Quasi sliding mode control (QSMC)

N.B. Ammar · S. Bouallègue (✉) · J. Haggège
Research Laboratory in Automatic Control (LA.R.A), National Engineering
School of Tunis (ENIT), University of Tunis El MANAR, BP 37,
1002 Le Belvédère, Tunis, Tunisia
e-mail: soufiene.bouallegue@issig.rnu.tn

N.B. Ammar
e-mail: nourelhouda.benammar@enit.rnu.tn

J. Haggège
e-mail: joseph.haggege@enit.rnu.tn

S. Bouallègue
High Institute of Industrial Systems of Gabès (ISSIG),
Salaheddine EL AYOUBI Street, 6011 Gabès, Tunisia

S. Vaidyanathan
Research and Development Centre, Vel Tech University, No. 42,
Avadi-Vel Tech Road, Avadi, Chennai 600062, Tamil Nadu, India
e-mail: sundarvtu@gmail.com

© Springer International Publishing AG 2017
S. Vaidyanathan and C.-H. Lien (eds.), *Applications of Sliding Mode Control
in Science and Engineering*, Studies in Computational Intelligence 709,
DOI 10.1007/978-3-319-55598-0_3

1 Introduction

An Unmanned Aerial Vehicle (UAV) refers to a flying machine without an on-board human pilot. These vehicles are increasingly popular platforms due to their ability to effectively carry out a wide range of applications from civil and military domains [5, 9, 13, 19]. Researchers have led to different designs for this type of aircrafts. A Quadrotor UAV is one of the Vertical Take-Off and Landing (VTOL) designs which are proven to have promising flying concepts due to their high maneuverability [3, 13, 19, 20]. The complex mechanical structure of the Quadrotor, its strongly nonlinear and coupled dynamics, its multiple inputs-outputs and the observation difficulty of its states allowed this VTOL aircraft to be a popular topic of research in the field of robotics and nonlinear control theory. So, modeling and control of this kind of nonlinear systems became increasingly difficult and hard tasks in the practical design and prototyping framework [3, 9, 10, 12].

Nowadays, several methods have been proposed to control Quadrotors. First, the linear control methods such as PID, Linear Quadratic Regulator (LQR) and Linear Quadratic Gaussian (LQG) approaches have been proposed in the literature and applied for the attitude stabilization and/or position tracking [7, 17]. However, these methods can impose limitations on application of Quadrotors for extended flight regions, i.e. aggressive maneuvers, where the system is no longer linear. Moreover, the stability of the closed-loop system can only be achieved for small regions around the equilibrium point, which are extremely hard to compute. In addition, the performances on tracking trajectories of these control laws are not satisfactory enough comparing with other more advanced methods. To overcome this problem, nonlinear control alternatives, such as the Backstepping [8, 14] and especially the sliding mode control [1, 6, 10, 16, 22, 30] approaches, are recently used in the VTOL aircrafts control framework.

Sliding Mode Control (SMC) is a particular type of robust Variable Structure Control (VSC) which evolved from the pioneering work in Russia in the early of 1960. The idea did not appear outside of Russia until the mid of 1970 when books of Itkis [15] and Utkin [23] where published in English. The ideas have successfully been applied to problem as diverse as automatic flight control, control of electric motors, space systems and robots [2, 4, 11, 18, 20, 24–29]. So, in this chapter, an improved SMC approach is proposed for the Quadrotor dynamics stabilization and trajectory tracking. Roll, pitch, yaw and position dynamics are separately controlled with Lyapunov-based SMC strategy. The chattering problem, which appeared in these designed sliding mode control laws, is handled thanks to a proposed Quasi Sliding Mode Control (QSMC) technique.

The reminder of this chapter is organized as follows. Section 2 presents the flight dynamics modeling of the Quadrotor UAV based on the well known Newton-Euler approach. A nonlinear model is derived and set in a state-space form. Section 3 is devoted to design nonlinear sliding mode controllers for the Quadrotor UAV flight stabilization and path tracking. Such designed SMC laws are then improved in terms of chattering phenomenon avoidance while replacing the well known sign function in the control expressions with a continuous term. Demonstrative simulation results, obtained for this proposed QSMC approach, are presented, compared and discussed in Sect. 4. Finally, Sect. 5 concludes this chapter.

2 Description and Dynamics Modeling of the Quadrotor UAV

Design and analysis of control systems are usually started by carefully considering mathematical models of physical systems. In this section, a complete dynamical model of the studied Quadrotor UAV is established using the Newton-Euler formalism.

2.1 Quadrotor Description and Aerodynamic Forces

A Quadrotor is an UAV with four rotors that are controlled independently. The movement of the Quadrotor results from changes in the speed of the rotors as shown in Fig. 1. In this chapter, the Quadrotor structure is assumed to be rigid and symmetrical. The center of gravity and the body fixed frame origin are coincided. The propellers are rigid and the thrust and drag forces are proportional to the square of propeller's speed. The studied Quadrotor is detailed with its body- and earth-frames $\mathcal{F}_b(b, x^b, y^b, z^b)$ and $\mathcal{F}_e(G, X^G, Y^G, Z^G)$, respectively.

Let us denote by m the total mass of the Quadrotor UAV and g the acceleration of the gravity. The following model partitions naturally into translational and rotational coordinates:

$$\boldsymbol{\xi} = (x, y, z) \in \mathbb{R}^3, \quad \boldsymbol{\eta} = (\phi, \theta, \psi) \in \mathbb{R}^3 \tag{1}$$

Fig. 1 Mechanical structure of the Quadrotor UAV and its related frames

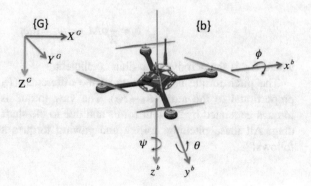

where $\boldsymbol{\xi} = (x, y, z)$ denotes the position vector of the center of mass of the Quadrotor in the fixed inertial frame, and $\boldsymbol{\eta} = (\phi, \theta, \psi)$ denotes the Quadrotor attitude given by the Euler pitch, roll and yaw angles, given as ϕ, θ and ψ, respectively. All those angles are bounded as follows: $-\frac{\pi}{2} < \phi < \frac{\pi}{2}$; $-\frac{\pi}{2} < \theta < \frac{\pi}{2}$ and $-\pi < \psi < \pi$.

Each motor of the Quadrotor produces a force F_i (i = 1, 2, 3 and 4) which is proportional to the square of the angular speed. Known that the motors are supposedly turning only in a fixed direction, the produced force F_i is always positive. The front and rear motors (M1 and M3) rotate counter-clockwise, while the left and right motors (M2 and M4) rotate clockwise. Gyroscopic effects and aerodynamic torques tend to cancel in trimmed flight. The total thrust F is the sum of individual thrusts of each motor.

The orientation of the Quadrotor is given by the rotation matrix $\boldsymbol{R}: \mathcal{F}_e \rightarrow \mathcal{F}_b$ which depends on the three Euler angles (ϕ, θ, ψ) and defined by the following equation:

$$\boldsymbol{R}(\phi, \theta, \psi) = \begin{bmatrix} c\psi c\theta & s\phi s\theta c\psi - s\psi c\theta & c\phi s\theta c\psi + s\psi s\phi \\ s\psi c\theta & s\phi s\theta s\psi + c\psi c\theta & c\phi s\theta s\psi - s\phi c\psi \\ -s\theta & s\phi c\theta & c\phi c\theta \end{bmatrix} \qquad (2)$$

where $c(.) = \cos(.)$ and $s(.) = \sin(.)$.

During its flight, the Quadrotor is subject to external forces like the gusts of wind, gravity, viscous friction and others self generated such as the thrust and drag forces. In addition, external torques are provided mainly by the thrust of rotors and the drag on the body and propellers. Moments generated by gyroscopic effects of motors are also considered. The thrust force generated by the ith rotor of Quadrotor is given by:

$$F_i = \frac{1}{2}\rho\Lambda C_T r^2 \omega_i^2 = b\omega_i^2 \qquad (3)$$

where ρ is the air density, r and Λ are the radius and the section of the propeller respectively, C_T is the aerodynamic thrust coefficient.

The aerodynamic drag torque, caused by the drag force at the propeller of the ith rotor and opposed the motor torque, is defined as follows:

$$\delta_i = \frac{1}{2}\rho\Lambda C_D r^2 \omega_i^2 = d\omega_i^2 \qquad (4)$$

where C_D is the aerodynamic drag coefficient.

The pitch torque is a function of the difference $(F_3 - F_1)$ and the roll one is proportional to the term $(F_4 - F_2)$. The yaw torque is the sum of all reactions torques generated by the four rotors and due to the shaft acceleration and propeller drag. All these pitching, rolling and yawing torques are defined respectively as follows:

$$\tau_\theta = l(F_3 - F_1) \tag{5}$$

$$\tau_\phi = l(F_4 - F_2) \tag{6}$$

$$\tau_\psi = C(F_1 - F_2 + F_3 - F_4) \tag{7}$$

where C is the proportional coefficient and l denotes the distance from the center of each rotor to the center of gravity of the Quadrotor.

Two gyroscopic effects torques, due to the motion of the propellers and the Quadrotor body, are additively provided. These moments are given respectively by:

$$M_{gp} = \sum_{i=1}^{4} \Omega \wedge \begin{bmatrix} 0 & 0 & J_r(-1)^{i+1}\omega_i \end{bmatrix}^T \tag{8}$$

$$M_{gb} = \Omega \wedge J\Omega \tag{9}$$

where Ω is the vector of angular velocities in the fixed earth frame and $J = diag\begin{bmatrix} I_x, I_y, I_z \end{bmatrix}$ is the inertia matrix of the Quadrotor, I_x, I_y and I_z denote the inertias of the x-axis, y-axis and z-axis of the Quadrotor, respectively, J_r denotes the z-axis inertia of the propellers' rotors.

The Quadrotor is controlled by independently varying the speed of the four rotors. Hence, these control inputs $u_{1,2,3,4}$ are defined as follows:

$$u = \begin{bmatrix} u_1 \\ u_2 \\ u_3 \\ u_4 \end{bmatrix} = \begin{bmatrix} F \\ \tau_\phi \\ \tau_\theta \\ \tau_\psi \end{bmatrix} = \begin{bmatrix} b & b & b & b \\ 0 & -lb & 0 & lb \\ -lb & 0 & lb & 0 \\ d & -d & d & -d \end{bmatrix} \begin{bmatrix} \omega_1^2 \\ \omega_2^2 \\ \omega_3^2 \\ \omega_4^2 \end{bmatrix} \tag{10}$$

where $b > 0$ and $d > 0$ are two parameters depending on the air density, the geometry and the lift and drag coefficients of the propeller as shown in Eqs. (3) and (4), and $\omega_{1,2,3,4}$ are the angular speeds of the four rotors, respectively.

From Eq. (10), it can be observed that the input u_1 denotes the total thrust force on the Quadrotor body in the z-axis, the inputs u_2 and u_3 represent the roll and pitch torques, respectively. The input u_4 represents a yawing torque.

2.2 Dynamical System Model

While using the Newton-Euler formalism for modeling, the Newton's laws lead to the following motion equations of the Quadrotor:

$$\begin{cases} m\ddot{\xi} = \boldsymbol{F}_{th} + \boldsymbol{F}_d + \boldsymbol{F}_g \\ J\dot{\Omega} = \boldsymbol{M} - \boldsymbol{M}_{gp} - \boldsymbol{M}_{gb} - \boldsymbol{M}_a \end{cases} \tag{11}$$

where $\boldsymbol{F}_{th} = \boldsymbol{R}(\phi, \theta, \psi) \left[0, 0, \sum\limits_{i=1}^{4} F_i \right]^T$ denotes the total thrust force of the four

rotors, $\boldsymbol{F}_d = diag(\kappa_1, \kappa_2, \kappa_3) \dot{\xi}^T$ is the air drag force which resists to the Quadrotor

motion, $\boldsymbol{F}_g = [0, 0, mg]^T$ is the gravity force, $\boldsymbol{M} = \left[\tau_\phi, \tau_\theta, \tau_\psi \right]^T$ represents the total

rolling, pitching and yawing torques, $\boldsymbol{M}_{gp}$ and $\boldsymbol{M}_{gb}$ are the gyroscopic torques and

$\boldsymbol{M}_a = diag(\kappa_4, \kappa_5, \kappa_6) \left[\dot{\phi}^2, \dot{\theta}^2, \dot{\psi}^2 \right]^T$ is the torque resulting from the aerodynamic

frictions.

Substituting the position vector and the forces expressions into Eq. (11), we obtain the following translational dynamics of the Quadrotor:

$$\begin{cases} \ddot{x} = \frac{1}{m} (c\phi c\psi s\theta + s\phi s\psi) u_1 - \frac{\kappa_1}{m} \dot{x} \\ \ddot{y} = \frac{1}{m} (c\phi s\psi s\theta - s\phi c\psi) u_1 - \frac{\kappa_2}{m} \dot{y} \\ \ddot{z} = \frac{1}{m} c\phi c\theta u_1 - g - \frac{\kappa_3}{m} \dot{z} \end{cases} \tag{12}$$

From the second part of Eq. (11), and while substituting each moment by its expression, we deduce the following rotational dynamics of the rotorcraft:

$$\begin{cases} \ddot{\phi} = \frac{(I_y - I_z)}{I_x} \dot{\theta} \dot{\psi} - \frac{J_r}{I_x} \Omega_r \dot{\theta} - \frac{\kappa_4}{I_x} \dot{\phi}^2 + \frac{1}{I_x} u_2 \\ \ddot{\theta} = \frac{(I_z - I_x)}{I_y} \dot{\phi} \dot{\psi} - \frac{J_r}{I_y} \Omega_r \dot{\phi} - \frac{\kappa_5}{I_y} \dot{\theta}^2 + \frac{1}{I_y} u_3 \\ \ddot{\psi} = \frac{(I_x - I_y)}{I_z} \dot{\theta} \dot{\phi} - \frac{\kappa_6}{I_z} \dot{\psi}^2 + \frac{1}{I_z} u_4 \end{cases} \tag{13}$$

where $\kappa_{1,2,\ldots,6} > 0$ are the drag coefficients, $\Omega_r = \omega_1 - \omega_2 + \omega_3 - \omega_4$ is the overall residual rotor angular velocity.

Taking $\boldsymbol{x} = (\phi, \dot{\phi}, \theta, \dot{\theta}, \psi, \dot{\psi}, x, \dot{x}, y, \dot{y}, z, \dot{z})^T \in \mathbb{R}^{12}$ as state vector, a state-space representation of the studied system is obtained as follows:

$$\dot{\boldsymbol{x}} = f(\boldsymbol{x}, \boldsymbol{u}) = \begin{cases} \dot{x}_1 = x_2 \\ \dot{x}_2 = a_1 x_4 x_6 + a_3 \Omega_r x_4 + a_2 x_2^2 + b_1 u_2 \\ \dot{x}_3 = x_4 \\ \dot{x}_4 = a_4 x_2 x_6 + a_6 \Omega_r x_2 + a_5 x_4^2 + b_2 u_3 \\ \dot{x}_5 = x_6 \\ \dot{x}_6 = a_7 x_2 x_4 + a_8 x_6^2 + b_3 u_4 \\ \dot{x}_7 = x_8 \\ \dot{x}_8 = a_9 x_8 + \frac{1}{m} (c\phi c\psi s\theta + s\phi s\psi) u_1 \\ \dot{x}_9 = x_{10} \\ \dot{x}_{10} = a_{10} x_{10} + \frac{1}{m} (c\phi s\theta s\psi - s\phi c\psi) u_1 \\ \dot{x}_{11} = x_{12} \\ \dot{x}_{12} = a_{11} x_{12} + \frac{c\phi c\theta}{m} u_1 - g \end{cases} \tag{14}$$

where:

$$a_1 = \frac{I_y - I_z}{I_x}, \quad a_2 = -\frac{\kappa_4}{I_x}, \quad a_3 = -\frac{J_r}{I_x}, \quad a_4 = \frac{(I_z - I_x)}{I_y}, \quad a_5 = -\frac{\kappa_5}{I_y}, \quad a_6 = -\frac{J_r}{I_y},$$

$$a_7 = \frac{(I_x - I_y)}{I_z}, \quad a_8 = -\frac{\kappa_6}{I_z}, \quad a_9 = -\frac{\kappa_1}{m}, \quad a_{10} = -\frac{\kappa_2}{m}, \quad a_{11} = -\frac{\kappa_3}{m}, \quad b_1 = \frac{1}{I_x},$$

$$b_2 = \frac{1}{I_y} \text{ and } b_3 = \frac{1}{I_z}.$$

3 Sliding Mode Control of the Quadrotor

3.1 Basic Concepts

The basic idea of the SMC approach is to attract the system states towards a surface, called sliding surface, suitably chosen, and design a stabilizing control law that keeps the system states on such a surface. For the choice of the sliding surface shape, the general form of Eq. (15) was proposed by Slotine and Li [22, 23]:

$$s(x, t) = \left(\frac{d}{dt} + \lambda_x \right)^{n-1} e \qquad (15)$$

where $x \in \mathbb{R}^n$ denotes the variable state, $e = x - x_d$ is the tracking error and x_d is the desired trajectory, λ_x is a positive constant that interprets the dynamics of the sliding manifold and presents a design parameter for the SMC method, and $n \in \mathbb{N}$ is the order of the controlled system.

Condition called attractiveness, is the condition under which the state trajectory will reach the sliding surface. There are two types of conditions of access to the sliding surface [23]. In this study, we use the Lyapunov based approach. It consists to make a positive scalar function of Eq. (16), called Lyapunov candidate function, for the system state variables and then choosing the control law that will decrease this function:

$$\dot{V}(x) < 0, \quad \text{with } V(x) > 0 \qquad (16)$$

In this case, the Lyapunov function can be chosen as:

$$V(x) = \frac{1}{2} s(x, t)^2 \qquad (17)$$

So, the derivative of this above function is negative when the following expression is checked:

$$s(\boldsymbol{x}, t)\dot{s}(\boldsymbol{x}, t) < 0 \tag{18}$$

The purpose is to force the system state trajectories to reach the sliding surface and to stay on this same surface despite the presence of uncertainty. The sliding control law contains two terms and is given as follows:

$$u(t) = u_{eq}(t) + u_D(t) \tag{19}$$

where $u_{eq}(t)$ denotes the equivalent control which is a way to determine the behavior of the system when an ideal sliding regime is established. It is calculated from the following invariance condition of the surface [22]:

$$\begin{cases} s(\boldsymbol{x}, t) = 0 \\ \dot{s}(\boldsymbol{x}, t) = 0 \end{cases} \tag{20}$$

and $u_D(t)$ is a discontinuous function calculated by checking the condition of the attractiveness. It is useful to compensate the uncertainties of the model and often defined as follows:

$$u_D(t) = -K\mathrm{sgn}(s(\boldsymbol{x}, t)) - K\mathrm{sgn}(s(\boldsymbol{x}, t)) \tag{21}$$

where K is a positive control parameter and sgn(.) is the mathematical signum function.

3.2 Sliding Mode Controllers Design

For the Quadrotor attitude control, we use the rotational motion model given by Eq. (13). The translational dynamics model of Eq. (12) is used to design the position controllers. Let also consider the state vector given by Eq. (14).

We begin by defining the tracking errors which represent the difference between the set-point and current values of the state:

$$\begin{cases} e_{i+1} = \dot{e}_i \\ e_i = x_i - x_{id}, i = 1, 2, \ldots, 11 \end{cases} \tag{22}$$

So, the sliding surfaces for all controlled dynamics are chosen based on the tracking errors and according to Eq. (15). They are given as follows:

$$\begin{cases} s_\phi = e_2 + \lambda_1 e_1 \\ s_\theta = e_4 + \lambda_2 e_3 \\ s_\psi = e_6 + \lambda_3 e_5 \\ s_x = e_8 + \lambda_4 e_7 \\ s_y = e_{10} + \lambda_5 e_9 \\ s_z = e_{12} + \lambda_6 e_{11} \end{cases} \tag{23}$$

where s_ϕ, s_θ, s_ψ, s_x, s_y and s_z denote the sliding manifolds of roll, pitch, yaw, and positions dynamics respectively, $\lambda_{1,2,...,6}$ are the SMC design parameters relatively of each mentioned dynamics of the Quadrotor.

Starting with the roll dynamics SMC of the Quadrotor, let consider the following Lyapunov function:

$$V(s_\phi) = \frac{1}{2}s_\phi^2 \tag{24}$$

While referring to Eqs. (16) and (18), we deduce the expression of the derivative roll surface given as:

$$\dot{s}(\phi,t) = -k_1\text{sgn}(s(\phi,t)) - k_2 s(\phi,t) \tag{25}$$

where $k_1 > 0$ and $k_2 > 0$ are the SMC design parameters of the controlled roll dynamics.

By replacing $\dot{x}_2$ variable with its expression and referring to the above equations, the control law u_2 is given by:

$$u_2 = \frac{1}{b_1}\left[-a_1 x_4 x_6 - a_3 \bar{\Omega}_r x_4 - a_2 x_2^2 + \ddot{x}_{1d} - \lambda_1 \dot{e}_1 - k_1\text{sgn}(s_\phi) - k_2 s_\phi\right] \tag{26}$$

So, when following exactly the same steps as for the roll SMC controller design, the control inputs u_3 and u_4, responsible of generating the pitch and yaw rotations respectively, are calculated as follows:

$$u_3 = \frac{1}{b_2}\left[-a_4 x_2 x_6 - a_6 \bar{\Omega}_r x_2 - a_5 x_4^2 + \ddot{x}_{3d} - \lambda_2 \dot{e}_3 - k_3\text{sgn}(s_\theta) - k_4 s_\theta\right] \tag{27}$$

$$u_4 = \frac{1}{b_3}\left[-a_7 x_2 x_4 - a_8 x_6^2 + \ddot{x}_{5d} - \lambda_3 \dot{e}_5 - k_5\text{sgn}(s_\psi) - k_6 s_\psi\right] \tag{28}$$

where (k_3, k_4) and (k_5, k_6) denote the SMC design positive parameters to be tuned for the pitch and yaw dynamics respectively.

Using the same method as for the Quadrotor attitude control, we deduce the SMC control laws u_1, u_x and u_y for the z, x and y positions, respectively. These control inputs are computed as follows:

$$u_1 = \frac{m}{c\phi c\theta}\left[-a_{11}x_{12} + \ddot{x}_{11d} - \lambda_6 \dot{e}_{11} - k_7\text{sgn}(s_z) - k_8 s_z + g\right] \tag{29}$$

$$u_x = \frac{m}{u_1}\left[-a_9 x_8 + \ddot{x}_{7d} - \lambda_5 \dot{e}_7 - k_9\text{sgn}(s_x) - k_{10} s_x\right] \tag{30}$$

$$u_y = \frac{m}{u_1}\left[-a_9 x_{10} + \ddot{x}_{9d} - \lambda_6 \dot{e}_9 - k_{11}\text{sgn}(s_y) - k_{12} s_y\right] \tag{31}$$

As given in Eqs. (26)–(31), these computed SMC laws for the stabilization of Quadrotor dynamics cannot be implemented in practical framework. This handicap is due to the well known chattering problem that occurs in the controlled system responses [21]. So, an improvement mechanism for chattering handling and suppression is proposed in the next section based on a QSMC method.

3.3 Chattering Problem Avoidance

The chattering problem in SMC approach is the one of the most common handicaps for applying to real-time prototyping. The main obstacle is an undesirable phenomenon of oscillation with finite frequency and amplitude on system responses. The chattering is harmful because it leads to low control accuracy, high wear of moving mechanical parts, and high thermal losses in electrical power circuits. In case that there exist fast dynamics which are neglected in the ideal model, the chattering may appear since an ideal sliding mode may not occur.

For practical implementation, it is imperative to avoid the control chattering by providing continuous/smooth control signals. Approximate smoothed implementations of sliding mode control technique have been suggested. The discontinuous sgn(.) term of the standard SMC law is replaced by a continuous approximation. In this chapter, we use one of the QSMC methods of the literature [4, 11, 21]. With this method, the sgn(.) term in Eqs. (26)–(31) is replaced with a continuous term, known as sigmoid function form, as follows [21]:

$$\text{sigm}(s) = \frac{s}{\|s\| + \alpha} \tag{32}$$

α is a small design constant that smoothes the discontinuity.

4 Simulation Results and Discussion

In this section, the proposed QSMC approach for the Quadrotor attitude stabilization and desired trajectories tracking is implemented and discussed. For the simulation, we use from [7] the physical parameters of the Quadrotor UAV as given in Table 1. We treat separately the problems of stabilization and tracking trajectories.

4.1 Quadrotor Dynamics Stabilization

Numerical simulations are performed for both SMC and QSMC structures and lead to the results of Figs. 2, 3, 4, 5, 6, 7, 8 and 9. The sliding mode control laws of

Table 1 Quadrotor model parameters

Model parameters	Values
Lift coefficient b	2.984×10^{-5} N s^2/rad^2
Drag coefficient d	3.30×10^{-7} N s^2/rad^2
Mass m	0.5 Kg
Arm length l	0.50 m
Motor inertia J_r	2.8385×10^{-5} N m s^{-2}/rad
Quadrotor inertia $\boldsymbol{J}$	$diag(0.005, 0.005, 0.010)$
Aerodynamic friction coeffs. $\kappa_{1,2,3}$	0.3729
Translational drag coeffs. $\kappa_{4,5,6}$	5.56×10^{-4}
Acceleration of the gravity g	9.81 m s^{-2}

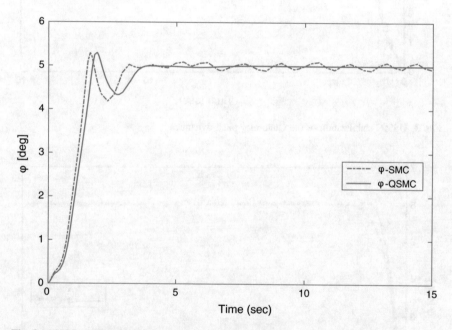

Fig. 2 QSMC stabilization of the Quadrotor roll dynamics

Eqs. (26)–(32) are implemented for the studied Quadrotor UAV. The initial position and angle values are set as [0; 0; 0] m and [0; 0; 0] degree. The purpose of the controller is to bring the states of the system to the following desired reference values [0; 0; 2] m and [5; 5; 6] degree, which allows us to have a comparative idea between the SMC and the QSMC.

Simulation results, which are illustrated in Figs. 2, 3, 4, 5, 6 and 7, show the resulted attitude, heading and altitude control with signum function and the mathematical sigmoid function of Eq. (32), introduced to avoid the SMC chattering problem.

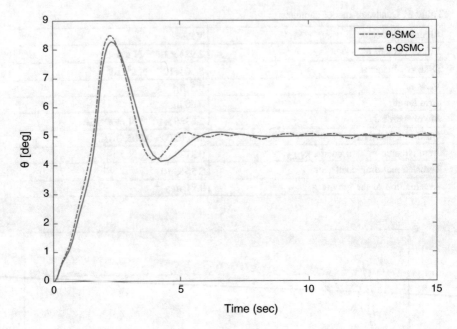

Fig. 3 QSMC stabilization of the Quadrotor pitch dynamics

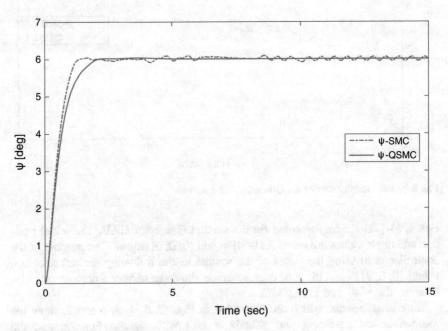

Fig. 4 QSMC stabilization of the Quadrotor yaw dynamics

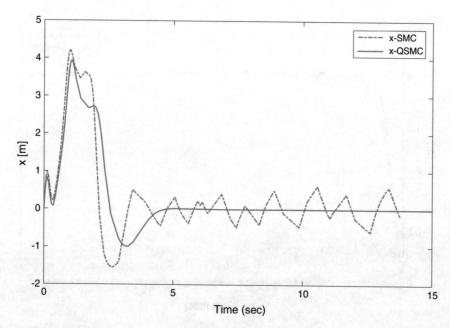

Fig. 5 QSMC stabilization of the Quadrotor x-position dynamics

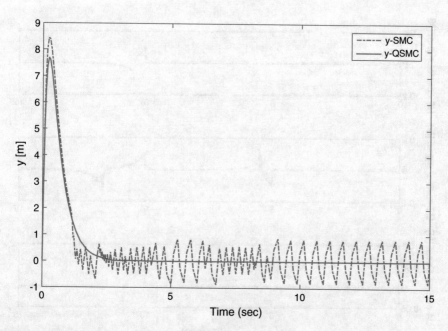

Fig. 6 QSMC stabilization of the Quadrotor y-position dynamics

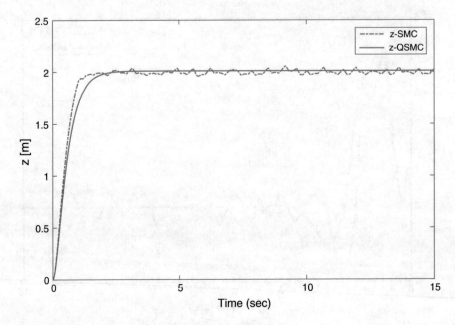

Fig. 7 QSMC stabilization of the Quadrotor z-position dynamics

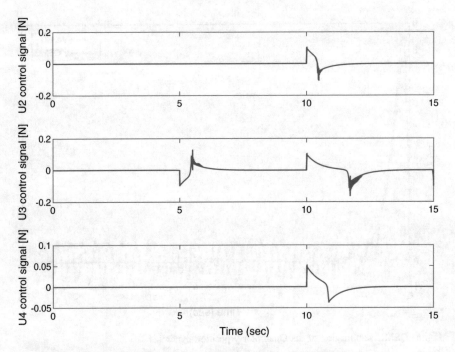

Fig. 8 Control signals evolution for the rotational dynamics of Quadrotor

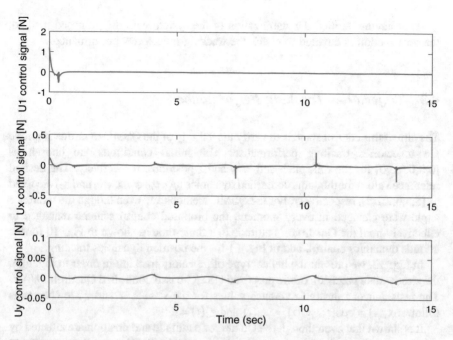

Fig. 9 Control signals evolution for the translational dynamics of Quadrotor

Table 2 SMC design parameters for simulation

Design parameters	Values
$\lambda_1; k_1; k_2$	10; 12; 2
$\lambda_2; k_3; k_4$	10; 12; 2
$\lambda_3; k_5; k_6$	10; 12; 2
$\lambda_4; k_7; k_8$	10; 12; 2
$\lambda_5; k_9; k_{10}$	10; 12; 2
$\lambda_6; k_{11}; k_{12}$	10; 12; 2
α	0.025

As shown in figures, the system is able to achieve the desired position but there is a clear chattering effect for the classical SMC based results. But, it can be clearly seen that when using the quasi sliding mode control algorithms (QSMC), the designed controllers can avoid the chattering phenomena and reduce it in the same time it maintain the stability of the system.

QSMC design parameters are manually adjusted to reach good performances such as the reducing of the chattering problems as well as convergence of the tracking errors and steady-state precision. All these parameters are chosen relatively for the altitude, roll, pitch and yaw dynamics of the studied Quadrotor UAV as given in Table 2.

The control signals, shown in Figs. 8 and 9, are smooth as desired and can be easily applied to a real-world model of the Quadrotor. It is noted that although the controlled states reach their steady states the stability does not appear affected.

After having verified the stabilization of the system with the proposed method, the next section is devoted to check the tracking trajectories performances.

4.2 Trajectories Tracking Performance

The aim of this control stage is to verify the validity of the QSMC to farther improve the trajectories tracking performance. The initial conditions are chosen as [0;0;0;0;0;0] for the roll, pitch and x, y and z positions, respectively. The desired reference values for the same controlled dynamics (ϕ_d, θ_d, ψ_d, x_d, y_d and z_d) are fixed as [5;5;6;2;2;2], respectively. For the QSMC methods and even though the reference angle were changed in every moment, the proposed control scheme managed to effectively hold the Quadrotor's attitude in a finite-time as shown in Fig. 10 for the attitude dynamics control, and in Fig. 11 for the position dynamics tracking.

In Fig. 12, we present the helical type of trajectory tracking in order to highlight the steady-state precision of the proposed QSMC-based controlled Quadrotor UAV. The instantaneous desired reference values for the state x, y and z are chosen as follows: $x_d(t) = \cos(t)$, $y_d(t) = \sin(t)$ and $z_d(t) = t$.

It is shown that even though the Quadrotor's attitude and position are affected by the abruptly changed reference angles, the designed QSMC controllers are able to drive all these state variables back to the new reference angle and position within some seconds. Moreover, the aerodynamic forces and moments are taken into account in the controllers design. Those demonstrate the robustness of the proposed control strategy and its effectiveness.

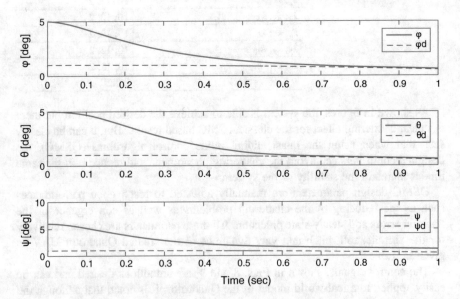

Fig. 10 QSMC-based attitude tracking of the Quadrotor

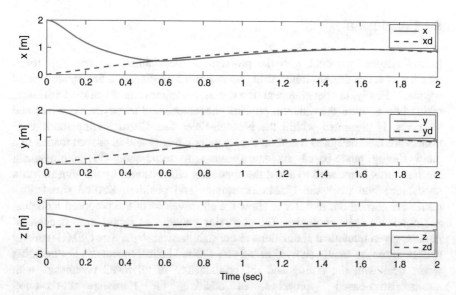

Fig. 11 QSMC-based positions tracking of the Quadrotor

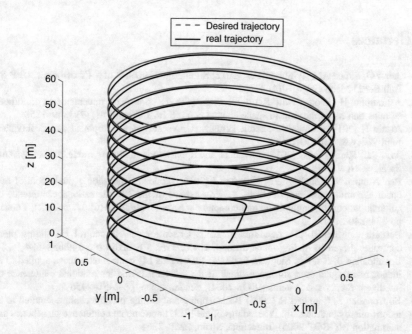

Fig. 12 Tracking of a helical type of desired trajectory

5 Conclusion

In this chapter, we deal with the problem of the stabilization and trajectories tracking of a Quadrotor aircraft using an improved nonlinear sliding mode control approach. Firstly, the development of a dynamic nonlinear model of the Quadrotor, taking into account the different physics phenomena and aerodynamic forces and moments, is presented within the Newton-Euler formalism. Sliding mode controllers are then designed based on the Lyapunov theory and improved thanks to a quasi sliding mode-based avoidance chattering technique. This improvement mechanism is introduced to avoid the chattering effect caused by the sign function to stabilize and track the Quadrotor attitude and position. Several simulations results are carried out in order to show the effectiveness of the proposed modeling and nonlinear SMC strategies. The chattering phenomena suppression is depicted through given numerical simulations in the stabilization stage. The QSMC tracking performances are finally showed based on a desired helical trajectory. Forthcoming works deal with the tuning and the optimization of all SMC parameters with metaheuristics-based approaches. In addition, the Hardware In-the-Loop (HIL) co-simulation of the designed QSMC laws will be also investigated.

References

1. Adr VG, Stoica AM, Whidborne JF (2012) Sliding mode control of a 4Y octorotor. UPB Sci Bull Ser D Mech Eng J 74(4):37–52
2. Ashrafiuon H, Scott Erwin R (2008) Sliding mode control of underactuated multibody systems and its application to shape change control. Int J Control 81(12):1849–1858
3. Austin R (2010) Unmanned aircraft systems: UAVs design, development and deployment. John Wiley & Sons, UK
4. Azar AT, Zhu Q (eds) (2015) Advances and applications in sliding mode control systems. Springer, New York
5. Ben Ammar N, Bouallègue S, Haggège J (2016) Modeling and sliding mode control of a quadrotor unmanned aerial vehicle. In: Proceedings of the 3th international conference on automation, control engineering and computer science (ACECS 2016), Hammamet, Tunisia, pp 834–840
6. Besnard L, Shtessel YB, Landrum B (2012) Quadrotor vehicle control via sliding mode controller driven by sliding mode disturbance observer. J Franklin Inst 349:658–684
7. Bouabdallah S, Noth A, Siegwart R (2004) PID versus LQ control techniques applied to an indoor micro quadrotor. In: Proceedings of the 2004 IEEE/RSJ international conference on intelligent robots and systems (IROS 2004), Sendai, Japan, pp 2451–2456
8. Bouabdallah S, Siegwart R (2005) Backstepping and sliding mode technique applied to an indoor micro quadrotor. In: Proceedings of the IEEE international conference on robotics and automation (ROBOT 2005), Barcelona, Spain, 2259–2264
9. Bouallègue S, Fessi R (2016) Rapid control prototyping and PIL co-simulation of a quadrotor UAV based on NI myRIO-1900 board. Int J Adv Comput Sci Appl 7(6):26–35
10. Bresciani T (2008) Modelling, identification and control of a quadrotor helicopter. Master thesis, department of automatic control, Lund University, ISSN 0280-5316, Sweden

11. Edwards C, Spurgeon SK (1998) Sliding mode control: theory and applications. CRC Press Taylor and Francis, London
12. Fantoni I, Lozano R (2002) Nonlinear control for underactuated mechanical systems. Springer, London
13. Guerrero J, Lozano R (eds) (2012) Flight formation control. Wiley-ISTE, UK
14. Islam S, Dias J, Seneviratne LD (2014) Adaptive tracking control for quadrotor unmanned flying vehicle. In: Proceedings of the international conference on advanced intelligent mechatronics (AIM 2014), Besançon, France, pp 441–445
15. Itkis U (1976) Control systems of variable structure. Wiley, Hoboken, New Jersey
16. Khalil HK (2002) Nonlinear systems, 3rd edn. Prentice Hall, Upper Saddle River, New Jersey
17. Khatoon S, Gupta D, Das LK (2014) PID and LQR control for a quadrotor: modeling and simulation. In: Proceedings of the 2014 international conference on advances in computing, communications and informatics (ICACCI 2014), New Delhi, pp 796–802
18. Lakhekar GV, Waghmare LM, Vaidyanathan S (2016) Diving autopilot design for underwater vehicles using an adaptive neuro-fuzzy sliding mode controller. Stud Comput Intell 635:477–503
19. Lozano R (ed) (2013) Unmanned aerial vehicles: embedded control. ISTE and Wiley, London UK and Hoboken USA
20. Nonami K, Kendoul F, Suzuki S, Wang W, Nakazawa D (2010) Autonomous flying robots: unmanned aerial vehicles and micro aerial vehicles. Springer, New York
21. O'Toole MD, Bouazza-Marouf K, Kerr D (2010) Chatter suppression in sliding mode control: strategies and tuning methods. In: Parenti-Castelli V, Schiehlen W (eds) ROMANSY 18 robot design, dynamics and control. Springer, New York
22. Slotine J-JE, Li W (1991) Applied nonlinear control. Prentice-Hall, Englewood Cliffs, New Jersey
23. Utkin VI (1992) Sliding mode in control and optimization. Springer, Heidelberg
24. Vaidyanathan S (2015) Sliding mode control of Rucklidge chaotic system for nonlinear double convection. Int J ChemTech Res 8(8):25–35
25. Vaidyanathan S (2015) Integral sliding mode control design for the global chaos synchronization of identical novel chemical chaotic reactor systems. Int J ChemTech Res 8 (11):684–699
26. Vaidyanathan S (2015) Sliding controller design for the global chaos synchronization of enzymes-substrates systems. Int J PharmTech Res 8(7):89–99
27. Vaidyanathan S (2015) Global chaos synchronization of chemical chaotic reactors via novel sliding mode control method. Int J ChemTech Res 8(7):209–221
28. Vaidyanathan S, Sampath S, Azar AT (2015) Global chaos synchronisation of identical chaotic systems via novel sliding mode control method and its application to Zhu system. Int J Model Ident Control 23(1):92–100
29. Young KD, Utkin VI, Özgüner U (1999) A control engineer's guide to sliding mode control. IEEE Trans Control Syst Technol 7(3):328–342
30. Zheng E-H, Xiong J-J, Luo J-L (2014) Second order sliding mode control for a quadrotor UAV. ISA Trans 53(4):1350–1356

Terminal Sliding Mode Controller Design for a Quadrotor Unmanned Aerial Vehicle

Rabii Fessi, Soufiene Bouallègue, Joseph Haggège
and Sundarapandian Vaidyanathan

Abstract This chapter deals with the modeling and the control of a Quadrotor type of Unmanned Aerial Vehicles (UAVs) using a Terminal Sliding Mode Control (TSMC) approach. The objectives of this proposed nonlinear control strategy are the stabilization and path tracking of the altitude and the attitude of such an aircraft. The TSMC structure is designed to overcome several problems occur with the classical SMC one such as the chattering phenomenon. With this TSMC approach, it is guaranteed that the output tracking error converges to zero in a finite time unlike the classical SMC. The main structural difference between all proposed SMC structures, i.e. classical and terminal variants, is defined at the sliding surface form that determines the states dynamics of the controlled Quadrotor by chosen the suitable parameters of this surface. High performances of the proposed TSMC controllers are showed through the tracking of a desired flight path. Demonstrative

R. Fessi · S. Bouallègue (✉) · J. Haggège
Research Laboratory in Automatic Control (LA.R.A), National Engineering
School of Tunis (ENIT), University of Tunis EL MANAR, BP 37,
Le Belvédère 1002, Tunis, Tunisia
e-mail: soufiene.bouallegue@issig.rnu.tn

R. Fessi
e-mail: rabii.fessi@gmail.com

J. Haggège
e-mail: joseph.haggege@enit.rnu.tn

R. Fessi
National Engineering School of Gabès (ENIG), University of Gabès,
Omar IBN EL KHATTAB Street, Gabès 6072, Tunisia

S. Bouallègue
High Institute of Industrial Systems of Gabès (ISSIG), Salaheddine
EL AYOUBI Street, 6011 Gabès, Tunisia

S. Vaidyanathan
Research and Development Centre, Vel Tech University, Avadi,
Chennai 600062, Tamil Nadu, India
e-mail: sundarvtu@gmail.com

© Springer International Publishing AG 2017
S. Vaidyanathan and C.-H. Lien (eds.), *Applications of Sliding Mode Control in Science and Engineering*, Studies in Computational Intelligence 709,
DOI 10.1007/978-3-319-55598-0_4

simulation results are carried out in order to show the effectiveness of the proposed normal SMC and TSMC approaches.

Keywords Quadrotor UAV · Modelling · Terminal sliding mode control (TSMC) · Attitude and altitude stabilization · Flight path tracking

1 Introduction

The Unmanned Aerial Vehicles (UAVs), particularly the Vertical Take-Off and Landing (VTOL) Quadrotors, are flying robots without pilot which are able to conduct missions in hostile and disturbed environments. Military and civilian missions, such as the inspection of dams and border monitoring, the prevention of forest fires and others, are the main applications of the Quadrotors [1, 5, 6, 13, 17, 18, 33]. Recently, these aircrafts have seen a great evolution in terms of the miniaturization, the modeling and especially the control design and prototyping [7]. This explains the interest shown by many researchers to study the flight dynamics and the control laws of these kinds of vehicles.

Sliding Mode Control (SMC) approach is a type of Variable Structure Control (VSC) strategies that uses a high speed switching control law to force the states trajectories to stay in a defined sliding surface [12, 16, 20–22, 30, 31]. It is a robust nonlinear control method which alters the dynamics of a system by using a dis-continuous control signal and forces the system to slide along a prescribed switching manifold. When the system states are in this switching surface, its dynamics in the closed-loop control are totally determined by this surface that usually has a linear form [21, 22]. This kind of control is well known for its robustness to stabilize such uncertain process with the presence of external dis-turbances and presents a good solution for the control and prototyping of the UAVs process.

In the literature, several works are realized based on the SMC concepts to stabilize and track a variety of real systems. Among these works, R. Xu and U. Özgüner synthesized a SMC control to drive an under-actuated system presented by a Quadrotor [30]. The same controller is implemented in [6] with a nonlinear observer design to stabilize an aircraft UAV. In [2–4, 15, 16, 20, 23–27], the authors applied the SMC approach on many industrial systems to solve accuracy and stability problems. However, this approach uses a commutation function to drive the states trajectories into an ideal hyper-plan surface. Consequentially, a chattering phenomenon has created at the control signal which effects directly on the stability of the controlled system. Moreover, the chosen sliding surface can guarantees asymptotic error convergence in the sliding manifold, but does not converge to zero in a finite-time [10–12, 16, 19].

Facing these structural problems, another class of SMC design, known as Terminal Sliding Mode Control (TSMC) approach, has been invented in 1993 by S.T. Venkataraman and S. Gulati to overcome these difficulties [28]. Compared to the classical SMC, or simply known as the Linear Sliding Mode Control (LSMC) approach, TSMC offers some superior properties such as the fast and finite-time convergence as well as the high performance on steady-state tracking precision in the controlled system responses [11, 12, 32, 33]. The main concept of TSM controller is to create a terminal attractor term in the sliding manifold, with a nonlinear form, that guarantees the rapid convergence in a finite time of the system states unlike the normal LSMC [9, 11, 20, 28].

Substantial research has been done and several works have widely realized with the recourse of the TSM control paradigm. They show the efficiency and the robustness of such a VSC approach to stabilize and tracking of various classes of uncertain and nonlinear systems. Among these works, we cite those developed in [33, 34] that implemented a TSM controller to drive the position and the attitude of a helicopter using inner and outer control loops. In [14], the authors synthesized a fuzzy TSM controller for the extracting of the maximum marine current energy. Another form of the TSM controllers is formulated in [19] for under water robot to eliminate the chattering problem in the reach phase by tuning the suitable values of odds numbers parameters. Many other works have been proposed for different mechanical and electrical types of physical process in [8, 10]. In this chapter, a TSMC approach is applied for the dynamics stabilization and tracking of a studied Quadrotor UAV. Such a nonlinear control technique is implemented and compared to the classical LSMC structure also applied for the same vehicle.

The remainder of this chapter is organized as follows. Section 2 presents the aerodynamic forces and torques of the Quadrotor in the VTOL flight mode. A dynamical model of the Quadrotor is then established thanks to the Newton-Euler formalism. In Sect. 3, a linear SMC approach is applied to stabilize and track the altitude and attitude dynamics of the studied drone. All mathematical developments are presented to describe the sliding manifold and derive the related control laws. Section 4 is devoted to the presentation of the proposed TSMC structure applied for the altitude, roll, pitch and yaw dynamics of the Quadrotor. The prescribed sliding manifold is given and then used to compute the related four TSM control laws. In Sect. 5, several simulations are carried out to validate the proposed control methods. Performances on steady-state precision and flight paths tracking for both SMC controllers are presented and discussed.

2 Modeling of the Quadrotor UAV

2.1 System Description and Aerodynamic Forces

The studied Quadrotor is detailed with its body-frame $R_B(O, x_b, y_b, z_b)$ and earth-frame $R_E(o, e_x, e_y, e_z)$ as shown in Fig. 1. Let denote by m the total mass of the Quadrotor, g the acceleration of the gravity and l the distance from the center of each rotor to the center of gravity (COG) of the drone.

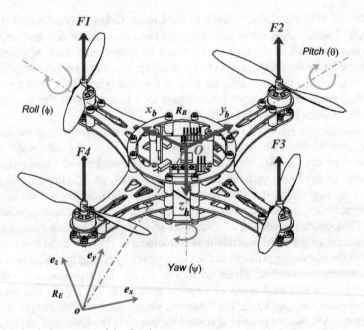

Fig. 1 Mechanical structure of the Quadrotor

The orientation of the Quadrotor is given by the rotation matrix $\mathbf{R}:R_E \to R_B$ which depends on the three Euler angles (ϕ, θ, ψ) and defined by the following equation:

$$\mathbf{R}(\phi, \theta, \psi) = \begin{pmatrix} c\psi c\theta & s\phi s\theta c\psi - s\psi c\theta & c\phi s\theta c\psi + s\psi s\phi \\ s\psi c\theta & s\phi s\theta s\psi + c\psi c\theta & c\phi s\theta s\psi - s\phi c\psi \\ -s\theta & s\phi c\theta & c\phi c\theta \end{pmatrix} \quad (1)$$

where $c(.) = \cos(.)$ and $s(.) = \sin(.)$.

Let consider the following model partitions of the Quadrotor naturally into translational and rotational coordinates:

$$\boldsymbol{\xi} = (x, y, z)^T \in \mathbb{R}^3, \quad \boldsymbol{\eta} = (\phi, \theta, \psi)^T \in \mathbb{R}^3 \quad (2)$$

where $\boldsymbol{\xi} = (x, y, z)^T$ denotes the position vector of the Quadrotor's COG relatively to its fixed earth-frame, $\boldsymbol{\eta} = (\phi, \theta, \psi)^T$ denotes the attitude of the Quadrotor given by the Euler angles for rolling $\phi \in [-\pi/2, \pi/2]$, pitching $\theta \in [-\pi/2, \pi/2]$ and yawing $\psi \in [-\pi, \pi]$ motions.

Let a vector $\boldsymbol{\nu} = (v_1, v_2, v_3)^T$ denote the linear velocity of the VTOL Quadrotor in the earth-frame R_E, while the vector $\boldsymbol{\vartheta} = (p, q, r)^T$ represents its angular velocity in R_B frame. The kinematic equations of rotational and translational movements are obtained, respectively, as follows [7, 13].

$$\begin{pmatrix} \dot{\phi} \\ \dot{\theta} \\ \dot{\psi} \end{pmatrix} = \begin{pmatrix} 1 & \sin\phi \tan\theta & \cos\phi \tan\theta \\ 0 & \cos\phi & -\sin\phi \\ 0 & \sin\phi \sec\theta & \cos\phi \sec\theta \end{pmatrix} \begin{pmatrix} p \\ q \\ r \end{pmatrix} \tag{3}$$

$$\nu_e = \mathbf{R}(\phi, \theta, \psi)\nu_B \tag{4}$$

where ν_e and ν_B are the linear velocities of the mass center expressed in the earth- and body-frames, respectively.

Each motor of the Quadrotor produces the force F_i which is proportional to the square of the angular speed. The trust force generated by the ith rotor of such a Quadrotor is given by:

$$F_i = \frac{1}{2}\rho\Lambda C_T r^2 \Omega_i^2 = b\Omega_i^2 \tag{5}$$

where ρ is the air density, r and Λ are the radius and the section of the propeller respectively, C_T is the aerodynamic thrust coefficient.

The aerodynamic drag torque, caused by the drag force at the propeller of the ith rotor and opposed the motor torque, is defined as follows:

$$\delta_i = \frac{1}{2}\rho\Lambda C_D r^2 \Omega_i^2 = d\Omega_i^2 \tag{6}$$

where C_D is the aerodynamic drag coefficient.

The pitch torque is a function of the difference $(F_3 - F_1)$ and the roll one is proportional to the term $(F_4 - F_2)$. The yaw torque is the sum of all reactions torques generated by the four rotors and is due to the shaft acceleration and the propeller drag. All these pitching, rolling and yawing torques are defined respectively as follows:

$$\Gamma_\theta = l(F_3 - F_1) \tag{7}$$

$$\Gamma_\phi = l(F_4 - F_2) \tag{8}$$

$$\Gamma_\psi = \mu(F_1 - F_2 + F_3 - F_4) \tag{9}$$

where μ is a proportional positive coefficient.

Two gyroscopic effects torques, due to the motion of the propellers and the Quadrotor body, are additively provided. These moments are given respectively by:

$$\mathbf{M}_{gp} = \sum_{i=1}^{4} \mathbf{\Omega} \wedge \left(0, 0, J_r(-1)^{i+1}\Omega_i\right)^T \tag{10}$$

$$\mathbf{M}_{gb} = \mathbf{\Omega} \wedge J\mathbf{\Omega} \tag{11}$$

where Ω is the vector of the angular velocities in the fixed earth-frame and $J = diag(I_{xx}, I_{yy}, I_{zz})$ is the inertia matrix of the Quadrotor, J_r denotes the z-axis inertia of the propellers' rotors.

The Quadrotor is controlled by independently varying the speed of its four rotors. Hence, these control inputs are defined as follows:

$$\begin{pmatrix} u_1 \\ u_2 \\ u_3 \\ u_4 \end{pmatrix} = \begin{pmatrix} F \\ \Gamma_\phi \\ \Gamma_\theta \\ \Gamma_\psi \end{pmatrix} = \begin{pmatrix} b & b & b & b \\ 0 & -lb & 0 & lb \\ -lb & 0 & lb & 0 \\ d & -d & d & -d \end{pmatrix} \begin{pmatrix} \Omega_1^2 \\ \Omega_2^2 \\ \Omega_3^2 \\ \Omega_4^2 \end{pmatrix} \tag{12}$$

where $\Omega_{1,2,3,4}$ are the angular speeds of the four rotors, respectively.

From Eq. (12), it can be observed that the input u_1 denotes the total thrust force on the Quadrotor body in the z-axis, the inputs u_2 and u_3 represent the roll and pitch torques, respectively. The input u_4 represents the yawing torque.

2.2 Dynamic System Model

While using the Newton-Euler formalism for modeling, the Newton's laws lead to the following motion equations of the studied Quadrotor:

$$\begin{cases} m\ddot{\xi} = F_{th} + F_d + F_g \\ J\dot{\Omega} = M - M_{gp} - M_{gb} - M_a \end{cases} \tag{13}$$

where $F_{th} = \mathbf{R}(\phi, \theta, \psi)\left(0, 0, \sum_{i=1}^{4} F_i\right)^T$ denotes the total thrust force of the four rotors, $F_d = diag(\kappa_1, \kappa_2, \kappa_3)\nu^T$ is the air drag force which resists to the Quadrotor motion, $F_g = (0, 0, mg)^T$ is the gravity force, $\mathbf{M} = \left(\Gamma_\phi, \Gamma_\theta, \Gamma_\psi\right)^T$ represents the total rolling, pitching and yawing torques, $\mathbf{M}_{gp}$ and $\mathbf{M}_{gb}$ are the gyroscopic torques and $\mathbf{M}_a = diag(\kappa_4, \kappa_5, \kappa_6)\vartheta^T$ is the torque resulting from aerodynamic frictions.

Substituting the position vector and the forces expressions into the Eq. (13), we have the following translational dynamics of the Quadrotor:

$$\begin{cases} \ddot{x} = \frac{1}{m}(c\phi c\psi s\theta + s\phi s\psi)u_1 - \frac{\kappa_1}{m}\dot{x} \\ \ddot{y} = \frac{1}{m}(c\phi s\psi s\theta - s\phi c\psi)u_1 - \frac{\kappa_2}{m}\dot{y} \\ \ddot{z} = \frac{1}{m}c\phi c\theta u_1 - g - \frac{\kappa_3}{m}\dot{z} \end{cases} \tag{14}$$

From the second part of Eq. (13), and while substituting each moment by its expression, we deduce the following rotational dynamics of the rotorcraft:

$$\begin{cases} \ddot{\phi} = qr\frac{I_y - I_z}{I_x} - \frac{J_r}{I_x}\Omega_r q - \frac{\kappa_4}{I_x}p + \frac{1}{I_x}u_2 \\ \ddot{\theta} = pr\frac{I_z - I_x}{I_y} + \frac{J_r}{I_y}\Omega_r p - \frac{\kappa_5}{I_y}q + \frac{1}{I_y}u_3 \\ \ddot{\psi} = pq\frac{I_x - I_y}{I_z} - \frac{\kappa_6}{I_z}r + \frac{1}{I_z}u_4 \end{cases} \tag{15}$$

According to these above established equations, i.e. Eqs. (14) and (15), $x = (\phi, \dot{\phi}, \theta, \dot{\theta}, \psi, \dot{\psi}, x, \dot{x}, y, \dot{y}, z, \dot{z})^T \in \mathbb{R}^{12}$ is retained as the state vector of the Quadrotor nonlinear model. Note that $\kappa_{1,2,...,6}$ are the aerodynamic friction and translational drag coefficients, $\Omega_r = \Omega_1 - \Omega_2 + \Omega_3 - \Omega_4$ is the overall residual rotor angular velocity.

3 Conventional Sliding Mode Control of the Quadrotor

3.1 Preliminaries and Problem Formulation

Sliding Mode Control (SMC) approach is a kind of robust Variable Structure Control (VSC) family that produces a switching control law in order to force the states trajectories of the controlled system to follow and maintain a defined linear sliding surface [16, 20–22, 30, 31].

In this framework, the sliding mode control law consists of two parts as given as follows:

$$u(t) = u_c(t) + u_{eq}(t) \tag{16}$$

where $u_c(t)$ is the corrective term which compensates the variations of the states trajectories from the time-varying sliding surface $s(x, t)$ in order to reach it, and the part $u_{eq}(t)$ is the equivalent control law which can be interpreted as the continuous control law that would maintain $\dot{s} = 0$ if the dynamics were exactly known.

Let consider a second-order nonlinear system given as follows:

$$\begin{cases} \dot{x}_1 = x_2 \\ \dot{x}_2 = f(x) + g(x)u + \delta(x) \end{cases} \tag{17}$$

where $x = [x_1, x_2]^T \in \mathbb{R}^2$ represents the system state vector, $f(x)$ and $g(x) \neq 0$ are two smooth nonlinear functions of x, u is the control input, and $\delta(x)$, which represents the uncertainties and the disturbances, is assumed to satisfy $|\delta(x)| \leq \delta_{max}$, where $\delta_{max} > 0$ is a constant.

The task of SMC approach is to design a control law $u(x)$ to stabilize the considered system of Eq. (17) around the origin $x = [0, 0]^T$ based on the requirements of the controlled system.

Usually, the sliding surface used in the conventional SMC framework for a nonlinear system with order n, can be chosen as follows [21]:

$$s(\boldsymbol{x}, t) = \left(\frac{d}{dt} + \lambda\right)^{n-1} e \qquad (18)$$

where λ is a design strictly positive constant and $e = x_1 - x_{1d}$ is the output tracking error and x_{1d} is the desired trajectory.

For the second-order system of Eq. (17), it is obvious that the sliding manifold is given as follows:

$$s(\boldsymbol{x}, t) = \lambda e + \dot{e} \qquad (19)$$

As given in [21], we obtain the equivalent SMC law for the system (17) by solving the equation $\dot{s}(\boldsymbol{x}, t) = \lambda \dot{e} + \ddot{e} = \lambda \dot{e} + \dot{x}_2 - \ddot{x}_{1d} = 0$ formally for the control input. The equivalent control law, $u_{eq}(t)$, is based on the nominal plant parameters with $\delta(\boldsymbol{x}, t) = 0$ and provides the main control action. So, we have:

$$u_{eq} = \frac{1}{g(\boldsymbol{x})} \left(\ddot{x}_{1d} - f(\boldsymbol{x}) - \lambda \dot{e}\right) \qquad (20)$$

A Lyapunov function is then defined to determine the corrective term $u_c(t)$ of the SMC control law that satisfied the following condition [22]:

$$V(s, t) = \frac{1}{2} s^2(t) > 0 \qquad (21)$$

The derivative time of $V(s, t)$ must be negative definite, as given by the relationship of Eq. (22) that defines the well known reaching condition, in order to satisfy the stability conditions of the controlled system and that the trajectory of the error will translate from reaching phase to sliding phase [3, 21, 22]:

$$\dot{V}(s, t) = s(t)\dot{s}(t) < 0, \quad s(t) \neq 0 \qquad (22)$$

Consequently, the corrective law is obtained with an exponential reaching law, which is defined as follows:

$$u_c = -k \operatorname{sgn}(s) \qquad (23)$$

where k is a design positive constant, $k > \delta_{max}$ to dominate the matching uncertainties and sgn(.) denotes the sign function defined as:

$$\operatorname{sgn}(s) = \begin{cases} 1, & s > 0 \\ 0, & s = 0 \\ -1, & s < 0 \end{cases} \qquad (24)$$

Finally, the complete expression of the SMC control law for the second-order nonlinear system (17) has the following form:

$$u = \frac{1}{g(x)}(\ddot{x}_{1d} - f(x) - \lambda\dot{e}) - k\,\mathrm{sgn}(s) \tag{25}$$

3.2 Controllers Design

In this subsection, the first order SMC approach, which the control law is given in Eq. (25), is implemented on the studied Quadrotor in order to stabilize its altitude and attitude dynamics.

We begin with the altitude controller design, which ensures the tracking of a reference trajectory of z state. The output error considered for this SMC design is defined as follows:

$$e_z = z_d - z \tag{26}$$

where z_d denotes the desired altitude trajectory.

The altitude control is mainly based on the dynamic equation given in the third part of Eq. (14) that creates the react aerodynamic forces on z state. So, the corrective and equivalent control laws are obtained respectively as follows:

$$u_{c1} = -k_z\,\mathrm{sgn}(s_z) \tag{27}$$

$$u_{eq1} = \frac{m}{c\phi c\theta}(\ddot{z}_d + g + \kappa_3\dot{z} + \lambda_z\dot{e}_z) \tag{28}$$

where s_z denotes the z altitude sliding manifold that is derived from Eq. (19) and given by:

$$s_z(t) = \lambda_z e_z + \dot{e}_z \tag{29}$$

According to Eq. (16), the complete first order SMC law for the stabilization of the Quadrotor altitude dynamics is deduced as follows:

$$u_1 = \frac{m}{c\phi c\theta}(\ddot{z}_d + g + \kappa_3\dot{z} + \lambda_z\dot{e}_z) - k_z\,\mathrm{sgn}(s_z) \tag{30}$$

where $\lambda_z > 0$ and $k_z > 0$ are the design parameters of the SMC of the Quadrotor altitude dynamics.

When following exactly the same steps of Eqs. (25)–(29), the first order SMC laws for the roll, pitch and yaw dynamics of Quadrotor UAV are given as follows:

$$u_2 = I_x \left(qr \frac{I_y - I_z}{I_x} + \frac{\kappa_4}{I_x} p + \lambda_\phi \dot{e}_\phi \right) - k_\phi \, \text{sgn}(s_\phi) \tag{31}$$

$$u_3 = I_y \left(pr \frac{I_z - I_x}{I_y} + \frac{\kappa_5}{I_y} q + \lambda_\theta \dot{e}_\theta \right) - k_\theta \, \text{sgn}(s_\theta) \tag{32}$$

$$u_4 = I_z \left(pq \frac{I_x - I_y}{I_z} + \frac{\kappa_6}{I_z} r + \lambda_\psi \dot{e}_\psi \right) - k_\psi \, \text{sgn}(s_\psi) \tag{33}$$

where (λ_ϕ, k_ϕ), $(\lambda_\theta, k_\theta)$, and (λ_ψ, k_ψ) are the design positive parameters for the SMC of the roll, pitch and yaw dynamics, respectively. The terms s_ϕ, s_θ, and s_ψ denote the sliding manifolds of the roll, pitch and yaw dynamics, respectively.

4 Terminal Sliding Mode Control of the Quadrotor

4.1 Preliminaries and Problem Formulation

In 1993, S.T. Venkataraman and S. Gulati have proposed an advanced variant of the traditional SMC approach, known as the Terminal Sliding Mode Control (TSMC) [28]. This nonlinear control structure is based on the conception of a terminal attractor that guarantees a finite time convergence, unlikely with the traditional SMC which attaints an asymptotical stability but in infinite time [9, 19, 29]. In TSMC framework, the sliding surface is designed thanks to a nonlinear term, i.e. the sliding manifold is formulated as an attractor and consequently, the state trajectories are attracted with this new manifold and converge in finite time [8, 10–12, 33, 34].

Consider once again the second-order nonlinear system of Eq. (17). A TSM sliding manifold can be selected in the form [11, 16]:

$$s(\boldsymbol{x}, t) = x_2 + \beta x_1^{q/p} \tag{34}$$

where $\beta > 0$ is a design constant, both p and q are positive odd integers that satisfy the conditions $(q < p)$ and $(1 < p/q < 2)$. All these parameters determine the TSM controller behavior.

As shown in [11], the second-order system (17) will reach the sliding mode surface $s = 0$ within finite-time and converge to zero along $s = 0$ within finite-time, if the sliding mode manifold is chosen as Eq. (34) and the control law is designed as follows:

$$u = -\frac{1}{g(\boldsymbol{x})} \left(f(\boldsymbol{x}) + \beta \frac{q}{p} x_1^{q/p-1} x_2 + (\delta_{\max} + \eta) \text{sgn}(s) \right) \tag{35}$$

where $\eta > 0$ is a design constant.

The finite-time and stability analysis of this particular structure of SMC approach are found in [16]. Several methods have been proposed in the literature to overcome the famous singularity problem of this kind of sliding mode control scheme [8, 10–12].

4.2 Controllers Design

Like for the above traditional SMC approach of the studied Quadrotor, this chapter subsection is devoted to the application of the proposed TSMC structure, given in Eq. (35), for the aircraft UAV altitude, roll, pitch and yaw dynamics.

We begin by the altitude dynamics control which is based on the differential equation of z state given in Eq. (14). In this case, let us note that $x_1 = z$ and $x_2 = \dot{z}$. According to the TSMC law of Eq. (35), we deduce the control signal form for the Quadrotor altitude dynamics as follows:

$$u_1 = \frac{m}{c\phi c\theta}\left(g + \beta_z \frac{q_z}{p_z} z^{q_z/p_z - 1}\dot{z} + (\delta_z + \eta_z)\mathrm{sgn}(s_z) \right) \tag{36}$$

where β_z, (p_z, q_z), δ_z and η_z are the positive design TSMC parameters relative to the controlled Quadrotor altitude dynamics, s_z denotes the sliding manifold of the altitude state as defined in Eq. (34).

The remains TSMC laws for the Quadrotors roll, pitch and yaw dynamics are synthesized by following exactly the same steps used in the altitude dynamics case, i.e. the form of Eq. (36). Therefore, we obtain respectively the following control laws:

$$u_2 = I_x\left(\frac{I_y - I_z}{I_x}qr + \beta_\phi \frac{q_\phi}{p_\phi} \phi^{q_\phi/p_\phi - 1}\dot{\phi} + (\delta_\phi + \eta_\phi)\mathrm{sgn}(s_\phi) \right) \tag{37}$$

$$u_3 = I_y\left(\frac{I_z - I_x}{I_y}pr + \beta_\theta \frac{q_\theta}{p_\theta} \theta^{q_\theta/p_\theta - 1}\dot{\theta} + (\delta_\theta + \eta_\theta)\mathrm{sgn}(s_\theta) \right) \tag{38}$$

$$u_4 = I_z\left(\frac{I_x - I_y}{I_z}pq + \beta_\psi \frac{q_\psi}{p_\psi} \psi^{q_\psi/p_\psi - 1}\dot{\psi} + (\delta_\psi + \eta_\psi)\mathrm{sgn}(s_\psi) \right) \tag{39}$$

where $\beta_\phi, \beta_\theta, \beta_\psi$, (p_ϕ, q_ϕ), (p_θ, q_θ), $(p_\psi, q_\psi)\delta_\phi$, δ_θ, δ_ψ, η_ϕ, η_θ and η_ψ are all positive design parameters for the TSM control of the Quadrotor roll, pitch and yaw dynamics, respectively. The terms s_ϕ, s_θ and s_ψ denote the sliding manifolds of the roll, pitch and yaw states, respectively, as mentioned in Eq. (34) with their general form.

Table 1 Quadrotor model parameters

Parameters	Values
Mass m	0.650 (kg)
Rotor distance to COG l	0.23 (m)
Lift coefficient b	2.9e-05 (N/rad/s)
Drag propellers coefficient d	3.23e-07 (Nm/rad/s)
Body inertias I_{xx}, I_{yy} and I_{zz}	0.0075, 0.0075, 0.013 (kg m^2)
Translational drag coefficients κ_1, κ_2 and κ_3	5.57e-04, 5.57e-04, 6.35e-04 (N/rad/s)
Aerodynamic friction coefficients κ_4, κ_5 and κ_6	5.57e-04, 5.57e-04, 6.35e-04 (N/m/s)
Acceleration of the gravity g	9.81 (m/s^2)

5   Simulation Results and Discussion

In this section, both TSMC and classical SMC approaches for the Quadrotor attitude and altitude stabilization are implemented and discussed. Controllers' performances on steady-state tracking precision are also shown. The used physical parameters of the Quadrotor UAV are given in Table 1.

5.1   Linear SMC Design

Numerical simulations are performed for this SMC structure and lead to the results of Figs. 2, 3 and 4. The sliding mode control laws of Eqs. (30)–(33) are implemented with the following design parameters $(\lambda_z = 1.7, k_z = 2.66)$, $(\lambda_\phi = 2.5, k_\phi = 1.99)$, $(\lambda_\theta = 2.5, k_\theta = 1.99)$ and $(\lambda_\psi = 5.24, k_\psi = 1.72)$ for the altitude, roll, pitch and yaw Quadrotor dynamics, respectively.

It can be observed from Fig. 2 that the controlled system states track its references trajectories in an acceptable time with the presence of external disturbances. The asymptotic stability objectives are guaranteed for all controlled dynamics. However, at the reach phase of control signals in Fig. 3, a chattering effect is produced due to the toggle effect of switching function that is effects on the accuracy and the robustness of system responses.

For the trajectories tracking performance, we give in Fig. 4 an example of path tracking illustration of the Quadrotor in 3D frame. Starting from an arbitrary initial position, the SMC-based controlled Quadrotor UAV can track with high performance the pre-specified desired trajectory.

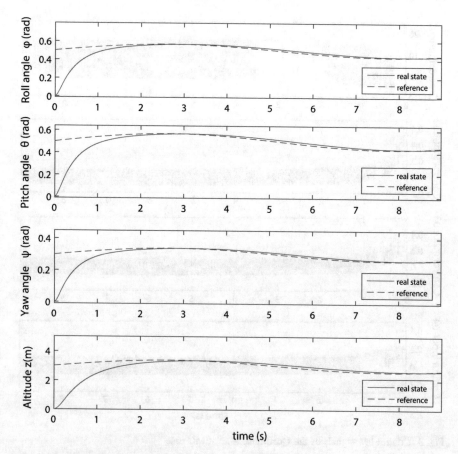

Fig. 2 System responses using the linear SMC controllers

5.2 Terminal SMC Design

Numerical simulations are performed for this TSMC structure and lead to the results of Figs. 5, 6 and 7. The terminal sliding mode control laws of Eqs. (36)–(39) are implemented for the studied Quadrotor UAV.

TSMC design parameters are manually adjusted to reach good performances such as the reducing of the chattering problems as well as the finite-time and the fastness convergence of the tracking errors and steady-state precision. All these parameters are chosen as $(\beta_z = 2, \beta_\phi = \beta_\theta = \beta_\psi = 1)$, $(q_z = 7, q_\phi = q_\theta = q_\psi = 5)$, $(p_z = p_\phi = p_\theta = p_\psi = 9)$ and $(\eta_z = 15, \eta_\phi = \eta_\theta = \eta_\psi = 5)$, relatively for the altitude, roll, pitch and yaw Quadrotor dynamics.

Compared to the results obtained for the linear SMC approach, we can clearly observe that the performances on steady-state tracking precision are improved and the chattering phenomenon on control signals is farther reduced. Indeed, the

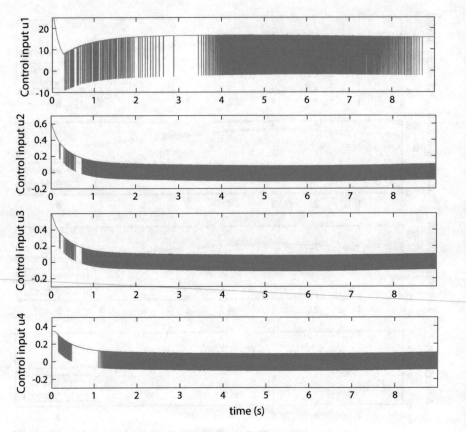

Fig. 3 Control law signals of the Quadrotor: linear SMC case

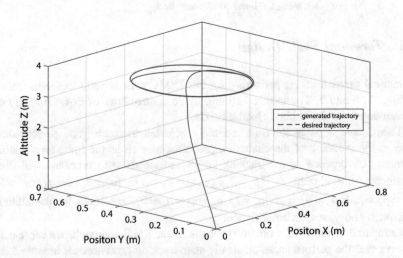

Fig. 4 Desired trajectories tracking: linear SMC case

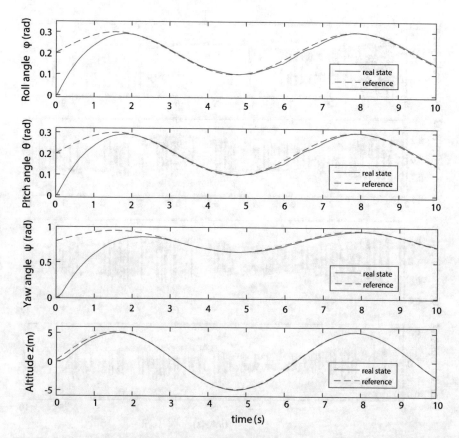

Fig. 5 System responses using the terminal SMC controllers

chattering effect is reduced due to remove the commutation function from the control law. The control signals depicted in Fig. 6 toggle up and down to force the controlled states to maintain on the sliding manifold with the presence of external disturbances. Stabilization and tracking objectives of the TSM controller are made with a high performances and the effectiveness of the proposed control approach is guaranteed.

On the other hand, the chattering problem stell exists in the proposed terminal SMC approach can be completely eliminated while replacing the signum function in the control laws, given in Eqs. (36)–(39), by a saturation function, as shown in the literature [11].

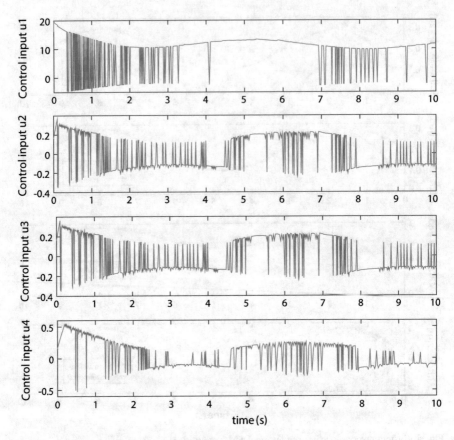

Fig. 6 Control law signals of the Quadrotor: terminal SMC case

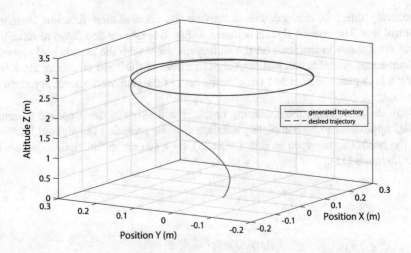

Fig. 7 Desired trajectories tracking: terminal SMC case

6 Conclusion

In this chapter, a nonlinear dynamical model of a Quadrotor UAV is firstly established using the Newton-Euler formalism. All aerodynamic forces and moments of the studied Quadrotor UAV are described within an inertial frame. Such an established dynamical model is then used to design both linear and terminal SMC structures for the stabilization and tracking of the Quadrotor altitude and attitude. Demonstrative simulation results are carried out and have shown the useful and the effectiveness of the proposed sliding mode control approaches for the stabilization and the trajectories tracking of the Quadrotor UAV. Compared to the results obtained for the linear SMC approach implementation, the performances on finite-time and fastness of the steady-state tracking precision as well as the chattering phenomenon reduction are further improved with the application of the terminal SMC algorithms. All used SMC design parameters, for both linear and terminal variants of the control approach, are manually adjusted thanks to trials-errors based procedure. Furthercoming works deal with the optimization of these design parameters based on advanced global metaheuristics approches.

References

1. Austin R (2010) Unmanned aircraft systems: UAVs design, development and deployment, 1st edn. Wiley, London
2. Bandyopadhyay B, Janardhanan S (2006) Discrete-time sliding mode control: a multirate output feedback approach, 1st edn. Springer, Heidelberg
3. Bandyopadhyay B, Shyam K (2015) Stabilization and control of fractional order systems: a sliding mode approach, 1st edn. Springer International Publishing, Switzerland. ISBN: 978-3-319-08620-0
4. Bartoszewicz A (ed) (2011). Sliding mode control. 1st edition, InTech, Rijeka, Croatia. ISBN: 978-953-307-162-6
5. Ben Ammar N, Bouallègue S, Haggège J (2016) Modeling and sliding mode control of a quadrotor unmanned aerial vehicle. In: Proceedings of the 3th international conference on automation, control engineering and computer science (ACECS 2016), Hammamet, Tunisia, pp 834–840
6. Bouadi H, Tadjine M (2007) Nonlinear observer design and sliding mode control of four rotors helicopter. I J Mech Aerosp Ind Mechatron Manuf Eng 1(7):329–334
7. Bouallègue S, Fessi R (2016) Rapid control prototyping and PIL co-simulation of a quadrotor UAV based on NI myRIO-1900 board. I J Adv Comput Sci Appl 7(6):26–35
8. Cao Q, Li S, Zhao D (2016) Full-order multi-input/multi-output terminal sliding mode control for robotic manipulators. I J Model Ident Control 15(1):17–27
9. Chen M, Wu QX, Cui RX (2013) Terminal sliding mode tracking control for a class of SISO uncertain nonlinear systems. ISA Trans 52(2):198–206
10. Feng Y, Yu X, Man Z (2002) Non-singular terminal sliding mode control of rigid manipulators. Automatica 38:2159–2167
11. Feng Y, Yu X, Han F (2013) On nonsingular terminal sliding mode control of nonlinear systems. Automatica 49:1715–1722
12. Feng Y, Han F, Yu X (2014) Chattering free full-order sliding mode control. Automatica 50:1310–1314

13. Fessi R, Bouallègue S (2016) Modeling and optimal LQG controller design for a quadrotor UAV. In: Proceedings of the 3th international conference on automation, control engineering and computer science (ACECS 2016), Hammamet, Tunisia, pp 264–270
14. Gu YJ, Yin XX, Liu HW, Wei Li W, Lin YG (2015) Fuzzy terminal sliding mode control for extracting maximum marine current energy. Energy 90(1):258–265
15. Lakhekar GV, Waghmare LM, Vaidyanathan S (2016) Diving autopilot design for underwater vehicles using an adaptive neuro-fuzzy sliding mode controller. Stud Comput Intell 635: 477–503
16. Liu J, Wang X (2012) Advanced sliding mode control for mechanical systems: design. Springer, Heidelberg, Analysis and MATLAB Simulation
17. Lozano R (ed) (2010) Unmanned aerial vehicles: embedded control, 1st edn. Wiley, New York
18. Nagati A, Saeedi S, Thibault C, Seto M, Li H (2013) Control and navigation framework for quadrotor helicopters. J Intell Rob Syst 70:1–12
19. Park K-B, Tsuji T (1999) Terminal sliding mode control of second-order nonlinear uncertain systems. Int J Robust Nonlinear Control 9(11):769–780
20. Qian D, Yi J (2015) Hierarchical sliding mode control for underactuated cranes: design. Springer, Heidelberg, Analysis and Simulation
21. Slotine J-JE, Li W (1991) Applied nonlinear control, 1st edn. Prentice Hall, New Jersey
22. Utkin V (1992) Sliding modes in control and optimization, 1st edn. Springer, Heidelberg
23. Vaidyanathan S (2015) Sliding mode control of rucklidge chaotic system for nonlinear double convection. Int J ChemTech Res 8(8):25–35
24. Vaidyanathan S (2015) Integral sliding mode control design for the global chaos synchronization of identical novel chemical chaotic reactor systems. Int J ChemTech Res 8 (11):684–699
25. Vaidyanathan S (2015) Sliding controller design for the global chaos synchronization of enzymes-substrates systems. Int J PharmTech Res 8(7):89–99
26. Vaidyanathan S (2015) Global chaos synchronization of chemical chaotic reactors via novel sliding mode control method. Int J ChemTech Res 8(7):209–221
27. Vaidyanathan S, Sampath S, Azar AT (2015) Global chaos synchronisation of identical chaotic systems via novel sliding mode control method and its application to Zhu system. Int J Model Ident Control 23(1):92–100
28. Venkataraman ST, Gulati S (1993) Control of nonlinear systems using terminal sliding modes. J Dyn Syst Meas Contr 115(3):554–560
29. Wu Y, Yu X, Man Z (1998) terminal sliding mode control design for uncertain dynamic systems. Syst Control Lett 34:281–287
30. Xu R, Özgüner U (2008) Sliding mode control of a class of underactuated systems. Automatica 44(1):233–241
31. Young KD, Özgüner U (eds) (1999) Variable structure systems, sliding mode and nonlinear control, 1st edn. Springer, London
32. Yu S, Yu X, Man Z (2000) Robust global terminal sliding mode control of SISO nonlinear uncertain systems. In: Proceedings of the 39th IEEE conference on decision and control (CDC 2000), Sydney, pp 2198–2203
33. Zheng E, Xiong J (2014) Quad-rotor unmanned helicopter control via novel robust terminal sliding mode controller and under-actuated system sliding mode controller. Int J Light Electron Opt 125(12):2817–2825
34. Zhou W, Zhu P, Wang C (2015) Position and attitude tracking control for a quadrotor UAV based on terminal sliding mode control. In: Proceedings of the 34th Chinese control conference (CCC 2015), Hangzhou, China, pp 3398–3404

Insensibility of the Second Order Sliding Mode Control via Measurement Noises: Application to a Robot Manipulator Surveillance Camera

Marwa Fathallah, Fatma Abdelhedi and Nabil Derbel

Abstract The Sliding Mode Control (SMC) is a widely spread approach thanks to its efficiency and robustness. However, it suffers from the undesirable chattering phenomenon which leads to the mechanical system damage. Then, this study takes into consideration the second order SMC which is known by its reliability regarding disturbances and nonlinear uncertainties, in order to improve the system stability and performances. In fact, such a high order SMC ensures a trade off between chattering reduction and disruption resistance. First, a second order sliding mode controller has been elaborated. Then, and in order to test the robustness of the proposed strategy, a measurement noise has been introduced. Simulation results show the efficiency of the proposed second order SMC applied to a robot manipulator system in a motion control task, which is used to ensure the displacement of a surveillance camera.

Keywords Sliding mode control · Second order sliding mode control · Robot manipulator · Measurement noise · Motion control

1 Introduction

In order to guarantee required performances of a closed loop system, the design of an adequate control law that can be able to overcome all control problems remains an interesting challenge. In this context, classic methods such as PID controllers are unable to achieve expected performances, mainly in the robotic field case, known by

M. Fathallah (✉) · F. Abdelhedi
Control & Energy Management Laboratory (Cemlab), Sfax Engineering School,
University of Sfax, Sfax, Tunisia
e-mail: fathallahmaroua@gmail.com

F. Abdelhedi
e-mail: fatma.abdelhedi@live.com

N. Derbel
Sfax Engineering School, University of Sfax, Sfax, Tunisia
e-mail: n.derbel@enis.rnu.tn

© Springer International Publishing AG 2017
S. Vaidyanathan and C.-H. Lien (eds.), *Applications of Sliding Mode Control
in Science and Engineering*, Studies in Computational Intelligence 709,
DOI 10.1007/978-3-319-55598-0_5

the presence of intense nonlinearities [1, 2]. For this reason, the formulation of robust control methods is highly recommended. One particular control method, namely the sliding mode control seems to be able to solve several control problems [3–5].

As one of the most influential nonlinear controllers, it has been prouved that SMC is able to submit two major control challenges: the system stability and robustness [6]. Such a variable structure controller (VSC) is able to deal with several difficulties encountered by different classes of systems [7], such as the power electronic systems, MIMO system, nonlinear system, discrete time models [8], etc. Basically, the SMC is presented as a discontinuous feedback control, where system trajectories are forced to reach and then to remain on a specific surface in the state space which is called the sliding surface. Then, system states have to evolve according to some specified sliding dynamics [9]. However, the undesirable chattering phenomenon is the main drawback of such a VSC method, whose impact is reflected by the presence of disrupting high switching frequencies in the system control input [8, 10]. Specifically, the problem consists of rapid and sudden changing control signals which lead to a low control accuracy, high wear and even damage the system mechanical parts. For this reason, several methods have been developed to overcome this problem and to improve the SMC performances [11, 12], such as intelligent approaches including the Fuzzy logic control [8, 13–15], the interpolation of the control inside the boundary layer [16, 17], as well as methods based on an approximation of the signum function (to be replaced for example by a saturation one in the control law expression [18]). Currently, nonlinear systems have attracted several researchers in various industrial fields, especially as the noise phenomenon is evident and imposes itself in any real dynamic process. In the light of what was said, and in order to mitigate these discruptions, the formulation of a robust controller which can prove a high capacity of insensitivity to these drawbacks becomes a requisite task. In this context, we propose a method which allow not only to test its robustness via disturbances and noises, but also to reduce the amplitude of the switching part of the controller (which eliminate the chattering drawback), by the use of a second order (SOSMC). In fact, by keeping main advantages of the classical sliding mode approach, the SOSMC presents an efficient approach to maintain the dynamic system stability, to achieve better performances even in the presence of parametric imprecisions. It can be also considered as a solution to remove the chattering phenomenon and to provide higher accuracy in practice [19, 20]. Besides, the adopted control strategy has been implemented on a surveillance camera system, inasmuch as the surveillance task is considered as an essential and important part of several industrial fields, in order to ensure the security of goods and human beings which present a major safety goal in the society [21]. Actually, the second order sliding mode approach is the most useful among the high order SMC thanks to its capability of reducing the chattering phenomenon and its relative simplicity of application, compared to the higher order controls. In this chapter, the presented work introduces the second order sliding mode control which can be exhibit on a 3 DOF surveillance camera system, where we focus on the manipulative arm managing the camera movements. Then, and in order to test and then to prove the effectiveness of the proposed strategy, an additive measurement noise has been introduced. This chapter is organized as follows. Section 2 gives a brief presentation of the mathematical model of a manipulator robot, on which the proposed control has been applied in order to illustrate the effectiveness of the proposed controller. In this context, the classical SMC has been introduced in Sect. 3 while the second order SMC has been developed in Sect. 4. After that and in order to test the robustness of the proposed controller against the disturbances, an additive measurement noise has been introduced in Sect. 5. The higher order controller has been applied to a 3 DOF robot manipulator surveillance camera and simulation results have been demonstrated in Sect. 6.

2 Mathematical Model of the System

A robotic manipulator system that serves to be equipped with a surveillance camera system presents the adopted dynamic model of this study. Using the Lagrangian formulation, the motion equations of a manipulator robot can be written as:

$$M(q)\ddot{q} + C(q,\dot{q})\dot{q} + G(q) = \tau \tag{1}$$

where:

- $q(t) \in \mathbb{R}^n$ is the measured articulation angles of the manipulator,
- $\dot{q} \in \mathbb{R}^n$ is the velocity vector,
- $\ddot{q} \in \mathbb{R}^n$ is the joint acceleration vector,
- $M(q) \in \mathbb{R}^{n \times n}$ is symmetric uniformly bounded and positive definite inertia matrix,
- $C(q,\dot{q})\dot{q} \in \mathbb{R}^n$ represents the Coriolis and the centrifugal forces,
- $G(q) \in \mathbb{R}^n$ is the vector of gravitational torques,
- $u = \tau \in \mathbb{R}^n$ denotes the control torques vector applied to each articulation.

3 Classical Sliding Mode Control

3.1 Preliminaries

State equations of system (1) can be described as:

$$\begin{cases} \dfrac{dx}{dt} = f(x) + g(x)u \\ \quad y = q \end{cases} \tag{2}$$

where $x = [q^T \ \dot{q}^T]^T \in \mathbb{R}^{2n}$, and $u = \tau \in \mathbb{R}^n$. The main objective of the trajectory tracking control is to achieve the appropriate input torque τ, that allows to the angular position q to follow the desired reference q_d.

The tracking error can be defined as follows:

$$e = q - q_d \tag{3}$$

The $2n$-dimensional state variable x is required to achieve the sliding condition:

$$S(x) = \dot{e} + 2\lambda e + \lambda^2 \int e \, dt = 0 \tag{4}$$

where λ is a positive constant.

The most important task is to conceive a switched control that pushes the system state to attain the sliding surface and to maintain it on the surface. In order to achieve this task, the Lyapunov theory with condition (5) are required to ensure the convergence to $S(x) = 0$ while meeting condition (5).

$$S^T(x)\dot{S}(x) < 0, \tag{5}$$

3.2 Control Design

The SMC is a Variable Structure Control strategy which is able to control several classes of systems. In general, this VSC formulation requires discontinuous feedback control laws whose role is to force the system trajectory to reach and then to remain on the particular sliding surface [22]. The control structure is composed of two parts (Fig. 1):

$$u = u_{eq} + \Delta u \tag{6}$$

Fig. 1 General SMC principal

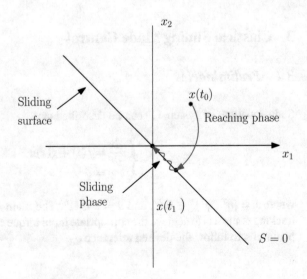

where u_{eq} represents the equivalent control that ensures the convergence to the sliding surface and then to remain there. The object of the corrective term Δu is to avoid all deviations from the sliding surface caused by the presence of perturbations. The expression of the equivalent control $u_{eq}(t)$ can be deduced from the following equation:

$$\dot{S}(x) = F(x) + G(x)u = 0 \tag{7}$$

Hence, the obtained equivalent control expression is presented as:

$$u_{eq} = -[G(x)]^{-1}F(x) \tag{8}$$

under the regularity of matrix $G(x)$. Besides, the discontinuous term Δu is defined as (9):

$$\Delta u = -[G(x)]^{-1}K\text{sign}(S) \tag{9}$$

where K is a positive definite diagonal matrix.

As it is mentioned above, the control law (6) suffers from the dangerous chattering phenomenon induced by the discontinuous term Δu [23, 24].

4   Second Order Sliding Mode Controller

Currently, the second order sliding mode approach can be considered as the most useful among the high order SMC thanks to its capability of reducing the chattering phenomenon and its relative simplicity of application [25]. Then, the sliding surface expression of the second order SMC approach has been modified as follows [26]:

$$\dot{S} = \sigma \quad \mapsto \quad \Delta u = [G(x)]^{-1}\sigma \tag{10}$$

The new description of the dynamic control behavior is defined as follows:

$$\begin{cases} \dot{S} = \sigma \\ \dot{\sigma} = -a_0 S - a_1 \sigma + v \end{cases} \tag{11}$$

where v is a new control variable.

System (10) is linear. Its characteristic equation which is a Hurwitz polynomial can be chosen as:

$$A(p) = (p + \mu)^{2n} \tag{12}$$

where μ is a positive scalar.

In order to ensure the stability, the dynamic control behaviour's representation can be reformulated as the following condensed form:

$$\dot{Z} = \phi Z + \Gamma v \tag{13}$$

where: $Z = \begin{pmatrix} \dot{S} \\ \dot{\sigma} \end{pmatrix}$, $\phi = \begin{pmatrix} O & I \\ -\mu^2 I & -2\mu I \end{pmatrix}$, and $\Gamma = \begin{pmatrix} 0 \\ I \end{pmatrix}$

O presents the null matrix and I is the identity matrix. Besides, the discontinuous term v is given by:

$$v = -K \operatorname{sign} (\Gamma^T P S) \tag{14}$$

where P and Q are two positive definite matrices providing the Lyapunov equation: $P\phi + \phi^T P = -Q$ and K is a diagonal definite positive matrix.

4.1 Stability Analysis

The chosen Lyapunov function for the system stability is chosen as:

$$V = \frac{1}{2}(\mu^2 S^T S + \sigma^T \sigma) \tag{15}$$

After computing the derivative of this function, the final expression of $\dot{V}$ can be written as follows:

$$\dot{V} = -2\mu \sigma^T \sigma - \sum_i \mu_i |\rho_i| \leq 0 \tag{16}$$

where $\rho = \Gamma^T P S$. Besides, this confirms the stability of the surveillance camera system.

5 Analysis of Measurement Noise Effects

This study addresses the trajectory tracking control of a 3-DOF robot manipulator based on the second order SMC in the presence of an additive disturbing noises [27]. As the measurement errors currently exists in the dynamical behavior of mechanical systems as external disturbances, and seen that it have a significant impact on the trajectory tracking performances, this research study has been devoted to synthesize such advanced control methods for improving the system control robustness in the presence of disturbances [28].

5.1 Preliminaries

There are two kinds of external disturbances which have been considered in the present section:

- The input disturbance: it is a perturbation input applied to the system. In this case, Eq. (1) becomes:

$$\ddot{q} = M^{-1}(q)\left[\tau - C(q,\dot{q}) - G(q)\right] + b_1(t) \tag{17}$$

where $b_1(t)$ is a bounded noise such that $\|b_1(t)\| \le d_1$, with d_1 is a positive constant.
- The measurement disturbance: it results from sensors and system devices. Then, the measured state can be expressed as:

$$x_m(t) = x(t)(1 + b_2(t)) = x(t) + \Delta x(t) \tag{18}$$

where b_2 is a multiplicative bounded measure noise such that:
$\|b_2(t)\| \le d_2$, and d_2 is a positive constant.

The controller uses the measured state $x_m(t)$ instead of the actual state $x(t)$. So the sliding control expression becomes:

$$\begin{cases} u = u_{eq_m} + \Delta u_m \\ u_{eq_m} = G(x_m)^{-1}F(x_m) \\ \Delta u_m = [G(x_m)]^{-1}\sigma \end{cases} \tag{19}$$

Therefore:

$$\dot{S} = F(x) + G(x)\left[-G(x_m)^{-1}\,F(x_m) + G(x_m)^{-1}\sigma \right] + b_1$$

However:

$$F(x) - G(x)G(x_m)^{-1}F(x_m) = \frac{\partial f}{\partial x}\Delta x - \sum \frac{\partial G}{\partial x_i}G(x_m)^{-1}F(x_m)\Delta x_i + O(\Delta x)^2$$

$$= M_1\Delta x + O(\Delta x)^2 \tag{20}$$

$$G(x)G(x_m)^{-1}\sigma = \sigma + \sum \frac{\partial G}{\partial x_i}G(x_m)^{-1}\sigma\Delta x_i + O(\Delta x)^2$$

$$= \sigma + M_2\Delta x + O(\Delta x)^2 \tag{21}$$

Thus:

$$\dot{S} = \sigma + M\Delta x + O(\Delta x)^2 \tag{22}$$

in which:

$$\begin{cases} M_1 = \dfrac{\partial f}{\partial x}\Delta x - \sum \dfrac{\partial G}{\partial x_i}G(x_m)^{-1}F(x_m)\Delta x_i \\ M_2 = \sum \dfrac{\partial G}{\partial x_i}G(x_m)^{-1}\sigma \Delta x_i \\ \quad M = M_1 + M_2 \end{cases} \tag{23}$$

5.2 Robustness Discussion

To verify the robustness of the second order sliding mode controller via uncertainties, different additive measurement errors have been introduced. The level of the

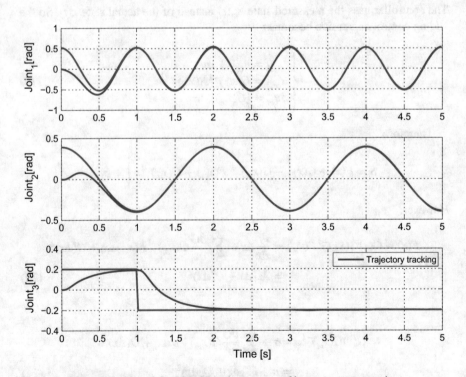

Fig. 2 Positions evolutions of the SOSMC in the presence of low measurement noises

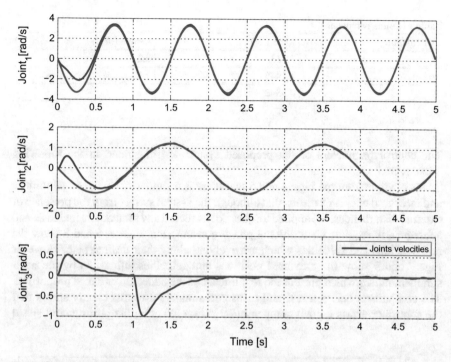

Fig. 3 Velocities evolutions of the SOSMC in the presence of low measurement noises

error effect has been varied from the lower impact to be gradually more intense, aiming to test the controller's ability to withstand such perturbations. Firstly, low noise has been induced (we fluctuate the noise from 5% to 20%), we notice that the positions and the velocities still evolve by following the sliding surface (Figs. 2 and 3) which improve the robustness and the insensibility of the SOSMC via disturbances. Then, the noise has been increased (almost 30%) and in this case, the perturbations start to affect tracking evolution of the system (Figs. 5 and 7).

6 Application to a Surveillance Camera

The considered surveillance camera system has been driven by a 3-DOF robot manipulator and controlled by the second order sliding mode approach [29].

The desired trajectory can be expressed by:

$$
q_d(t) = \begin{bmatrix} q_{d1}(t) \\ q_{d2}(t) \\ q_{d3}(t) \end{bmatrix} = \frac{\pi}{2} \times \begin{bmatrix} \sin \pi t \\ \sin 2\pi t \\ \sin 4\pi t \end{bmatrix} \tag{24}
$$

Table 1 Joints parameters

Articulation	Mass (kg)	Length (m)
q_1	2.7132	0.2
q_2	1.1446	0.15
q_3	0.3392	0.1

The control parameters of the proposed system are illustrated in the following Table 1.

It's obvious from the Figs. 4 and 5 that there is a similarity between real torques and measured ones, this means that although the system suffers from the presence of external disturbances, the applied torques do not record any remarkable increase, and subsequently requires no additional power consumption. Figures 6 and 8 show the real positions and velocities which are evolving along their desired path. However Figs. 7 and 9 show the measured positions and velocities influenced by the measurement noises which present the real impact of the added disturbance proportion. The comparison between real and measured simulation results clearly shows that the controller exerts a great compensation action affecting the position and speed

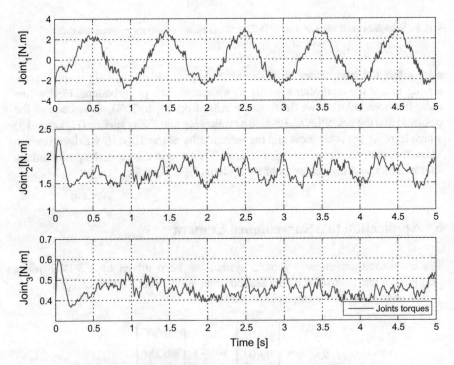

Fig. 4 Torques evolutions of the SOSMC

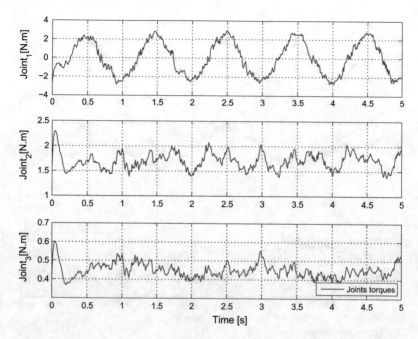

Fig. 5 Measured torques evolutions of the SOSMC

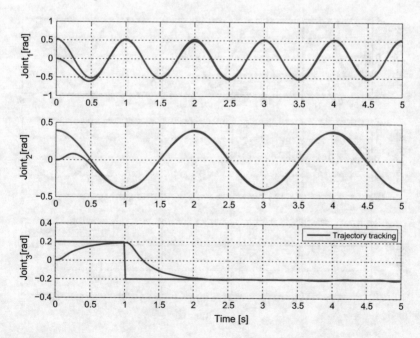

Fig. 6 Positions evolutions of the SOSMC

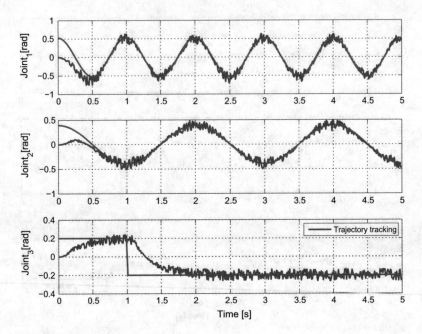

Fig. 7 Measured positions evolutions of the SOSMC

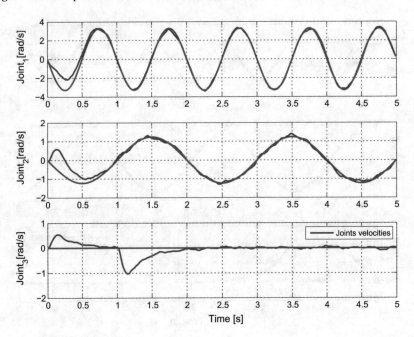

Fig. 8 Velocities evolutions of the SOSMC

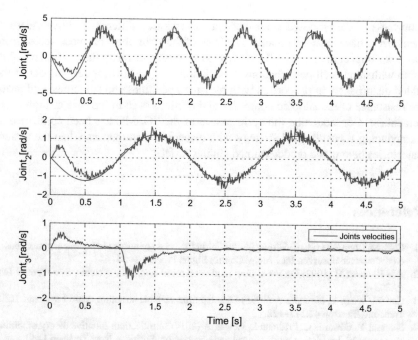

Fig. 9 Measured velocities evolutions of the SOSMC

evolution. This compensation becomes lower while increasing the external disturbance effect, and it becomes unable to manage high imposed disruptions, i.e. starting
from the presence of 30 of errors applied to the system. Nevertheless, at this level the
adopted controller has still resist to deviations, and seen that this percentage is relatively high, it can be then considered as a sufficient control, and proves its robustness
against external disturbances.

A clear behavioral difference between the first and the second order SMC
approaches can be illustrated from the simulation results, especially in terms of
torque inputs, where the superiority of the higher order controller in ensuring a suc-
cessful chattering reduction and an improved robustness via perturbations has been
proved.

7 Conclusion

A second order SMC design has been suggested in this paper, aiming to ensure the
motion control of a 3-DOF manipulator robot. The adopted robotic model serves to
manage a surveillance camera movements. At the beginning, and in order to reduce
the undesirable chattering phenomenon induced by the classical sliding mode controller, a reformulated SOSMC has been developed. Then, and in order to test the

robustness of the proposed controller via uncertainties, different additive measurement errors have been introduced. The level of the error effect has been varied from the lower impact to be gradually more intense, aiming to verify the controller's ability to withstand such perturbations. Simulation results show the real impact of the added disturbance, in proportion with measured positions and velocities, and prove the resistance of the adopted controller to deviations even in the case of high noise percentage. Consequently, the performance of the SOSMC has been confirmed as a convincing robust control strategy with respect to disturbances when the system trajectories belong the predetermined sliding surface.

References

1. Chen M, Chen WH (2010) Sliding mode control for a class of uncertain nonlinear system based on disturbance observer. Int J Adapt Control Signal Process 24:51–64
2. Sira-Ramirez H (1992) On the sliding mode control of nonlinear systems. Syst Control Lett 19:303–312
3. Gao W, Wang Y, Homaifa A (1995) Discrete-time variable structure control systems. IEEE Trans Ind Electron 42:117–122
4. Shtessel Y, Edwards C, Fridman L, Levant A (2014) Introduction: intuitive theory of sliding mode control. In: Sliding mode control and observation. Springer, New York, pp 1–42
5. Vaidyanathan S, Azar AT (2015) Hybrid synchronization of identical chaotic systems using sliding mode control and an application to Vaidyanathan chaotic systems. Stud Comput Intell 576:549–569
6. Piltan F, Sulaiman N, Marhaban MH (2011) Design of FPGA based sliding mode controller for robot manipulator. Int J Robot Autom 2:183–204
7. Xu L, Yao B (2001) Adaptative robust precision motion control of linear motors with negligible electrical dynamics: theory and experiments. IEEE Trans Mechatron 6:444–452
8. Shafiei SE (2010) Sliding mode control of robot manipulators via intelligent approaches. INTECH Open Access Publisher
9. Perruquetti W, Barbot JP (2002) Sliding mode control in engineering. CRC Press Publisher
10. Levant A (2010) Chattering analysis. IEEE Trans Autom Control 55:1380–1389
11. Vaidyanathan S (2016) Anti-synchronization of 3-cells cellular neural network attractors via integral sliding mode control. Int J PharmTech Res 9:193–205
12. Kapoor N, Ohri J (2013) Integrating a few actions for chattering reduction and error convergence in sliding mode controller in robotic manipulator. Int J Eng Res Technol 2
13. Vaidyanathan S (2015) Global chaos synchronization of 3-cells cellular neural network attractors via integral sliding mode control. Int J PharmTech Res 8:118–130
14. Vaidyanathan S (2015) Global chaos control of 3-cells cellular neural network attractor via integral sliding mode control. Int J PharmTech Res 8:211–221
15. Golea N, Debbache G, Golea A (2012) Neural network-based adaptive sliding mode control for uncertain non-linear mimo systems. Int J Model Ident Control 16:334–344
16. Kapoor N, Ohri J (2013a) Fuzzy sliding mode controller (FSMC) with global stabilization and saturation function for tracking control of a robotic manipulator. J Control Syst Eng 1:50
17. Slotine JJE, Li W (1991) Applied nonlinear control. Prentice Hall, Englewood Cliffs, NJ
18. Zinober ASI, ElGhezawi OE, Billings SA (1982) Multivariable structure adaptive model following control systems. IEE Proc D (Control Theor Appl) IET Digital Libr 129:6–12
19. Vaidyanathan S (2016) Global chaos regulation of a symmetric nonlinear gyro system via integral sliding mode control. Int J ChemTech Res 9:462–469
20. Fridman L, Levant A (2002) Higher order sliding modes. Sliding Mode Control Eng 11:53–102

21. Vaidyanathan S (2016) A highly chaotic system with four quadratic nonlinearities, its analysis, control and synchronization via integral sliding mode control. Int J Control Theory Appl 9:279–297
22. Vaidyanathan S, Sampath S (2016) A novel hyperchaotic system with two quadratic nonlinearities, its analysis and synchronization via integral sliding mode control. Int J Control Theory Appl 9:221–235
23. Bartolini G, Pydynowski P (1996) An improved, chattering free, VSC scheme for uncertain dynamical systems. IEEE Trans Autom Control 41:1220–1226
24. Bartolini G, Ferrara A, Usani E (1998) Chattering avoidance by second-order sliding mode control. IEEE Trans Autom Control 43(2):241–246
25. Ferrara A, Magnani L (2007) Motion control of rigid robot manipulators via first and second order sliding modes. J Intell Robot Syst 48:23–36
26. Abdelhedi F, Bouteraa Y, Derbel N (2014) Distributed second order sliding mode control for synchronization of robot manipulator. In: Second international conference on automation, control, engineering and computer science
27. Vaidyanathan S, Sampath S (2016) Hybrid synchronization of identical chaotic systems via novel sliding control method with application to Sampath four-scroll chaotic system. Int J Control Theory Appl 9:321–337
28. Vaidyanathan S, Sampath S (2016) Anti-synchronization of identical chaotic systems via novel sliding control method with application to Vaidyanathan-Madhavan chaotic system. Int J Control Theory Appl 9:85–100
29. Vaidyanathan S, Sampath S, Azar AT (2015) Global chaos synchronisation of identical chaotic systems via novel sliding mode control method and its application to Zhu system. Int J Model Ident Control 23:92–100

Robust Control of a Photovoltaic Battery System via Fuzzy Sliding Mode Approach

Alessandro Baldini, Lucio Ciabattoni, Riccardo Felicetti,
Francesco Ferracuti, Alessandro Freddi, Andrea Monteriù
and Sundarapandian Vaidyanathan

Abstract In this chapter we propose a novel fuzzy sliding mode approach to manage the power flow of a Photovoltaic (PV) battery system. In particular, due to the inner stochastic nature and intermittency of the solar production and in order to face the irradiance rapid changes, a robust and fast controller is needed. Sliding Mode Control (SMC) is a well-known approach to control systems under heavy uncertain conditions. However, one of the major drawbacks of this control technique is the high frequency chattering generated by the switching control term. In the proposed solution, we introduce a fuzzy inference system to set the controller parameters (boundary layer and gains) according to the measured irradiance. A comparison of the designed Fuzzy Sliding Mode Control (FSMC) with two popular controllers (PI and Backstepping) is performed. In particular, FSMC shows better performances in

A. Baldini · L. Ciabattoni (✉) · R. Felicetti · F. Ferracuti · A. Monteriù
Dipartimento di Ingegneria dell'Informazione, Università Politecnica delle Marche,
Via Brecce Bianche, 60131 Ancona, Italy
e-mail: l.ciabattoni@univpm.it

A. Baldini
e-mail: a.baldini@univpm.it; alessandro.baldini.91@gmail.com

R. Felicetti
e-mail: r.felicetti@univpm.it; felicetti.riccardo@gmail.com

F. Ferracuti
e-mail: f.ferracuti@univpm.it

A. Monteriù
e-mail: a.monteriu@univpm.it

A. Freddi
SMARTEST Research Centre, Università degli Studi eCampus, Via Isimbardi 10,
22060 Novedrate (CO), Italy
e-mail: alessandro.freddi@uniecampus.it; a.freddi@univpm.it

S. Vaidyanathan
Research and Development Centre, Vel Tech University, Chennai 600062,
Tamil Nadu, India
e-mail: sundarvtu@gmail.com

© Springer International Publishing AG 2017
S. Vaidyanathan and C.-H. Lien (eds.), *Applications of Sliding Mode Control
in Science and Engineering*, Studies in Computational Intelligence 709,
DOI 10.1007/978-3-319-55598-0_6

115

terms of steady state chattering and transient response, as confirmed by IAE, ISE and ITAE performance indexes.

Keywords Sliding mode control · Fuzzy sliding mode · Photovoltaic system · Battery system

1 Introduction

As concerns about climate change, rising fossil fuel prices, and energy security demand, there is a growing interest around the world in renewable energy resources. However, since most renewable energy sources are intermittent in nature, the integration of a significant portion of renewable energy resources into the power grid infrastructure is a challenging task.

Traditional electricity grid connections were designed as unidirectional point (i.e., large conventional power plants) to multipoint (substations or loads) links. In contrast to this traditional scenario, in recent years there have been a large expansion of renewable energy plants, i.e., small, distributed and stochastic generators. The integration of Distributed Energy Resources (DERs) may have a great impact on the grid operation, and calls for new methods and technologies.

Nowadays, less attractive Photovoltaic (PV) feed-in-tariffs and incentives to promote self-consumption, request to explore new operation modes for PV in order to reach grid parity, which has been predicted to become a reality in the next years in the European Union, as suggested in [3, 9, 11, 15]. By increasing the self consumed local generated energy, the grid parity could be achieved earlier and solar power will finally make economic sense becoming cheaper (over the lifetime of the system) than to buy it from electricity [12, 13]. In particular, there are two main tasks to integrate DERs, both locally and globally: integrating them into the electricity network, and into the energy market. One solution for decreasing the problems caused by the variable output of some distributed generation, is to add energy storages into the systems (centralised or distributed energy storages, as proposed in [1, 41, 42]). The main reasons of storage absence in the grid can be found in its expensive cost and in the reduced generation control capabilities of the classical power stations, making apparently not profitable the implementation of new storage technologies on a large scale. However, the current technological and energetic situation encourages the investment in storage systems. In fact, a massive use of renewable energy generation implicates a strong hourly mismatch between demand and generation patterns. Due to the aforementioned mismatch, typical self-consumption percentage of domestic PV systems is usually limited between 25% and 35%, as shown in [4]. Obviously, it could be possible to increase this percentage by reducing the size of the PV installation. However, this choice is not desirable because it reduces the share of renewable energy content in the whole energy mix of the household. Thus, the only viable solution to reach both goals is the usage of electrical energy storage systems to buffer solar energy and, therewith, to temporarily decouple generation and con-

sumption. Lithium-ion batteries are well suited in this application, as presented in
[5], but they require reliable, robust and high performance control systems in order
to preserve each battery from damages and increase its lifetime.

In this paper, we propose and develop a smart controller for a PV storage system
capable to reduce the variability of the electricity exported to the grid, to manage the
power flow, and to indirectly use the PV electricity such that to maximize the self
consumption. Sliding Mode Control (SMC), see [7, 8, 18], is an effective approach
to control system design under heavy uncertainty conditions and, in addition, it is
accurate, robust and very simple to implement, as shown in [31, 32, 35, 38, 39]. In
case of the control of a PV storage system, the conventional SMC can guarantee the
robustness for the uncertainty of the system and external disturbance, as described
in [6], but the high frequency chattering, which is generated by the switching term
of the control input, may cause electrical grid stability issues as well as damages to
the energy storage device. In order to improve the transient response while reducing
chattering phenomenon, in this paper, we propose a smart controller based on the
SMC technique coupled with a fuzzy logic algorithm, namely Fuzzy Sliding Mode
Control (FSMC). A comparison of the designed fuzzy sliding mode controller with
two popular controllers, namely PI and Backstepping, is proposed and the corresponding results shown.

The chapter is organized as follows. The nonlinear state space model of the PV
battery dynamics is presented in Sect. 2. The developed SMC algorithm is reported
in Sect. 3 as well as the tuning procedure, and the designed fuzzy inference system.
Simulation results are reported in Sect. 4 where the proposed fuzzy sliding mode controller is compared with a PI controller and a Backstepping controller. The chapter
ends with some comments on the performance of the proposed solution.

2 Photovoltaic Battery Model

In this section, we derive the mathematical model of a photovoltaic battery system, and its state space representation suited to controller design. To this aim, we
firstly introduce the so called "one diode and three parameters PV cell model" [43],
obtained from the electrical circuit shown in Fig. 1.

By applying the Kirchhoff law, the current of the photovoltaic cell I_g can be
obtained by:

$$I_g = I_{ph} - I_d \tag{1}$$

where I_d is the diode current which is proportional to the reverse saturation current
I_0:

$$I_d = I_0 \cdot \left(e^{\frac{V_g/n_s + R_s \cdot I_g}{A \cdot V_T}} - 1 \right) \tag{2}$$

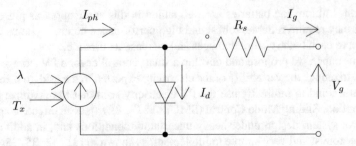

Fig. 1 PV one diode and three parameters cell circuital model

with n_s the number of PV cells connected in series, A the ideality factor, that is a constant value related to the cell manufacturing technology (see Table 1), while V_T is the so called thermal voltage (due to its exclusive dependence by the temperature).

The thermal voltage can be derived by:

$$V_T = \frac{K \cdot T_x}{q} \tag{3}$$

where T_x is the actual photovoltaic cell temperature, K is the Boltzmann constant and q is the electron charge value.

The reverse saturation (or leakage) current I_0 (see Eq. (2)) of the diode can be computed according to:

$$I_0 = I_0(T_0) \cdot \left(\frac{T_x}{T_0}\right)^{\frac{3}{A}} e^{-\frac{q \cdot E_g}{A \cdot K} \cdot \left(\frac{1}{T_x} - \frac{1}{T_0}\right)} \tag{4}$$

where E_g is the energy band gap related to the cell's manufacturing technology (see Table 1), $I_0(T_0)$ is the leakage current at nominal temperature T_0, that is at Standard Test Conditions (STCs), which can be obtained by:

Table 1 Ideality factor (A) and band gap (E_g) for different PV cells technologies

Technology	Formula	Ideality factor	Band gap
Monocrystalline silicon	Si-mono	1.2	
Polycrystalline silicon	Si-poly	1.3	
Amorphous silicon	a-Si-H	1.8	
Amorphous silicon	a-Si-H tandem	3.3	
Amorphous silicon	a-Si-H triple	5.0	
Cadmium telluride	CdTe	1.5	1.5
	CTs	1.5	
Gallium arsenide	AsGa	1.3	1.43

$$I_0(T_0) = \frac{I_{sc}(T_0)}{e^{\frac{q \cdot V_{oc}(T_0)}{A \cdot K \cdot T_0}} - 1} \tag{5}$$

with $V_{oc}(T_0)$ the open circuit voltage at STCs.

The photocurrent I_{ph} (see Eq. (1)) depends on both irradiance and temperature:

$$I_{ph} = I_{ph}(T_0) \cdot (1 + K_0 \cdot (T_x - T_0)) \tag{6}$$

with the photocurrent at STCs $I_{ph}(T_0)$ which can be obtained by:

$$I_{ph}(T_0) = I_{sc}(T_0) \cdot \frac{G}{G_0} \tag{7}$$

where $I_{sc}(T_0)$ and G_0 are the short circuit current and irradiance at STCs, respectively, and G is the actual irradiance.

The effects of the loss efficiency of the PV cell and, consequently, of the overall PV module, is modeled by a series resistance R_s which can be estimated as the slope of the $I - V$ function for $V = V_{oc}$ as follows:

$$R_s = \frac{A \cdot K}{q \cdot I_{sc}(T_0)} \cdot log \left| \frac{I_{ph} - I_{sc}(T_0)}{I_0} \right| \tag{8}$$

In order to ensure that the operation point of the circuit of Fig. 1 coincides, or is as closest as possible to the maximum power point, a specific solution, called Maximum Power Point Tracker (MPPT), is employed. Usually, the MPPT is achieved by interposing a power converter (DC-to-DC converter) between the PV generator and the load (battery). Controlling the reference signal of the converter (i.e., the duty cycle), it is thus possible to extract the maximum power for charging the battery.

Finally, the solar generation model consists of a PV array module, a DC-to-DC buck converter and a battery, as shown in Fig. 2.

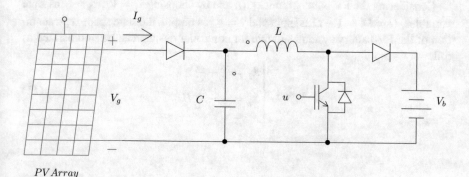

PV Array

Fig. 2 PV battery circuital model

The buck converter is used to drive a low voltage load from the high voltage PV array, thus the PV battery circuital mathematical equations can be summarized as:

$$\begin{cases} C \cdot \dot{V}_g = I_g - I_L \\ L \cdot \dot{I}_L = V_g - (1 - U) \cdot V_b \end{cases} \tag{9}$$

where C is the capacity, L the inductance, I_L the current across the inductance, V_g and I_g are the voltage and current of the photovoltaic array, respectively, V_b the battery voltage and U is the switched control signal which is a binary variable, i.e., "0" for switch opened and "1" for switch closed.

Although the proposed mathematical model can be theoretically suitable for controller design, it is necessary to take into account the following average-state system in order to apply the controller in real power electronics systems:

$$\begin{cases} C \cdot \dot{\bar{V}}_g = \bar{I}_g - \bar{I}_L \\ L \cdot \dot{\bar{I}}_L = \bar{V}_g - D' \cdot \bar{V}_b \end{cases} \tag{10}$$

where D is the duty cycle and $D' = 1 - D$. This model has a continuous input and a continuous output, and thus it results suitable for a continuous controller. The desired continuous input, in real power electronics systems (e.g., inverters), represents the pulse-width modulation (PWM) signal.

In order to develop an effective controller, the following assumptions can be made:

Assumption 2.1 $\bar{I}_g, \bar{V}_g, \bar{I}_L$ and $\bar{V}_b$ are measurable.

Assumption 2.2 C and L are known constants.

Assumption 2.3 $\bar{V}_b$ is considered as a constant value due to its slow dynamics.

Assumption 2.4 $\bar{V}_g, \bar{I}_g$ and $\dot{\bar{I}}_g$ are bounded.

Considering the PV battery model (10) and by choosing $x_1 = \bar{V}_g, x_2 = \bar{I}_L$ as state variables, $U = D' = 1 - D$ as input and $y = \bar{V}_g$ as output, the state space representation of the PV battery system, suitable for controller design, can be summarized as follows:

$$\Sigma : \begin{cases} \dot{x}_1 = \dfrac{\bar{I}_g}{C} - \dfrac{x_2}{C} \\ \dot{x}_2 = \dfrac{x_1}{L} - \dfrac{\bar{V}_b}{L} U \\ y = x_1 \end{cases} \tag{11}$$

3 Control Design

3.1 Sliding Mode Control Recap

Consider the following system

$$
\begin{cases}
\dot{x}_1 & = x_2 \\
\dot{x}_2 & = x_3 \\
\vdots & = \vdots \\
\dot{x}_{n-1} & = x_n \\
\dot{x}_n & = \alpha(x) + \gamma(x)u
\end{cases}
\tag{12}
$$

where u is the control input, $x^T = (x_1, x_2, \ldots, x_n)^T \in D \subset \mathbb{R}^n$ is the state vector and $\alpha(x)$ and $\gamma(x)$ are nonlinear functions, where $\gamma(x) \neq 0$. Without loss of generality, we can suppose $\gamma(x) \geq \gamma_0 > 0$. Our goal is to stabilize the system with a control law $u = u(x)$, such as $x_i \to 0$ for $i = 1, 2, \ldots, n$. Let us define the following sliding surface

$$
S = k_1 x_1 + \cdots + k_{n-1} x_{n-1} + x_n
$$

with the coefficients $k_1, \ldots, k_{n-1}$ chosen such that the polynomial $p(t) = k_1 + k_2 t + \cdots + k_{n-1} t^{n-2} + t^{n-1}$ is a Hurwitz one. On this surface ($S = 0$), the motion is governed by

$$
x_n = -(k_1 x_1 + \cdots + k_{n-1} x_{n-1})
\tag{13}
$$

so the resulting system is asymptotically stable and $x_1 \to 0$ (and $x_2, \ldots, x_n \to 0$ too). It is possible to notice that the Eq. (13) can be rewritten as the differential equation

$$
x_1^{(n-1)} = -(k_1 x_1 + \cdots + k_{n-1} x_1^{(n-2)})
$$

and it follows that $x_1(t) \to 0$ (and then $\dot{x}_1 \to 0$), thanks to the choice of the parameters k_i. Consequently $x_2 = \dot{x}_1 \to 0$ and $x_3, \ldots, x_n \to 0$. If the state of the system (12) moves on the surface, then the dynamic of the state $x^T = (x_1, \ldots, x_n)^T$ converges asymptotically to the origin. So, the first step to design the sliding mode controller is to choose a proper sliding surface.

Then we must impose $S = 0$ by the action of a control law $u = u(x)$, where u is the control input. From another point of view, we must ensure that the closed-loop system is asymptotically stable. It is usually needed to reach the convergence of the system in finite time, if possible. We can use Lyapunov theory of the non linear system theory.

Choosing $V = \frac{S^2}{2}$ as a candidate Lyapunov function, it follows that

$$
\begin{aligned}
\dot{V} = S\dot{S} &= S(k_1\dot{x}_1 + \cdots + k_{n-1}\dot{x}_{n-1} + \dot{x}_n) \\
&= S(k_1 x_2 + \cdots + k_{n-1}x_n + \alpha(x) + \gamma(x)u)
\end{aligned}
\tag{14}
$$

By choosing $u = -\frac{1}{\gamma(x)}(k_1 x_2 + \cdots + k_{n-1}x_n + \alpha(x)) + v$, we obtain $\dot{V} = S \cdot v$. Taking $v = -\beta(x)sign(S)$, with $\beta(x) \geq \beta_0 > 0$, yields

$$
\dot{V} = -\beta(x)Ssign(S) = -\beta(x)|S| \leq -\beta_0|S|
\tag{15}
$$

Therefore, the trajectory reaches the surface $S = 0$ in finite time and once on the surface it cannot leave it, because $\dot{V} \leq -\beta_0|S| < 0 \quad \forall x \in D - \{0\}$.

This approach has the following negative effect: suppose that the state x of the system is in a point $x' \in D$ where $S > 0$. The control law forces the system to reach the surface, with $sign(S) = 1$. Once the system reaches the surface, it remains on the surface for a while (instantaneously) and then leave that, due to physical phenomena like inertia, hysteresis or delays, as discussed in [19, 24, 27]. Suppose now that the state $(x_1, \ldots, x_n)$ of the system is in a point $x'' \in D$ where $S < 0$ (reached under the effect of the control law). This time $sign(S) = -1$, so the control input, with an inverse action if compared to previous, moves the system towards the sliding surface again. The final result of this actions is known as chattering problem, examined also in [23, 25, 26]. It is the main problem of the sliding mode control and lots of solutions are proposed in literature (e.g. in [17, 22, 29]) to reduce its impact to the system.

To avoid the well known chattering problem, let us introduce a boundary layer: we avoid impose $S = 0$, instead $|S| \leq \epsilon$ is desired, with $\epsilon > 0$ arbitrarily chosen. In this way, the control law can be separated in two parts. When $|S| > \epsilon$, the control input is the same of the previous case, otherwise, if $|S| \leq \epsilon$, the control law can be freely modified. The standard solution is choosing a linear function; this control law is simple and permits to avoid any discontinuity of the control input, in other words, $sign(S)$ is replaced by $sat(S/\epsilon)$. A deeper explanation of chattering problem and its mathematical consequences can be found in [2].

Rarely we can dispose of a system in form (12), and, moreover, we are usually interesting to a tracking problem. Consider now the system

$$
\begin{cases}
\dot{x} = f(x) + g(x)u \\
y = h(x)
\end{cases}
\tag{16}
$$

where $x \in D \subset \mathbb{R}^n$ is the state of the system, while $f(x)$, $g(x)$ and $h(x)$ are sufficiently smooth vectorial non-linear functions. The system in exam has same form of the circuital model considered in the photovoltaic panel. To design a sliding mode controller that permit to track the system we must do two steps: the first one is to transform the system (16) into a new form, like (12), by a change of variables. The second step is to resolve the tracking problem. For doing the transformation of the variables we can use the theory developed for the Feedback Linearization, which is largely explained in the literature [10, 14]. By introducing the Lie derivative

$$L_{T(x)}W(x) = \frac{\partial W(x)}{\partial x}T(x)$$

where $W(x)$ and $T(x)$ are vector fields, we can define the relative degree ρ for a system as follows:

$$L_g h(x) = L_g L_f h(x) = \cdots = L_g L_f^{\rho-2} h(x) = 0, \qquad L_g L_f^{\rho-1} h(x) \geq a > 0 \quad \forall x \in D$$

From another point view, the relative degree ρ represents the number of times we have to derive the output until the input will appear. The control objective is to make the output y asymptotically track the given reference signal $r(t)$, where $r(t)$ and its derivatives $\dot{r}(t), \ddot{r}(t), \ldots, r^{(\rho)}(t)$ are known and bounded for all $t \geq 0$ and $r^{(\rho)}(t)$ is a piecewise continuous function of t. Applying the input-output feedback linearization the system can be transformed in the following normal form:

$$\Sigma' \begin{cases} \dot{\eta} = f_0(\eta, \xi) \\ \dot{\xi}_1 = \xi_2 \\ \dot{\xi}_2 = \xi_3 \\ \vdots = \vdots \\ \dot{\xi}_{\rho-1} = \xi_\rho \\ \dot{\xi}_\rho = L_f^\rho h(x) + L_g L_f^{\rho-1} h(x)u \\ y = \xi_1 \end{cases} \tag{17}$$

by the change of variables

$$\begin{bmatrix} \eta \\ --- \\ \xi \end{bmatrix} = \begin{bmatrix} \phi(x) \\ --- \\ \psi(x) \end{bmatrix} = \begin{bmatrix} \phi_1(x) \\ \vdots \\ \phi_{n-\rho}(x) \\ --- \\ h(x) \\ \vdots \\ L_f^{(\rho-1)} h(x) \end{bmatrix} = T(x)$$

where $\frac{\partial \phi_i}{\partial x} g(x) = 0$ for $1 \leq i \leq n - \rho$ $\forall x \in D$. The mapping rule $T(x) : \mathbb{R}^n \to \mathbb{R}^n$ must be a diffeomorphism. It is compose by the two applications $\psi(x)$ and $\phi(x)$. Without elaborating on the theory we observe that the construction of $\psi(x)$ follows mechanical rules, and it involves exactly ρ equations. The transformation $\phi(x)$, in contrast, involves $n - \rho$ equations, and the mapped variables are not controllable by the control input. This can be easily seen on the explicit system (17), where $f_0(\eta, \xi)$ is defined as zero dynamics.

We define the following reference for the system

$$R = \begin{bmatrix} r \\ \vdots \\ r^{(\rho-1)} \end{bmatrix}$$

Applying the change of variables $e = \xi - R$ we obtain

$$
\Sigma'_e \begin{cases}
\dot{\eta} = f_0(\eta, \xi) \\
\dot{e}_1 = e_2 \\
\dot{e}_2 = e_3 \\
\vdots = \vdots \\
\dot{e}_{\rho-1} = e_\rho \\
\dot{e}_\rho = L_f^\rho h(x) + L_g L_f^{\rho-1} h(x) u - r^{(\rho)}(t)
\end{cases} \tag{18}
$$

Supposing that $\dot{\eta} = f_0(\eta, \xi)$ is ISS,[1] it is sufficient to stabilize the origin $\xi = 0$. This is a linear system in the controllable canonical form, so we can use a linear control such $e_\rho = -(k_1 e_1 + \cdots + k_{\rho-1} e_{\rho-1})$, with $k_1, \ldots, k_{\rho-1}$ such that $s^{\rho-1} + k_{\rho-1} s^{\rho-2} + \cdots + k_1$ is Hurwitz. Then, we can choose the sliding surface

$$
s = (k_1 e_1 + \cdots + k_{\rho-1} e_{\rho-1}) + e_\rho = 0
$$

Defining

$$
u = -\frac{1}{L_g L_f^{(\rho-1)} h(x)} [(k_1 e_2 + \cdots + k_{\rho-1} e_\rho) + L_f^{(\rho)} h(x) - r^{(\rho)}(t)] + v
$$

we cancel the known terms of $\dot{s}$ and stabilize the system with the switching component:

$$
v = -\beta(x) \text{sign}(s)
$$

with $\beta(x) > \beta_0 > 0$ sufficiently large to guarantee $\dot{V} = s\dot{s} \le -k|s|$, so $s \to 0$ asymptotically. We observe that the sliding mode theory is widely developed in literature, so mathematical proofs and insights are well known.

3.2   Sliding Mode Control Design for the PV Battery System

Let us apply the theory presented in the previous section on the photovoltaic battery system. Starting from the state space system (11), the first step is to perform a state transformation by defining:

$$
h(x) = y = x_1
$$

$$
L_f h(x) = \frac{\bar{I}_g}{C} - \frac{x_2}{C}
$$

---

[1] Input to State Stable.

$$L_g h(x) = 0$$

$$L_f^2 h(x) = \begin{bmatrix} \frac{1}{C} \frac{\partial \bar{I}_g}{\partial x_1} & -\frac{1}{C} \end{bmatrix} \cdot \begin{bmatrix} \frac{\bar{I}_g}{C} - \frac{x_2}{C} \\ \frac{x_1}{L} \end{bmatrix} = \frac{1}{C} \frac{\partial \bar{I}_g}{\partial x_1} \cdot \frac{\bar{I}_g}{C} - \frac{x_2}{C} - \frac{1}{C} \cdot \frac{x_1}{L} = \frac{\dot{\bar{I}}_g}{C} - \frac{x_1}{LC}$$

$$L_g L_f h(x) = \frac{\bar{V}_b}{LC} \neq 0$$

The relative degree is $\rho = 2$, so the system is completely linearizable with transformation

$$T = \begin{bmatrix} h(x) \\ L_f h(x) \end{bmatrix} = \begin{bmatrix} x_1 \\ \frac{\bar{I}_g}{C} - \frac{x_2}{C} \end{bmatrix} = \begin{bmatrix} \xi_1 \\ \xi_2 \end{bmatrix}$$

We define $r(t)$ as the reference for $y = x_1 = \xi_1$, as shown previously. By introducing the following errors

$$\begin{cases} e_1 = \xi_1 - r \\ e_2 = \xi_2 - \dot{r} \end{cases}$$

so the error dynamic is

$$\begin{cases} \dot{e}_1 = \xi_2 - \dot{r} \\ \dot{e}_2 = \frac{\bar{I}_g}{C} - \frac{x_1}{LC} + \frac{\bar{V}_b}{LC} U - \ddot{r} \end{cases}$$

that can be rewritten as

$$\begin{cases} \dot{e}_1 = e_2 \\ \dot{e}_2 = \frac{\dot{\bar{I}}_g}{C} - \frac{x_1}{LC} + \frac{\bar{V}_b}{LC} U - \ddot{r} \end{cases}$$

Defining the sliding surface $S = k_1 e_1 + e_2$, we can apply the theory introduced above and set the control law as

$$\begin{aligned} U &= -\frac{1}{L_g L_f^{(\rho-1)} h(x)} [(k_1 e_2 + \cdots + k_{\rho-1} e_\rho) + L_f^{(\rho)} h(x) - r^{(\rho)}(t)] + v \\ &= -\frac{1}{L_g L_f h(x)} [k_1 e_2 + L_f^{(2)} h(x) - \ddot{r}(t)] + v \\ &= -\frac{LC}{\bar{V}_b} [k_1 e_2 + \frac{\dot{\bar{I}}_g}{C} - \frac{x_1}{LC} - \ddot{r}] + v \end{aligned}$$

where v is the equivalent part of the control law. To impose the convergence we define the following Lyapunov function

$$V = \frac{1}{2} S^2$$

so, the time derivative of the Lyapunov function is

$$\dot{V} = S\dot{S} = S[k_1\dot{e}_1 + \dot{e}_2]$$
$$= S[k_1 e_2 + \frac{\dot{I}_g}{C} - \frac{x_1}{LC} + \frac{\bar{V}_b}{LC}U - \ddot{r}(t)]$$

To obtain the asymptotic convergence of the system, we must impose $\dot{V} < 0$. Defining the control input in the form $U = U_n + U_{eq}$, where U_n is the nominal part and U_{eq} is the equivalent part:

$$U = -\frac{LC}{\bar{V}_b}\left[k_1 e_2 + \frac{\dot{I}_g}{C} - \frac{x_1}{LC} - \ddot{r}(t)\right] + v$$

the Lyapunov function becomes

$$\dot{V} = S\left[\frac{\bar{V}_b}{LC}v\right]$$

so, choosing $v = -\beta(x)sign(S)$, with $\beta(x) > 0 \forall x \in D$ we obtain $\dot{V} < 0$. In conclusion, by imposing

$$U = -\frac{LC}{\bar{V}_b}\left[k_1 e_2 + \frac{\dot{I}_g}{C} - \frac{x_1}{LC} - \ddot{r}(t)\right] + v$$

$$v = -K\frac{LC}{\bar{V}_b} \cdot sign(S)$$

$$\beta(x) = K\frac{LC}{\bar{V}_b}$$

it follows that $\dot{V} = -K|S|$, with $k > 0$, so we reach the surface in finite time. To reduce the well known chattering problem the $sign(\cdot)$ can be replaced by $sat(\cdot)$ function. In particular $sign(s)$ becomes $sat(\frac{s}{\epsilon})$, where ϵ defines the boundary layer. The final control law is:

$$U = -\frac{LC}{\bar{V}_b}\left[k_1 e_2 + \frac{\dot{I}_g}{C} - \frac{x_1}{LC} - \ddot{r}(t)\right] + v$$

$$v = -K\frac{LC}{\bar{V}_b} \cdot sat\left(\frac{s}{\epsilon}\right) \tag{19}$$

$$\beta(x) = K\frac{LC}{\bar{V}_b}$$

where $k_1, \beta(x), \epsilon > 0$.

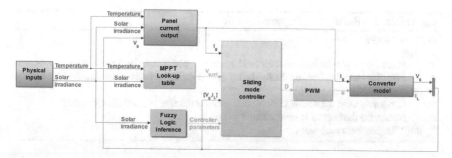

Fig. 3 Fuzzy sliding mode control scheme

3.3 Fuzzy Sliding Mode

The designed sliding mode approach allows us to synthesize a robust control law. Anyway, the designed control performances strongly depend on the system parameters and coefficients. In this context, the control law designed and optimized on a specific set of parameters P_1 does not have the same performances on another set of parameters P_2, even if stability (or tracking in general) is being maintained. To improve the global performances of our controller we thus introduce a fuzzy sliding mode. The approach is mainly composed by two steps: (1) the design of an optimal sliding mode control law for different operating points of the system; (2) the use of a fuzzy inference to define the proper control input by combining the different laws previously computed. The whole scheme of the controller is depicted in Fig. 3.

An assumption is needed, although many installed PV systems are yet equipped with irradiance sensors:

Assumption 3.1 The value of the solar irradiance λ can be directly measured.

3.3.1 Step 1: The Optimization Procedure

For each considered configuration of the system parameters P_i an optimal sliding mode control law L_i needs to be computed. Since the most important parameter in a PV application is the solar irradiance λ, an optimal control law L_i (i.e., with an optimal set of parameters) is computed for different irradiance levels. In order to perform the task we employed an evolutionary computation technique, Artificial Bee Colonies (ABC). In ABC algorithm, a possible solution of the optimized problem is described by the position of a food source and the nectar amount of a food source represents the fitness of the associated solution. Three different agents perform the search: scout bees (random search), employer bees (greedy search) and onlooker bees (probabilistic search). The basic pseudo algorithm is described below:

Algorithm 1 ABC algorithm pseudo-code

1: Initialize Population
2: **repeat**
3:     Place the employed bees on their food sources to perform a greedy search
4:     Place the onlooker bees on the food sources depending on their nectar amounts (probabilistic search)
5:     Randomly send the scouts to the search area for discovering new food sources
6:     Memorize the best food source found so far
7: **until** Requirements are met

3.3.2   Step 2: The Fuzzy Inference

A Fuzzy Inference System (FIS) is finally used to merge the control parameters depending on the measured irradiance level. A zero-order Takagi Sugeno fuzzy inference system has been chosen. The FIS has one input (i.e., the parameter λ) which is partitioned into eight fuzzy sets corresponding to different degrees of solar irradiance: very high, high, more than medium, medium, lower than medium, low, less than low, extremely low. (reported in Table 2 and depicted in Fig. 4).

The membership functions of the input variable consist of trapezoidal functions chosen by experiments. The trapezoidal fuzzy set A in the universe of discourse U (the real line $\mathbf{R}$) with the membership function μ_A is parametrized by four real scalar parameters: (a; b; c; d) with $a < b \leq c < d$. This representation can be interpreted as a mathematical membership function as follow

$$\mu_A(x) = \begin{cases} 0, & x < a \\ \frac{x-a}{b-a}, & a < x < b \\ 1, & b < x < c \\ \frac{d-x}{d-c}, & c < x < d \\ 0, & x > d \end{cases} \tag{20}$$

Table 2 Linguistic terms and their corresponding trapezoidal fuzzy sets for solar irradiance

Input variable	Linguistic terms	a	b	c	d
λ	Very high	0	0	0.025	0.0375
	High	0.025	0.0375	0.0625	0.0875
	More than medium	0.0625	0.0875	0.125	0.175
	Medium	0.125	0.175	0.25	0.35
	Lower than medium	0.25	0.35	0.45	0.55
	Low	0.45	0.55	0.65	0.75
	Less than low	0.65	0.75	0.85	0.95
	Extremely low	0.85	0.95	1.025	1.075

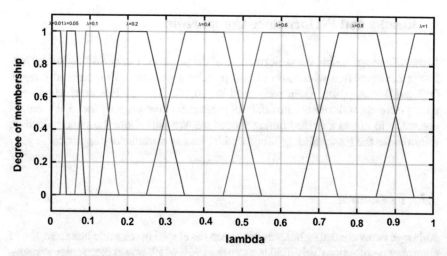

Fig. 4 The input membership function

The control parameters computed through ABC algorithm for each λ_i are k_1, β and ϵ defined in (19). In particular, ϵ and $\beta(x)$ lead to a saturation function with a non-linear slope inside the boundary layer, while the parameter k_1 allows us to change the slope of the sliding surface. Therefore, the control law becomes:

$$U = -\frac{LC}{\bar{V}_b}\left(FSMC_{k_1}(\lambda)e_2 + \frac{\dot{I}_g}{C} - \frac{x_1}{LC} - \ddot{r}(t)\right) - FSMC_{\beta(x)}(\lambda)\cdot sat\left(\frac{S}{FSMC_\epsilon(\lambda)}\right)$$

(21)

where $FSMC_{k_1}(\lambda), FSMC_{\beta(x)}(\lambda)$ and $FSMC_\epsilon(\lambda)$ are respectively the fuzzy output of the parameters k_1, β end ϵ. The rules and output values of the FIS are reported in Table 3.

Table 3 Rule set and output values of fuzzy inference system

Input	Outputs		
λ	$FSMC_{k_1}(\lambda)$	$FSMC_{\beta(x)}(\lambda)$	$FSMC_\epsilon(\lambda)$
Very high	248.8907	0.8593	2664.8393
High	164.1434	1	1965.7679
More than medium	157.4098	0.9874	1430.2668
Medium	118.6231	0.511	724.3177
Lower than medium	124.2885	1	1000.2
Low	234.6996	1	1135.7
Less than low	1000	0.9457	281.1894
Extremely low	394.7043	0.5723	11.0502

4   Results and Performance Comparison

In the following section, the performances of the proposed Fuzzy Sliding Mode Controller are compared with those of two popular approaches: Proportional-Integral (PI) and Backstepping, extensively used in [16, 21, 37, 40]. However, when comparing different controllers with different parameters to be properly tuned, it becomes necessary to assess a unified tuning procedure. We thus chose the same procedure followed for the Fuzzy Sliding Mode design, i.e., an evolutionary approach.

4.1   PI Control

Although many control techniques have been developed in scientific literature, the PI controller is still extensively used in more than 90% of PV power electronics systems. It can guarantee acceptable performances in transient and steady-state response and disturbance rejection for a wide range of application.

Figures 5 and 6 show a tracking error not negligible: there is a slow transient until the controlled variable reaches its reference. On the other side, in the central interval the system cannot reach a stable state and keeps oscillating: this is an undesired condition, whose main cause is the PWM delay. In fact, as evident from Fig. 8, the duty cycle imposed by the controller is unstable and presents heavy variations along the entire simulation time. Due to the high gains, necessary to minimize the IAE of the tracking errors, the controller does not find the right output to avoid the oscillations in the central period, that is characterized by strong solar irradiance.

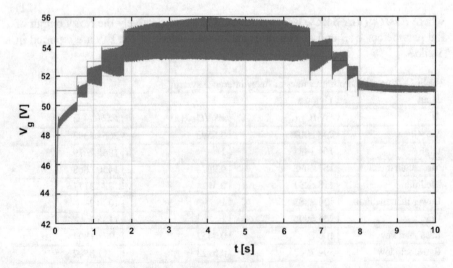

Fig. 5   The controlled variable $\bar{V}_g$

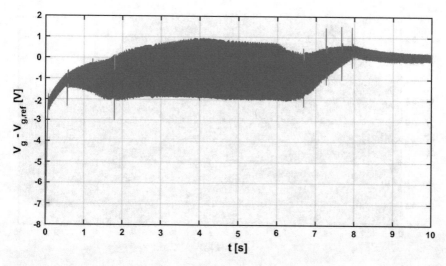

Fig. 6 Error of the controlled variable $\bar{V}_g$

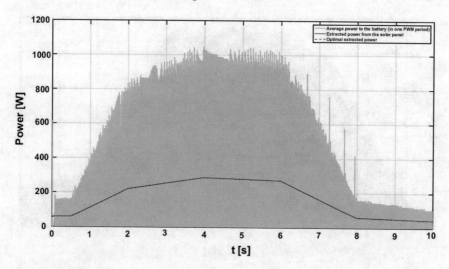

Fig. 7 Power extraction

Figure 9 shows the same problem for the current that flows along the inductance, leading to analogous conclusions. As a result, the amount of power extracted from the panel is near to the ideal amount of power (Fig. 7), but the power given to the battery shows significant oscillations, therefore the power output, as it is, may damage the battery.

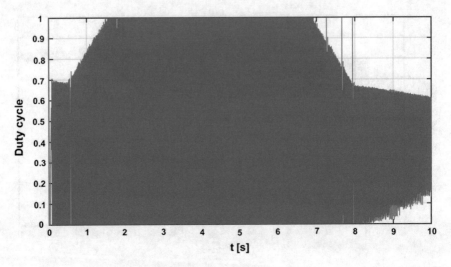

Fig. 8 Output of the PI controller (duty cycle)

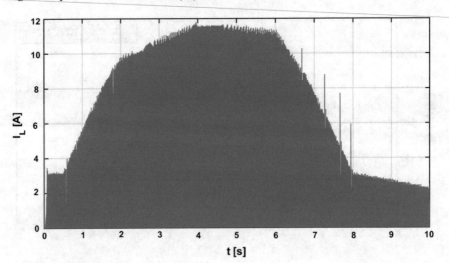

Fig. 9 Current in the inductance

4.2 Backstepping Control

The results obtained with a more complex non-linear control are shown below. Back-stepping is a recursive procedure able to design a controller for a class of systems known as parametric strict-feedback system, as described in [34, 36]. The design problem of the full system is decomposed into a sequence of design problems for lower order subsystems by considering some of the state variables as "virtual controls" and designing for them intermediate control laws (see e.g., [20, 28, 30, 33]). The controller expression used in this paper is the following:

$$S1 = x_1 - x_{1d}$$

$$S_2 = x_2 - C\left(\frac{\bar{I}_g}{C} - \dot{x}_{1d} + K_1 S_1\right)$$

$$u = \frac{L}{\bar{V}_b}\left(\frac{x_1}{L} - \dot{\bar{I}}_g + C\ddot{x}_{1d} - CK_1\left(-\frac{S_2}{C} - K_1 S_1\right) - \frac{S_1}{C} + K_2 S_2\right) \quad (22)$$

Even in this case, as depicted in Figs. 10 and 11, Backstepping shows a large tracking error: compared to the PI controller, the initial transient is faster, but there is still an offset between the measured voltage and the reference. In the central simulation interval the system oscillates as seen in the previous case. As highlighted in Fig. 13, the duty cycle imposed by the controller presents strong oscillations along the entire simulation time. The controller does not find the right output to avoid the oscillations in the central period, due to its high gain. The current on the inductance shows the same pattern, as depicted in Fig. 14. The amount of power extracted from the panel, depicted in Fig. 12, is comparable to that obtained with the PI, and it is very close to the ideal amount of power. The power fed to the battery shows stronger oscillations than the PI controller. The effort of using such a non linear technique is not justified and shows only marginal improvement in comparison with the PI controller.

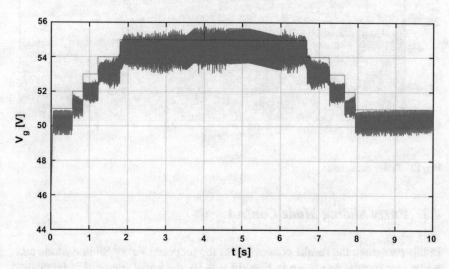

Fig. 10 The controlled variable $\bar{V}_g$

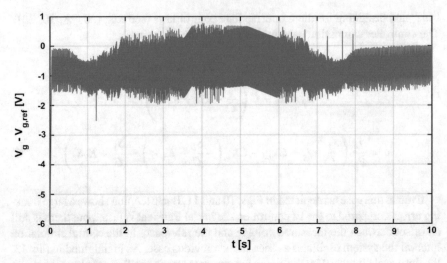

Fig. 11 Error of the controlled variable $\bar{V}_g$

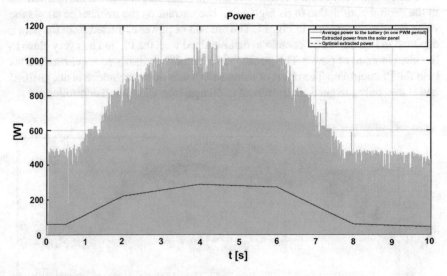

Fig. 12 Power extraction

4.3 Fuzzy Sliding Mode Control

In this paragraph the results obtained with the proposed Fuzzy Sliding Mode controller, are shown. As shown in Figs. 15 and 16, the initial transient is faster than the previous controllers one. In the central interval, the system shows very small

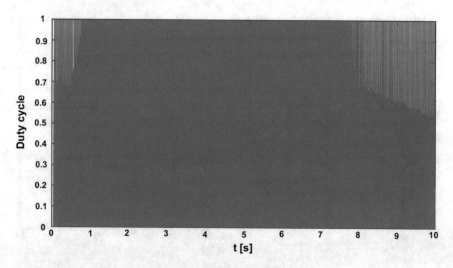

Fig. 13 Output of the backstepping controller (duty cycle)

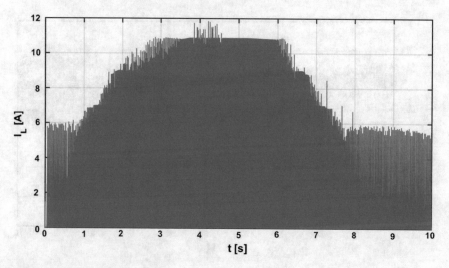

Fig. 14 Current in the inductance

ripples and the tracking error is less than $0.15V$. Figure 18 shows that the duty cycle imposed by the controller is much more stable than the PI and Backstepping ones. Thus, in presence of strong variation of solar irradiance, the output power given to the battery is more regular and instants with no power transmission to the battery are completely absent. This is confirmed by Fig. 19: in the central part, the current

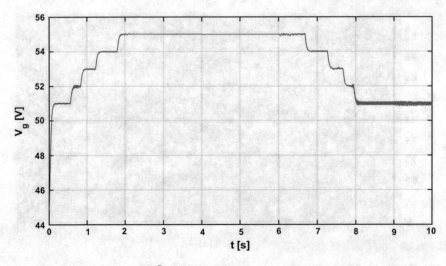

Fig. 15 The controlled variable $\bar{V}_g$

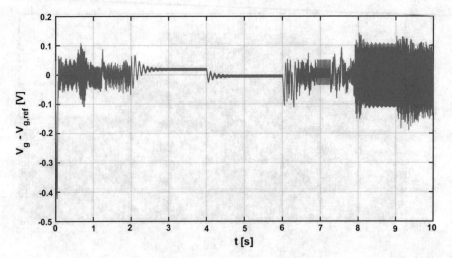

Fig. 16 Error of the controlled variable $\bar{V}_g$

flowing in the inductance doesn't drop to zero, thus the converter operates in continuous conduction mode. The amount of power extracted from the panel is almost equal to the theoretical optimum (computed by the MPPT), as shown in Fig. 18, and the stability of the power transmitted to the battery is drastically improved (Fig. 17)

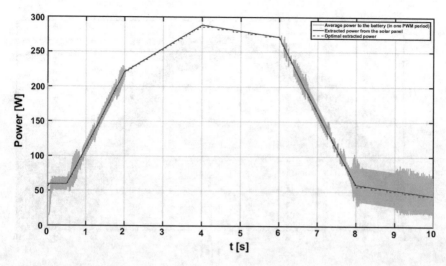

Fig. 17 Power extraction

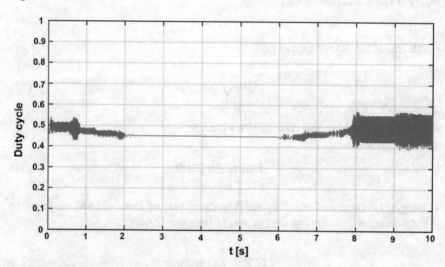

Fig. 18 Output of the fuzzy sliding mode controller (duty cycle)

4.4 Performance Comparison

A proper evaluation of the controller performances needs to be performed through a quantitative comparison. In particular, three performance index measurements have been considered in the comparison process:

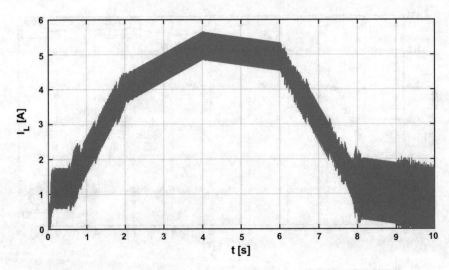

Fig. 19 Current in the inductance

- Integral of the Squared Error (ISE):

$$ISE = \int_0^T [e(t)]^2 dt \qquad (23)$$

- Integral of the Absolute value of Error (IAE):

$$IAE = \int_0^T |e(t)|\, dt \qquad (24)$$

- Integral of Time multiplied by the Absolute value of Error (ITAE), to penalize long duration transients as well as steady state oscillations:

$$ITAE = \int_0^T t \cdot |e(t)|\, dt \qquad (25)$$

The values have been computed for the three controllers previously described. Tables 4 and 5 show the results of the voltage and the exported power tracking performance respectively (Fig. 20).

Table 4 IAE, ISE and ITAE of tracking error $\left(\bar{V}_g - V_{g,ref} \right)$

Controller	IAE	ISE	ITAE
PI	8.1159	11.9174	30.0147
Backstepping	8.0077	9.3133	38.1581
Fuzzy sliding mode	0.3144	0.0234	1.89631

Table 5 IAE, ISE and ITAE of power error $\left(P_{max} - P_{avg} \right)$

Controller	IAE	ISE	ITAE
PI	1.8114×10^3	5.8258×10^5	8.0039×10^3
Backstepping	1.9814×10^3	6.6056×10^5	8.9555×10^3
Fuzzy sliding mode	0.0667×10^3	0.0114×10^5	0.4273×10^3

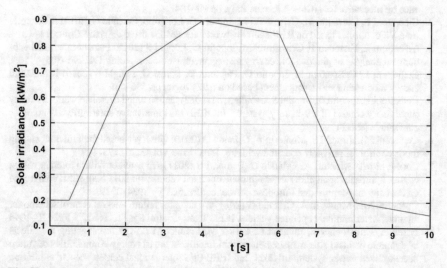

Fig. 20 Solar irradiance

5 Concluding Remarks

In this chapter we proposed a novel fuzzy sliding mode approach to manage power flow of a photovoltaic (PV) battery system. In particular, due to the inner stochastic nature and intermittency of the solar production, a robust and fast controller is needed to face rapid irradiance changes. In this context Sliding Mode Control (SMC) is a popular approach to control systems under heavy uncertain conditions. The solution proposed in this paper to avoid one of the major drawbacks of SMC (i.e., the chattering) is the introduction of a Fuzzy Inference System which varies the controller parameters (boundary layer and gains) according to the measured irradiance.

A simulation comparison of the designed Fuzzy Sliding Mode Control (FSMC) with two popular controllers (PI and Backstepping) has been proposed. FSMC shows better performances in terms of steady state chattering and transient response, as confirmed by performance indexes IAE, ISE and ITAE. In particular, as confirmed by simulations, in presence of strong variation of solar irradiance, FSMC drastically improves the stability of the power transmitted to the battery, thus preserving it from damages.

References

1. Ciabattoni L, Cimini G, Grisostomi M, Ippoliti G, Longhi S, Mainardi E (2013) Supervisory control of PV-battery systems by online tuned neural networks. In: IEEE international conference on mechatronics (ICM), Vicenza, Italy, pp 99–104
2. Ciabattoni L, Corradini M, Grisostomi M, Ippoliti G, Longhi S, Orlando G (2013) A discrete-time VS controller based on RBF neural networks for PMSM drives. Asian J Control 1–13
3. Ciabattoni L, Ferracuti F, Grisostomi M, Ippoliti G, Longhi S (2015) Fuzzy logic based economical analysis of photovoltaic energy management. Neurocomputing 170:296–305
4. Ciabattoni L, Grisostomi M, Ippoliti G, Pagnotta D, Foresi G, Longhi S (2015) Residential energy monitoring and management based on fuzzy logic, pp 536–539
5. Ciabattoni L, Ippoliti G, Longhi S, Cavalletti M (2013) Online tuned neural networks for fuzzy supervisory control of PV-battery systems. In: IEEE PES innovative smart grid technologies conference (ISGT)
6. Corradini ML, Longhi S, Monteriù A, Orlando G (2010) Observer-based fault tolerant sliding mode control for remotely operated vehicles. IFAC Proc Vol 43(20):173–178
7. Corradini ML, Monteri A, Orlando G, Pettinari S (2011) An actuator failure tolerant robust control approach for an underwater remotely operated vehicle. In: 2011 50th IEEE conference on decision and control and European control conference, pp 3934–3939
8. Corradini ML, Monteriu A, Orlando G (2011) An actuator failure tolerant control scheme for an underwater remotely operated vehicle. IEEE Trans Control Syst Technol 19(5):1036–1046
9. Fazeli A, Christopher E, Johnson C, Gillott M, Sumner M (2011) Investigating the effects of dynamic demand side management within intelligent smart energy communities of future decentralized power system. In: 2011 2nd IEEE PES international conference and exhibition on innovative smart grid technologies (ISGT Europe), pp 1 –8
10. Freddi A, Lanzon A, Longhi S (2011) A feedback linearization approach to fault tolerance in quadrotor vehicles. IFAC Proc Vol 44(1):5413–5418
11. Kanchev H, Lu D, Colas F, Lazarov V, Francois B (2011) Energy management and operational planning of a microgrid with a PV-based active generator for smart grid applications. IEEE Trans Ind Electron 58(10):4583–4592
12. Lewis D (2009) Solar grid parity—[power solar]. Eng Technol 4(9):50–53
13. Lopez-Polo A, Haas R, Panzer C, Auer H (2012) Prospects for grid-parity of photovoltaics due to effective promotion schemes in major countries. In: Power and energy engineering conference (APPEEC), 2012 Asia-Pacific, pp 1–4
14. Mancini A, Caponetti F, Monteriù A, Frontoni E, Zingaretti P, Longhi S (2007) Safe flying for an UAV helicopter. In: Mediterranean conference on control and automation. IEEE, Athens, Greece, pp 1–6
15. Palensky P, Dietrich D (2011) Demand side management: demand response, intelligent energy systems, and smart loads. IEEE Trans Ind Inf 7(381):388
16. Rasappan S, Vaidyanathan S (2013) Hybrid synchronization of n-scroll chaotic Chua circuits using adaptive backstepping control design with recursive feedback. Malays J Math Sci 7(2):219–246

17. Sampath S, Vaidyanathan S (2016) Hybrid synchronization of identical chaotic systems via novel sliding control method with application to Sampath four-scroll chaotic system. Int J Control Theory Appl 9(1):221–235
18. Utkin VI, Guldner J, Shi J (1999) Sliding mode control in electromechanical systems. In: Systems and control. CRC Press LLC, Florida, USA
19. Vaidyanathan S (2014) Global chaos synchronisation of identical Li-Wu chaotic systems via sliding mode control. Int J Model Ident Control 22(2):170–177
20. Vaidyanathan S (2015) Adaptive chaotic synchronization of enzymes-substrates system with ferroelectric behaviour in brain waves. Int J PharmTech Res 8(5):964–973
21. Vaidyanathan S (2015) Analysis, control, and synchronization of a 3-D novel jerk chaotic system with two quadratic nonlinearities. Kyungpook Math J 55(3):563–586
22. Vaidyanathan S (2015) Anti-synchronization of brusselator chemical reaction systems via integral sliding mode control. Int J ChemTech Res 8(11):700–713
23. Vaidyanathan S (2015) Global chaos synchronization of chemical chaotic reactors via novel sliding mode control method. Int J ChemTech Res 8(7):209–221
24. Vaidyanathan S (2015) Global chaos synchronization of Rucklidge chaotic systems for double convection via sliding mode control. Int J ChemTech Res 8(8):61–72
25. Vaidyanathan S (2015) Integral sliding mode control design for the global chaos synchronization of identical novel chemical chaotic reactor systems. Int J ChemTech Res 8(11):684–699
26. Vaidyanathan S (2015) A novel chemical chaotic reactor system and its output regulation via integral sliding mode control. Int J ChemTech Res 8(11):669–683
27. Vaidyanathan S (2015) Sliding controller design for the global chaos synchronization of enzymes-substrates systems. Int J PharmTech Res 8(7):89–99
28. Vaidyanathan S (2016) Analysis, adaptive control and synchronization of a novel 4-D hyperchaotic hyperjerk system via backstepping control method. Arch Control Sci 26(3):281–308
29. Vaidyanathan S (2016) Anti-synchronization of duffing double-well chaotic oscillators via integral sliding mode control. Int J ChemTech Res 9(2):297–304
30. Vaidyanathan S (2016) Anti-synchronization of enzymes-substrates biological systems via adaptive backstepping control. Int J PharmTech Res 9(2):193–205
31. Vaidyanathan S (2016) Global chaos control of the generalized Lotka-Volterra three-species system via integral sliding mode control. Int J PharmTech Res 9(4):399–412
32. Vaidyanathan S (2016) A highly chaotic system with four quadratic nonlinearities, its analysis, control and synchronization via integral sliding mode control. Int J Control Theory Appl 9(1):279–297
33. Vaidyanathan S (2016) A novel 3-D jerk chaotic system with two quadratic nonlinearities and its adaptive backstepping control. Int J Control Theory Appl 9(1):199–219
34. Vaidyanathan S (2016) A novel hyperchaotic hyperjerk system with two nonlinearities, its analysis, adaptive control and synchronization via backstepping control method. Int J Control Theory Appl 9(1):257–278
35. Vaidyanathan S, Boulkroune A (2016) A novel hyperchaotic system with two quadratic nonlinearities, its analysis and synchronization via integral sliding mode control. Int J Control Theory Appl 9(1):321–337
36. Vaidyanathan S, Madhavan K, Idowu BA (2016) Backstepping control design for the adaptive stabilization and synchronization of the Pandey jerk chaotic system with unknown parameters. Int J Control Theory Appl 9(1):299–319
37. Vaidyanathan S, Rasappan S (2014) Global chaos synchronization of n-scroll Chua circuit and Lur'e system using backstepping control design with recursive feedback. Arab J Sci Eng 39(4):3351–3364
38. Vaidyanathan S, Volos C (2016) Advances and applications in chaotic systems, vol 636. Springer, Berlin, Germany
39. Vaidyanathan S, Volos C (2016) Advances and applications in nonlinear control systems, vol 635. Springer, Berlin, Germany
40. Vaidyanathan S, Volos C, Pham VT, Madhavan K, Idowu BA (2014) Adaptive backstepping control, synchronization and circuit simulation of a 3-D novel jerk chaotic system with two hyperbolic sinusoidal nonlinearities. Arch Control Sci 24(3):375–403

41. Vallve X, Graillot A, Gual S, Colin H (2007) Micro storage and demand side management in distributed PV grid-connected installations. In: 9th international conference on electrical power quality and utilisation, 2007. EPQU 2007, pp 1–6
42. Veldman E, Gibescu M, Slootweg J, Kling W (2009) Technical benefits of distributed storage and load management in distribution grids. In: 2009 IEEE Bucharest PowerTech, pp 1–8
43. Xiao W, Dunford WG, Capel A (2004) A novel modeling method for photovoltaic cells. In: IEEE 35th annual power electronics specialists conference, vol 3, pp 1950–1956

Particle Swarm Optimization Based Sliding Mode Control Design: Application to a Quadrotor Vehicle

Alessandro Baldini, Lucio Ciabattoni, Riccardo Felicetti,
Francesco Ferracuti, Alessandro Freddi, Andrea Monteriù
and Sundarapandian Vaidyanathan

Abstract In this chapter, a design method for determining the optimal sliding mode controller parameters for a quadrotor dynamic model using the Particle Swarm Optimization algorithm is presented. In particular, due to the effort to determine optimal or near optimal sliding mode parameters, which depend on the nature of the considered dynamic model, a population based solution is proposed to tune the parameters. The proposed population based-method tunes the controller parameters (boundary layers and gains) according to a fitness function that measures the controller performances. A comparison of the designed sliding mode control with two popular controllers (PID and Backstepping) applied to a quadrotor dynamic model is proposed. In particular sliding mode control shows better performances in terms of steady state and transient response, as confirmed by performance indexes IAE, ISE, ITAE and ITSE.

A. Baldini · L. Ciabattoni (✉) · R. Felicetti · F. Ferracuti · A. Monteriù
Dipartimento di Ingegneria dell'Informazione, Università Politecnica delle Marche,
Via Brecce Bianche, 60131 Ancona, Italy
e-mail: l.ciabattoni@univpm.it

A. Baldini
e-mail: a.baldini@univpm.it

R. Felicetti
e-mail: r.felicetti@univpm.it

F. Ferracuti
e-mail: f.ferracuti@univpm.it

A. Monteriù
e-mail: a.monteriu@univpm.it

A. Freddi
Università degli studi eCampus, Via Isimbardi 10, Novedrate (CO), Italy
e-mail: alessandro.freddi@uniecampus.it

S. Vaidyanathan
Research and Development Centre, Vel Tech University,
Avadi, Chennai 600062, Tamil Nadu, India
e-mail: sundarvtu@gmail.com

© Springer International Publishing AG 2017
S. Vaidyanathan and C.-H. Lien (eds.), *Applications of Sliding Mode Control in Science and Engineering*, Studies in Computational Intelligence 709,
DOI 10.1007/978-3-319-55598-0_7

143

Keywords Sliding mode · Quadrotor · Particle swarm optimization

1 Introduction

In recent years an ever growing interest in Unmanned Aerial Vehicles (UAVs) has been shown among the research community, including industry, governments and academia [20]. This popularity is mainly due to their ability to carry out a wide range of applications e.g., military and rescue missions, wild fire surveillance, law enforcement, mapping, aerial multimedia productions, power plants inspection. On the one hand the potential to remove human pilots from danger and size/cost effectiveness of modern UAVs are incredibly attractive; on the other hand the need to obtain comparable (or even better) performances with respect to manned aircrafts in terms of capabilities, efficiency and flexibility poses many engineering challenges.

The quadrotor UAV is a four-rotor Vertical Take-Off and Landing (VTOL) aircraft, which has all advantages of VTOL aircrafts along with an increased payload capability, a stability in hover inherent to its design as well as an increased maneuverability. Furthermore, its most important advantage, when compared to conventional aircraft, is its reduced mechanical complexity. Nowadays, the emerging and existing motions of the quadrotor UAV include:

- the precession motion, however, neutralized by designing the front and rear rotors to spin in the opposite direction to the left and right propellers so eliminating any reactive torque around the vertical coordinate axis;
- the hover motion, obtained by making the rotational velocity of each propellers are same;
- the roll and pitch motions, attained by applying a difference of rotational velocity between the opposing rotors forcing the vehicle to tilt towards the slowest propeller;
- the yaw motion, generated by making the rotational velocity of neighboring rotors different from others, thus forcing the vehicle to tilt towards the two slower propellers;
- the vertical motion, generated by increasing or decreasing the rotational velocity of all rotors by the same amount;
- the horizontal motion, realized by making the vehicle roll or pitch firstly, so as to shift the direction of the thrust vector and then produce a forward component.

In practical missions, the stability of the aircraft is easily affected by abruptly changed commands. The flight controller design should be capable to offer to the aircraft an accurate and robust control which is crucial in the flight process.

In this context, Sliding Mode Control (SMC) is a popular approach to control systems under heavy uncertainty conditions, and is accurate, robust and simple to implement [11–13, 33, 36–40, 42, 43]. The main drawbacks of a classical first-order SMC (1-SMC) are principally related to the so-called chattering effect, undesired steady-state vibrations of the system variables. A major cause of chattering has been identified as the presence of unmodelled parasitic dynamics in the switching devices [6]. To mitigate the chattering's effects, many solutions have been developed over the last three decades [5, 17, 29]. In particular three main approaches have been developed:

- the use of a continuous approximation of the relay (e.g., the saturation function);
- the use of asymptotic state-observer to confine chattering in the observer dynamics bypassing the plant;
- the use of Higher-Order Sliding Mode control algorithms (HOSM).

The SMC literature is therefore wide for the flight controller design of a quadrotor vehicles, e.g., [14, 28, 31, 46]. SMC is extensively proposed in literature also in the context of fault tolerant systems [10, 16]. A SMC requires the tuning of different parameters that, unfortunately, are quite difficult to tune properly due to time delays and nonlinearities of real systems [2, 4, 6, 42]. Over the years, several heuristic methods have been proposed for tuning the SMCs [15, 22, 45]. In general, it is often hard to determine optimal or near optimal sliding mode parameters. For these reasons, advanced tuning methods are needed to increase the capabilities of SMCs. In this context, Artificial Intelligence (AI) techniques have been proposed in the control theory literature to increase controller performances and to proper tune the Proportional-Integral-Derivative (PID) controller parameters [3, 21, 43]. Many heuristic algorithms methods have recently received much interest for dealing with nonlinear optimization problems. One of the modern random search method is the Particle Swarm Optimization (PSO) method introduced by [24]. PSO originated from the study of a simulator for a social system, and it demonstrated to be robust in solving continuous nonlinear optimization problems [24].

In this paper, we have developed a SMC tuned by PSO in order to search optimal SMC parameters. As the novelty, we propose a procedure for selecting the parameters of SMC, in order to maximize its performance. The SMC tuned by PSO is tested on the quadrotor dynamic model and compared with PID and backstepping controllers. All the controllers are tuned by PSO and four indexes, i.e., Integral of the Squared Error (ISE), Integral of the Absolute value of Error (IAE), Integral of Time multiplied by the Squared Error value (ITSE) and Integral of Time multiplied by the Absolute value of Error (ITAE), are considered to compare the performances of all controllers.

The chapter is organized as follows. Section 2 shows the quadrotor dynamic model considered to test the controllers. Section 3 describes the designing of sliding mode, backstepping and PID techniques for the quadrotor model. In Sect. 4 we present the PSO algorithm, the swarm intelligence-based technique used to solve the tuning problem of SMC. Results on the testing model are reported in Sect. 5. The chapter ends with comments on the performance of the proposed solution.

2 Quadrotor Dynamics

A quadrotor consists in four DC motors on which propellers are fixed, and arranged to the extremities of a X-shaped frame, where all the arms make an angle of 90° with one another. The speed of rotation of the motors (i.e.: the lift force associated to the propeller attached to that motor) can be individually changed, thus modifying the attitude of the vehicle allowing the quadrotor to translate into the space.

In this section we first use the Newton-Euler approach to model the quadrotor as a rigid body subject to forces and moments and provide the most relevant aerodynamic effects, similarly to [30] and [19]. Then we simplify the model in order to derive a set of equations more suitable for control purposes, as commonly found in the literature, see for instance [7, 27].

2.1 Kinematics

The quadrotor vehicle is modelled using the convention shown in Fig. 1. Two frames are used to study the system motion, in particular a frame integral with the earth $\{R\}$ (O, x, y, z), which is supposed to be inertial, and a body-fixed frame $\{R_B\}$ $\{O_B, x_B, y_B, z_B\}$, where O_B is fixed to the center of mass of the quadrotor. $\{R_B\}$ is related to $\{R\}$ by a position vector $\boldsymbol{\xi} = \begin{bmatrix} x & y & z \end{bmatrix}^T$, describing the position of the center of gravity in $\{R_B\}$ relative to $\{R\}$ and by a vector of three independent angles $\boldsymbol{\eta} = \begin{bmatrix} \psi & \theta & \phi \end{bmatrix}^T$, which represent the orientation of the body-fixed frame $\{R_B\}$

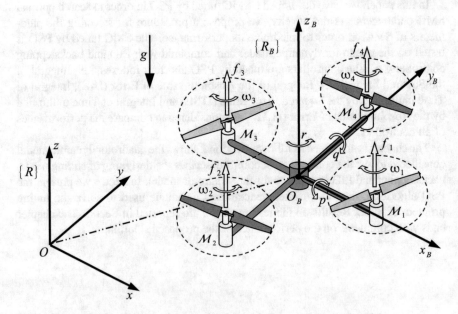

Fig. 1 The quadrotor scheme used for the development of the mathematical model

$\{O_B, x_B, y_B, z_B\}$ with respect to the earth frame $\{R\}$ (O, x, y, z). The adopted notation, usually called yaw, pitch and roll, is based on the assumption that the earth frame $\{R\}$ (O, x, y, z) can reach the same orientation of the body-fixed frame $\{R_B\}$ $\{O_B, x_B, y_B, z_B\}$ by first performing a rotation of an angle ψ around the z-axis (yaw), then a rotation of an angle θ around the new y-axis (pitch) and finally a rotation of an angle ϕ around the new x-axis (roll). All the rotations are right-handed with $\left(-\frac{\pi}{2} < \phi < \frac{\pi}{2}\right)$, $\left(-\frac{\pi}{2} < \theta < \frac{\pi}{2}\right)$ and ψ is unrestricted. In this way ξ and η fully describe, respectively, the translational and the rotational movement of the rotorcraft with respect to the earth frame.

Given a vector v_B, expressed using the coordinates of the body frame, the vector v expressed in the coordinates of the earth frame is:

$$v = Rv_B \tag{1}$$

where R is the rotation transformation from vectors read in the body reference frame to vectors read in the earth reference frame given by

$$R \doteq \begin{bmatrix} C_\theta C_\psi & C_\psi S_\theta S_\phi - C_\phi S_\psi & C_\phi C_\psi S_\theta + S_\phi S_\psi \\ C_\theta S_\psi & S_\theta S_\phi S_\psi + C_\phi C_\psi & C_\phi S_\theta S_\psi - C_\psi S_\phi \\ -S_\theta & C_\theta S_\phi & C_\theta C_\phi \end{bmatrix} \tag{2}$$

In a similar way, given the angular velocity vector $\omega = \begin{bmatrix} p & q & r \end{bmatrix}^T$, where p, q and r represent the instantaneous angular velocities around the x_B-axis, y_B-axis and z_B-axis respectively, it is related to the rate of change of the yaw, pitch and roll angles by:

$$\omega = W\dot{\eta} \tag{3}$$

where

$$W \doteq \begin{bmatrix} 1 & 0 & -S_\theta \\ 0 & C_\phi & S_\phi C_\theta \\ 0 & -S_\phi & C_\phi C_\theta \end{bmatrix} \tag{4}$$

and $S_{(.)}$ and $C_{(.)}$ represent sin (.) and cos (.) respectively.

2.2 Rigid Body Dynamics

The quadrotor is modelled as a rigid body subject to actuation and external forces. The Newton-Euler approach is adopted in order to derive the rigid body dynamics of the quadrotor model, as it represents the most straightforward approach for modelling purposes [18].

Denote with m the whole mass of the rotorcraft and define the inertia matrix in the body frame as

$$I \doteq \begin{bmatrix} I_x & -I_{xy} & -I_{xz} \\ -I_{yx} & I_y & -I_{yz} \\ -I_{zx} & -I_{zy} & I_z \end{bmatrix} \tag{5}$$

where I_x, I_y, I_z are the moments of inertia about x_B, y_B, z_B axes, and $I_{xy} = I_{yx}, I_{xz} = I_{zx}$, $I_{yz} = I_{zy}$ are the products of inertia. Let

- $f_B^o = [F_x, F_y, F_z]^T$ be the vector of forces acting on the quadrotor read in the body frame;
- $m_B^o = [K, M, N]^T$ be the vector of moments acting on the quadrotor read in the body frame;
- $v_B^o = [u, v, w]^T$ be the linear velocities along the x_B, y_B, z_B axes read in the body frame;
- $r_g = [x_g, y_g, z_g]^T$ be the vector of offsets between the body frame and the center of gravity of the quadrotor read in the body frame;
- $v = [u, v, w, p, q, r]^T$ be the generalized velocity vector read in the body frame;
- $\tau = [F_x, F_y, F_z, K, M, N]^T$ be the generalized force vector read in the body frame.

Then the rigid body dynamics can be expressed in a vectorial form as

$$M\dot{v} + C(v)v = \tau \tag{6}$$

where M is the rigid body system inertia matrix and $C(v)$ is the rigid body Coriolis and Centripetal matrix.

The representation of the rigid body inertia matrix is unique

$$M = \begin{bmatrix} m & 0 & 0 & 0 & mz_g & -my_g \\ 0 & m & 0 & -mz_g & 0 & mx_g \\ 0 & 0 & m & my_g & -mx_g & 0 \\ 0 & -mz_g & my_g & I_x & -I_{xy} & -I_{xz} \\ mz_g & 0 & -mx_g & -I_{yx} & I_y & -I_{yz} \\ -my_g & mx_g & 0 & -I_{zx} & -I_{zy} & I_z \end{bmatrix} \tag{7}$$

The rigid body system inertia matrix is built such that it takes into account Newton second law and the Euler fictitious force. It is thus necessary to include the Coriolis and centrifugal forces in matrix $C(v)$. The rigid body Coriolis and centrifugal matrix $C(v)$ can always be represented in a skew-symmetrical form, which is not unique. We use the following representation:

$$C(v) = \begin{bmatrix} C_{11} & C_{12} \\ C_{21} & C_{22} \end{bmatrix} \tag{8}$$

with

$$C_{11} = \begin{bmatrix} 0 & 0 & 0 \\ 0 & 0 & 0 \\ 0 & 0 & 0 \end{bmatrix} \tag{9a}$$

$$C_{12} = \begin{bmatrix} m(rz_g + qy_g) & m(w - qx_g) & -m(v + rx_g) \\ -m(w + py_g) & m(rz_g + px_g) & m(u - ry_g) \\ m(v - pz_g) & -m(u + qz_g) & m(qy_g + px_g) \end{bmatrix} \tag{9b}$$

$$C_{21} = \begin{bmatrix} -m(rz_g + qy_g) & m(w + py_g) & m(pz_g - v) \\ m(qx_g - w) & -m(rz_g + px_g) & m(u + qz_g) \\ m(v + rx_g) & m(ry_g - u) & -m(qy_g + px_g) \end{bmatrix} \tag{9c}$$

$$C_{22} = \begin{bmatrix} 0 & (-pI_{zx} - qI_{zy} + rI_z) & (pI_{yx} - qI_y + rI_{yz}) \\ (pI_{zx} + qI_{zy} - rI_z) & 0 & (pI_x - qI_{xy} - rI_{zx}) \\ (-pI_{yx} + qI_y - rI_{yz}) & (-pI_x + qI_{xy} + rI_{xz}) & 0 \end{bmatrix} \tag{9d}$$

2.3 Forces and Moments

2.3.1 Force of Gravity

The weight force is applied to the center of gravity and directed along the negative z-axis in the earth frame

$$F_g = R^T \begin{bmatrix} 0 \\ 0 \\ -mg \end{bmatrix} \tag{10}$$

where R^T is the rotation transformation from vectors read in the earth reference frame to vectors read in the body reference frame. Since the point of application is the center of gravity, the gravity force does not generate any moment.

2.3.2 Thrust

The thrust force f_j, where $j = 1, 2, 3, 4$, is applied to the center of the j-th motor, distant l from the center of mass, and directed along the positive z_B-axis. Labeling as f_1, f_2, f_3 and f_4 the upward lifting forces generated by the propellers, the force due to the control inputs, expressed into the body frame, has a non zero component in the z_B direction only:

$$f_j = \begin{bmatrix} 0 \\ 0 \\ f_j \end{bmatrix} \tag{11}$$

The expression of the thrust can be derived using the actuator disk and blade elements theory well known in literature, which lead to

$$f_j = C_T \rho A r^2 \Omega_j^2 \doteq c\Omega_j^2 \tag{12}$$

where C_T is the thrust coefficient, ρ is the air density, A is the rotor disk area, r is the rotor radius and Ω_j is the rotational speed of motor j.

Denoting with

$$l_j = \begin{bmatrix} l_{j,x} \\ l_{j,y} \\ l_{j,z} \end{bmatrix} \tag{13}$$

the vector whose components represent the offset between the center of rotor j with respect to the center of gravity, the moment around the body frame axis due to the force f_j is described by

$$m_j = f_j \times l_j \tag{14}$$

2.3.3 Propeller Drag

The aerodynamic forces acting on the blade of each rotor, in a direction orthogonal to the thrust force (blade reaction to rotation), generate and additional moment which must be included into the model. This moment is called drag moment and it is directed along the z_B axis:

$$q_j = \begin{bmatrix} 0 \\ 0 \\ q_j \end{bmatrix} \tag{15}$$

The expression of the drag moment can be derived similarly to that used to derive the thrust force:

$$q_j = C_Q \rho A r^3 \Omega_j |\Omega_j| \doteq d\Omega_j |\Omega_j| \tag{16}$$

where C_Q is the drag coefficient and Ω_j is multiplied by its magnitude to preserve the sing of rotation for counter-rotating rotors.

2.3.4 Blade Gyroscopic Effect

The quadrotor vehicle is propelled by four blades which rotate at high speed. Even if the inertia of the propeller is small when compared to that of the body, the high rotation speed causes a gyroscopic effect which can affect the motion of the vehicle when the propellers spin with different velocities.

The gyroscopic torque m_g caused by rotations of the vehicle with rotating rotors can be expressed as

$$m_g = J_R \left(\omega \times \begin{bmatrix} 0 \\ 0 \\ 1 \end{bmatrix} \right) \left(\Omega_1 + \Omega_3 - \Omega_2 - \Omega_4 \right) \tag{17}$$

where ω is the angular velocity of the body read into the body frame, Ω_j is the rotational speed of motor j and J_R is the propeller inertia.

2.4 Model Equation Set

The vehicle model can now be expressed using the rigid body dynamics, the forces and moments acting on the quadrotor and the kinematic equations described in the previous sections. The complete model in vector form is

$$\dot{\xi} = R v_B^o \tag{18a}$$

$$\dot{\eta} = W^{-1} \omega \tag{18b}$$

$$M\dot{v} + C(v)v = \begin{bmatrix} f_B^o \\ m_B^o \end{bmatrix} = \begin{bmatrix} \sum_{j=1}^4 f_j + F_g \\ \sum_{j=1}^4 \left(q_j + m_j \right) - m_g \end{bmatrix} \tag{18c}$$

where f_j is defined in (12), F_g is defined in (10), q_j defined in (16), m_j is defined in (14) and m_g is defined in (17).

Equations 18 describe the dynamics of a quadrotor for a wide range of flight conditions, however it can be simplified when certain assumptions are made, in order to obtain simpler equations for simulation and control purposes. These assumptions are reasonable for the majority of quadrotors studied in the literature.

Usually quadrotor vehicles are built symmetrically with respect to the body axes, with the center of gravity close to the geometric center and the motors having a distance l from the center of gravity. According to this design, Eq. 5 can be written as

$$I = \begin{bmatrix} I_x & 0 & 0 \\ 0 & I_y & 0 \\ 0 & 0 & I_z \end{bmatrix} \tag{19}$$

and it is possible to consider $r_g = 0$, $l_1 = -l_3 = [l \ 0 \ 0]^T$ and $-l_2 = l_4 = [0 \ l \ 0]^T$. Moreover quadrotors are usually adopted for low flight speed missions with conservative attitude, thus the roll and pitch angles are small and matrix W simplifies to the identity matrix I. The simplified model for control purposes then becomes [7]

$$\begin{cases} \dot{x}_1 = x_2 \\ \dot{x}_2 = a_1 x_4 x_6 + a_2 \Omega x_4 + b_1 U_2 \\ \dot{x}_3 = x_4 \\ \dot{x}_4 = a_3 x_2 x_6 + a_4 \Omega x_2 + b_2 U_3 \\ \dot{x}_5 = x_6 \\ \dot{x}_6 = a_5 x_2 x_4 + b_3 U_4 \\ \dot{x}_7 = x_8 \\ \dot{x}_8 = -g + cos(x_1)cos(x_3)\frac{1}{m}U_1 \\ \dot{x}_9 = x_{10} \\ \dot{x}_{10} = \frac{1}{m}u_x U_1 \\ \dot{x}_{11} = x_{12} \\ \dot{x}_{12} = \frac{1}{m}u_y U_1 \end{cases} \tag{20}$$

where

$$\begin{aligned} u_x &= (cosx_1 sinx_3 cosx_5 + sinx_1 sinx_5) \\ u_y &= (cosx_1 sinx_3 sinx_5 - sinx_1 cosx_5) \end{aligned} \tag{21}$$

$$\begin{aligned} U_1 &= c(\Omega_1^2 + \Omega_2^2 + \Omega_3^2 + \Omega_4^2) \\ U_2 &= c(\Omega_4^2 - \Omega 2^2) \\ U_3 &= c(\Omega_3^2 - \Omega_1^2) \\ U_4 &= d(\Omega_2^2 + \Omega_4^2 - \Omega_1^2 - \Omega_3^2) \\ \Omega &= \Omega_2 + \Omega_4 - \Omega_1 - \Omega_3 \end{aligned} \tag{22}$$

$$\begin{aligned} x_1 &= \phi & x_7 &= z \\ x_2 &= \dot{x}_1 = \dot{\phi} & x_8 &= \dot{x}_7 = \dot{z} \\ x_3 &= \theta & x_9 &= x \\ x_4 &= \dot{x}_3 = \dot{\theta} & x_{10} &= \dot{x}_9 = \dot{x} \\ x_5 &= \psi & x_{11} &= y \\ x_6 &= \dot{x}_5 = \dot{\psi} & x_{12} &= \dot{x}_{11} = \dot{y} \end{aligned} \tag{23}$$

$$\begin{aligned} a_1 &= \frac{I_y - I_z}{I_x} & b_1 &= \frac{l}{I_x} \\ a_2 &= \frac{-J_R}{I_x} & b_2 &= \frac{l}{I_y} \\ a_3 &= \frac{I_z - I_x}{I_y} & b_3 &= \frac{l}{I_z} \\ a_4 &= \frac{J_R}{I_y} \\ a_5 &= \frac{I_x - I_y}{I_z} \end{aligned} \tag{24}$$

3 Control Design

The following section shows the sliding mode, backstepping and PID techniques designed to control a quadrotor, whose mathematical model is described in Sect. 2.4. In this paragraph, the design of sliding mode controller is described in details, whereas backstepping and PID controllers are designed according to [7].

3.1 Sliding Mode Control

In this subsection we present the theory of sliding mode for a simple class of non-linear problems (Sect. 3.1.1) and then extends it to the case of the quadrotor vehicle (Sect. 3.1.2) by focusing on the tracking problem.

3.1.1 Basics of Sliding Mode Control

Let us consider a simple nonlinear system

$$\Sigma \begin{cases} \dot{x}_1 = x_2 \\ \dot{x}_2 = h(x) + g(x)u \end{cases} \tag{25}$$

where $(x_1, x_2) \in D \subset \mathbb{R}^2$ is the state space, $h(x)$ and $g(x)$ are nonlinear functions and $g(x) \neq 0$. Without loss of generality, we can suppose $g(x) \geq g_0 > 0$. This system can be stabilized as reported in [25]. Let us define the following sliding surface $S = a_1 x_1 + x_2 = 0$, where $a_1 > 0$ can be freely chosen. On this surface, the motion is governed by $\dot{x}_1 = -a_1 x_1$, with $a_1 > 0$, so the resulting system is asymptotically stable and $x_1 \to 0$ (and $x_2 \to 0$ too). We note, in fact, that the equation $\dot{x}_1 = -a_1 x_1$ is a differential equation with solution $x_1(t) = x_1(t_0)e^{(t-t_0)}$. In other words, if the state of the system (see Eq. (20)) is on the surface, then the dynamic of x_1 converges to 0. It follows that x_2 must be constant, but from $S = a_1 x_1 + x_2 = 0$ we obtain $x_2 = -a_1 x_1 = 0$. This is the reason why such a surface S is called "sliding surface": the state of the system (x_1, x_2) slides on a surface S, designed on the state space $D \subset \mathbb{R}^2$, and reaches the point $(0, 0)$ asymptotically. In order to design a sliding mode controller, the first step is to choose a proper sliding surface. Then we must impose $S = 0$ by the action of a control law $u = u(x)$, where u is the control input. From another point of view we can say that the closed-loop system must be asymptotically stable. It is usually needed, if possible, that the convergence of the system can be reached in finite time. For doing this we can use Lyapunov theory of the non linear system theory. The derivative of S is

$$\dot{S} = a_1 \dot{x}_1 + \dot{x}_2 = a_1 x_2 + h(x) + g(x)u \tag{26}$$

We define $\rho(x)$ such that

$$\left|\frac{a_1 x_2 + h(x)}{g(x)}\right| \le \rho(x) \qquad \forall x \in \mathbb{R}^2 \tag{27}$$

Choosing $V = \frac{S^2}{2}$ as a candidate Lyapunov function, it follows that

$$\begin{aligned}\dot{V} = S\dot{S} &= S[a_1 x_2 + h(x) + g(x)u] = \\ &= S[a_1 x_2 + h(x)] + Sg(x)u \le g(x)|S|\rho(x) + g(x)Su\end{aligned} \tag{28}$$

Choosing $u = -\beta(x)sign(S)$, with $\beta(x) \ge \rho(x) + \beta_0$ and $\beta_0 > 0$, yields

$$\dot{V} \le g(x)|S|\rho(x) - g(x)[\rho(x) + \beta_0]S \cdot sign(S) = -g(x)\beta_0|S| \le -g_0\beta_0|S| \tag{29}$$

Therefore, the trajectory reaches the surface $S = 0$ in finite time and, once on the surface, it cannot leave it because $\dot{V} \le -g_0\beta_0|S|$, hence it is definitive negative.

This approach has the following negative effect: suppose that the state (x_1, x_2) of the system is in a point $x' \in D$ where $S > 0$. The control law forces the system to reach the surface with $sign(S) = 1$. Once the system reaches the surface, it remains on the surface for a while (instantaneously) and then leaves it, due to physical problems, like inertia. Suppose now that the state (x_1, x_2) of the system is in a point $x'' \in D$ where $S < 0$ (reached under the effect of the control law). This time $sign(S) = -1$, so the control input, with an inverse action if compared to the previous, can move the system to the sliding surface again. The final result of this "zig-zag" action is the well-known chattering problem, which represents the main problem of the sliding mode control.

To avoid chattering, let us introduce a boundary layer: we do not impose that $S = 0$, but $|S| \le \varepsilon$, with $\varepsilon > 0$ arbitrarily chosen. In this way the control law can be separated in two parts. When $|S| > \varepsilon$, the control input is the same of the previous case, otherwise, if $|S| \le \varepsilon$, the control law can be freely modified. The standard solution is to choose a linear function: this control law is simple and permits to avoid any discontinuity of the control input, in other words, $sign(S)$ is replaced by $sat(S/\varepsilon)$. A deeper explanation of chattering problem and its mathematical consequences can be found in literature [6].

$$sign(S) = \begin{cases} +1 & S > 0 \\ 0 & S = 0 \\ -1 & S < 0 \end{cases} \qquad sat(\frac{S}{\varepsilon}) = \begin{cases} +1 & |S| > \varepsilon \\ \frac{S}{\varepsilon} & |S| < \varepsilon \end{cases} \tag{30}$$

The control law that grants stabilization is finally

$$u = -\beta(x)sat(\frac{S}{\varepsilon}) \tag{31}$$

This control law solves the stabilization problem. For the application to the quadrotor vehicle, we are interested to the tracking problem, so we define $x_{1d}(t)$, $x_{2d}(t) = \dot{x}_{1d}(t)$ as references for the states x_1, x_2. It is possible to solve the tracking problem such that $x_1(t) \to x_{1d}(t)$ and $x_2(t) \to x_{2d}(t)$ for $t \to \infty$. Consider the following tracking errors $e_1(t) = x_1(t) - x_{1d}(t)$ and $e_2(t) = x_2(t) - x_{2d}(t)$, the error system is then

$$\Sigma_e \begin{cases} \dot{e}_1 = \dot{x}_1 - \dot{x}_{1d} = e_2 \\ \dot{e}_2 = \dot{x}_2 - \dot{x}_{2d} = h(x) + g(x)u - \ddot{x}_{1d} \end{cases} \tag{32}$$

We note that this system has the same structure of the previous one by naming $h'(x) = h(x) - \ddot{x}_{1d}$[1] and the same steps can be followed if we know a bound of $\ddot{x}_{1d}$, that implies a known bound for $\frac{a_1 x_2 + h'(x)}{g(x)}$ at the same conditions. The sliding surface is $S = a_1 e_1 + e_2 = 0$ and its derivative is

$$\dot{S} = a_1 \dot{e}_1 + \dot{e}_2 = a_1 e_2 + h'(x) + g(x)u \tag{33}$$

We define $\rho(x)$ such that

$$\left| \frac{a_1 e_2 + h'(x)}{g(x)} \right| \le \rho(x) \qquad \forall x \in \mathbb{R}^2 \tag{34}$$

Choosing $V = \frac{S^2}{2}$ as a candidate Lyapunov function it follows that

$$\dot{V} = S\dot{S} = S[a_1 e_2 + h'(x) + g(x)u] \le g(x)|S|\rho(x) + g(x)Su \tag{35}$$

Once again, by choosing $u = -\beta(x)sign(S)$, with $\beta(x) \ge \rho(x) + \beta_0$ and $\beta_0 > 0$, yields

$$\dot{V} \le g(x)|S|\rho(x) - g(x)[\rho(x) + \beta_0]S \cdot sign(S) = -g(x)\beta_0|S| \le -g_0\beta_0|S| \tag{36}$$

The trajectory reaches the surface $S = 0$ (in finite time) and once on the surface it cannot leave it, because $\dot{V} \le -g_0\beta_0|S|$. It follows that $e_1 \to 0$ and $e_2 \to 0$, or rather, $x_1 \to x_{1d}$ and $x_2 \to x_{2d}$, that means $\dot{x}_1 \to \dot{x}_{1d}$. To avoid the problem of chattering, $sign(\cdot)$ function is substituted by $sat(\cdot)$ function, as discussed previously, choosing a boundary layer with an arbitrary $\varepsilon > 0$. The control law that grants tracking is finally

$$u = -\beta(x)sat(\frac{S}{\varepsilon}) \tag{37}$$

where $\beta(x)$ and the sliding function S are now related to the tracking error system.

---

[1] For the sake of clarity, we note that h' is time dependent due to $\ddot{x}_{1d}$, so it is actually $h'(x, t)$. Therefore in the following steps, we assert relations $\forall t$.

3.1.2 Sliding Mode Control Applied to the Quadrotor Vehicle

The SMC presented here is not the general theory, but it's only valid in the simple case when the system has the structure discussed above. It is natural to apply this theory to a quadrotor helicopter, because the considered model is composed by six couples of equations in this form, but only four are considered to synthesize the control laws because the system is underactuated.[2] This follows by the nature of mechanical system, where state variables are the linear and angular positions and speeds, while the equations involve also linear and angular accelerations. Before doing this we must note the following assumption:

1. Let us assume the state of the system (20) to be known $\forall t$.

   This assumption is usually true for quadrotor vehicles, since all the state variables are measured (or precisely estimated) by using different sensor fusion algorithms (see [27]). Considering the quadrotor system described by Eqs. (20)–(24), where $b_i > 0$, $i = 1, 2, 3, 4$, the first couple of equations is

$$\Sigma_{1,2} \begin{cases} \dot{x}_1 = x_2 \\ \dot{x}_2 = a_1 x_4 x_6 + a_2 \Omega x_4 + b_1 U_2 \end{cases}$$

so we can apply the SMC theory by naming $h(x) = a_1 x_4 x_6 + a_2 \Omega x_4$ and $g(x) = g_0 = b_1 > 0$. The sliding surface is

$$S_1 = K_1 e_1 + e_2$$

It is possible to choose $\rho_2(x) = \left| \frac{K_1 e_2 + h'(x)}{g(x)} \right|$, because the terms are known

$$\left| \frac{K_1 e_2 + h'(x)}{g(x)} \right| \leq \rho_2(x)$$

Following the procedure it is possible to set

$$\beta_2 = \left| \frac{K_1 e_2 + a_1 x_4 x_6 + a_2 \Omega x_4 - \ddot{x}_{1d}}{b_1} \right| + \sigma_1 = \frac{|K_1 e_2 + \frac{I_y - I_z}{I_x} x_4 x_6 + \frac{-J_R}{I_x} \Omega x_4 - \ddot{x}_{1d}|}{b_1} + \sigma_1$$

where $\sigma_1 > 0$ is a project parameter. The control law is

$$U_2 = -\beta_2 sat(\frac{S_1}{\varepsilon_1})$$

---

[2] The remaining two couples of equations are used in the outer loop, which permits to calculate the reference angles required to track a desired position.

The second couple of equations is

$$\Sigma_{3,4} \begin{cases} \dot{x}_3 = x_4 \\ \dot{x}_4 = a_3 x_2 x_6 + a_4 \Omega x_2 + b_2 U_3 \end{cases}$$

again, we can apply the SMC theory by defining $h(x) = a_3 x_2 x_6 + a_4 \Omega x_2$ and $g(x) = g_0 = b_2 > 0$. The sliding surface is $S_3 = K_2 e_3 + e_4$, where $e_3 = x_3 - x_{3d}$, $e_4 = x_4 - x_{4d} = \dot{x}_3 - \dot{x}_{3d}$. Following the procedure it is possible to set

$$\beta_3 = \left| \frac{K_2 e_4 + a_3 x_2 x_6 + a_4 \Omega x_2 - \ddot{x}_{3d}}{b_2} \right| + \sigma_2 = \frac{|K_2 e_4 + \frac{I_z - I_x}{I_y} x_2 x_6 + \frac{J_R}{I_y} \Omega x_2 - \ddot{x}_{3d}|}{b_2} + \sigma_2$$

where $\sigma_2 > 0$ is a project parameter. The control law is

$$U_3 = -\beta_3 sat(\frac{S_3}{\varepsilon_2})$$

The third couple of equations is

$$\Sigma_{5,6} \begin{cases} \dot{x}_5 = x_6 \\ \dot{x}_6 = a_5 x_2 x_4 + b_3 U_4 \end{cases}$$

The SMC theory can applied by defining $h(x) = a_5 x_2 x_4$ and $g(x) = g_0 = b_3 > 0$. The sliding surface is $S_5 = K_3 e_5 + e_6$, where $e_5 = x_5 - x_{5d}$, $e_6 = x_6 - x_{6d} = \dot{x}_5 - \dot{x}_{5d}$. Following the procedure it is possible to set

$$\beta_4 = \left| \frac{K_3 e_6 + a_5 x_2 x_4 - \ddot{x}_{5d}}{b_3} \right| + \sigma_3 = \frac{|K_3 e_6 + \frac{I_x - I_y}{I_z} x_2 x_4 - \ddot{x}_{5d}|}{b_3} + \sigma_3$$

The control law is

$$U_4 = -\beta_4 sat(\frac{S_5}{\varepsilon_3})$$

where $\sigma_3 > 0$ is a project parameter. The last couple of equations is

$$\Sigma_{7,8} \begin{cases} \dot{x}_7 = x_8 \\ \dot{x}_8 = -g + cos(x_1)cos(x_3)\frac{1}{m}U_1 \end{cases}$$

We note that $-\pi/2 < x_1 < \pi/2$ and $-\pi/2 < x_3 < \pi/2$, otherwise the system looses its physical meaning. The SMC theory can be applied by defining $h(x) = -g$ and $g(x) = cos(x_1)cos(x_3)\frac{1}{m} > g_0 > 0$. The sliding surface is $S_7 = K_4 e_7 + e_8$, where $e_7 = x_7 - x_{7d}$, $e_8 = x_8 - x_{8d} = \dot{x}_7 - \dot{x}_{7d}$. Following the procedure it is possible to set

$$\beta_1 = \left| \frac{K_4 e_8 - g - \ddot{x}_{7d}}{cos(x_1)cos(x_3)\frac{1}{m}} \right| + \sigma_4 = \frac{|K_4 e_8 - g - \ddot{x}_{7d}|}{cos(x_1)cos(x_3)\frac{1}{m}} + \sigma_4$$

where $\sigma_3 > 0$ is a project parameter. The control law is

$$U_1 = -\beta_1 sat(\frac{S_7}{\varepsilon_4})$$

In conclusion, the overall control laws of the tracking problem for the quadrotor system are

$$\begin{cases} U_2 = -\beta_2 sat(\frac{S_1}{\varepsilon_1}) \\ U_3 = -\beta_3 sat(\frac{S_3}{\varepsilon_2}) \\ U_4 = -\beta_4 sat(\frac{S_5}{\varepsilon_3}) \\ U_1 = -\beta_1 sat(\frac{S_7}{\varepsilon_4}) \end{cases}$$

where

$$\begin{cases} \beta_2 = \frac{|K_1 e_2 + \frac{I_y - I_z}{I_x} x_4 x_6 + \frac{-J_R}{I_x}\Omega x_4 - \ddot{x}_{1d}|}{b_1} + \sigma_1 \\ \beta_3 = \frac{|K_2 e_4 + \frac{I_z - I_x}{I_y} x_2 x_6 + \frac{J_R}{I_y}\Omega x_2 - \ddot{x}_{3d}|}{b_2} + \sigma_2 \\ \beta_4 = \frac{|K_3 e_6 + \frac{I_x - I_y}{I_z} x_2 x_4 - \ddot{x}_{5d}|}{b_3} + \sigma_3 \\ \beta_1 = \frac{|K_4 e_8 - g - \ddot{x}_{7d}|}{cos(x_1)cos(x_3)\frac{1}{m}} + \sigma_4 \end{cases}$$

with S_1, S_3, S_5, S_7 sliding surfaces of the tracking error system, and $\sigma_1, \sigma_2, \sigma_3, \sigma_4 > 0$.

Twelve parameters have to be set by the automation designer to control a quadrotor by sliding mode: K_i, ε_i and σ_i, $i = 1, \ldots, 4$. Finally, we note three important points

- **Remark 1**: The controllers are designed for tracking of variables $x_1, \ldots, x_8$, which means

$$\begin{aligned}
\phi &\to \phi_d & x_1 &\to x_{1d} \\
\dot{\phi} &\to \dot{\phi}_d & x_2 &\to x_{2d} = \dot{x}_{1d} \\
\theta &\to \theta_d & x_3 &\to x_{3d} \\
\dot{\theta} &\to \dot{\theta}_d & x_4 &\to x_{4d} = \dot{x}_{3d} \\
\psi &\to \psi_d & x_5 &\to x_{5d} \\
\dot{\psi} &\to \dot{\psi}_d & x_6 &\to x_{6d} = \dot{x}_{5d} \\
z &\to z_d & x_7 &\to x_{7d} \\
\dot{z} &\to \dot{z}_d & x_8 &\to x_{8d} = \dot{x}_{7d}
\end{aligned}$$

This means that the controller is designed to track the quote z and the attitude, together with their time derivative. We can note from Eq. (20) that the first 8 equations are

$$\Sigma_{\phi,\theta,\psi,z} \begin{cases} \dot{x}_1 = x_2 \\ \dot{x}_2 = a_1 x_4 x_6 + a_2 \Omega x_4 + b_1 U_2 \\ \dot{x}_3 = x_4 \\ \dot{x}_4 = a_3 x_2 x_6 + a_4 \Omega x_2 + b_2 U_3 \\ \dot{x}_5 = x_6 \\ \dot{x}_6 = a_5 x_2 x_4 + b_3 U_4 \\ \dot{x}_7 = x_8 \\ \dot{x}_8 = -g + cos(x_1)cos(x_3)\dfrac{1}{m}U_1 \end{cases} \tag{38}$$

so the state variables $x_9, x_{10}, x_{11}, x_{12}$ do not appear in the equations of interest. In conclusion, we want that the controller sets the desired attitude and quote, but we neglect the longitudinal and lateral motion.

- **Remark 2**: The quadrotor system is considered as a group of coupled equations. The stability for each couple of equation is given with the control law design. For completeness we note that the stability of the system $\Sigma_{\phi,\theta,\psi,z}$ is given by the Lyapunov function

$$V = \frac{1}{2}S_1^2 + \frac{1}{2}S_3^2 + \frac{1}{2}S_5^2 + \frac{1}{2}S_7^2$$
$$= V_1 + V_3 + V_5 + V_7$$

in fact with the control law designed above we have $\dot{V}_i = S_i \dot{S}_i < k_i |S_i|$, for $i = 1, 3, 5, 7$ and $k_1, k_3, k_5, k_7 > 0$. It follows that $V \to 0$ (in finite time) and each $V_i \to 0$, so $x_i \to 0$ with $i = 1, 3, 5, 7$ and $x_j \to 0$ with $j = 2, 4, 6, 8$.

- **Remark 3**: The variables controlled by the control laws U_1, U_2, U_3, U_4 are $\phi, \dot{\phi}, \theta, \dot{\theta}, \psi, \dot{\psi}, z, \dot{z}$, so the position x, y in the plane, and their derivatives, are not considered, as discussed in remark 1. The aim of the SMC is the tracking of the variables $x, \dot{x}, y, \dot{y}, z, \dot{z}, \psi, \dot{\psi}$, which correspond to the state variables $x_9, x_{10}, x_{11}, x_{12}, x_7, x_8, x_5, x_6$. The solution is to create an external loop with desired variables, with the aim of calculating the correct tracking values for the controller. To achieve that aim, it is necessary to consider the last 4 equations of the system, that link linear accelerations to angles.

$$\Sigma_{x,y} \begin{cases} \dot{x}_9 = x_{10} \\ \dot{x}_{10} = \dfrac{1}{m}u_x U_1 = (cosx_1 sinx_3 cosx_5 + sinx_1 sinx_5)\dfrac{U_1}{m} \\ \dot{x}_{11} = x_{12} \\ \dot{x}_{12} = \dfrac{1}{m}u_y U_1 = (cosx_1 sinx_3 sinx_5 - sinx_1 cosx_5)\dfrac{U_1}{m} \end{cases} \tag{39}$$

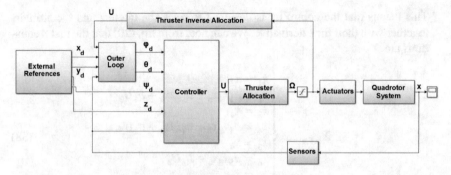

Fig. 2 Block diagram of an quadrotor system with sliding mode controller

Considering the physical variables of $x_9, x_{10}, x_{11}, x_{12}$ and a small-angle approximation for θ and ϕ, so $\cos(\gamma) \approx 1$ and $\sin(\gamma) \approx \gamma$ for any variable γ, it follows that

$$\begin{pmatrix} \ddot{x} \\ \ddot{y} \end{pmatrix} = \begin{pmatrix} (\theta cos\psi + \phi sin\psi)\frac{U_1}{m} \\ (\theta sin\psi - \phi cos\psi)\frac{U_1}{m} \end{pmatrix} = \frac{U_1}{m} \begin{pmatrix} sin\psi & cos\psi \\ -cos\psi & sin\psi \end{pmatrix} \begin{pmatrix} \phi \\ \theta \end{pmatrix} \tag{40}$$

We observe that the matrix is invertible $\forall \, \psi$, because its determinant is 1, so the angular references for the controller are given by

$$\begin{pmatrix} \phi \\ \theta \end{pmatrix} = \frac{m}{U_1} \begin{pmatrix} sin\psi & cos\psi \\ -cos\psi & sin\psi \end{pmatrix}^{-1} \begin{pmatrix} \ddot{x} \\ \ddot{y} \end{pmatrix} \tag{41}$$

Finally, the block diagram of an quadrotor system with sliding mode controller is shown in Fig. 2. It is worth to note as the external reference signals are the desired quote z_d and attitude of the quadrotor (e.g., x_d, y_d, ψ_d), whereas the controller acts on the quote z_d and the angles ϕ_d, θ_d and ψ_d. The variables ϕ_d and θ_d are calculated by x_d and y_d as shown in Eq. (41).

3.2 Backstepping

Backstepping is a recursive procedure able to design a controller for a class of systems known as parametric strict-feedback system. The design problem of the full system is decomposed into a sequence of design problems for lower order subsystems by considering some of the state variables as "virtual controls" and designing for them intermediate control laws [26, 32, 34, 35, 41, 44]. Using the backstepping approach, the control law can be synthesized as follow [7]

$$z_1 = x_{1d} - x_1$$
$$z_2 = x_2 - \dot{x}_{1d} - \alpha_1 z_1$$
$$z_3 = x_{3d} - x_3$$
$$z_4 = x_4 - \dot{x}_{3d} - \alpha_3 z_3$$
$$z_5 = x_{5d} - x_5$$
$$z_6 = x_6 - \dot{x}_{5d} - \alpha_5 z_5$$
$$z_7 = x_{7d} - x_7$$
$$z_8 = x_8 - \dot{x}_{7d} - \alpha_7 z_7$$

where

$$u_1 = \frac{m}{\cos(x_1)\cos(x_3)}(z_7 + g - \alpha_7(z_8 + \alpha_7 z_7) - \alpha_8 z_8 + \ddot{x}_{7d})$$
$$u_2 = \frac{1}{b_1}(z_1 - a_1 x_4 x_6 - a_2 x_4 \Omega - \alpha_1(z_2 + \alpha_1 z_1) - \alpha_2 z_2 + \ddot{x}_{1d})$$
$$u_3 = \frac{1}{b_2}(z_3 - a_3 x_2 x_6 - a_4 x_2 \Omega - \alpha_3(z_4 + \alpha_3 z_3) - \alpha_4 z_4 + \ddot{x}_{3d})$$
$$u_4 = \frac{1}{b_3}(z_5 - a_5 x_2 x_4 - \alpha_5(z_6 + \alpha_5 z_5) - \alpha_6 z_6 + \ddot{x}_{5d})$$

$$\begin{pmatrix} x_{1d} \\ x_{3d} \end{pmatrix} = \frac{m}{U_1} \begin{pmatrix} \sin\psi & -\cos\psi \\ \cos\psi & \sin\psi \end{pmatrix} \begin{pmatrix} \ddot{x}_{9d} - \ddot{x}_9 - \dot{x}_9 + \dot{x}_{9d} - 2x_9 + 2x_{9d} \\ \ddot{y}_{9d} - \ddot{y}_9 - \dot{y}_9 + \dot{y}_{9d} - 2y_9 + 2y_{9d} \end{pmatrix} \quad (42)$$

Eight parameters have be set by the automation designer to control a quadrotor by backstepping: α_i, $i = 1, \ldots, 8$.

3.3 PID

Though many control techniques have been developed in scientific literature, the Proportional-Integral-Derivative (PID) controller is still extensively used in more than 90% of industrial processes [1]. It can guarantee acceptable performances in transient and steady-state response and disturbance rejection for a wide range of application. Using the PID technique, the control law can be synthesized as follow

$$e_1 = x_{1d} - x_1$$
$$U_2 = (P_\phi + I_\phi \frac{1}{s} + D_\phi \frac{N_\phi}{1 + N_\phi \frac{1}{s}}) e_1$$

$$e_3 = x_{3d} - x_3$$

$$U_3 = (P_\theta + I_\theta \frac{1}{s} + D_\theta \frac{N_\theta}{1 + N_\theta \frac{1}{s}})e_3$$

$$e_5 = x_{5d} - x_5$$

$$U_4 = (P_\psi + I_\psi \frac{1}{s} + D_\psi \frac{N_\psi}{1 + N_\psi \frac{1}{s}})e_5$$

$$e_7 = x_{7d} - x_7$$

$$U_1 = (P_z + I_z \frac{1}{s} + D_z \frac{N_z}{1 + N_z \frac{1}{s}})e_7$$

Twelve parameters have be set by the automation designer for PID controlling of a quadrotor: the proportional, integral and derivative gain of each controller. In particular P_ϕ, P_θ, P_ψ, P_z, I_ϕ, I_θ, I_ψ, I_z, D_ϕ, D_θ, D_ψ and D_z.

4 Particle Swarm Optimization

The first version of particle swarm optimization was described in [24]. The algorithm starts by positioning uniformly and randomly the initial particles, and assigning them initial velocities. The velocities are randomly initialized in the range $[v_{min}, v_{max}]$. Then the objective function is evaluated at each particle location and after the best function value and the best location are evaluated. New velocities are chosen, based on the current velocity $(\vec{v}(t))$, the particles' individual best locations $(\vec{p}(t)$, particle best) and the best locations of the population $(\vec{g}(t)$, global best). The velocity is updated as described in Eq. (43).

$$\vec{v}(t+1) = \lambda \vec{v}(t) + r_1 \text{rand}(0,1)(\vec{p}(t) - \vec{x}(t)) + \qquad (43)$$
$$+ r_2 \text{rand}(0,1)(\vec{g}(t) - \vec{x}(t))$$

The parameter λ is the inertia weight, whereas r_1 and r_2 set the significance of $\vec{p}(t)$ and $\vec{g}(t)$, respectively. Then, iteratively, each particle is moved to a new position calculated by the velocity updated at each time step t. The new position is the sum of the previous position and the new velocity as described in Eq. (44).

$$\vec{x}(t+1) = \vec{x}(t) + \vec{v}(t+1) \qquad (44)$$

Table 1 Tuning parameters

Controllers	Parameters
Sliding mode	$K_1, \varepsilon_1, \sigma_1, K_2, \varepsilon_2, \sigma_2, K_3, \varepsilon_3, \sigma_3, K_4, \varepsilon_4, \sigma_4$
Backstepping	$\alpha_1, \alpha_2, \alpha_3, \alpha_4, \alpha_5, \alpha_6, \alpha_7, \alpha_8$
PID	$P_\phi, I_\phi, D_\phi, P_\theta, I_\theta, D_\theta, P_\psi, I_\psi, D_\psi, P_z, I_z, D_z$

The iterations proceed until the algorithm reaches a stopping criterion. Main steps of the basic PSO are listed in the Algorithm 1 [8, 9].

Algorithm 1 PSO algorithm pseudo-code

1: Initialize Particles
2: **repeat**
3:     Calculate fitness values of particles
4:     Modify the best particles in the swarm
5:     Choose the best particle
6:     Calculate the velocities of particles
7:     Update the particle positions
8: **until** Requirements are met

In this work, the IAE index is considered as fitness function and the parameters shown in Table 1 are tuned through PSO algorithm.

5 Experimental Results

The proposed controllers have been developed and tested on Simulink environment using the quadrotor dynamic model Sect. 2.4. The controller parameters shown in Table 1 are tuned through PSO algorithm. In the experiment reported in this section, we considered the basic version of the technique, as described in Sect. 4. Furthermore the values of the common parameters (i.e., λ, r_1, r_2) used in the PSO algorithm were chosen according to [23]. In particular the population size is 10 and the number of runs is 50. The IAE index is considered as fitness function and four performance index measurements have been considered in the comparison process, i.e., ISE, IAE, ITSE and ISAE defined as follows:

- the integral of the squared error (ISE):

$$\text{ISE} = \int_0^T [e(t)]^2 dt \tag{45}$$

- the integral of the absolute value of error (IAE):

$$IAE = \int_0^T |e(t)| \, dt \qquad (46)$$

- the integral of time multiplied by the absolute value of error (ITAE), to penalize long duration transients as well as steady state oscillations:

$$ITAE = \int_0^T t \cdot |e(t)| \, dt \qquad (47)$$

- the integral of time multiplied by the squared error value (ITSE), to penalize long duration transients as well as steady state oscillations:

$$ITSE = \int_0^T t \cdot [e(t)]^2 \, dt \qquad (48)$$

The simulated task was to perform an 8-shape trajectory. In particular, the quadrotor has to take-off at the position x_d, y_d, $z_d = (3, 0, 0)$, where $x_d = x_9$, $y_d = x_{11}$, $z_d = x_7$, follow an 8-shape path and land at the initial position. During the path, the quadrotor has to keep the yaw angle $\psi_d = x_5 = 0$ rad. Figure 3 shows the task accomplished by the quadrotor.

Figure 4 shows the actual path following for each controller, whereas Figs. 5, 6 and 7 depict the error positions of the quadrotor performed during the task. In particular, Figs. 5, 6 and 7 show the error along x, y and z ($e_9(t)$, $e_{11}(t)$, $e_7(t)$), respectively. PID controller gives the highest error along z due to the overshoots and sliding mode controller achieves the best performances, whereas along x and y the errors are comparable for all controllers.

The simulated performances are showed in Fig. 8, where Fig. 8a, b, c and d depict the indexes IAE, ISE, ITAE and ITSE respectively. The figures show that PID tech-

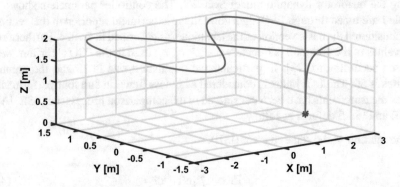

Fig. 3 Desired path: the quadrotor has to take-off at the position $(3, 0, 0)$, follow an 8-shape path and land at the initial position

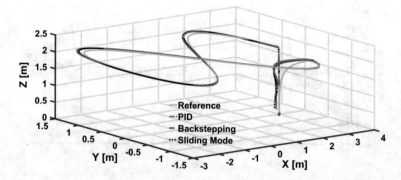

Fig. 4 Path following: *dash–dotted blue line* refers to PID controller, *dashed red line* refers to backstepping controller and *dotted black line* refers to sliding mode controller

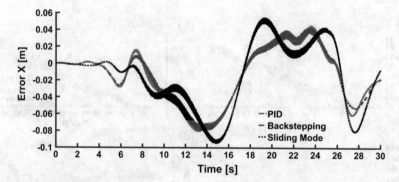

Fig. 5 Control results: error output $e_9(t)$ related to x. *Dash–dotted blue line* refers to PID controller, *dashed red line* refers to backstepping controller and *dotted black line* refers to sliding mode controller

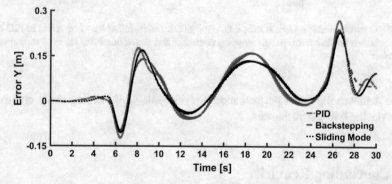

Fig. 6 Control results: error output $e_{11}(t)$ related to y. *Dash–dotted blue line* refers to PID controller, *dashed red line* refers to backstepping controller and *dotted black line* refers to sliding mode controller

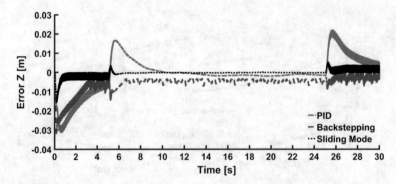

Fig. 7 Control results: error output $e_7(t)$ related to z. *Dash–dotted blue line* refers to PID controller, *dashed red line* refers to backstepping controller and *dotted black line* refers to sliding mode controller

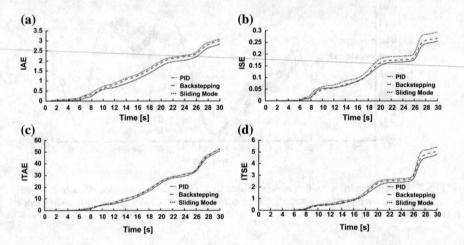

Fig. 8 Control results: **a** IAE; **b** ISE; **c** ITAE; **d** ITSE. *Dash–dotted blue line* refers to PID controller, *dashed red line* refers to backstepping controller and *dotted black line* refers to sliding mode controller

nique achieves the worst performances for this task, while sliding mode controller achieves the best performances.

6 Concluding Remarks

In this paper we proposed a particle swarm optimization method for choosing the control design parameters of well-known control algorithms applied to the tracking problem of a quadrotor vehicle. In detail, we compared the results among three different control laws, PID, backstepping and sliding mode. The sliding mode control showed the best performances in the presented case. The combination of sliding mode control and particle swarm optimization allows to solve the problem of control (stabilization and tracking) of several nonlinear systems, by limiting at the same time the design effort (i.e., parameters choice) and achieving near optimal results. As a drawback, the proposed approach requires a intensive use of process simulation, however this is quite common in control problems where hardware in the loop is usually adopted.

References

1. Ang KH, Chong G, Li Y (2005) Pid control system analysis, design, and technology. IEEE Trans Control Syst Technol 13(4):559–576
2. Azar AT, Vaidyanathan S (2015) Chaos modeling and control systems design, vol 581. Springer, Berlin, Germany
3. Azar AT, Vaidyanathan S (2015) Computational intelligence applications in modeling and control, vol 575. Springer, Berlin, Germany
4. Azar AT, Vaidyanathan S (2016) Advances in chaos theory and intelligent control, vol 337. Springer, Berlin, Germany
5. Boiko I, Fridman L, Pisano A, Usai E (2007) Analysis of chattering in systems with second-order sliding modes. IEEE Trans Autom Control 52(11):2085–2102
6. Bondarev AG, Bondarev SA, Kostyleva NE, Utkin VI (1985) Sliding modes in systems with asymptotic state observers. Autom Remote Control 46:679–684
7. Bouabdallah S, Siegwart R (2005) Backstepping and sliding-mode techniques applied to an indoor micro quadrotor. In: Proceedings of the 2005 IEEE international conference on robotics and automation, pp 2247–2252
8. Cavanini L, Ciabattoni L, Ferracuti F, Ippoliti G, Longhi S (2016) Microgrid sizing via profit maximization: a population based optimization approach. In: 2016 IEEE 14th international conference on industrial informatics (INDIN)
9. Ciabattoni L, Ferracuti F, Ippoliti G, Longhi S (2016) Artificial bee colonies based optimal sizing of microgrid components: a profit maximization approach. In 2016 IEEE congress on evolutionary computation (IEEE CEC 2016)
10. Corradini M, Longhi S, Monteri A, Orlando G (2010) Observer-based fault tolerant sliding mode control for remotely operated vehicles. In: IFAC proceedings volumes (IFAC-PapersOnline), pp 173–178
11. Corradini ML, Longhi S, Monteriù A, Orlando G (2010) Observer-based fault tolerant sliding mode control for remotely operated vehicles. IFAC Proc Vol 43(20):173–178
12. Corradini ML, Monteri A, Orlando G, Pettinari S (2011) An actuator failure tolerant robust control approach for an underwater remotely operated vehicle. In: 2011 50th IEEE conference on decision and control and European control conference, pp 3934–3939
13. Corradini ML, Monteriu A, Orlando G (2011) An actuator failure tolerant control scheme for an underwater remotely operated vehicle. IEEE Trans Control Syst Technol 19(5):1036–1046
14. Derafa L, Fridman L, Benallegue A, Ouldali A (2010). Super twisting control algorithm for the four rotors helicopter attitude tracking problem. In: 2010 11th international workshop on variable structure systems (VSS), pp 62–67
15. Erbatur K, Kaynak O (2001) Use of adaptive fuzzy systems in parameter tuning of sliding-mode controllers. IEEE/ASME Trans Mechatron 6(4):474–482
16. Fasano A, Ferracuti F, Freddi A, Longhi S, Monteri A (2015) A virtual thruster-based failure tolerant control scheme for underwater vehicles. IFAC-PapersOnLine 48(16):146–151
17. Ferrara A, Incremona GP, Stocchetti V (2014) Networked sliding mode control with chattering alleviation. In: 53rd IEEE conference on decision and control, pp 5542–5547

18. Fossen TI (2011) Handbook of marine craft hydrodynamics and motion control. Wiley
19. Freddi A (2012) Model-based diagnosis and control of unmanned aerial vehicles: application to the quadrotor system. PhD thesis, Università Politecnica delle Marche
20. Freddi A, Longhi S, Monteriù A (2009) A model-based fault diagnosis system for unmanned aerial vehicles. In: IFAC proceedings volumes (IFAC-PapersOnline), pp 71–76
21. Gaing Z-L (2004) A particle swarm optimization approach for optimum design of PID controller in AVR system. IEEE Trans Energy Convers 19(2):384–391
22. Ha QP (1996) Robust sliding mode controller with fuzzy tuning. Electron Lett 32(17):1626–1628
23. Karaboga D, Akay B (2009) A comparative study of artificial bee colony algorithm. Appl Math Comput 214(1):108–132
24. Kennedy J, Eberhart R (1995) Particle swarm optimization. In: IEEE International conference on neural networks, 1995, vol 4, pp 1942–1948
25. Khalil HK, Grizzle J (1996) Nonlinear systems. Prentice Hall, New Jersey
26. Krstic M, Kokotovic PV, Kanellakopoulos I (1995) Nonlinear and adaptive control design. Wiley, New York, NY, USA
27. Lanzon A, Freddi A, Longhi S (2014) Flight control of a quadrotor vehicle subsequent to a rotor failure. J Guidance Control Dyn 37(2):580–591
28. Lee D, Kim HJ, Sastry S (2009) Feedback linearization vs. adaptive sliding mode control for a quadrotor helicopter. Int J Control Autom Syst 7(3):419–428
29. Levant A (2010) Chattering analysis. IEEE Trans Autom Control 55(6):1380–1389
30. Mahony R, Kumar V, Corke P (2012) Multirotor aerial vehicles: modeling, estimation, and control of quadrotor. IEEE Robot Autom Mag 19(3):20–32
31. Mercado D, Castillo P, Castro R, Lozano R (2014) 2-sliding mode trajectory tracking control and ekf estimation for quadrotors. IFAC Proc Vol 47(3):8849–8854
32. Rasappan S, Vaidyanathan S (2013) Hybrid synchronization of n-scroll chaotic chua circuits using adaptive backstepping control design with recursive feedback. Malays J Math Sci 7(2):219–246
33. Vaidyanathan S (2014) Global chaos synchronisation of identical Li-Wu chaotic systems via sliding mode control. Int J Modell Ident Control 22(2):170–177
34. Vaidyanathan S (2015) Adaptive chaotic synchronization of enzymes-substrates system with ferroelectric behaviour in brain waves. Int J PharmTech Res 8(5):964–973
35. Vaidyanathan S (2015) Analysis, control, and synchronization of a 3-d novel jerk chaotic system with two quadratic nonlinearities. Kyungpook Math J 55(3):563–586
36. Vaidyanathan S (2015) Global chaos synchronization of chemical chaotic reactors via novel sliding mode control method. Int J ChemTech Res 8(7):209–221
37. Vaidyanathan S (2015) Global chaos synchronization of rucklidge chaotic systems for double convection via sliding mode control. Int J ChemTech Res 8(8):61–72
38. Vaidyanathan S (2015) Integral sliding mode control design for the global chaos synchronization of identical novel chemical chaotic reactor systems. Int J ChemTech Res 8(11):684–699
39. Vaidyanathan S (2015) A novel chemical chaotic reactor system and its output regulation via integral sliding mode control. Int J ChemTech Res 8(11):669–683
40. Vaidyanathan S (2015) Sliding controller design for the global chaos synchronization of enzymes-substrates systems. Int J PharmTech Res 8(7):89–99
41. Vaidyanathan S, Rasappan S (2014) Global chaos synchronization of n-scroll Chua circuit and Lur'e system using backstepping control design with recursive feedback. Arab J Sci Eng 39(4):3351–3364
42. Vaidyanathan S, Volos C (2016) Advances and applications in chaotic systems, vol 636. Springer, Berlin, Germany
43. Vaidyanathan S, Volos C (2016) Advances and applications in nonlinear control systems, vol 635. Springer, Berlin, Germany
44. Vaidyanathan S, Volos C, Pham VT, Madhavan K, Idowu BA (2014) Adaptive backstepping control, synchronization and circuit simulation of a 3-d novel jerk chaotic system with two hyperbolic sinusoidal nonlinearities. Arch Control Sci 24(3):375–403

45. Wai RJ (2007) Fuzzy sliding-mode control using adaptive tuning technique. IEEE Trans Ind Electron 54(1):586–594
46. Zheng E-H, Xiong J-J, Luo J-L (2014) Second order sliding mode control for a quadrotor UAV. ISA Trans 53(4):1350–1356. Disturbance estimation and mitigation

Global Stabilization of Nonlinear Systems via Novel Second Order Sliding Mode Control with an Application to a Novel Highly Chaotic System

Sundarapandian Vaidyanathan

Abstract Sliding mode control is an important method used to solve various problems in control systems engineering. In robust control systems, the sliding mode control is often adopted due to its inherent advantages of easy realization, fast response and good transient performance as well as insensitivity to parameter uncertainties and disturbance. In this work, we derive a novel second order sliding mode control method for the global stabilization of any nonlinear system. The global stabilization result is derived using novel second order sliding mode control method and established using Lyapunov stability theory. Chaos in nonlinear dynamics occurs widely in physics, chemistry, biology, ecology, secure communications, cryptosystems and many scientific branches. Synchronization of chaotic systems is an important research problem in chaos theory. As an application of the general result, the problem of global chaos control of a novel highly chaotic system is studied and a new sliding mode controller is derived. The Lyapunov exponents of the novel chaotic system are obtained as $L_1 = 12.8393$, $L_2 = 0$ and $L_3 = -33.1207$. The large value of the maximal Lyapunov exponent (MLE) shows that the novel chaotic system is highly chaotic. The Kaplan-Yorke dimension of the novel chaotic system is obtained as $D_{KY} = 2.3877$. We show that the novel highly chaotic system has three unstable equilibrium points. Numerical simulations using MATLAB have been shown to depict the phase portraits of the novel highly chaotic system and the global chaos control of the state trajectories of the novel highly chaotic system.

Keywords Chaos · Chaotic systems · Chaos control · Sliding mode control · Lyapunov exponents

S. Vaidyanathan (✉)
Research and Development Centre, Vel Tech University,
Avadi, Chennai 600062, Tamil Nadu, India
e-mail: sundarcontrol@gmail.com

© Springer International Publishing AG 2017
S. Vaidyanathan and C.-H. Lien (eds.), *Applications of Sliding Mode Control
in Science and Engineering*, Studies in Computational Intelligence 709,
DOI 10.1007/978-3-319-55598-0_8

171

1 Introduction

Chaos theory describes the quantitative study of unstable aperiodic dynamic behavior in deterministic nonlinear dynamical systems. For the motion of a dynamical system to be chaotic, the system variables should contain some nonlinear terms and the system must satisfy three properties: boundedness, infinite recurrence and sensitive dependence on initial conditions [4–6, 125–127].

Chaos theory has applications in several fields such as memristors [28, 30, 31, 34–36, 137, 140, 142], fuzzy logic [7, 49, 111, 146], communication systems [13, 14, 143], cryptosystems [10, 12], electromechanical systems [15, 58], lasers [8, 21, 145], encryption [22, 23, 147], electrical circuits [1, 2, 16, 120, 138], chemical reactions [75, 76, 78, 79, 81, 83, 85–87, 90, 92, 96, 112], oscillators [93, 94, 97, 98, 139], tokamak systems [91, 99], neurology [80, 88, 89, 95, 105, 141], ecology [77, 82, 106, 108], etc.

The problem of global control of a chaotic system is to device feedback control laws so that the closed-loop system is globally asymptotically stable. The problem of global chaos synchronization of chaotic systems is to find feedback control laws so that the master and slave systems are globally and asymptotically synchronized with respect to their states. There are many techniques available in the control literature for the regulation and synchronization of chaotic systems such as active control [17, 44, 45, 54, 118, 129], adaptive control [46–48, 51, 53, 64, 104, 115, 119, 120], backstepping control [37–41, 57, 122, 130, 135, 136], sliding mode control [19, 26, 56, 63, 65, 66, 73, 103, 107, 123], etc.

Some classical paradigms of 3-D chaotic systems in the literature are Lorenz system [24], Rössler system [42], ACT system [3], Sprott systems [50], Chen system [11], Lü system [25], Cai system [9], Tigan system [60], etc.

Many new chaotic systems have been discovered in the recent years such as Zhou system [148], Zhu system [149], Li system [20], Sundarapandian systems [52, 55], Vaidyanathan systems [67–72, 74, 84, 100, 102, 109, 110, 113, 114, 116, 117, 121, 124, 128, 131–134], Pehlivan system [27], Sampath system [43], Tacha system [59], Pham systems [29, 32, 33, 35], Akgul system [2], etc.

In this research work, we derive a general result for the global stabilization of nonlinear systems using second order sliding mode control (SMC) [61, 62]. The sliding mode control approach is recognized as an efficient tool for designing robust controllers for linear or nonlinear control systems operating under uncertainty conditions.

A major advantage of sliding mode control is low sensitivity to parameter variations in the plant and disturbances affecting the plant, which eliminates the necessity of exact modeling of the plant. In the sliding mode control, the control dynamics will have two sequential modes, viz. the reaching mode and the sliding mode. Basically, a sliding mode controller design consists of two parts: hyperplane design and controller design. A hyperplane is first designed via the pole-placement approach and a controller is then designed based on the sliding condition. The stability of the overall system is guaranteed by the sliding condition and by a stable hyperplane.

This work is organized as follows. In Sect. 2, we discuss the problem statement for the global stabilization of nonlinear systems. Then we derive a general result for the global stabilization of nonlinear systems using novel second order sliding mode control. In Sect. 3, we describe the novel highly chaotic system and its phase portraits. In Sect. 4, we describe the qualitative properties of the novel highly chaotic system. The Lyapunov exponents of the novel chaotic system are obtained as $L_1 = 12.8393$, $L_2 = 0$ and $L_3 = -33.1207$. The Kaplan-Yorke dimension of the novel chaotic system is obtained as $D_{KY} = 2.3877$. We show that the novel highly chaotic system has three unstable equilibrium points. In Sect. 5, we describe the second order sliding mode controller design for the global chaos control of the novel highly chaotic system and its numerical simulations. Section 6 contains the conclusions of this work.

2 Second Order Sliding Mode Control for Nonlinear Systems

We consider a general nonlinear system given by

$$\dot{\mathbf{x}} = A\mathbf{x} + f(\mathbf{x}) + \mathbf{u} \tag{1}$$

where $\mathbf{x} \in \mathbf{R}^n$ denotes the state of the system, $A \in \mathbf{R}^{n \times n}$ denotes the matrix of system parameters and $f(\mathbf{x}) \in \mathbf{R}^n$ contains the nonlinear parts of the system. Also, $\mathbf{u}$ represents the sliding mode controller to be designed.

Then the global stabilization problem for the system (1) can be stated as follows: Find a controller $\mathbf{u}(\mathbf{x})$ so as to render the state $\mathbf{x}(t)$ to be globally asymptotically stable for all values of $\mathbf{x}(0) \in \mathbf{R}^n$, i.e.

$$\lim_{t \to \infty} \|\mathbf{x}(t)\| = 0 \quad \text{for all} \quad \mathbf{x}(0) \in \mathbf{R}^n \tag{2}$$

We start the controller design by setting

$$\mathbf{u}(t) = -f(\mathbf{x}) + Bv(t) \tag{3}$$

In Eq. (3), $B \in \mathbf{R}^n$ is chosen such that (A, B) is completely controllable.

By substituting (3) into (1), we get the closed-loop system dynamics

$$\dot{\mathbf{x}} = A\mathbf{x} + Bv \tag{4}$$

Next, we start the sliding controller design by defining the sliding variable as

$$s(\mathbf{x}) = C\mathbf{x} = c_1 x_1 + c_2 x_2 + \cdots + c_n x_n, \tag{5}$$

where $C \in \mathbf{R}^{1 \times n}$ is a constant vector to be determined.

The sliding manifold S is defined as the hyperplane

$$S = \{\mathbf{x} \in \mathbf{R}^n \; : \; s(\mathbf{x}) = C\mathbf{x} = 0\} \tag{6}$$

We shall assume that a sliding motion occurs on the hyperplane S.

In the second order sliding mode control, the following equations must be satisfied:

$$s = 0 \tag{7a}$$

$$\dot{s} = CA\mathbf{x} + CB v = 0 \tag{7b}$$

We assume that

$$CB \neq 0 \tag{8}$$

The sliding motion is influenced by equivalent control derived from (7b) as

$$v_{\text{eq}}(t) = -(CB)^{-1} CA\mathbf{x}(t) \tag{9}$$

By substituting (9) into (4), we obtain the equivalent error dynamics in the sliding phase as follows:

$$\dot{\mathbf{x}} = A\mathbf{x} - (CB)^{-1}CA\mathbf{x} = E\mathbf{x}, \tag{10}$$

where

$$E = \left[I - B(CB)^{-1}C\right] A \tag{11}$$

We note that E is independent of the control and has at most $(n-1)$ non-zero eigenvalues, depending on the chosen switching surface, while the associated eigenvectors belong to $\ker(C)$.

Since (A, B) is controllable, we can use sliding control theory [61, 62] to choose B and C so that E has any desired $(n-1)$ stable eigenvalues.

This shows that the dynamics (10) is globally asymptotically stable.

Finally, for the sliding controller design, we apply a novel sliding control law, viz.

$$\dot{s} = -ks - qs^2 \operatorname{sgn}(s) \tag{12}$$

In (12), $\operatorname{sgn}(\cdot)$ denotes the *sign* function and the SMC constants $k > 0, q > 0$ are found in such a way that the sliding condition is satisfied and that the sliding motion will occur.

By combining Eqs. (7b), (9) and (12), we finally obtain the sliding mode controller $v(t)$ as

$$v(t) = -(CB)^{-1} \left[C(kI + A)\mathbf{x} + qs^2 \operatorname{sgn}(s)\right] \tag{13}$$

Next, we establish the main result of this section.

Theorem 1 *The second order sliding mode controller defined by (3) achieves global stabilization for all the states of the system (1) where v is defined by the novel sliding mode control law (13), $B \in \mathbf{R}^{n \times 1}$ is such that (A, B) is controllable, $C \in \mathbf{R}^{1 \times n}$ is such that $CB \neq 0$ and the matrix E defined by (11) has $(n-1)$ stable eigenvalues.*

Proof Upon substitution of the control laws (3) and (13) into the state dynamics (1), we obtain the closed-loop system dynamics as

$$\dot{\mathbf{x}} = A\mathbf{x} - B(CB)^{-1} \left[C(kI + A)\mathbf{x} + qs^2 \operatorname{sgn}(s) \right] \tag{14}$$

We shall show that the error dynamics (14) is globally asymptotically stable by considering the quadratic Lyapunov function

$$V(\mathbf{x}) = \frac{1}{2} s^2(\mathbf{x}) \tag{15}$$

The sliding mode motion is characterized by the equations

$$s(\mathbf{x}) = 0 \quad \text{and} \quad \dot{s}(\mathbf{x}) = 0 \tag{16}$$

By the choice of E, the dynamics in the sliding mode given by Eq. (10) is globally asymptotically stable.

When $s(\mathbf{x}) \neq 0$, $V(\mathbf{x}) > 0$.

Also, when $s(\mathbf{x}) \neq 0$, differentiating V along the error dynamics (14) or the equivalent dynamics (12), we get

$$\dot{V}(\mathbf{x}) = s\dot{s} = -ks^2 - qs^3 \operatorname{sgn}(s) < 0 \tag{17}$$

Hence, by Lyapunov stability theory [18], the error dynamics (14) is globally asymptotically stable for all $\mathbf{x}(0) \in \mathbf{R}^n$.

This completes the proof.   ∎

3 A Novel Highly Chaotic System

In this work, we propose a novel highly chaotic system described by

$$\begin{aligned}
\dot{x}_1 &= a(x_2 - x_1) + dx_2 x_3 \\
\dot{x}_2 &= bx_1 - x_2 - x_1 x_3 \\
\dot{x}_3 &= x_1 x_2 - cx_3 + px_2^2
\end{aligned} \tag{18}$$

where x_1, x_2, x_3 are the states and a, b, c, d, p are constant, positive, parameters.

In this chapter, we show that the system (18) is chaotic when the parameters take the values

$$a = 14, \quad b = 18, \quad c = 6, \quad d = 98, \quad p = 12 \tag{19}$$

For numerical simulations, we take the initial state of the system (18) as

$$x_1(0) = 0.2, \quad x_2(0) = 0.2, \quad x_3(0) = 0.2 \tag{20}$$

The Lyapunov exponents of the system (18) for the parameter values (19) and the initial state (20) are determined by Wolf's algorithm [144] as

$$L_1 = 12.8393, \quad L_2 = 0, \quad L_3 = -33.1207 \tag{21}$$

Since $L_1 > 0$, we conclude that the system (18) is chaotic.

Since $L_1 + L_2 + L_3 < 0$, we deduce that the system (18) is dissipative.

Hence, the limit sets of the system (18) are ultimately confined into a specific limit set of zero volume, and the asymptotic motion of the novel highly chaotic system (18) settles onto a strange attractor of the system.

From (21), we see that the maximal Lyapunov exponent (MLE) of the chaotic system (18) is $L_1 = 12.8393$, which is very large.

Thus, we conclude that the proposed novel system (18) is highly chaotic.

Also, the Kaplan-Yorke dimension of the novel highly chaotic system (18) is calculated as

$$D_{KY} = 2 + \frac{L_1 + L_2}{|L_3|} = 2.3877, \tag{22}$$

which shows the high complexity of the system (18).

Figure 1 shows the 3-D phase portrait of the highly chaotic system (18) in $\mathbf{R}^3$.

Figures 2, 3 and 4 show the 2-D projections of the highly chaotic system (18) in (x_1, x_2), (x_2, x_3) and (x_1, x_3) planes, respectively.

4 Qualitative Properties of the Novel Highly Chaotic System

4.1 Dissipativity

In vector notation, the novel highly chaotic system (18) can be expressed as

$$\dot{\mathbf{x}} = f(\mathbf{x}) = \begin{bmatrix} f_1(x_1, x_2, x_3) \\ f_2(x_1, x_2, x_3) \\ f_3(x_1, x_2, x_3) \end{bmatrix}, \tag{23}$$

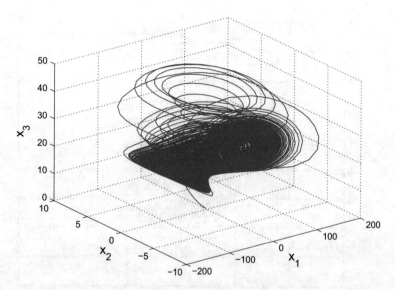

Fig. 1   3-D phase portrait of the novel highly chaotic system in $\mathbf{R}^3$

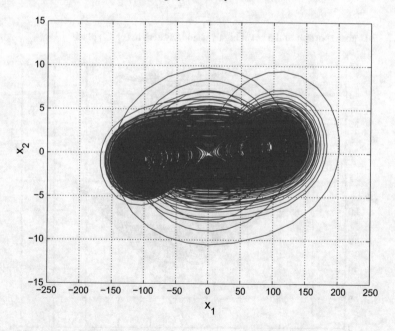

Fig. 2   2-D phase portrait of the novel highly chaotic system in (x_1, x_2) plane

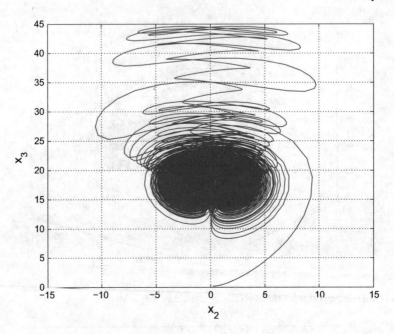

Fig. 3 2-D phase portrait of the novel highly chaotic system in (x_2, x_3) plane

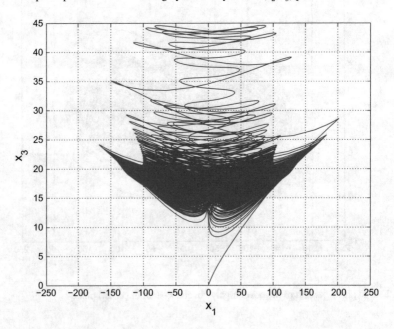

Fig. 4 2-D phase portrait of the novel highly chaotic system in (x_1, x_3) plane

where

$$
\begin{cases}
f_1(x_1, x_2, x_3) = a(x_2 - x_1) + dx_2 x_3 \\
f_2(x_1, x_2, x_3) = bx_1 - x_2 - x_1 x_3 \\
f_3(x_1, x_2, x_3) = x_1 x_2 - cx_3 + px_2^2
\end{cases} \tag{24}
$$

We take the parameter values as in the chaotic case (19).

Let Ω be any region in $\mathbf{R}^3$ with a smooth boundary and also, $\Omega(t) = \Phi_t(\Omega)$, where Φ_t is the flow of f. Furthermore, let $V(t)$ denote the volume of $\Omega(t)$.

By Liouville's theorem, we know that

$$
\dot{V}(t) = \int_{\Omega(t)} (\nabla \cdot f) \, dx_1 \, dx_2 \, dx_3 \tag{25}
$$

The divergence of the novel chaotic system (23) is calculated as

$$
\nabla \cdot f = \frac{\partial f_1}{\partial x_1} + \frac{\partial f_2}{\partial x_2} + \frac{\partial f_3}{\partial x_3} = -(a + c + 1) = -\mu < 0 \tag{26}
$$

where $\mu = a + c + 1 = 21 > 0$.

Inserting the value of $\nabla \cdot f$ from (26) into (25), we get

$$
\dot{V}(t) = \int_{\Omega(t)} (-\mu) \, dx_1 \, dx_2 \, dx_3 = -\mu V(t) \tag{27}
$$

Integrating the first order linear differential equation (27), we get

$$
V(t) = \exp(-\mu t) V(0) \tag{28}
$$

Since $\mu > 0$, it follows from Eq. (28) that $V(t) \to 0$ exponentially as $t \to \infty$.

This shows that the novel chaotic system (18) is dissipative. Hence, the limit sets of the system (18) are ultimately confined into a specific limit set of zero volume, and the asymptotic motion of the novel chaotic system (18) settles onto a strange attractor of the system.

4.2 Equilibrium Points

The equilibrium points of the novel highly chaotic system (18) are obtained by solving the equations

$$\begin{cases} f_1(x_1, x_2, x_3) = a(x_2 - x_1) + dx_2 x_3 = 0 \\ f_2(x_1, x_2, x_3) = bx_1 - x_2 - x_1 x_3 = 0 \\ f_3(x_1, x_2, x_3) = x_1 x_2 - cx_3 + px_2^2 = 0 \end{cases} \qquad (29)$$

We take the parameter values as in the chaotic case (19).

Solving the system (29), we find that the system (18) has three equilibrium points given by

$$E_0 = \begin{bmatrix} 0 \\ 0 \\ 0 \end{bmatrix}, \quad E_1 = \begin{bmatrix} 111.8950 \\ 0.8814 \\ 17.9921 \end{bmatrix}, \quad E_2 = \begin{bmatrix} -111.8950 \\ -0.8814 \\ 17.9921 \end{bmatrix} \qquad (30)$$

To test the stability type of the equilibrium points, we calculate the Jacobian of the system (18) at any $\mathbf{x} \in \mathbf{R}^3$ as

$$J(\mathbf{x}) = \begin{bmatrix} -a & a + dx_3 & dx_2 \\ b - x_3 & -1 & -x_1 \\ x_2 & x_1 + 2px_2 & -c \end{bmatrix} = \begin{bmatrix} -14 & 14 + 98x_3 & 98x_2 \\ 18 - x_3 & -1 & -x_1 \\ x_2 & x_1 + 24x_2 & -6 \end{bmatrix} \qquad (31)$$

The matrix $J_0 = J(E_0)$ has the eigenvalues

$$\lambda_1 = -24.6537, \quad \lambda_2 = -6, \quad \lambda_3 = 9.6537 \qquad (32)$$

This shows that the equilibrium point E_0 is a saddle point, which is unstable. The matrix $J_1 = J(E_1)$ has the eigenvalues

$$\lambda_1 = -25.54, \quad \lambda_{2,3} = 2.27 \pm 122.52i \qquad (33)$$

This shows that the equilibrium point E_1 is a saddle-focus, which is unstable.

The matrix $J_2 = J(E_2)$ has the same eigenvalues as $J_1 = J(E_1)$. Hence, we conclude that the equilibrium point E_2 is also a saddle-focus, which is unstable.

Thus, all three equilibrium points of the novel highly chaotic system (18) are unstable.

4.3 Rotation Symmetry About the x_3-Axis

It is easy to see that the system (18) is invariant under the change of coordinates

$$(x_1, x_2, x_3) \mapsto (-x_1, -x_2, x_3) \qquad (34)$$

Thus, the 3-D novel chaotic system (18) has rotation symmetry about the x_3-axis. Hence, it follows that any non-trivial trajectory of the system (18) must have a twin trajectory

4.4 Invariance

It is easy to see that the x_3-axis is invariant under the flow of the 3-D novel highly chaotic system (18). The invariant motion along the x_3-axis is characterized by

$$\dot{x}_3 = -cx_3, \quad (c > 0) \tag{35}$$

which is globally exponentially stable.

4.5 Lyapunov Exponents and Kaplan-Yorke Dimension

We take the parameters of the system (18) as

$$a = 14, \quad b = 18, \quad c = 6, \quad d = 98, \quad p = 12 \tag{36}$$

Also, we take the initial state of the system (18) as

$$x_1(0) = 0.2, \quad x_2(0) = 0.2, \quad x_3(0) = 0.2 \tag{37}$$

The Lyapunov exponents of the system (18) for the parameter values (36) and the initial state (37) are determined by Wolf's algorithm [144] as

$$L_1 = 12.8393, \quad L_2 = 0, \quad L_3 = -33.1207 \tag{38}$$

Figure 5 describes the MATLAB plot for the Lyapunov exponents of the novel chaotic system (18).

From (21), we see that the maximal Lyapunov exponent (MLE) of the chaotic system (18) is $L_1 = 12.8393$, which is very large. Thus, novel system (18) is highly chaotic. Also, the Kaplan-Yorke dimension of the novel highly chaotic system (18) is calculated as

$$D_{KY} = 2 + \frac{L_1 + L_2}{|L_3|} = 2.3877, \tag{39}$$

which shows the high complexity of the system (18).

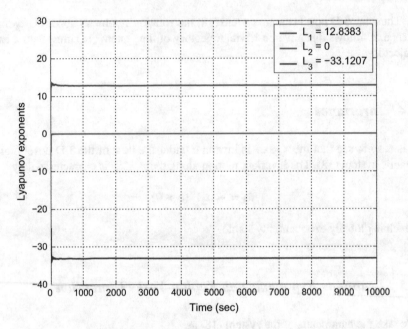

Fig. 5 Lyapunov exponents of the novel highly chaotic system

5 Sliding Mode Controller Design for the Global Chaos Control of the Novel Highly Chaotic System

In this section, we describe the sliding mode controller design for the global chaos control of the novel highly chaotic system by applying the novel sliding mode control method described by Theorem 1 in Sect. 2.

Thus, we consider the controlled novel highly chaotic system given by

$$\begin{aligned}
\dot{x}_1 &= a(x_2 - x_1) + dx_2x_3 + u_1 \\
\dot{x}_2 &= bx_1 - x_2 - x_1x_3 + u_2 \\
\dot{x}_3 &= x_1x_2 - cx_3 + px_2^2 + u_3
\end{aligned} \tag{40}$$

where x_1, x_2, x_3 are the states and u is the sliding mode control to be designed.

We take the parameter values as in the chaotic case (19), i.e.

$$a = 14, \quad b = 18, \quad c = 6, \quad d = 98, \quad p = 12 \tag{41}$$

In matrix form, we can write the system dynamics (40) as

$$\dot{x} = Ax + f(\mathbf{x}) + \mathbf{u} \tag{42}$$

The matrices in (42) are given by

$$A = \begin{bmatrix} -a & a & 0 \\ b & -1 & 0 \\ 0 & 0 & -c \end{bmatrix} \text{ and } f(\mathbf{x}) = \begin{bmatrix} dx_2x_3 \\ -x_1x_3 \\ x_1x_2 + px_2^2 \end{bmatrix} \tag{43}$$

We follow the procedure given in Sect. 2 for the construction of the novel sliding controller to achieve global chaos stabilization of the chaotic system (42).

First, we set $\mathbf{u}$ as

$$\mathbf{u}(t) = -f(\mathbf{x}) + Bv(t) \tag{44}$$

where B is selected such that (A, B) is completely controllable.

A simple choice of B is

$$B = \begin{bmatrix} 1 \\ 1 \\ 1 \end{bmatrix} \tag{45}$$

It can be easily checked that (A, B) is completely controllable.

Next, we find a sliding variable $s = C\mathbf{x}$ such that the matrix $E = [I - B(CB)^{-1}C]A$ has two stable eigenvalues.

A simple calculation gives

$$s(\mathbf{x}) = C\mathbf{x} = \begin{bmatrix} 1 & 8 & 1 \end{bmatrix} \mathbf{x} = x_1 + 8x_2 + x_3 \tag{46}$$

We also note that the matrix $E = [I - B(CB)^{-1}C]A$ has the eigenvalues

$$\lambda_1 = -28.6140, \quad \lambda_2 = -4.3860, \quad \lambda_3 = 0 \tag{47}$$

Next, we take the sliding mode gains as

$$k = 6, \quad q = 0.2 \tag{48}$$

From Eq. (13) in Sect. 2, we obtain the novel sliding control v as

$$v(t) = -13.6x_1 - 5.4x_2 - 0.02s^2 \, \text{sgn}(s) \tag{49}$$

As an application of Theorem 1 to the novel highly chaotic system (40), we obtain the main result of this section as follows.

Theorem 2 *The novel highly chaotic system (40) is globally and asymptotically sta-bilized for all initial conditions* $\mathbf{x}(0) \in \mathbf{R}^3$ *with the sliding controller* $\mathbf{u}$ *defined by (44), where* $f(\mathbf{x})$ *is defined by (43),* B *is defined by (45) and* v *is defined by (49).* ∎

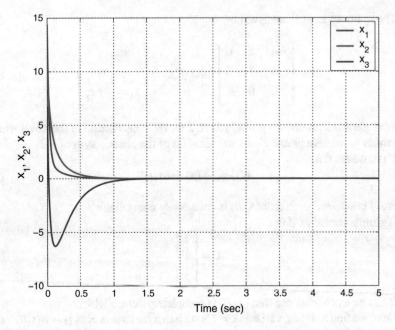

Fig. 6 Time-history of the controlled states x_1, x_2, x_3

For numerical simulations, we use MATLAB for solving the systems of differential equations using the classical fourth-order Runge-Kutta method with step size $h = 10^{-8}$.

The parameter values of the novel highly chaotic system (40) are taken as in the chaotic case (47).

The sliding mode gains are taken as $k = 6$ and $q = 0.2$.

We take the initial state of the chaotic system (40) as

$$x_1(0) = 14.4, \quad x_2(0) = 10.7, \quad x_3(0) = 6.8 \tag{50}$$

Figure 6 shows the time-history of the controlled states x_1, x_2, x_3.

6 Conclusions

In robust control systems, the sliding mode control is commonly used due to its inherent advantages of easy realization, fast response and good transient performance as well as insensitivity to parameter uncertainties and disturbance. In this work, we derived a novel second order sliding mode control method for the global stabilization of nonlinear systems. We proved the main result using Lyapunov stability theory. As an application of the general result, the problem of global chaos control of a novel highly chaotic system was studied and a new second order sliding mode controller has been derived. Numerical simulations using MATLAB were shown to depict the phase portraits of the novel highly chaotic system and the second order sliding mode controller design for the global chaos control of the novel highly chaotic system.

References

1. Akgul A, Hussain S, Pehlivan I (2016) A new three-dimensional chaotic system, its dynamical analysis and electronic circuit applications. Optik 127(18):7062–7071
2. Akgul A, Moroz I, Pehlivan I, Vaidyanathan S (2016) A new four-scroll chaotic attractor and its engineering applications. Optik 127(13):5491–5499
3. Arneodo A, Coullet P, Tresser C (1981) Possible new strange attractors with spiral structure. Commun Math Phys 79(4):573–576
4. Azar AT, Vaidyanathan S (2015) Chaos modeling and control systems design. Springer, Berlin, Germany
5. Azar AT, Vaidyanathan S (2016) Advances in chaos theory and intelligent control. Springer, Berlin, Germany
6. Azar AT, Vaidyanathan S (2017) Fractional order control and synchronization of chaotic systems. Springer, Berlin, Germany
7. Boulkroune A, Bouzeriba A, Bouden T (2016) Fuzzy generalized projective synchronization of incommensurate fractional-order chaotic systems. Neurocomputing 173:606–614
8. Burov DA, Evstigneev NM, Magnitskii NA (2017) On the chaotic dynamics in two coupled partial differential equations for evolution of surface plasmon polaritons. Commun Nonlinear Sci Numer Simul 46:26–36
9. Cai G, Tan Z (2007) Chaos synchronization of a new chaotic system via nonlinear control. J Uncertain Syst 1(3):235–240
10. Chai X, Chen Y, Broyde L (2017) A novel chaos-based image encryption algorithm using DNA sequence operations. Opt Lasers Eng 88:197–213
11. Chen G, Ueta T (1999) Yet another chaotic attractor. Int J Bifurcat Chaos 9(7):1465–1466
12. Chenaghlu MA, Jamali S, Khasmakhi NN (2016) A novel keyed parallel hashing scheme based on a new chaotic system. Chaos Solitons Fractals 87:216–225
13. Fallahi K, Leung H (2010) A chaos secure communication scheme based on multiplication modulation. Commun Nonlinear Sci Numer Simul 15(2):368–383
14. Fontes RT, Eisencraft M (2016) A digital bandlimited chaos-based communication system. Commun Nonlinear Sci Numer Simul 37:374–385
15. Fotsa RT, Woafo P (2016) Chaos in a new bistable rotating electromechanical system. Chaos Solitons Fractals 93:48–57
16. Kacar S (2016) Analog circuit and microcontroller based RNG application of a new easy realizable 4D chaotic system. Optik 127(20):9551–9561
17. Karthikeyan R, Sundarapandian V (2014) Hybrid chaos synchronization of four-scroll systems via active control. J Electr Eng 65(2):97–103
18. Khalil HK (2002) Nonlinear Syst. Prentice Hall, New York, USA
19. Lakhekar GV, Waghmare LM, Vaidyanathan S (2016) Diving autopilot design for underwater vehicles using an adaptive neuro-fuzzy sliding mode controller. In: Vaidyanathan S, Volos C (eds) Advances and applications in nonlinear control systems. Springer, Berlin, Germany, pp 477–503
20. Li D (2008) A three-scroll chaotic attractor. Phys Lett A 372(4):387–393
21. Liu H, Ren B, Zhao Q, Li N (2016) Characterizing the optical chaos in a special type of small networks of semiconductor lasers using permutation entropy. Opt Commun 359:79–84
22. Liu W, Sun K, Zhu C (2016) A fast image encryption algorithm based on chaotic map. Opt Lasers Eng 84:26–36

23. Liu X, Mei W, Du H (2016) Simultaneous image compression, fusion and encryption algorithm based on compressive sensing and chaos. Opt Commun 366:22–32
24. Lorenz EN (1963) Deterministic periodic flow. J Atmos Sci 20(2):130–141
25. Lü J, Chen G (2002) A new chaotic attractor coined. Int J Bifurcat Chaos 12(3):659–661
26. Moussaoui S, Boulkroune A, Vaidyanathan S (2016) Fuzzy adaptive sliding-mode control scheme for uncertain underactuated systems. In: Vaidyanathan S, Volos C (eds) Advances and applications in nonlinear control systems. Springer, Berlin, Germany, pp 351–367
27. Pehlivan I, Moroz IM, Vaidyanathan S (2014) Analysis, synchronization and circuit design of a novel butterfly attractor. J Sound Vib 333(20):5077–5096
28. Pham VT, Volos C, Jafari S, Wang X, Vaidyanathan S (2014) Hidden hyperchaotic attractor in a novel simple memristive neural network. Optoelectron Adv Mater Rapid Commun 8(11–12):1157–1163
29. Pham VT, Volos CK, Vaidyanathan S (2015) Multi-scroll chaotic oscillator based on a first-order delay differential equation. In: Azar AT, Vaidyanathan S (eds) Chaos modeling and control systems design. Studies in computational intelligence, vol 581. Springer, Germany, pp 59–72
30. Pham VT, Volos CK, Vaidyanathan S, Le TP, Vu VY (2015) A memristor-based hyperchaotic system with hidden attractors: dynamics, synchronization and circuital emulating. J Eng Sci Technol Rev 8(2):205–214
31. Pham VT, Jafari S, Vaidyanathan S, Volos C, Wang X (2016) A novel memristive neural network with hidden attractors and its circuitry implementation. Sci China Technol Sci 59(3):358–363
32. Pham VT, Jafari S, Volos C, Giakoumis A, Vaidyanathan S, Kapitaniak T (2016) A chaotic system with equilibria located on the rounded square loop and its circuit implementation. IEEE Trans Circuits Syst II: Express Briefs 63(9):878–882
33. Pham VT, Jafari S, Volos C, Vaidyanathan S, Kapitaniak T (2016) A chaotic system with infinite equilibria located on a piecewise linear curve. Optik 127(20):9111–9117
34. Pham VT, Vaidyanathan S, Volos CK, Hoang TM, Yem VV (2016) Dynamics, synchronization and SPICE implementation of a memristive system with hidden hyperchaotic attractor. In: Azar AT, Vaidyanathan S (eds) Advances in chaos theory and intelligent control. Springer, Berlin, Germany, pp 35–52
35. Pham VT, Vaidyanathan S, Volos CK, Jafari S, Kuznetsov NV, Hoang TM (2016) A novel memristive time-delay chaotic system without equilibrium points. Eur Phys J: Spec Top 225(1):127–136
36. Pham VT, Vaidyanathan S, Volos CK, Jafari S, Wang X (2016) A chaotic hyperjerk system based on memristive device. In: Vaidyanathan S, Volos C (eds) Advances and applications in chaotic systems. Springer, Berlin, Germany, pp 39–58
37. Rasappan S, Vaidyanathan S (2012) Global chaos synchronization of WINDMI and Coullet chaotic systems by backstepping control. Far East J Math Sci 67(2):265–287
38. Rasappan S, Vaidyanathan S (2012) Hybrid synchronization of n-scroll Chua and Lur'e chaotic systems via backstepping control with novel feedback. Arch Control Sci 22(3):343–365
39. Rasappan S, Vaidyanathan S (2012) Synchronization of hyperchaotic Liu system via backstepping control with recursive feedback. Commun Comput Inf Sci 305:212–221
40. Rasappan S, Vaidyanathan S (2013) Hybrid synchronization of n-scroll chaotic Chua circuits using adaptive backstepping control design with recursive feedback. Malay J Math Sci 7(2):219–246
41. Rasappan S, Vaidyanathan S (2014) Global chaos synchronization of WINDMI and Coullet chaotic systems using adaptive backstepping control design. Kyungpook Math J 54(1):293–320
42. Rössler OE (1976) An equation for continuous chaos. Phys Lett A 57(5):397–398
43. Sampath S, Vaidyanathan S, Volos CK, Pham VT (2015) An eight-term novel four-scroll chaotic system with cubic nonlinearity and its circuit simulation. J Eng Sci Technol Rev 8(2):1–6

44. Sarasu P, Sundarapandian V (2011) Active controller design for generalized projective synchronization of four-scroll chaotic systems. Int J Syst Signal Control Eng Appl 4(2):26–33
45. Sarasu P, Sundarapandian V (2011) The generalized projective synchronization of hyperchaotic Lorenz and hyperchaotic Qi systems via active control. Int J Soft Comput 6(5):216–223
46. Sarasu P, Sundarapandian V (2012) Adaptive controller design for the generalized projective synchronization of 4-scroll systems. Int J Syst Signal Control Eng Appl 5(2):21–30
47. Sarasu P, Sundarapandian V (2012) Generalized projective synchronization of three-scroll chaotic systems via adaptive control. Eur J Sci Res 72(4):504–522
48. Sarasu P, Sundarapandian V (2012) Generalized projective synchronization of two-scroll systems via adaptive control. Eur J Sci Res 72(4):146–156
49. Shirkhani N, Khanesar M, Teshnehlab M (2016) Indirect model reference fuzzy control of SISO fractional order nonlinear chaotic systems. Proc Comput Sci 102:309–316
50. Sprott JC (1994) Some simple chaotic flows. Phys Rev E 50(2):647–650
51. Sundarapandian V (2013) Adaptive control and synchronization design for the Lu-Xiao chaotic system. Lect Notes Electr Eng 131:319–327
52. Sundarapandian V (2013) Analysis and anti-synchronization of a novel chaotic system via active and adaptive controllers. J Eng Sci Technol Rev 6(4):45–52
53. Sundarapandian V, Karthikeyan R (2011) Anti-synchronization of hyperchaotic Lorenz and hyperchaotic Chen systems by adaptive control. Int J Syst Signal Control Eng Appl 4(2):18–25
54. Sundarapandian V, Karthikeyan R (2012) Hybrid synchronization of hyperchaotic Lorenz and hyperchaotic Chen systems via active control. Int J Syst Signal Control Eng Appl 7(3):254–264
55. Sundarapandian V, Pehlivan I (2012) Analysis, control, synchronization, and circuit design of a novel chaotic system. Math Comput Modell 55(7–8):1904–1915
56. Sundarapandian V, Sivaperumal S (2011) Sliding controller design of hybrid synchronization of four-wing Chaotic systems. Int J Soft Comput 6(5):224–231
57. Suresh R, Sundarapandian V (2013) Global chaos synchronization of a family of n-scroll hyperchaotic Chua circuits using backstepping control with recursive feedback. Far East J Math Sci 73(1):73–95
58. Szmit Z, Warminski J (2016) Nonlinear dynamics of electro-mechanical system composed of two pendulums and rotating hub. Proc Eng 144:953–958
59. Tacha OI, Volos CK, Kyprianidis IM, Stouboulos IN, Vaidyanathan S, Pham VT (2016) Analysis, adaptive control and circuit simulation of a novel nonlinear finance system. Appl Math Comput 276:200–217
60. Tigan G, Opris D (2008) Analysis of a 3D chaotic system. Chaos Solitons Fractals 36:1315–1319
61. Utkin VI (1977) Variable structure systems with sliding modes. IEEE Trans Autom Control 22(2):212–222
62. Utkin VI (1993) Sliding mode control design principles and applications to electric drives. IEEE Trans Ind Electron 40(1):23–36
63. Vaidyanathan S (2011) Analysis and synchronization of the hyperchaotic Yujun systems via sliding mode control. Adv Intell Syst Comput 176:329–337
64. Vaidyanathan S (2012) Anti-synchronization of Sprott-L and Sprott-M chaotic systems via adaptive control. Int J Control Theory Appl 5(1):41–59
65. Vaidyanathan S (2012) Global chaos control of hyperchaotic Liu system via sliding control method. Int J Control Theory Appl 5(2):117–123
66. Vaidyanathan S (2012) Sliding mode control based global chaos control of Liu-Liu-Liu-Su chaotic system. Int J Control Theory Appl 5(1):15–20
67. Vaidyanathan S (2013) A new six-term 3-D chaotic system with an exponential nonlinearity. Far East J Math Sci 79(1):135–143
68. Vaidyanathan S (2013) Analysis and adaptive synchronization of two novel chaotic systems with hyperbolic sinusoidal and cosinusoidal nonlinearity and unknown parameters. J Eng Sci Technol Rev 6(4):53–65

69. Vaidyanathan S (2014) A new eight-term 3-D polynomial chaotic system with three quadratic nonlinearities. Far East J Math Sci 84(2):219–226
70. Vaidyanathan S (2014) Analysis and adaptive synchronization of eight-term 3-D polynomial chaotic systems with three quadratic nonlinearities. Eur Phys J: Spec Top 223(8):1519–1529
71. Vaidyanathan S (2014) Analysis, control and synchronisation of a six-term novel chaotic system with three quadratic nonlinearities. Int J Modell Ident Control 22(1):41–53
72. Vaidyanathan S (2014) Generalised projective synchronisation of novel 3-D chaotic systems with an exponential non-linearity via active and adaptive control. Int J Modell Ident Control 22(3):207–217
73. Vaidyanathan S (2014) Global chaos synchronisation of identical Li-Wu chaotic systems via sliding mode control. Int J Modell Ident Control 22(2):170–177
74. Vaidyanathan S (2015) A 3-D novel highly chaotic system with four quadratic nonlinearities, its adaptive control and anti-synchronization with unknown parameters. J Eng Sci Technol Rev 8(2):106–115
75. Vaidyanathan S (2015) A novel chemical chaotic reactor system and its adaptive control. Int J ChemTech Res 8(7):146–158
76. Vaidyanathan S (2015) A novel chemical chaotic reactor system and its output regulation via integral sliding mode control. Int J ChemTech Res 8(11):669–683
77. Vaidyanathan S (2015) Active control design for the anti-synchronization of Lotka-Volterra biological systems with four competitive species. Int J PharmTech Res 8(7):58–70
78. Vaidyanathan S (2015) Adaptive control design for the anti-synchronization of novel 3-D chemical chaotic reactor systems. Int J ChemTech Res 8(11):654–668
79. Vaidyanathan S (2015) Adaptive control of a chemical chaotic reactor. Int J PharmTech Res 8(3):377–382
80. Vaidyanathan S (2015) Adaptive control of the FitzHugh-Nagumo chaotic neuron model. Int J PharmTech Res 8(6):117–127
81. Vaidyanathan S (2015) Adaptive synchronization of chemical chaotic reactors. Int J ChemTech Res 8(2):612–621
82. Vaidyanathan S (2015) Adaptive synchronization of generalized Lotka-Volterra three-species biological systems. Int J PharmTech Res 8(5):928–937
83. Vaidyanathan S (2015) Adaptive synchronization of novel 3-D chemical chaotic reactor systems. Int J ChemTech Res 8(7):159–171
84. Vaidyanathan S (2015) Analysis, properties and control of an eight-term 3-D chaotic system with an exponential nonlinearity. Int J Modell Ident Control 23(2):164–172
85. Vaidyanathan S (2015) Anti-synchronization of brusselator chemical reaction systems via adaptive control. Int J ChemTech Res 8(6):759–768
86. Vaidyanathan S (2015) Anti-synchronization of brusselator chemical reaction systems via integral sliding mode control. Int J ChemTech Res 8(11):700–713
87. Vaidyanathan S (2015) Anti-synchronization of chemical chaotic reactors via adaptive control method. Int J ChemTech Res 8(8):73–85
88. Vaidyanathan S (2015) Anti-synchronization of the FitzHugh-Nagumo chaotic neuron models via adaptive control method. Int J PharmTech Res 8(7):71–83
89. Vaidyanathan S (2015) Chaos in neurons and synchronization of Birkhoff-Shaw strange chaotic attractors via adaptive control. Int J PharmTech Res 8(6):1–11
90. Vaidyanathan S (2015) Dynamics and control of brusselator chemical reaction. Int J ChemTech Res 8(6):740–749
91. Vaidyanathan S (2015) Dynamics and control of Tokamak system with symmetric and magnetically confined plasma. Int J ChemTech Res 8(6):795–802
92. Vaidyanathan S (2015) Global chaos synchronization of chemical chaotic reactors via novel sliding mode control method. Int J ChemTech Res 8(7):209–221
93. Vaidyanathan S (2015) Global chaos synchronization of Duffing double-well chaotic oscillators via integral sliding mode control. Int J ChemTech Res 8(11):141–151
94. Vaidyanathan S (2015) Global chaos synchronization of the forced Van der Pol chaotic oscillators via adaptive control method. Int J PharmTech Res 8(6):156–166

95. Vaidyanathan S (2015) Hybrid chaos synchronization of the FitzHugh-Nagumo chaotic neuron models via adaptive control method. Int J PharmTech Res 8(8):48–60
96. Vaidyanathan S (2015) Integral sliding mode control design for the global chaos synchronization of identical novel chemical chaotic reactor systems. Int J ChemTech Res 8(11):684–699
97. Vaidyanathan S (2015) Output regulation of the forced Van der Pol chaotic oscillator via adaptive control method. Int J PharmTech Res 8(6):106–116
98. Vaidyanathan S (2015) Sliding controller design for the global chaos synchronization of forced Van der Pol chaotic oscillators. Int J PharmTech Res 8(7):100–111
99. Vaidyanathan S (2015) Synchronization of Tokamak systems with symmetric and magnetically confined plasma via adaptive control. Int J ChemTech Res 8(6):818–827
100. Vaidyanathan S (2016) A novel 3-D conservative chaotic system with a sinusoidal nonlinearity and its adaptive control. Int J Control Theory Appl 9(1):115–132
101. Vaidyanathan S (2016) A novel 3-D jerk chaotic system with three quadratic nonlinearities and its adaptive control. Arch Control Sci 26(1):19–47
102. Vaidyanathan S (2016) A novel 3-D jerk chaotic system with two quadratic nonlinearities and its adaptive backstepping control. Int J Control Theory Appl 9(1):199–219
103. Vaidyanathan S (2016) Anti-synchronization of 3-cells cellular neural network attractors via integral sliding mode control. Int J PharmTech Res 9(1):193–205
104. Vaidyanathan S (2016) Generalized projective synchronization of Vaidyanathan Chaotic system via active and adaptive control. In: Vaidyanathan S, Volos C (eds) Advances and applications in nonlinear control systems. Springer, Berlin, Germany, pp 97–116
105. Vaidyanathan S (2016) Global chaos control of the FitzHugh-Nagumo chaotic neuron model via integral sliding mode control. Int J PharmTech Res 9(4):413–425
106. Vaidyanathan S (2016) Global chaos control of the generalized Lotka-Volterra three-species system via integral sliding mode control. Int J PharmTech Res 9(4):399–412
107. Vaidyanathan S (2016) Global chaos regulation of a symmetric nonlinear gyro system via integral sliding mode control. Int J ChemTech Res 9(5):462–469
108. Vaidyanathan S (2016) Hybrid synchronization of the generalized Lotka-Volterra three-species biological systems via adaptive control. Int J PharmTech Res 9(1):179–192
109. Vaidyanathan S (2016) Mathematical analysis, adaptive control and synchronization of a ten-term novel three-scroll chaotic system with four quadratic nonlinearities. Int J Control Theory Appl 9(1):1–20
110. Vaidyanathan S, Azar AT (2015) Analysis, control and synchronization of a nine-term 3-D novel chaotic system. In: Azar AT, Vaidyanathan S (eds) Chaos modelling and control systems design. Studies in computational intelligence, vol 581. Springer, Germany, pp 19–38
111. Vaidyanathan S, Azar AT (2016) Takagi-Sugeno fuzzy logic controller for Liu-Chen four-scroll chaotic system. Int J Intell Eng Inform 4(2):135–150
112. Vaidyanathan S, Boulkroune A (2016) A novel 4-D hyperchaotic chemical reactor system and its adaptive control. In: Vaidyanathan S, Volos C (eds) Advances and applications in chaotic systems. Springer, Berlin, Germany, pp 447–469
113. Vaidyanathan S, Madhavan K (2013) Analysis, adaptive control and synchronization of a seven-term novel 3-D chaotic system. Int J Control Theory Appl 6(2):121–137
114. Vaidyanathan S, Pakiriswamy S (2016) A five-term 3-D novel conservative chaotic system and its generalized projective synchronization via adaptive control method. Int J Control Theory Appl 9(1):61–78
115. Vaidyanathan S, Pakiriswamy S (2013) Generalized projective synchronization of six-term Sundarapandian chaotic systems by adaptive control. Int J Control Theory Appl 6(2):153–163
116. Vaidyanathan S, Pakiriswamy S (2015) A 3-D novel conservative chaotic system and its generalized projective synchronization via adaptive control. J Eng Sci Technol Rev 8(2):52–60
117. Vaidyanathan S, Pakiriswamy S (2016) Adaptive control and synchronization design of a seven-term novel chaotic system with a quartic nonlinearity. Int J Control Theory Appl 9(1):237–256
118. Vaidyanathan S, Rajagopal K (2011) Hybrid synchronization of hyperchaotic Wang-Chen and hyperchaotic Lorenz systems by active non-linear control. Int J Syst Signal Control Eng Appl 4(3):55–61

119. Vaidyanathan S, Rajagopal K (2012) Global chaos synchronization of hyperchaotic Pang and hyperchaotic Wang systems via adaptive control. Eur J Sci Res 7(1):28–37
120. Vaidyanathan S, Rajagopal K (2016) Adaptive control, synchronization and LabVIEW implementation of Rucklidge chaotic system for nonlinear double convection. Int J Control Theory Appl 9(1):175–197
121. Vaidyanathan S, Rajagopal K (2016) Analysis, control, synchronization and LabVIEW implementation of a seven-term novel chaotic system. Int J Control Theory Appl 9(1):151–174
122. Vaidyanathan S, Rasappan S (2014) Global chaos synchronization of n-scroll Chua circuit and Lur'e system using backstepping control design with recursive feedback. Arab J Sci Eng 39(4):3351–3364
123. Vaidyanathan S, Sampath S (2011) Global chaos synchronization of hyperchaotic Lorenz systems by sliding mode control. Commun Comput Inf Sci 205:156–164
124. Vaidyanathan S, Volos C (2015) Analysis and adaptive control of a novel 3-D conservative no-equilibrium chaotic system. Arch Control Sci 25(3):333–353
125. Vaidyanathan S, Volos C (2016) Advances and applications in chaotic systems. Springer, Berlin, Germany
126. Vaidyanathan S, Volos C (2016) Advances and applications in nonlinear control systems. Springer, Berlin, Germany
127. Vaidyanathan S, Volos C (2017) Advances in memristors, memristive devices and systems. Springer, Berlin, Germany
128. Vaidyanathan S, Volos C, Pham VT, Madhavan K, Idowu BA (2014) Adaptive backstepping control, synchronization and circuit simulation of a 3-D novel jerk chaotic system with two hyperbolic sinusoidal nonlinearities. Arch Control Sci 24(3):375–403
129. Vaidyanathan S, Azar AT, Rajagopal K, Alexander P (2015) Design and SPICE implementation of a 12-term novel hyperchaotic system and its synchronisation via active control. Int J Modell Ident Control 23(3):267–277
130. Vaidyanathan S, Idowu BA, Azar AT (2015) Backstepping controller design for the global chaos synchronization of Sprott's jerk systems. In: Azar AT, Vaidyanathan S (eds) Chaos modeling and control systems design. Springer, Berlin, Germany, pp 39–58
131. Vaidyanathan S, Rajagopal K, Volos CK, Kyprianidis IM, Stouboulos IN (2015) Analysis, adaptive control and synchronization of a seven-term novel 3-D chaotic system with three quadratic nonlinearities and its digital implementation in LabVIEW. J Eng Sci Technol Rev 8(2):130–141
132. Vaidyanathan S, Volos CK, Kyprianidis IM, Stouboulos IN, Pham VT (2015) Analysis, adaptive control and anti-synchronization of a six-term novel jerk chaotic system with two exponential nonlinearities and its circuit simulation. J Eng Sci Technol Rev 8(2):24–36
133. Vaidyanathan S, Volos CK, Pham VT (2015) Analysis, adaptive control and adaptive synchronization of a nine-term novel 3-D chaotic system with four quadratic nonlinearities and its circuit simulation. J Eng Sci Technol Rev 8(2):181–191
134. Vaidyanathan S, Volos CK, Pham VT (2015) Global chaos control of a novel nine-term chaotic system via sliding mode control. In: Azar AT, Zhu Q (eds) Advances and applications in sliding mode control systems. Studies in computational intelligence, vol 576. Springer, Germany, pp 571–590
135. Vaidyanathan S, Volos CK, Rajagopal K, Kyprianidis IM, Stouboulos IN (2015) Adaptive backstepping controller design for the anti-synchronization of identical WINDMI chaotic systems with unknown parameters and its SPICE implementation. J Eng Sci Technol Rev 8(2):74–82
136. Vaidyanathan S, Madhavan K, Idowu BA (2016) Backstepping control design for the adaptive stabilization and synchronization of the Pandey jerk chaotic system with unknown parameters. Int J Control Theory Appl 9(1):299–319
137. Volos CK, Kyprianidis IM, Stouboulos IN, Tlelo-Cuautle E, Vaidyanathan S (2015) Memristor: a new concept in synchronization of coupled neuromorphic circuits. J Eng Sci Technol Rev 8(2):157–173

138. Volos CK, Pham VT, Vaidyanathan S, Kyprianidis IM, Stouboulos IN (2015) Synchronization phenomena in coupled Colpitts circuits. J Eng Sci Technol Rev 8(2):142–151
139. Volos CK, Pham VT, Vaidyanathan S, Kyprianidis IM, Stouboulos IN (2016) Synchronization phenomena in coupled hyperchaotic oscillators with hidden attractors using a nonlinear open loop controller. In: Vaidyanathan S, Volos C (eds) Advances and applications in chaotic systems. Springer, Berlin, Germany, pp 1–38
140. Volos CK, Pham VT, Vaidyanathan S, Kyprianidis IM, Stouboulos IN (2016) The case of bidirectionally coupled nonlinear circuits via a memristor. In: Vaidyanathan S, Volos C (eds) Advances and applications in nonlinear control systems. Springer, Berlin, Germany, pp 317–350
141. Volos CK, Prousalis D, Kyprianidis IM, Stouboulos I, Vaidyanathan S, Pham VT (2016) Synchronization and anti-synchronization of coupled Hindmarsh-Rose neuron models. Int J Control Theory Appl 9(1):101–114
142. Volos CK, Vaidyanathan S, Pham VT, Maaita JO, Giakoumis A, Kyprianidis IM, Stouboulos IN (2016) A novel design approach of a nonlinear resistor based on a memristor emulator. In: Azar AT, Vaidyanathan S (eds) Advances in chaos theory and intelligent control. Springer, Berlin, Germany, pp 3–34
143. Wang B, Zhong SM, Dong XC (2016) On the novel chaotic secure communication scheme design. Commun Nonlinear Sci Numer Simul 39:108–117
144. Wolf A, Swift JB, Swinney HL, Vastano JA (1985) Determining Lyapunov exponents from a time series. Physica D 16:285–317
145. Wu T, Sun W, Zhang X, Zhang S (2016) Concealment of time delay signature of chaotic output in a slave semiconductor laser with chaos laser injection. Opt Commun 381:174–179
146. Xu G, Liu F, Xiu C, Sun L, Liu C (2016) Optimization of hysteretic chaotic neural network based on fuzzy sliding mode control. Neurocomputing 189:72–79
147. Xu H, Tong X, Meng X (2016) An efficient chaos pseudo-random number generator applied to video encryption. Optik 127(20):9305–9319
148. Zhou W, Xu Y, Lu H, Pan L (2008) On dynamics analysis of a new chaotic attractor. Phys Lett A 372(36):5773–5777
149. Zhu C, Liu Y, Guo Y (2010) Theoretic and numerical study of a new chaotic system. Intell Inf Manage 2:104–109

Complete Synchronization of Chaotic Systems via Novel Second Order Sliding Mode Control with an Application to a Novel Three-Scroll Chaotic System

Sundarapandian Vaidyanathan

Abstract Chaos in nonlinear dynamics occurs widely in physics, chemistry, biology, ecology, secure communications, cryptosystems and many scientific branches. Synchronization of chaotic systems is an important research problem in chaos theory. Sliding mode control is an important method used to solve various problems in control systems engineering. In robust control systems, the sliding mode control is often adopted due to its inherent advantages of easy realization, fast response and good transient performance as well as insensitivity to parameter uncertainties and disturbance. This work derives a new result for the complete synchronization of identical chaotic systems via novel second order sliding mode control method. The main control result is established by Lyapunov stability theory. As an application of the general result, the problem of global chaos synchronization of novel three-scroll chaotic systems is studied and a new sliding mode controller is derived. The Lyapunov exponents of the novel three-scroll chaotic system are obtained as $L_1 = 2.0469$, $L_2 = 0$ and $L_3 = -3.5533$. The Kaplan-Yorke dimension of the novel chaotic system is obtained as $D_{KY} = 2.5761$. The large value of D_{KY} shows the high complexity of the novel three-scroll chaotic system. Numerical simulations using MATLAB have been shown to depict the phase portraits of the novel three-scroll chaotic system and the global chaos synchronization of three-scroll chaotic systems.

Keywords Chaos · Chaotic systems · Chaos control · Chaos synchronization · Sliding mode control

S. Vaidyanathan (✉)
Research and Development Centre, Vel Tech University, Avadi,
Chennai 600062, Tamil Nadu, India
e-mail: sundarcontrol@gmail.com

© Springer International Publishing AG 2017
S. Vaidyanathan and C.-H. Lien (eds.), *Applications of Sliding Mode Control in Science and Engineering*, Studies in Computational Intelligence 709,
DOI 10.1007/978-3-319-55598-0_9

193

1 Introduction

Chaos theory describes the quantitative study of unstable aperiodic dynamic behavior in deterministic nonlinear dynamical systems. For the motion of a dynamical system to be chaotic, the system variables should contain some nonlinear terms and the system must satisfy three properties: boundedness, infinite recurrence and sensitive dependence on initial conditions [4–6, 128–130].

Some classical paradigms of 3-D chaotic systems in the literature are Lorenz system [25], Rössler system [45], ACT system [3], Sprott systems [53], Chen system [11], Lü system [26], Cai system [9], Tigan system [63], etc.

The problem of global control of a chaotic system is to device feedback control laws so that the closed-loop system is globally asymptotically stable. Global chaos control of various chaotic systems has been investigated via various methods in the control literature [4, 5, 128, 129].

The synchronization of chaotic systems was first researched by Yamada and Fujisaka [16] with subsequent work by Pecora and Carroll [28, 29]. The problem of complete synchronization of chaotic systems is to design a controller for a pair of chaotic systems called the *master* and *slave* systems such that the output of the slave system tracks the output of the master system asymptotically [4, 5, 128, 129].

There are many techniques available in the control literature for the regulation and synchronization of chaotic systems such as active control [18, 47, 48, 57, 121, 132], adaptive control [49–51, 54, 56, 67, 107, 118, 122, 123], backstepping control [40–44, 60, 125, 133, 138, 139], sliding mode control [20, 27, 59, 66, 68, 69, 76, 106, 110, 126], etc.

Many new chaotic systems have been discovered in the recent years such as Zhou system [151], Zhu system [152], Li system [21], Sundarapandian systems [55, 58], Vaidyanathan systems [70–75, 77, 87, 103–105, 112, 113, 116, 117, 119, 120, 124, 127, 131, 134–137], Pehlivan system [30], Sampath system [46], Tacha system [62], Pham systems [32, 35, 36, 38], Akgul system [2], etc.

Chaos theory has applications in several fields such as memristors [31, 33, 34, 37–39, 140, 143, 145], fuzzy logic [7, 52, 114, 149], communication systems [13, 14, 146], cryptosystems [10, 12], electromechanical systems [15, 61], lasers [8, 22, 148], encryption [23, 24, 150], electrical circuits [1, 2, 17, 123, 141], chemical reactions [78, 79, 81, 82, 84, 86, 88–90, 93, 95, 99, 115], oscillators [96, 97, 100, 101, 142], tokamak systems [94, 102], neurology [83, 91, 92, 98, 108, 144], ecology [80, 85, 109, 111], etc.

In this research work, we derive a general result for the complete synchronization of identical chaotic systems using sliding mode control (SMC) theory [64, 65]. The sliding mode control approach is recognized as an efficient tool for designing robust controllers for linear or nonlinear control systems operating under uncertainty conditions.

A major advantage of sliding mode control is low sensitivity to parameter variations in the plant and disturbances affecting the plant, which eliminates the necessity of exact modeling of the plant. In the sliding mode control, the control dynamics will have two sequential modes, viz. the reaching mode and the sliding mode. Basically, a sliding mode controller design consists of two parts: hyperplane design and controller design. A hyperplane is first designed via the pole-placement approach and a controller is then designed based on the sliding condition. The stability of the overall system is guaranteed by the sliding condition and by a stable hyperplane.

This work is organized as follows. In Sect. 2, we derive a general result for the complete synchronization of identical chaotic systems using novel second order sliding mode control. In Sect. 3, we describe a novel three-scroll chaotic system and its phase portraits. We also describe the qualitative properties of the novel three-scroll chaotic system. The Lyapunov exponents of the novel chaotic system are obtained as $L_1 = 2.0469$, $L_2 = 0$ and $L_3 = -3.5533$. The Kaplan-Yorke dimension of the novel chaotic system is obtained as $D_{KY} = 2.5761$. The large value of D_{KY} shows the high complexity of the novel three-scroll chaotic system. In Sect. 4, we describe the second order sliding mode controller design for the complete synchronization of identical novel three-scroll chaotic systems using novel sliding mode control and its numerical simulations. Section 5 contains the conclusions of this work.

2 Complete Synchronization of Chaotic Systems

In this section, we derive new results for the complete synchronization of identical chaotic systems via novel sliding mode control.

As the *master* system, we consider the chaotic system given by

$$\dot{\mathbf{x}} = A\mathbf{x} + f(\mathbf{x}) \tag{1}$$

where $\mathbf{x} \in \mathbf{R}^n$ denotes the state of the system, $A \in \mathbf{R}^{n \times n}$ denotes the matrix of system parameters and $f(\mathbf{x}) \in \mathbf{R}^n$ contains the nonlinear parts of the system.

As the *slave* system, we consider the controlled identical chaotic system given by

$$\dot{\mathbf{y}} = A\mathbf{y} + f(\mathbf{y}) + \mathbf{u} \tag{2}$$

where $\mathbf{y} \in \mathbf{R}^n$ denotes the state of the system and $\mathbf{u}$ is the sliding control to be determined.

The complete synchronization error is defined as

$$\mathbf{e} = \mathbf{y} - \mathbf{x} \tag{3}$$

The error dynamics is easily obtained as

$$\dot{\mathbf{e}} = A\mathbf{e} + \psi(\mathbf{x}, \mathbf{y}) + \mathbf{u}, \tag{4}$$

where

$$\psi(\mathbf{x}, \mathbf{y}) = f(\mathbf{y}) - f(\mathbf{x}) \tag{5}$$

Thus, the complete synchronization problem between the systems (1) and (2) can be stated as follows: Find a controller $\mathbf{u}(\mathbf{x}, \mathbf{y})$ so as to render the synchronization error $\mathbf{e}(t)$ to be globally asymptotically stable for all values of $\mathbf{e}(0) \in \mathbf{R}^n$, i.e.

$$\lim_{t \to \infty} ||\mathbf{e}(t)|| = 0 \text{ for all } \mathbf{e}(0) \in \mathbf{R}^n \tag{6}$$

Next, we describe the novel sliding mode controller design for achieving complete synchronization of the chaotic systems (1) and (2).

First, we start the design by setting the control as

$$\mathbf{u}(t) = -\psi(\mathbf{x}, \mathbf{y}) + Bv(t) \tag{7}$$

In Eq. (7), $B \in \mathbf{R}^n$ is chosen such that (A, B) is completely controllable.

By substituting (7) into (4), we get the closed-loop error dynamics

$$\dot{\mathbf{e}} = A\mathbf{e} + Bv \tag{8}$$

The system (8) is a linear time-invariant control system with single input v.

Next, we start the sliding controller design by defining the sliding variable as

$$s(\mathbf{e}) = C\mathbf{e} = c_1 e_1 + c_2 e_2 + \cdots + c_n e_n, \tag{9}$$

where $C \in \mathbf{R}^{1 \times n}$ is a constant vector to be determined.

The sliding manifold S is defined as the hyperplane

$$S = \{\mathbf{e} \in \mathbf{R}^n \ : \ s(\mathbf{e}) = C\mathbf{e} = 0\} \tag{10}$$

We shall assume that a sliding motion occurs on the hyperplane S.

In second order sliding mode control, the following equations must be satisfied:

$$s = 0 \tag{11a}$$

$$\dot{s} = CA\mathbf{e} + CBv = 0 \tag{11b}$$

We assume that

$$CB \neq 0 \tag{12}$$

The sliding motion is influenced by equivalent control derived from (11b) as

$$v_{eq}(t) = -(CB)^{-1} CA\mathbf{e}(t) \tag{13}$$

By substituting (13) into (8), we obtain the equivalent error dynamics in the sliding phase as follows:

$$\dot{\mathbf{e}} = A\mathbf{e} - (CB)^{-1} CA\mathbf{e} = E\mathbf{e}, \tag{14}$$

where

$$E = \left[I - B(CB)^{-1}C\right]A \tag{15}$$

We note that E is independent of the control and has at most $(n-1)$ non-zero eigenvalues, depending on the chosen switching surface, while the associated eigenvectors belong to $\ker(C)$.

Since (A, B) is controllable, we can use sliding control theory [64, 65] to choose B and C so that E has any desired $(n-1)$ stable eigenvalues.

This shows that the dynamics (14) is globally asympotically stable.

Finally, for the sliding controller design, we apply a novel sliding control law, viz.

$$\dot{s} = -ks - qs^2 \, \text{sgn}(s) \tag{16}$$

In (16), $\text{sgn}(\cdot)$ denotes the *sign* function and the SMC constants $k > 0, q > 0$ are found in such a way that the sliding condition is satisfied and that the sliding motion will occur.

By combining Eqs. (11b), (13) and (16), we finally obtain the sliding mode controller $v(t)$ as

$$v(t) = -(CB)^{-1}\left[C(kI + A)\mathbf{e} + qs^2 \, \text{sgn}(s)\right] \tag{17}$$

Next, we establish the main result of this section.

Theorem 1 *The second order sliding mode controller defined by (7) achieves complete synchronization between the identical chaotic systems (1) and (2) for all initial conditions* $\mathbf{x}(0), \mathbf{y}(0)$ *in* $\mathbf{R}^n$, *where v is defined by the novel sliding mode control law (17),* $B \in \mathbf{R}^{n \times 1}$ *is such that* (A, B) *is controllable,* $C \in \mathbf{R}^{1 \times n}$ *is such that* $CB \neq 0$ *and the matrix E defined by (15) has* $(n-1)$ *stable eigenvalues.*

Proof Upon substitution of the control laws (7) and (17) into the error dynamics (4), we obtain the closed-loop error dynamics as

$$\dot{\mathbf{e}} = A\mathbf{e} - B(CB)^{-1}\left[C(kI + A)\mathbf{e} + qs^2 \, \text{sgn}(s)\right] \tag{18}$$

We shall show that the error dynamics (18) is globally asymptotically stable by considering the quadratic Lyapunov function

$$V(\mathbf{e}) = \frac{1}{2}s^2(\mathbf{e}) \tag{19}$$

The sliding mode motion is characterized by the equations

$$s(\mathbf{e}) = 0 \quad \text{and} \quad \dot{s}(\mathbf{e}) = 0 \tag{20}$$

By the choice of E, the dynamics in the sliding mode given by Eq. (14) is globally asymptotically stable.

When $s(\mathbf{e}) \neq 0$, $V(\mathbf{e}) > 0$.

Also, when $s(\mathbf{e}) \neq 0$, differentiating V along the error dynamics (18) or the equivalent dynamics (16), we get

$$\dot{V}(\mathbf{e}) = s\dot{s} = -ks^2 - qs^3 \operatorname{sgn}(s) < 0 \tag{21}$$

Hence, by Lyapunov stability theory [19], the error dynamics (18) is globally asymptotically stable for all $\mathbf{e}(0) \in \mathbf{R}^n$.

This completes the proof.   ∎

3 A Novel Three-Scroll Chaotic System

In this work, we propose a novel three-scroll chaotic system described by

$$\begin{aligned}
\dot{x}_1 &= a(x_2 - x_1) + cx_1x_3 \\
\dot{x}_2 &= bx_1 + px_2 - x_1x_3 \\
\dot{x}_3 &= 2p - dx_1^2 + x_1x_2 + 3x_3
\end{aligned} \tag{22}$$

where x_1, x_2, x_3 are the states and a, b, c, d, p are constant, positive, parameters.

In this chapter, we show that the system (22) is chaotic when the parameters take the values

$$a = 40, \quad b = 55, \quad c = 0.2, \quad d = 0.6, \quad p = 12 \tag{23}$$

For numerical simulations, we take the initial state of the system (22) as

$$x_1(0) = 0.1, \quad x_2(0) = 0.1, \quad x_3(0) = 0.1 \tag{24}$$

The Lyapunov exponents of the system (22) for the parameter values (23) and the initial state (24) are determined by Wolf's algorithm [147] as

$$L_1 = 2.0469, \quad L_2 = 0, \quad L_3 = -3.5533 \tag{25}$$

Since $L_1 > 0$, we conclude that the system (22) is chaotic.

Since $L_1 + L_2 + L_3 < 0$, we deduce that the system (22) is dissipative.

Hence, the limit sets of the system (22) are ultimately confined into a specific limit set of zero volume, and the asymptotic motion of the novel highly chaotic system (22) settles onto a strange attractor of the system.

From (25), we see that the maximal Lyapunov exponent (MLE) of the chaotic system (22) is $L_1 = 2.0469$.

Also, the Kaplan-Yorke dimension of the novel three-scroll chaotic system (22) is calculated as

$$D_{KY} = 2 + \frac{L_1 + L_2}{|L_3|} = 2.5761, \tag{26}$$

which shows the high complexity of the system (22).

Figure 1 shows the 3-D phase portrait of the novel three-scroll chaotic system (22) in $\mathbf{R}^3$.

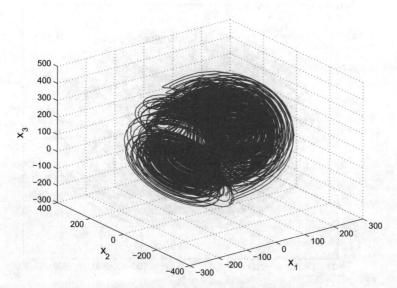

Fig. 1   3-D phase portrait of the novel three-scroll chaotic system in $\mathbf{R}^3$

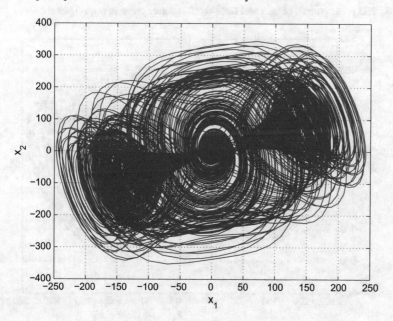

Fig. 2   2-D phase portrait of the novel three-scroll chaotic system in (x_1, x_2) plane

Figures 2, 3 and 4 show the 2-D projections of the novel three-scroll chaotic system (22) in (x_1, x_2), (x_2, x_3) and (x_1, x_3) planes, respectively.

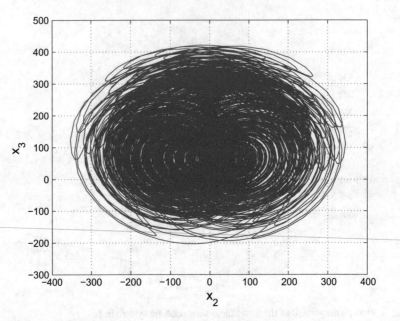

Fig. 3 2-D phase portrait of the novel three-scroll chaotic system in (x_2, x_3) plane

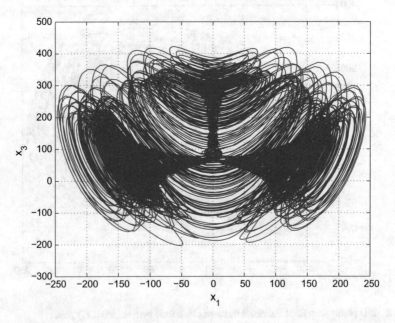

Fig. 4 2-D phase portrait of the novel three-scroll chaotic system in (x_1, x_3) plane

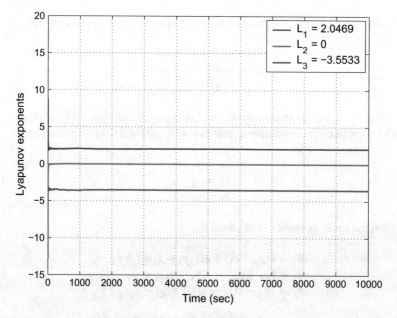

Fig. 5 Lyapunov exponents of the novel three-scroll chaotic system

Figure 5 shows the Lyapunov exponents of the novel three-scroll chaotic system (22).

4 Second Order Sliding Mode Controller Design for the Complete Synchronization of Novel Three-Scroll Chaotic Systems

In this section, we describe the second order sliding mode controller design for the synchronization of identical novel three-scroll chaotic systems by applying the novel second order sliding mode control method described in Sect. 2.

As the master system, we consider the three-scroll system given by

$$
\begin{aligned}
\dot{x}_1 &= a(x_2 - x_1) + cx_1x_3 \\
\dot{x}_2 &= bx_1 + px_2 - x_1x_3 \\
\dot{x}_3 &= 2p - dx_1^2 + x_1x_2 + 3x_3
\end{aligned}
\tag{27}
$$

where x_1, x_2, x_3 are the state variables and a, b, c, d, p are positive parameters.

As the slave system, we consider the three-scroll chaotic system given by

$$
\begin{aligned}
\dot{y}_1 &= a(y_2 - y_1) + c y_1 y_3 + u_1 \\
\dot{y}_2 &= b y_1 + p y_2 - y_1 y_3 + u_2 \\
\dot{y}_3 &= 2p - d y_1^2 + y_1 y_2 + 3 y_3 + u_3
\end{aligned}
\tag{28}
$$

where y_1, y_2, y_3 are the state variables and u_1, u_2, u_3 are sliding mode controls. The complete synchronization error between (27) and (28) is defined as

$$
\begin{aligned}
e_1 &= y_1 - x_1 \\
e_2 &= y_2 - x_2 \\
e_3 &= y_3 - x_3
\end{aligned}
\tag{29}
$$

Then the error dynamics is obtained as

$$
\begin{aligned}
\dot{e}_1 &= a(e_2 - e_1) + c(y_1 y_3 - x_1 x_3) + u_1 \\
\dot{e}_2 &= b e_1 + p e_2 - y_1 y_3 + x_1 x_3 + u_2 \\
\dot{e}_3 &= 3 e_3 - d(y_1^2 - x_1^2) + y_1 y_2 - x_1 x_2 + u_3
\end{aligned}
\tag{30}
$$

In matrix form, we can write the error dynamics (30) as

$$
\dot{\mathbf{e}} = A \mathbf{e} + \psi(\mathbf{x}, \mathbf{y}) + \mathbf{u}
\tag{31}
$$

The matrices in (31) are given by

$$
A = \begin{bmatrix} -a & a & 0 \\ b & p & 0 \\ 0 & 0 & 3 \end{bmatrix} \quad \text{and} \quad \psi(\mathbf{x}, \mathbf{y}) = \begin{bmatrix} c(y_1 y_3 - x_1 x_3) \\ -y_1 y_3 + x_1 x_3 \\ -d(y_1^2 - x_1^2) + y_1 y_2 - x_1 x_2 \end{bmatrix}
\tag{32}
$$

We follow the procedure given in Sect. 2 for the construction of the novel sliding controller to achieve complete synchronization of the identical three-scroll chaotic systems (27) and (28).

First, we set **u** as

$$
\mathbf{u}(t) = -\psi(\mathbf{x}, \mathbf{y}) + B v(t)
\tag{33}
$$

where B is selected such that (A, B) is completely controllable.

A simple choice of B is

$$
B = \begin{bmatrix} 1 \\ 1 \\ 1 \end{bmatrix}
\tag{34}
$$

It can be easily checked that (A, B) is completely controllable.

The three-scroll chaotic system (27) displays a strange attractor when

$$a = 40, \quad b = 55, \quad c = 0.2, \quad d = 0.6, \quad p = 12 \tag{35}$$

Next, we take the sliding variable as

$$s(\mathbf{e}) = C\mathbf{e} = \begin{bmatrix} 1 & -3 & 1 \end{bmatrix} \mathbf{e} = e_1 - 3e_2 + e_3 \tag{36}$$

Next, we take the sliding mode gains as

$$k = 6, \quad q = 0.2 \tag{37}$$

From Eq. (17) in Sect. 2, we obtain the novel sliding control v as

$$v(t) = -199e_1 - 14e_2 + 9e_3 + 0.2s^2 \, \mathrm{sgn}(s) \tag{38}$$

As an application of Theorem 1 to the identical novel three-scroll chaotic systems, we obtain the following main result of this section.

Theorem 2 *The identical novel three-scroll chaotic systems (27) and (28) are globally and asymptotically synchronized for all initial conditions* $\mathbf{x}(0), \mathbf{y}(0) \in \mathbf{R}^3$ *with the second order sliding controller* $\mathbf{u}$ *defined by (33), where* $\psi(\mathbf{x}, \mathbf{y})$ *is defined by (32), B is defined by (34) and v is defined by (38).* ∎

For numerical simulations, we use MATLAB for solving the systems of differential equations using the classical fourth-order Runge-Kutta method with step size $h = 10^{-8}$.

The parameter values of the novel three-scroll chaotic systems are taken as in the chaotic case (35). The sliding mode gains are taken as $k = 6$ and $q = 0.2$.

As an initial condition for the master system (27), we take

$$x_1(0) = 8.9, \quad x_2(0) = 1.5, \quad x_3(0) = 14.3 \tag{39}$$

As an initial condition for the slave system (28), we take

$$y_1(0) = 5.1, \quad y_2(0) = 9.7, \quad y_3(0) = 6.8 \tag{40}$$

Figures 6, 7 and 8 show the complete synchronization of the states of the identical novel three-scroll chaotic systems (27) and (28). Figure 9 shows the time-history of the complete synchronization errors e_1, e_2, e_3.

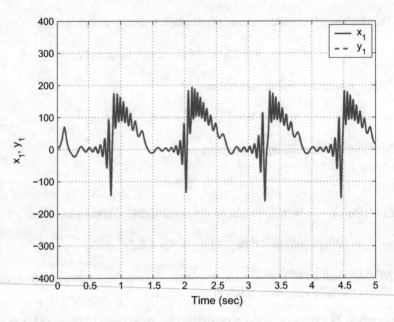

Fig. 6 Complete synchronization of the states x_1 and y_1

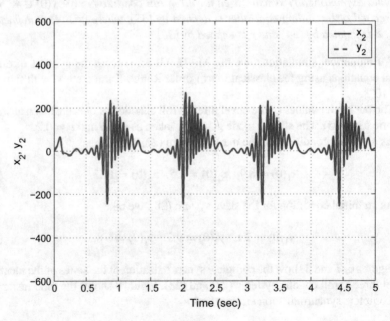

Fig. 7 Complete synchronization of the states x_2 and y_2

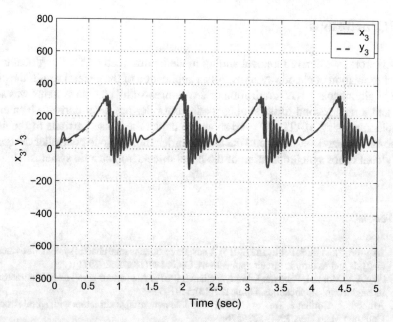

Fig. 8 Complete synchronization of the states x_3 and y_3

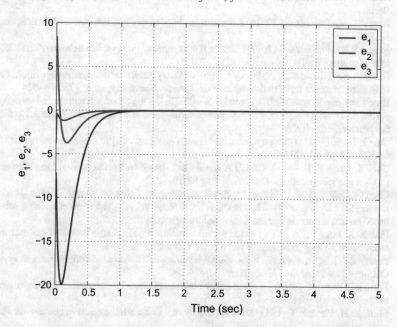

Fig. 9 Time-history of the complete synchronization errors e_1, e_2, e_3

5 Conclusions

In this work, we derived a novel sliding mode control method for the global chaos synchronization of chaotic systems. As an application of the general result, the problem of global chaos synchronization of novel three-scroll chaotic systems was studied and a new second order sliding mode controller has been derived. Numerical simulations using MATLAB were shown to depict the phase portraits of the novel three-scroll chaotic systems and the second order sliding mode controller design for the global chaos synchronization of the novel three-scroll chaotic systems.

References

1. Akgul A, Hussain S, Pehlivan I (2016) A new three-dimensional chaotic system, its dynamical analysis and electronic circuit applications. Optik 127(18):7062–7071
2. Akgul A, Moroz I, Pehlivan I, Vaidyanathan S (2016) A new four-scroll chaotic attractor and its engineering applications. Optik 127(13):5491–5499
3. Arneodo A, Coullet P, Tresser C (1981) Possible new strange attractors with spiral structure. Commun Math Phys 79(4):573–576
4. Azar AT, Vaidyanathan S (2015) Chaos modeling and control systems design. Springer, Berlin
5. Azar AT, Vaidyanathan S (2016) Advances in chaos theory and intelligent control. Springer, Berlin
6. Azar AT, Vaidyanathan S (2017) Fractional order control and synchronization of chaotic systems. Springer, Berlin
7. Boulkroune A, Bouzeriba A, Bouden T (2016) Fuzzy generalized projective synchronization of incommensurate fractional-order chaotic systems. Neurocomputing 173:606–614
8. Burov DA, Evstigneev NM, Magnitskii NA (2017) On the chaotic dynamics in two coupled partial differential equations for evolution of surface plasmon polaritons. Commun Nonlinear Sci Numer Simul 46:26–36
9. Cai G, Tan Z (2007) Chaos synchronization of a new chaotic system via nonlinear control. J Uncertain Syst 1(3):235–240
10. Chai X, Chen Y, Broyde L (2017) A novel chaos-based image encryption algorithm using DNA sequence operations. Opt Lasers Eng 88:197–213
11. Chen G, Ueta T (1999) Yet another chaotic attractor. Int J Bifurc Chaos 9(7):1465–1466
12. Chenaghlu MA, Jamali S, Khasmakhi NN (2016) A novel keyed parallel hashing scheme based on a new chaotic system. Chaos Solitons Fractals 87:216–225
13. Fallahi K, Leung H (2010) A chaos secure communication scheme based on multiplication modulation. Commun Nonlinear Sci Numer Simul 15(2):368–383
14. Fontes RT, Eisencraft M (2016) A digital bandlimited chaos-based communication system. Commun Nonlinear Sci Numer Simul 37:374–385
15. Fotsa RT, Woafo P (2016) Chaos in a new bistable rotating electromechanical system. Chaos Solitons Fractals 93:48–57
16. Fujisaka H, Yamada T (1983) Stability theory of synchronized motion in coupled-oscillator systems. Progress Theoret Phys 63:32–47
17. Kacar S (2016) Analog circuit and microcontroller based RNG application of a new easy realizable 4D chaotic system. Optik 127(20):9551–9561
18. Karthikeyan R, Sundarapandian V (2014) Hybrid chaos synchronization of four-scroll systems via active control. J Electr Eng 65(2):97–103
19. Khalil HK (2002) Nonlinear systems. Prentice Hall, New York

20. Lakhekar GV, Waghmare LM, Vaidyanathan S (2016) Diving autopilot design for underwater vehicles using an adaptive neuro-fuzzy sliding mode controller. In: Vaidyanathan S, Volos C (eds) Advances and applications in nonlinear control systems. Springer, Berlin, pp 477–503
21. Li D (2008) A three-scroll chaotic attractor. Phys Lett A 372(4):387–393
22. Liu H, Ren B, Zhao Q, Li N (2016) Characterizing the optical chaos in a special type of small networks of semiconductor lasers using permutation entropy. Opt Commun 359:79–84
23. Liu W, Sun K, Zhu C (2016) A fast image encryption algorithm based on chaotic map. Opt Lasers Eng 84:26–36
24. Liu X, Mei W, Du H (2016) Simultaneous image compression, fusion and encryption algorithm based on compressive sensing and chaos. Opt Commun 366:22–32
25. Lorenz EN (1963) Deterministic periodic flow. J Atmos Sci 20(2):130–141
26. Lü J, Chen G (2002) A new chaotic attractor coined. Int J Bifurc Chaos 12(3):659–661
27. Moussaoui S, Boulkroune A, Vaidyanathan S (2016) Fuzzy adaptive sliding-mode control scheme for uncertain underactuated systems. In: Vaidyanathan S, Volos C (eds) Advances and applications in nonlinear control systems. Springer, Berlin, pp 351–367
28. Pecora LM, Carroll TL (1991) Synchronizing chaotic circuits. IEEE Trans Circuits Syst 38:453–456
29. Pecora LM, Carroll TL (1990) Synchronization in chaotic systems. Phys Rev Lett 64:821–824
30. Pehlivan I, Moroz IM, Vaidyanathan S (2014) Analysis, synchronization and circuit design of a novel butterfly attractor. J Sound Vib 333(20):5077–5096
31. Pham VT, Volos C, Jafari S, Wang X, Vaidyanathan S (2014) Hidden hyperchaotic attractor in a novel simple memristive neural network. Optoelectron Adv Mater Rapid Commun 8(11–12):1157–1163
32. Pham VT, Volos CK, Vaidyanathan S (2015) Multi-scroll chaotic oscillator based on a first-order delay differential equation. In: Azar AT, Vaidyanathan S (eds) Chaos modeling and control systems design. Studies in computational intelligence, vol 581. Springer, Germany, pp 59–72
33. Pham VT, Volos CK, Vaidyanathan S, Le TP, Vu VY (2015) A memristor-based hyperchaotic system with hidden attractors: dynamics, synchronization and circuital emulating. J Eng Sci Technol Rev 8(2):205–214
34. Pham VT, Jafari S, Vaidyanathan S, Volos C, Wang X (2016) A novel memristive neural network with hidden attractors and its circuitry implementation. Sci China Technol Sci 59(3):358–363
35. Pham VT, Jafari S, Volos C, Giakoumis A, Vaidyanathan S, Kapitaniak T (2016) A chaotic system with equilibria located on the rounded square loop and its circuit implementation. IEEE Trans Circuits Syst II: Express Briefs 63(9):878–882
36. Pham VT, Jafari S, Volos C, Vaidyanathan S, Kapitaniak T (2016) A chaotic system with infinite equilibria located on a piecewise linear curve. Optik 127(20):9111–9117
37. Pham VT, Vaidyanathan S, Volos CK, Hoang TM, Yem VV (2016) Dynamics, synchronization and SPICE implementation of a memristive system with hidden hyperchaotic attractor. In: Azar AT, Vaidyanathan S (eds) Advances in chaos theory and intelligent control. Springer, Berlin, pp 35–52
38. Pham VT, Vaidyanathan S, Volos CK, Jafari S, Kuznetsov NV, Hoang TM (2016) A novel memristive time-delay chaotic system without equilibrium points. Eur Phys J Spec Top 225(1):127–136
39. Pham VT, Vaidyanathan S, Volos CK, Jafari S, Wang X (2016) A chaotic hyperjerk system based on memristive device. In: Vaidyanathan S, Volos C (eds) Advances and applications in chaotic systems. Springer, Berlin, pp 39–58
40. Rasappan S, Vaidyanathan S (2012) Global chaos synchronization of WINDMI and Coullet chaotic systems by backstepping control. Far East J Math Sci 67(2):265–287
41. Rasappan S, Vaidyanathan S (2012) Hybrid synchronization of n-scroll Chua and Lur'e chaotic systems via backstepping control with novel feedback. Arch Control Sci 22(3):343–365

42. Rasappan S, Vaidyanathan S (2012) Synchronization of hyperchaotic Liu system via back-stepping control with recursive feedback. Commun Comput Inf Sci 305:212–221
43. Rasappan S, Vaidyanathan S (2013) Hybrid synchronization of n-scroll chaotic Chua circuits using adaptive backstepping control design with recursive feedback. Malays J Math Sci 7(2):219–246
44. Rasappan S, Vaidyanathan S (2014) Global chaos synchronization of WINDMI and Coullet chaotic systems using adaptive backstepping control design. Kyungpook Math J 54(1):293–320
45. Rössler OE (1976) An equation for continuous chaos. Phys Lett A 57(5):397–398
46. Sampath S, Vaidyanathan S, Volos CK, Pham VT (2015) An eight-term novel four-scroll chaotic system with cubic nonlinearity and its circuit simulation. J Eng Sci Technol Rev 8(2):1–6
47. Sarasu P, Sundarapandian V (2011) Active controller design for generalized projective synchronization of four-scroll chaotic systems. Int J Syst Signal Control Eng Appl 4(2):26–33
48. Sarasu P, Sundarapandian V (2011) The generalized projective synchronization of hyperchaotic Lorenz and hyperchaotic Qi systems via active control. Int J Soft Comput 6(5):216–223
49. Sarasu P, Sundarapandian V (2012) Adaptive controller design for the generalized projective synchronization of 4-scroll systems. Int J Syst Signal Control Eng Appl 5(2):21–30
50. Sarasu P, Sundarapandian V (2012) Generalized projective synchronization of three-scroll chaotic systems via adaptive control. Eur J Sci Res 72(4):504–522
51. Sarasu P, Sundarapandian V (2012) Generalized projective synchronization of two-scroll systems via adaptive control. Eur J Sci Res 72(4):146–156
52. Shirkhani N, Khanesar M, Teshnehlab M (2016) Indirect model reference fuzzy control of SISO fractional order nonlinear chaotic systems. Procedia Comput Sci 102:309–316
53. Sprott JC (1994) Some simple chaotic flows. Phys Rev E 50(2):647–650
54. Sundarapandian V (2013) Adaptive control and synchronization design for the Lu-Xiao chaotic system. Lect Notes Electr Eng 131:319–327
55. Sundarapandian V (2013) Analysis and anti-synchronization of a novel chaotic system via active and adaptive controllers. J Eng Sci Technol Rev 6(4):45–52
56. Sundarapandian V, Karthikeyan R (2011) Anti-synchronization of hyperchaotic Lorenz and hyperchaotic Chen systems by adaptive control. Int J Syst Signal Control Eng Appl 4(2):18–25
57. Sundarapandian V, Karthikeyan R (2012) Hybrid synchronization of hyperchaotic Lorenz and hyperchaotic Chen systems via active control. Int J Syst Signal Control Eng Appl 7(3):254–264
58. Sundarapandian V, Pehlivan I (2012) Analysis, control, synchronization, and circuit design of a novel chaotic system. Math Comput Model 55(7–8):1904–1915
59. Sundarapandian V, Sivaperumal S (2011) Sliding controller design of hybrid synchronization of four-wing Chaotic systems. Int J Soft Comput 6(5):224–231
60. Suresh R, Sundarapandian V (2013) Global chaos synchronization of a family of n-scroll hyperchaotic Chua circuits using backstepping control with recursive feedback. Far East J Math Sci 73(1):73–95
61. Szmit Z, Warminski J (2016) Nonlinear dynamics of electro-mechanical system composed of two pendulums and rotating hub. Procedia Eng 144:953–958
62. Tacha OI, Volos CK, Kyprianidis IM, Stouboulos IN, Vaidyanathan S, Pham VT (2016) Analysis, adaptive control and circuit simulation of a novel nonlinear finance system. Appl Math Comput 276:200–217
63. Tigan G, Opris D (2008) Analysis of a 3D chaotic system. Chaos Solitons Fractals 36:1315–1319
64. Utkin VI (1977) Variable structure systems with sliding modes. IEEE Trans Autom Control 22(2):212–222
65. Utkin VI (1993) Sliding mode control design principles and applications to electric drives. IEEE Trans Industr Electron 40(1):23–36

66. Vaidyanathan S (2011) Analysis and synchronization of the hyperchaotic Yujun systems via sliding mode control. Adv Intell Syst Comput 176:329–337
67. Vaidyanathan S (2012) Anti-synchronization of Sprott-L and Sprott-M chaotic systems via adaptive control. Int J Control Theory Appl 5(1):41–59
68. Vaidyanathan S (2012) Global chaos control of hyperchaotic Liu system via sliding control method. Int J Control Theory Appl 5(2):117–123
69. Vaidyanathan S (2012) Sliding mode control based global chaos control of Liu-Liu-Liu-Su chaotic system. Int J Control Theory Appl 5(1):15–20
70. Vaidyanathan S (2013) A new six-term 3-D chaotic system with an exponential nonlinearity. Far East J Math Sci 79(1):135–143
71. Vaidyanathan S (2013) Analysis and adaptive synchronization of two novel chaotic systems with hyperbolic sinusoidal and cosinusoidal nonlinearity and unknown parameters. J Eng Sci Technol Rev 6(4):53–65
72. Vaidyanathan S (2014) A new eight-term 3-D polynomial chaotic system with three quadratic nonlinearities. Far East J Math Sci 84(2):219–226
73. Vaidyanathan S (2014) Analysis and adaptive synchronization of eight-term 3-D polynomial chaotic systems with three quadratic nonlinearities. Eur Phys J Spec Top 223(8):1519–1529
74. Vaidyanathan S (2014) Analysis, control and synchronisation of a six-term novel chaotic system with three quadratic nonlinearities. Int J Model Ident Control 22(1):41–53
75. Vaidyanathan S (2014) Generalised projective synchronisation of novel 3-D chaotic systems with an exponential non-linearity via active and adaptive control. Int J Model Ident Control 22(3):207–217
76. Vaidyanathan S (2014) Global chaos synchronisation of identical Li-Wu chaotic systems via sliding mode control. Int J Model Ident Control 22(2):170–177
77. Vaidyanathan S (2015) A 3-D novel highly chaotic system with four quadratic nonlinearities, its adaptive control and anti-synchronization with unknown parameters. J Eng Sci Technol Rev 8(2):106–115
78. Vaidyanathan S (2015) A novel chemical chaotic reactor system and its adaptive control. Int J ChemTech Res 8(7):146–158
79. Vaidyanathan S (2015) A novel chemical chaotic reactor system and its output regulation via integral sliding mode control. Int J ChemTech Res 8(11):669–683
80. Vaidyanathan S (2015) Active control design for the anti-synchronization of Lotka-Volterra biological systems with four competitive species. Int J PharmTech Res 8(7):58–70
81. Vaidyanathan S (2015) Adaptive control design for the anti-synchronization of novel 3-D chemical chaotic reactor systems. Int J ChemTech Res 8(11):654–668
82. Vaidyanathan S (2015) Adaptive control of a chemical chaotic reactor. Int J PharmTech Res 8(3):377–382
83. Vaidyanathan S (2015) Adaptive control of the FitzHugh-Nagumo chaotic neuron model. Int J PharmTech Res 8(6):117–127
84. Vaidyanathan S (2015) Adaptive synchronization of chemical chaotic reactors. Int J ChemTech Res 8(2):612–621
85. Vaidyanathan S (2015) Adaptive synchronization of generalized Lotka-Volterra three-species biological systems. Int J PharmTech Res 8(5):928–937
86. Vaidyanathan S (2015) Adaptive synchronization of novel 3-D chemical chaotic reactor systems. Int J ChemTech Res 8(7):159–171
87. Vaidyanathan S (2015) Analysis, properties and control of an eight-term 3-D chaotic system with an exponential nonlinearity. Int J Model Ident Control 23(2):164–172
88. Vaidyanathan S (2015) Anti-synchronization of brusselator chemical reaction systems via adaptive control. Int J ChemTech Res 8(6):759–768
89. Vaidyanathan S (2015) Anti-synchronization of brusselator chemical reaction systems via integral sliding mode control. Int J ChemTech Res 8(11):700–713
90. Vaidyanathan S (2015) Anti-synchronization of chemical chaotic reactors via adaptive control method. Int J ChemTech Res 8(8):73–85

91. Vaidyanathan S (2015) Anti-synchronization of the FitzHugh-Nagumo chaotic neuron models via adaptive control method. Int J PharmTech Res 8(7):71–83

92. Vaidyanathan S (2015) Chaos in neurons and synchronization of Birkhoff-Shaw strange chaotic attractors via adaptive control. Int J PharmTech Res 8(6):1–11

93. Vaidyanathan S (2015) Dynamics and control of brusselator chemical reaction. Int J ChemTech Res 8(6):740–749

94. Vaidyanathan S (2015) Dynamics and control of Tokamak system with symmetric and magnetically confined plasma. Int J ChemTech Res 8(6):795–802

95. Vaidyanathan S (2015) Global chaos synchronization of chemical chaotic reactors via novel sliding mode control method. Int J ChemTech Res 8(7):209–221

96. Vaidyanathan S (2015) Global chaos synchronization of Duffing double-well chaotic oscillators via integral sliding mode control. Int J ChemTech Res 8(11):141–151

97. Vaidyanathan S (2015) Global chaos synchronization of the forced Van der Pol chaotic oscillators via adaptive control method. Int J PharmTech Res 8(6):156–166

98. Vaidyanathan S (2015) Hybrid chaos synchronization of the FitzHugh-Nagumo chaotic neuron models via adaptive control method. Int J PharmTech Res 8(8):48–60

99. Vaidyanathan S (2015) Integral sliding mode control design for the global chaos synchronization of identical novel chemical chaotic reactor systems. Int J ChemTech Res 8(11):684–699

100. Vaidyanathan S (2015) Output regulation of the forced Van der Pol chaotic oscillator via adaptive control method. Int J PharmTech Res 8(6):106–116

101. Vaidyanathan S (2015) Sliding controller design for the global chaos synchronization of forced Van der Pol chaotic oscillators. Int J PharmTech Res 8(7):100–111

102. Vaidyanathan S (2015) Synchronization of Tokamak systems with symmetric and magnetically confined plasma via adaptive control. Int J ChemTech Res 8(6):818–827

103. Vaidyanathan S (2016) A novel 3-D conservative chaotic system with a sinusoidal nonlinearity and its adaptive control. Int J Control Theory Appl 9(1):115–132

104. Vaidyanathan S (2016) A novel 3-D jerk chaotic system with three quadratic nonlinearities and its adaptive control. Arch Control Sci 26(1):19–47

105. Vaidyanathan S (2016) A novel 3-D jerk chaotic system with two quadratic nonlinearities and its adaptive backstepping control. Int J Control Theory Appl 9(1):199–219

106. Vaidyanathan S (2016) Anti-synchronization of 3-cells cellular neural network attractors via integral sliding mode control. Int J PharmTech Res 9(1):193–205

107. Vaidyanathan S (2016) Generalized projective synchronization of Vaidyanathan Chaotic system via active and adaptive control. In: Vaidyanathan S, Volos C (eds) Advances and applications in nonlinear control systems. Springer, Berlin, pp 97–116

108. Vaidyanathan S (2016) Global chaos control of the FitzHugh-Nagumo chaotic neuron model via integral sliding mode control. Int J PharmTech Res 9(4):413–425

109. Vaidyanathan S (2016) Global chaos control of the generalized Lotka-Volterra three-species system via integral sliding mode control. Int J PharmTech Res 9(4):399–412

110. Vaidyanathan S (2016) Global chaos regulation of a symmetric nonlinear gyro system via integral sliding mode control. Int J ChemTech Res 9(5):462–469

111. Vaidyanathan S (2016) Hybrid synchronization of the generalized Lotka-Volterra three-species biological systems via adaptive control. Int J PharmTech Res 9(1):179–192

112. Vaidyanathan S (2016) Mathematical analysis, adaptive control and synchronization of a ten-term novel three-scroll chaotic system with four quadratic nonlinearities. Int J Control Theory Appl 9(1):1–20

113. Vaidyanathan S, Azar AT (2015) Analysis, control and synchronization of a nine-term 3-D novel chaotic system. In: Azar AT, Vaidyanathan S (eds) Chaos modelling and control systems design. Studies in computational intelligence, vol 581. Springer, Germany, pp 19–38

114. Vaidyanathan S, Azar AT (2016) Takagi-Sugeno fuzzy logic controller for Liu-Chen four-scroll chaotic system. Int J Intel Eng Inform 4(2):135–150

115. Vaidyanathan S, Boulkroune A (2016) A novel 4-D hyperchaotic chemical reactor system and its adaptive control. In: Vaidyanathan S, Volos C (eds) Advances and applications in chaotic systems. Springer, Berlin, pp 447–469

116. Vaidyanathan S, Madhavan K (2013) Analysis, adaptive control and synchronization of a seven-term novel 3-D chaotic system. Int J Control Theory Appl 6(2):121–137
117. Vaidyanathan S, Pakiriswamy S (2016) A five-term 3-D novel conservative chaotic system and its generalized projective synchronization via adaptive control method. Int J Control Theory Appl 9(1):61–78
118. Vaidyanathan S, Pakiriswamy S (2013) Generalized projective synchronization of six-term Sundarapandian chaotic systems by adaptive control. Int J Control Theory Appl 6(2):153–163
119. Vaidyanathan S, Pakiriswamy S (2015) A 3-D novel conservative chaotic system and its generalized projective synchronization via adaptive control. J Eng Sci Technol Rev 8(2):52–60
120. Vaidyanathan S, Pakiriswamy S (2016) Adaptive control and synchronization design of a seven-term novel chaotic system with a quartic nonlinearity. Int J Control Theory Appl 9(1):237–256
121. Vaidyanathan S, Rajagopal K (2011) Hybrid synchronization of hyperchaotic Wang-Chen and hyperchaotic Lorenz systems by active non-linear control. Int J Syst Signal Control Eng Appl 4(3):55–61
122. Vaidyanathan S, Rajagopal K (2012) Global chaos synchronization of hyperchaotic Pang and hyperchaotic Wang systems via adaptive control. Eur J Sci Res 7(1):28–37
123. Vaidyanathan S, Rajagopal K (2016) Adaptive control, synchronization and LabVIEW implementation of Rucklidge chaotic system for nonlinear double convection. Int J Control Theory Appl 9(1):175–197
124. Vaidyanathan S, Rajagopal K (2016) Analysis, control, synchronization and LabVIEW implementation of a seven-term novel chaotic system. Int J Control Theory Appl 9(1):151–174
125. Vaidyanathan S, Rasappan S (2014) Global chaos synchronization of n-scroll Chua circuit and Lur'e system using backstepping control design with recursive feedback. Arab J Sci Eng 39(4):3351–3364
126. Vaidyanathan S, Sampath S (2011) Global chaos synchronization of hyperchaotic Lorenz systems by sliding mode control. Commun Comput Inf Sci 205:156–164
127. Vaidyanathan S, Volos C (2015) Analysis and adaptive control of a novel 3-D conservative no-equilibrium chaotic system. Arch Control Sci 25(3):333–353
128. Vaidyanathan S, Volos C (2016) Advances and applications in chaotic systems. Springer, Berlin
129. Vaidyanathan S, Volos C (2016) Advances and applications in nonlinear control systems. Springer, Berlin
130. Vaidyanathan S, Volos C (2017) Advances in memristors. Memristive devices and systems. Springer, Berlin
131. Vaidyanathan S, Volos C, Pham VT, Madhavan K, Idowu BA (2014) Adaptive backstepping control, synchronization and circuit simulation of a 3-D novel jerk chaotic system with two hyperbolic sinusoidal nonlinearities. Arch Control Sci 24(3):375–403
132. Vaidyanathan S, Azar AT, Rajagopal K, Alexander P (2015) Design and SPICE implementation of a 12-term novel hyperchaotic system and its synchronisation via active control. Int J Model Ident Control 23(3):267–277
133. Vaidyanathan S, Idowu BA, Azar AT (2015) Backstepping controller design for the global chaos synchronization of Sprott's jerk systems. In: Azar AT, Vaidyanathan S (eds) Chaos modeling and control systems design. Springer, Berlin, pp 39–58
134. Vaidyanathan S, Rajagopal K, Volos CK, Kyprianidis IM, Stouboulos IN (2015) Analysis, adaptive control and synchronization of a seven-term novel 3-D chaotic system with three quadratic nonlinearities and its digital implementation in LabVIEW. J Eng Sci Technol Rev 8(2):130–141
135. Vaidyanathan S, Volos CK, Kyprianidis IM, Stouboulos IN, Pham VT (2015) Analysis, adaptive control and anti-synchronization of a six-term novel jerk chaotic system with two exponential nonlinearities and its circuit simulation. J Eng Sci Technol Rev 8(2):24–36
136. Vaidyanathan S, Volos CK, Pham VT (2015) Analysis, adaptive control and adaptive synchronization of a nine-term novel 3-D chaotic system with four quadratic nonlinearities and its circuit simulation. J Eng Sci Technol Rev 8(2):181–191

137. Vaidyanathan S, Volos CK, Pham VT (2015) Global chaos control of a novel nine-term chaotic system via sliding mode control. In: Azar AT, Zhu Q (eds) Advances and applications in sliding mode control systems. Studies in computational intelligence, vol 576. Springer, Germany, pp 571–590

138. Vaidyanathan S, Volos CK, Rajagopal K, Kyprianidis IM, Stouboulos IN (2015) Adaptive backstepping controller design for the anti-synchronization of identical WINDMI chaotic systems with unknown parameters and its SPICE implementation. J Eng Sci Technol Rev 8(2):74–82

139. Vaidyanathan S, Madhavan K, Idowu BA (2016) Backstepping control design for the adaptive stabilization and synchronization of the Pandey jerk chaotic system with unknown parameters. Int J Control Theory Appl 9(1):299–319

140. Volos CK, Kyprianidis IM, Stouboulos IN, Tlelo-Cuautle E, Vaidyanathan S (2015) Memristor: a new concept in synchronization of coupled neuromorphic circuits. J Eng Sci Technol Rev 8(2):157–173

141. Volos CK, Pham VT, Vaidyanathan S, Kyprianidis IM, Stouboulos IN (2015) Synchronization phenomena in coupled Colpitts circuits. J Eng Sci Technol Rev 8(2):142–151

142. Volos CK, Pham VT, Vaidyanathan S, Kyprianidis IM, Stouboulos IN (2016) Synchronization phenomena in coupled hyperchaotic oscillators with hidden attractors using a nonlinear open loop controller. In: Vaidyanathan S, Volos C (eds) Advances and applications in chaotic systems. Springer, Berlin, pp 1–38

143. Volos CK, Pham VT, Vaidyanathan S, Kyprianidis IM, Stouboulos IN (2016) The case of bidirectionally coupled nonlinear circuits via a memristor. In: Vaidyanathan S, Volos C (eds) Advances and applications in nonlinear control systems. Springer, Berlin, pp 317–350

144. Volos CK, Prousalis D, Kyprianidis IM, Stouboulos I, Vaidyanathan S, Pham VT (2016) Synchronization and anti-synchronization of coupled Hindmarsh-Rose neuron models. Int J Control Theory Appl 9(1):101–114

145. Volos CK, Vaidyanathan S, Pham VT, Maaita JO, Giakoumis A, Kyprianidis IM, Stouboulos IN (2016) A novel design approach of a nonlinear resistor based on a memristor emulator. In: Azar AT, Vaidyanathan S (eds) Advances in chaos theory and intelligent control. Springer, Berlin, pp 3–34

146. Wang B, Zhong SM, Dong XC (2016) On the novel chaotic secure communication scheme design. Commun Nonlinear Sci Numer Simul 39:108–117

147. Wolf A, Swift JB, Swinney HL, Vastano JA (1985) Determining Lyapunov exponents from a time series. Phys D 16:285–317

148. Wu T, Sun W, Zhang X, Zhang S (2016) Concealment of time delay signature of chaotic output in a slave semiconductor laser with chaos laser injection. Opt Commun 381:174–179

149. Xu G, Liu F, Xiu C, Sun L, Liu C (2016) Optimization of hysteretic chaotic neural network based on fuzzy sliding mode control. Neurocomputing 189:72–79

150. Xu H, Tong X, Meng X (2016) An efficient chaos pseudo-random number generator applied to video encryption. Optik 127(20):9305–9319

151. Zhou W, Xu Y, Lu H, Pan L (2008) On dynamics analysis of a new chaotic attractor. Phys Lett A 372(36):5773–5777

152. Zhu C, Liu Y, Guo Y (2010) Theoretic and numerical study of a new chaotic system. Intel Inf Manage 2:104–109

Novel Second Order Sliding Mode Control Design for the Anti-synchronization of Chaotic Systems with an Application to a Novel Four-Wing Chaotic System

Sundarapandian Vaidyanathan

Abstract Chaos in nonlinear dynamics occurs widely in physics, chemistry, biology, ecology, secure communications, cryptosystems and many scientific branches. Synchronization of chaotic systems is an important research problem in chaos theory. Sliding mode control is an important method used to solve various problems in control systems engineering. In robust control systems, the sliding mode control is often adopted due to its inherent advantages of easy realization, fast response and good transient performance as well as insensitivity to parameter uncertainties and disturbance. This work derives a new result for the anti-synchronization of identical chaotic systems via novel second order sliding mode control method. The main control result is established by Lyapunov stability theory. As an application of the general result, the problem of anti-synchronization of novel four-wing chaotic systems is studied and a new sliding mode controller is derived. The Lyapunov exponents of the novel four-wing chaotic system are obtained as $L_1 = 0.8312$, $L_2 = 0$ and $L_3 = -27.4625$. The Kaplan-Yorke dimension of the novel chaotic system is obtained as $D_{KY} = 2.0303$. We show that the novel four-wing chaotic system has five unstable equilibrium points. We also show that the novel four-wing chaotic system has rotation symmetry about the x_3-axis. Numerical simulations using MATLAB have been shown to depict the phase portraits of the novel four-wing chaotic system and the global anti-synchronization of the novel four-wing chaotic systems.

Keywords Chaos · Chaotic systems · Chaos control · Anti-synchronization · Four-wing system · Sliding mode control

S. Vaidyanathan (✉)
Research and Development Centre, Vel Tech University,
Avadi, Chennai 600062, Tamil Nadu, India
e-mail: sundarcontrol@gmail.com

© Springer International Publishing AG 2017
S. Vaidyanathan and C.-H. Lien (eds.), *Applications of Sliding Mode Control in Science and Engineering*, Studies in Computational Intelligence 709,
DOI 10.1007/978-3-319-55598-0_10

213

1 Introduction

Chaos theory describes the quantitative study of unstable aperiodic dynamic behavior in deterministic nonlinear dynamical systems. For the motion of a dynamical system to be chaotic, the system variables should contain some nonlinear terms and the system must satisfy three properties: boundedness, infinite recurrence and sensitive dependence on initial conditions [4–6, 128–130].

Some classical paradigms of 3-D chaotic systems in the literature are Lorenz system [25], Rössler system [45], ACT system [3], Sprott systems [53], Chen system [11], Lü system [26], Cai system [9], Tigan system [63], etc.

The problem of global control of a chaotic system is to device feedback control laws so that the closed-loop system is globally asymptotically stable. Global chaos control of various chaotic systems has been investigated via various methods in the control literature [4, 5, 128, 129].

The synchronization of chaotic systems was first researched by Yamada and Fujisaka [16] with subsequent work by Pecora and Carroll [27, 29].

The anti-synchronization of systems is a phenomenon that the state vectors of the synchronized systems (master and slave systems) have the same absolute values but opposite signs. Thus, the problem of anti-synchronization of chaotic systems is to design a controller for a pair of chaotic systems called the *master* and *slave* systems such that the sum of the states of master and slave systems converges to zero asymptotically [4, 5, 128, 129].

There are many techniques available in the control literature for the regulation and synchronization of chaotic systems such as active control [18, 47, 48, 57, 121, 132], adaptive control [49–51, 54, 56, 67, 107, 118, 122, 123], backstepping control [40–44, 60, 125, 133, 138, 139], sliding mode control [20, 28, 59, 66, 68, 69, 76, 106, 110, 126], etc.

Many new chaotic systems have been discovered in the recent years such as Zhou system [151], Zhu system [152], Li system [21], Sundarapandian systems [55, 58], Vaidyanathan systems [70–75, 77, 87, 103–105, 112, 113, 116, 117, 119, 120, 124, 127, 131, 134–137], Pehlivan system [30], Sampath system [46], Tacha system [62], Pham systems [32, 35, 36, 38], Akgul system [2], etc.

Chaos theory has applications in several fields such as memristors [31, 33, 34, 37–39, 140, 143, 145], fuzzy logic [7, 52, 114, 149], communication systems [13, 14, 146], cryptosystems [10, 12], electromechanical systems [15, 61], lasers [8, 22, 148], encryption [23, 24, 150], electrical circuits [1, 2, 17, 123, 141], chemical reactions [78, 79, 81, 82, 84, 86, 88–90, 93, 95, 99, 115], oscillators [96, 97, 100, 101, 142], tokamak systems [94, 102], neurology [83, 91, 92, 98, 108, 144], ecology [80, 85, 109, 111], etc.

In this research work, we derive a general result for the anti-synchronization of identical chaotic systems using sliding mode control (SMC) theory [64, 65]. The sliding mode control approach is recognized as an efficient tool for designing robust controllers for linear or nonlinear control systems operating under uncertainty conditions.

A major advantage of sliding mode control is low sensitivity to parameter variations in the plant and disturbances affecting the plant, which eliminates the necessity of exact modeling of the plant. In the sliding mode control, the control dynamics will have two sequential modes, viz. the reaching mode and the sliding mode. Basically, a sliding mode controller design consists of two parts: hyperplane design and controller design. A hyperplane is first designed via the pole-placement approach and a controller is then designed based on the sliding condition. The stability of the overall system is guaranteed by the sliding condition and by a stable hyperplane.

This work is organized as follows. In Sect. 2, we derive a general result for the anti-synchronization of identical chaotic systems using novel second order sliding mode control. In Sect. 3, we describe a novel four-wing chaotic system and its phase portraits. We also describe the qualitative properties of the novel four-wing chaotic system. In Sect. 4, we describe the second order sliding mode controller design for the anti-synchronization of identical novel four-wing chaotic systems using novel sliding mode control and its numerical simulations. Section 5 contains the conclusions of this work.

2 Anti-synchronization of Chaotic Systems

In this section, we derive new results for the anti-synchronization of identical chaotic systems via novel sliding mode control.

As the *master* system, we consider the chaotic system given by

$$\dot{\mathbf{x}} = A\mathbf{x} + f(\mathbf{x}) \tag{1}$$

where $\mathbf{x} \in \mathbf{R}^n$ denotes the state of the system, $A \in \mathbf{R}^{n \times n}$ denotes the matrix of system parameters and $f(\mathbf{x}) \in \mathbf{R}^n$ contains the nonlinear parts of the system.

As the *slave* system, we consider the controlled identical chaotic system given by

$$\dot{\mathbf{y}} = A\mathbf{y} + f(\mathbf{y}) + \mathbf{u} \tag{2}$$

where $\mathbf{y} \in \mathbf{R}^n$ denotes the state of the system and $\mathbf{u}$ is the sliding control to be determined.

The anti-synchronization error is defined as

$$\mathbf{e} = \mathbf{y} + \mathbf{x} \tag{3}$$

The error dynamics is easily obtained as

$$\dot{\mathbf{e}} = A\mathbf{e} + \psi(\mathbf{x}, \mathbf{y}) + \mathbf{u}, \tag{4}$$

where

$$\psi(\mathbf{x}, \mathbf{y}) = f(\mathbf{x}) + f(\mathbf{y}) \tag{5}$$

Thus, the anti-synchronization problem between the systems (1) and (2) can be stated as follows: Find a controller $\mathbf{u}(\mathbf{x}, \mathbf{y})$ so as to render the anti-synchronization error $\mathbf{e}(t)$ to be globally asymptotically stable for all values of $\mathbf{e}(0) \in \mathbf{R}^n$, i.e.

$$\lim_{t \to \infty} ||\mathbf{e}(t)|| = 0 \text{ for all } \mathbf{e}(0) \in \mathbf{R}^n \tag{6}$$

Next, we describe the novel sliding mode controller design for achieving anti-synchronization of the chaotic systems (1) and (2).

First, we start the design by setting the control as

$$\mathbf{u}(t) = -\psi(\mathbf{x}, \mathbf{y}) + Bv(t) \tag{7}$$

In Eq. (7), $B \in \mathbf{R}^n$ is chosen such that (A, B) is completely controllable.

By substituting (7) into (4), we get the closed-loop error dynamics

$$\dot{\mathbf{e}} = A\mathbf{e} + Bv \tag{8}$$

The system (8) is a linear time-invariant control system with single input v.

Next, we start the sliding controller design by defining the sliding variable as

$$s(\mathbf{e}) = C\mathbf{e} = c_1 e_1 + c_2 e_2 + \cdots + c_n e_n, \tag{9}$$

where $C \in \mathbf{R}^{1 \times n}$ is a constant vector to be determined.

The sliding manifold S is defined as the hyperplane

$$S = \{\mathbf{e} \in \mathbf{R}^n \ : \ s(\mathbf{e}) = C\mathbf{e} = 0\} \tag{10}$$

We shall assume that a sliding motion occurs on the hyperplane S.

In second order sliding mode control, the following equations must be satisfied:

$$s = 0 \tag{11a}$$

$$\dot{s} = CA\mathbf{e} + CBv = 0 \tag{11b}$$

We assume that

$$CB \neq 0 \tag{12}$$

The sliding motion is influenced by equivalent control derived from (11b) as

$$v_{eq}(t) = -(CB)^{-1} CA\mathbf{e}(t) \tag{13}$$

By substituting (13) into (8), we obtain the equivalent error dynamics in the sliding phase as follows:

$$\dot{\mathbf{e}} = A\mathbf{e} - (CB)^{-1} CA\mathbf{e} = E\mathbf{e}, \tag{14}$$

where

$$E = \left[I - B(CB)^{-1}C\right]A \qquad (15)$$

We note that E is independent of the control and has at most $(n-1)$ non-zero eigenvalues, depending on the chosen switching surface, while the associated eigenvectors belong to $\ker(C)$.

Since (A, B) is controllable, we can use sliding control theory [64, 65] to choose B and C so that E has any desired $(n-1)$ stable eigenvalues.

This shows that the dynamics (14) is globally asympotically stable.

Finally, for the sliding controller design, we apply a novel sliding control law, *viz.*

$$\dot{s} = -ks - qs^2 \operatorname{sgn}(s) \qquad (16)$$

In (16), $\operatorname{sgn}(\cdot)$ denotes the *sign* function and the SMC constants $k > 0, q > 0$ are found in such a way that the sliding condition is satisfied and that the sliding motion will occur.

By combining equations (11b), (13) and (16), we finally obtain the sliding mode controller $v(t)$ as

$$v(t) = -(CB)^{-1}\left[C(kI + A)\mathbf{e} + qs^2 \operatorname{sgn}(s)\right] \qquad (17)$$

Next, we establish the main result of this section.

Theorem 1 *The second order sliding mode controller defined by (7) achieves antisynchronization between the identical chaotic systems (1) and (2) for all initial conditions $\mathbf{x}(0), \mathbf{y}(0)$ in $\mathbf{R}^n$, where v is defined by the novel sliding mode control law (17), $B \in \mathbf{R}^{n \times 1}$ is such that (A, B) is controllable, $C \in \mathbf{R}^{1 \times n}$ is such that $CB \neq 0$ and the matrix E defined by (15) has $(n-1)$ stable eigenvalues.*

Proof Upon substitution of the control laws (7) and (17) into the error dynamics (4), we obtain the closed-loop error dynamics as

$$\dot{\mathbf{e}} = A\mathbf{e} - B(CB)^{-1}\left[C(kI + A)\mathbf{e} + qs^2 \operatorname{sgn}(s)\right] \qquad (18)$$

We shall show that the error dynamics (18) is globally asymptotically stable by considering the quadratic Lyapunov function

$$V(\mathbf{e}) = \frac{1}{2}s^2(\mathbf{e}) \qquad (19)$$

The sliding mode motion is characterized by the equations

$$s(\mathbf{e}) = 0 \quad \text{and} \quad \dot{s}(\mathbf{e}) = 0 \qquad (20)$$

By the choice of E, the dynamics in the sliding mode given by Eq. (14) is globally asymptotically stable.

When $s(\mathbf{e}) \neq 0$, $V(\mathbf{e}) > 0$.

Also, when $s(\mathbf{e}) \neq 0$, differentiating V along the error dynamics (18) or the equivalent dynamics (16), we get

$$\dot{V}(\mathbf{e}) = s\dot{s} = -ks^2 - qs^3 \, \mathrm{sgn}(s) < 0 \tag{21}$$

Hence, by Lyapunov stability theory [19], the error dynamics (18) is globally asymptotically stable for all $\mathbf{e}(0) \in \mathbf{R}^n$.

This completes the proof.   ∎

3 A Novel Four-Wing Chaotic System

In this work, we propose a novel four-wing chaotic system described by

$$\begin{aligned}
\dot{x}_1 &= a(x_2 - x_1) + bx_2x_3 \\
\dot{x}_2 &= dx_1 - x_2 - px_2^3 + 4x_1x_3 \\
\dot{x}_3 &= cx_3 - x_1x_2
\end{aligned} \tag{22}$$

where x_1, x_2, x_3 are the states and a, b, c, d, p are constant, positive, parameters.

In this chapter, we show that the system (22) is chaotic when the parameters take the values

$$a = 3, \quad b = 15, \quad c = 5, \quad d = 0.1, \quad p = 10 \tag{23}$$

For numerical simulations, we take the initial state of the system (22) as

$$x_1(0) = 1, \quad x_2(0) = 1, \quad x_3(0) = 1 \tag{24}$$

The Lyapunov exponents of the system (22) for the parameter values (23) and the initial state (24) are determined by Wolf's algorithm [147] as

$$L_1 = 0.8312, \quad L_2 = 0, \quad L_3 = -27.4625 \tag{25}$$

Since $L_1 > 0$, we conclude that the system (22) is chaotic.

Since $L_1 + L_2 + L_3 < 0$, we deduce that the system (22) is dissipative.

Hence, the limit sets of the system (22) are ultimately confined into a specific limit set of zero volume, and the asymptotic motion of the novel highly chaotic system (22) settles onto a strange attractor of the system.

From (25), we see that the maximal Lyapunov exponent (MLE) of the chaotic system (22) is $L_1 = 0.8312$.

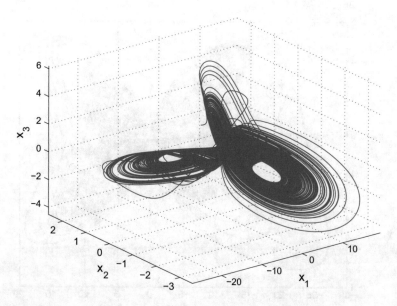

Fig. 1 3-D phase portrait of the novel four-wing chaotic system in $\mathbf{R}^3$

Also, the Kaplan-Yorke dimension of the novel chaotic system (22) is calculated as

$$D_{KY} = 2 + \frac{L_1 + L_2}{|L_3|} = 2.0303, \tag{26}$$

which is fractional.

Figure 1 shows the 3-D phase portrait of the novel three-scroll chaotic system (22) in $\mathbf{R}^3$.

Figures 2, 3 and 4 show the 2-D projections of the novel four-wing chaotic system (22) in (x_1, x_2), (x_2, x_3) and (x_1, x_3) planes, respectively.

Figure 5 shows the Lyapunov exponents of the novel four-wing chaotic system (22).

We also note that the novel four-wing chaotic system (22) is invariant under the coordinates transformation

$$(x_1, x_2, x_3) \mapsto (-x_1, -x_2, x_3) \tag{27}$$

This shows that the novel four-wing chaotic system (22) has rotation symmetry about the x_3-axis. Thus, every non-trivial trajectory of the system (22) must have a twin trajectory.

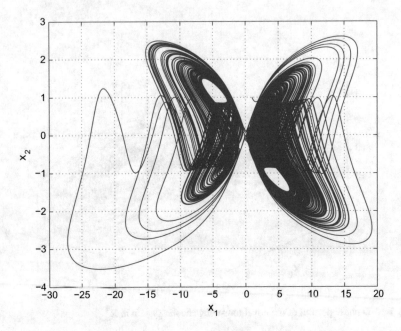

Fig. 2 2-D phase portrait of the novel four-wing chaotic system in (x_1, x_2) plane

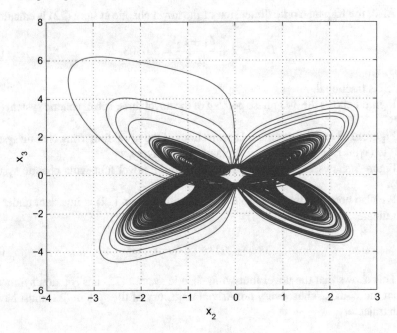

Fig. 3 2-D phase portrait of the novel four-wing chaotic system in (x_2, x_3) plane

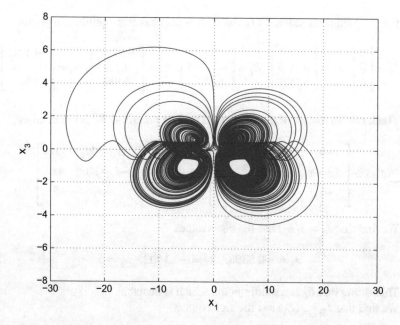

Fig. 4 2-D phase portrait of the novel four-wing chaotic system in (x_1, x_3) plane

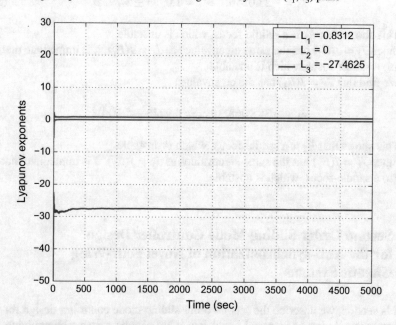

Fig. 5 Lyapunov exponents of the novel four-wing chaotic system

For the parameter values (23), the system (22) has five equilibrium points

$$
E_0 = \begin{bmatrix} 0 \\ 0 \\ 0 \end{bmatrix}, \ E_1 = \begin{bmatrix} 3.1461 \\ 0.8536 \\ 0.5371 \end{bmatrix}, \ E_2 = \begin{bmatrix} -3.1461 \\ -0.8536 \\ 0.5371 \end{bmatrix}, \ E_3 = \begin{bmatrix} 4.1921 \\ -1.1264 \\ -0.9444 \end{bmatrix}, \ E_4 = \begin{bmatrix} -4.1921 \\ 1.1264 \\ -0.9444 \end{bmatrix}
$$

The Jacobian matrix of the system (22) at any point $\mathbf{x} \in \mathbf{R}^3$ is obtained as

$$
J(x) = \begin{bmatrix} -a & a + bx_3 & bx_2 \\ d + 4x_3 & -1 - 3px_2^2 & 4x_1 \\ -x_2 & -x_1 & c \end{bmatrix} = \begin{bmatrix} -3 & 3 + 15x_3 & 15x_2 \\ 0.1 + 4x_3 & -1 - 30x_2^2 & 4x_1 \\ -x_2 & -x_1 & 5 \end{bmatrix} \tag{28}
$$

We find that $J_0 = J(E_0)$ has the eigenvalues

$$
\lambda_1 = -0.8598, \quad \lambda_2 = -3.1402, \quad \lambda_3 = 5 \tag{29}
$$

This shows that E_0 is a saddle-point, which is unstable.
We find that $J_1 = J(E_1)$ has the eigenvalues

$$
\lambda_1 = -23.0556, \quad \lambda_{2,3} = 1.0983 \pm 3.7922i \tag{30}
$$

This shows that E_1 is a saddle-focus, which is unstable.
Since $J_2 = J(E_2)$ has the same eigenvalues as $J_1 = J(E_1)$, it is immediate that E_2 is also a saddle-focus, which is unstable.
We find that $J_3 = J(E_3)$ has the eigenvalues

$$
\lambda_1 = -38.9068, \quad \lambda_{2,3} = 0.9218 \pm 5.1001i \tag{31}
$$

This shows that E_3 is a saddle-focus, which is unstable.
Since $J_4 = J(E_4)$ has the same eigenvalues as $J_3 = J(E_3)$, it is immediate that E_4 is also a saddle-focus, which is unstable.

4 Second Order Sliding Mode Controller Design for the Anti-synchronization of Novel Four-Wing Chaotic Systems

In this section, we describe the second order sliding mode controller design for the anti-synchronization of identical novel four-wing chaotic systems by applying the novel second order sliding mode control method described in Sect. 2.

As the master system, we consider the novel four-wing chaotic system given by

$$\begin{aligned}
\dot{x}_1 &= a(x_2 - x_1) + bx_2x_3 \\
\dot{x}_2 &= dx_1 - x_2 - px_2^3 + 4x_1x_3 \\
\dot{x}_3 &= cx_3 - x_1x_2
\end{aligned} \tag{32}$$

where x_1, x_2, x_3 are the state variables and a, b, c, d, p are positive parameters.

As the slave system, we consider the three-scroll chaotic system given by

$$\begin{aligned}
\dot{y}_1 &= a(y_2 - y_1) + by_2y_3 + u_1 \\
\dot{y}_2 &= dy_1 - y_2 - py_2^3 + 4y_1y_3 + u_2 \\
\dot{y}_3 &= cy_3 - y_1y_2 + u_3
\end{aligned} \tag{33}$$

where y_1, y_2, y_3 are the state variables and u_1, u_2, u_3 are sliding mode controls.

The anti-synchronization error between (32) and (33) is defined as

$$\begin{aligned}
e_1 &= y_1 + x_1 \\
e_2 &= y_2 + x_2 \\
e_3 &= y_3 + x_3
\end{aligned} \tag{34}$$

Then the anti-synchronization error dynamics is obtained as

$$\begin{aligned}
\dot{e}_1 &= a(e_2 - e_1) + b(y_2y_3 + x_2x_3) + u_1 \\
\dot{e}_2 &= de_1 - e_2 - p(y_2^3 + x_2^3) + 4(y_1y_3 + x_1x_3) + u_2 \\
\dot{e}_3 &= ce_3 - y_1y_2 - x_1x_2 + u_3
\end{aligned} \tag{35}$$

In matrix form, we can write the error dynamics (35) as

$$\dot{\mathbf{e}} = A\mathbf{e} + \psi(\mathbf{x}, \mathbf{y}) + \mathbf{u} \tag{36}$$

The matrices in (36) are given by

$$A = \begin{bmatrix} -a & a & 0 \\ d & -1 & 0 \\ 0 & 0 & c \end{bmatrix} \quad \text{and} \quad \psi(\mathbf{x}, \mathbf{y}) = \begin{bmatrix} b(y_2y_3 + x_2x_3) \\ -p(y_2^3 + x_2^3) + 4(y_1y_3 + x_1x_3) \\ -y_1y_2 - x_1x_2 \end{bmatrix} \tag{37}$$

We follow the procedure given in Sect. 2 for the construction of the novel sliding controller to achieve anti-synchronization of the identical four-wing chaotic systems (32) and (33).

First, we set **u** as

$$\mathbf{u}(t) = -\psi(\mathbf{x}, \mathbf{y}) + Bv(t) \tag{38}$$

where B is selected such that (A, B) is completely controllable.

A simple choice of B is

$$B = \begin{bmatrix} 1 \\ 1 \\ 1 \end{bmatrix} \tag{39}$$

It can be easily checked that (A, B) is completely controllable.

The four-wing chaotic system (32) displays a strange attractor when

$$a = 3, \quad b = 15, \quad c = 5, \quad d = 0.1, \quad p = 10 \tag{40}$$

Next, we take the sliding variable as

$$s(\mathbf{e}) = C\mathbf{e} = \begin{bmatrix} 9 & -1 & -9 \end{bmatrix} \mathbf{e} = 9e_1 - e_2 - 9e_3 \tag{41}$$

Next, we take the sliding mode gains as

$$k = 6, \quad q = 0.2 \tag{42}$$

From Eq. (17) in Sect. 2, we obtain the novel sliding control v as

$$v(t) = 26.9e_1 + 22e_2 - 99e_3 + 0.2s^2 \operatorname{sgn}(s) \tag{43}$$

As an application of Theorem 1 to the identical novel four-wing chaotic systems, we obtain the following main result of this section.

Theorem 2 *The identical novel four-wing chaotic systems (32) and (33) are globally and asymptotically anti-synchronized for all initial conditions $\mathbf{x}(0), \mathbf{y}(0) \in \mathbf{R}^3$ with the second order sliding controller $\mathbf{u}$ defined by (38), where $\psi(\mathbf{x}, \mathbf{y})$ is defined by (37), B is defined by (39) and v is defined by (43).* ∎

For numerical simulations, we use MATLAB for solving the systems of differential equations using the classical fourth-order Runge-Kutta method with step size $h = 10^{-8}$.

The parameter values of the novel four-wing chaotic systems are taken as in the chaotic case (40), *viz.*

$$a = 3, \quad b = 15, \quad c = 5, \quad d = 0.1, \quad p = 10 \tag{44}$$

The sliding mode gains are taken as $k = 6$ and $q = 0.2$.
As an initial condition for the master system (32), we take

$$x_1(0) = 7.5, \quad x_2(0) = 12.7, \quad x_3(0) = 24.3 \tag{45}$$

As an initial condition for the slave system (33), we take

$$y_1(0) = 16.1, \quad y_2(0) = 4.8, \quad y_3(0) = 6.4 \tag{46}$$

Figures 6, 7 and 8 show the anti-synchronization of the states of the identical novel four-wing chaotic systems (32) and (33).

Figure 9 shows the time-history of the anti-synchronization errors e_1, e_2, e_3.

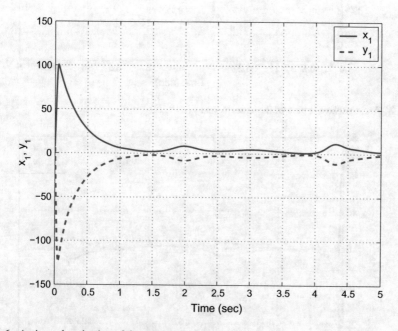

Fig. 6 Anti-synchronization of the states x_1 and y_1

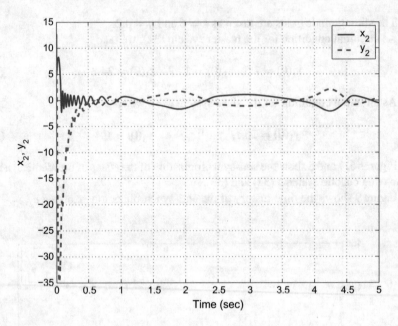

Fig. 7 Anti-synchronization of the states x_2 and y_2

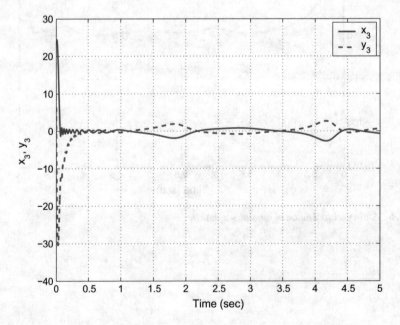

Fig. 8 Anti-synchronization of the states x_3 and y_3

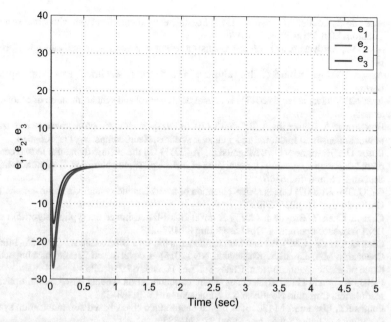

Fig. 9 Time-history of the anti-synchronization errors e_1, e_2, e_3

5 Conclusions

In this work, we derived a novel second-order sliding mode control method for the anti-synchronization of chaotic systems. The main result was established using Lyapunov stability theory. As an application of the general result, the problem of anti-synchronization of novel four-wing chaotic systems was studied and a new sliding mode controller was derived. The Lyapunov exponents of the novel four-wing chaotic system were obtained as $L_1 = 0.8312$, $L_2 = 0$ and $L_3 = -27.4625$. The Kaplan-Yorke dimension of the novel chaotic system was calculated as $D_{KY} = 2.0303$. We showed that the novel four-wing chaotic system has five unstable equilibrium points. We also established that the novel four-wing chaotic system has rotation symmetry about the x_3-axis. Numerical simulations using MATLAB were shown to depict the phase portraits of the novel four-wing chaotic system and the global anti-synchronization of the novel four-wing chaotic systems.

References

1. Akgul A, Hussain S, Pehlivan I (2016) A new three-dimensional chaotic system, its dynamical analysis and electronic circuit applications. Optik 127(18):7062–7071
2. Akgul A, Moroz I, Pehlivan I, Vaidyanathan S (2016) A new four-scroll chaotic attractor and its engineering applications. Optik 127(13):5491–5499

3. Arneodo A, Coullet P, Tresser C (1981) Possible new strange attractors with spiral structure. Commun Math Phys 79(4):573–576
4. Azar AT, Vaidyanathan S (2015) Chaos modeling and control systems design. Springer, Berlin
5. Azar AT, Vaidyanathan S (2016) Advances in chaos theory and intelligent control. Springer, Berlin
6. Azar AT, Vaidyanathan S (2017) Fractional order control and synchronization of chaotic systems. Springer, Berlin
7. Boulkroune A, Bouzeriba A, Bouden T (2016) Fuzzy generalized projective synchronization of incommensurate fractional-order chaotic systems. Neurocomputing 173:606–614
8. Burov DA, Evstigneev NM, Magnitskii NA (2017) On the chaotic dynamics in two coupled partial differential equations for evolution of surface plasmon polaritons. Commun Nonlinear Sci Numer Simul 46:26–36
9. Cai G, Tan Z (2007) Chaos synchronization of a new chaotic system via nonlinear control. J Uncertain Syst 1(3):235–240
10. Chai X, Chen Y, Broyde L (2017) A novel chaos-based image encryption algorithm using DNA sequence operations. Opt Lasers Eng 88:197–213
11. Chen G, Ueta T (1999) Yet another chaotic attractor. Int J Bifurc Chaos 9(7):1465–1466
12. Chenaghlu MA, Jamali S, Khasmakhi NN (2016) A novel keyed parallel hashing scheme based on a new chaotic system. Chaos Solitons Fractals 87:216–225
13. Fallahi K, Leung H (2010) A chaos secure communication scheme based on multiplication modulation. Commun Nonlinear Sci Numer Simul 15(2):368–383
14. Fontes RT, Eisencraft M (2016) A digital bandlimited chaos-based communication system. Commun Nonlinear Sci Numer Simul 37:374–385
15. Fotsa RT, Woafo P (2016) Chaos in a new bistable rotating electromechanical system. Chaos Solitons Fractals 93:48–57
16. Fujisaka H, Yamada T (1983) Stability theory of synchronized motion in coupled-oscillator systems. Progr Theor Phys 63:32–47
17. Kacar S (2016) Analog circuit and microcontroller based RNG application of a new easy realizable 4D chaotic system. Optik 127(20):9551–9561
18. Karthikeyan R, Sundarapandian V (2014) Hybrid chaos synchronization of four-scroll systems via active control. J Electr Eng 65(2):97–103
19. Khalil HK (2002) Nonlinear systems. Prentice Hall, New York
20. Lakhekar GV, Waghmare LM, Vaidyanathan S (2016) Diving autopilot design for underwater vehicles using an adaptive neuro-fuzzy sliding mode controller. In: Vaidyanathan S, Volos C (eds) Advances and applications in nonlinear control systems. Springer, Berlin, pp 477–503
21. Li D (2008) A three-scroll chaotic attractor. Phys Lett A 372(4):387–393
22. Liu H, Ren B, Zhao Q, Li N (2016) Characterizing the optical chaos in a special type of small networks of semiconductor lasers using permutation entropy. Opt Commun 359:79–84
23. Liu W, Sun K, Zhu C (2016) A fast image encryption algorithm based on chaotic map. Opt Lasers Eng 84:26–36
24. Liu X, Mei W, Du H (2016) Simultaneous image compression, fusion and encryption algorithm based on compressive sensing and chaos. Opt Commun 366:22–32
25. Lorenz EN (1963) Deterministic periodic flow. J Atmos Sci 20(2):130–141
26. Lü J, Chen G (2002) A new chaotic attractor coined. Int J Bifurc Chaos 12(3):659–661
27. Pl M, Carroll TL (1991) Synchronizing chaotic circuits. IEEE Trans Circuits Syst 38:453–456
28. Moussaoui S, Boulkroune A, Vaidyanathan S (2016) Fuzzy adaptive sliding-mode control scheme for uncertain underactuated systems. In: Vaidyanathan S, Volos C (eds) Advances and applications in nonlinear control systems. Springer, Berlin, pp 351–367
29. Pecora LM, Carroll TL (1990) Synchronization in chaotic systems. Phys Rev Lett 64:821–824
30. Pehlivan I, Moroz IM, Vaidyanathan S (2014) Analysis, synchronization and circuit design of a novel butterfly attractor. J Sound Vib 333(20):5077–5096
31. Pham VT, Volos C, Jafari S, Wang X, Vaidyanathan S (2014) Hidden hyperchaotic attractor in a novel simple memristive neural network. Optoelectr Adv Mater Rapid Commun 8(11–12):1157–1163

32. Pham VT, Volos CK, Vaidyanathan S (2015) Multi-scroll chaotic oscillator based on a first-order delay differential equation. In: Azar AT, Vaidyanathan S (eds) Chaos modeling and control systems design. Studies in computational intelligence, vol 581. Springer, Germany, pp 59–72
33. Pham VT, Volos CK, Vaidyanathan S, Le TP, Vu VY (2015) A memristor-based hyperchaotic system with hidden attractors: dynamics, synchronization and circuital emulating. J Eng Sci Technol Rev 8(2):205–214
34. Pham VT, Jafari S, Vaidyanathan S, Volos C, Wang X (2016) A novel memristive neural network with hidden attractors and its circuitry implementation. Sci Chin Technol Sci 59(3):358–363
35. Pham VT, Jafari S, Volos C, Giakoumis A, Vaidyanathan S, Kapitaniak T (2016) A chaotic system with equilibria located on the rounded square loop and its circuit implementation. IEEE Trans Circuits Syst II: Express Briefs 63(9):878–882
36. Pham VT, Jafari S, Volos C, Vaidyanathan S, Kapitaniak T (2016) A chaotic system with infinite equilibria located on a piecewise linear curve. Optik 127(20):9111–9117
37. Pham VT, Vaidyanathan S, Volos CK, Hoang TM, Yem VV (2016) Dynamics, synchronization and SPICE implementation of a memristive system with hidden hyperchaotic attractor. In: Azar AT, Vaidyanathan S (eds) Advances in chaos theory and intelligent control. Springer, Berlin, pp 35–52
38. Pham VT, Vaidyanathan S, Volos CK, Jafari S, Kuznetsov NV, Hoang TM (2016) A novel memristive time-delay chaotic system without equilibrium points. Eur Phys J: Spec Topics 225(1):127–136
39. Pham VT, Vaidyanathan S, Volos CK, Jafari S, Wang X (2016) A chaotic hyperjerk system based on memristive device. In: Vaidyanathan S, Volos C (eds) Advances and applications in chaotic systems. Springer, Berlin, pp 39–58
40. Rasappan S, Vaidyanathan S (2012) Global chaos synchronization of WINDMI and Coullet chaotic systems by backstepping control. Far East J Math Sci 67(2):265–287
41. Rasappan S, Vaidyanathan S (2012) Hybrid synchronization of n-Scroll Chua and Lur'e chaotic systems via backstepping control with novel feedback. Arch Control Sci 22(3):343–365
42. Rasappan S, Vaidyanathan S (2012) Synchronization of hyperchaotic Liu system via backstepping control with recursive feedback. Commun Comput Inf Sci 305:212–221
43. Rasappan S, Vaidyanathan S (2013) Hybrid synchronization of n-scroll chaotic Chua circuits using adaptive backstepping control design with recursive feedback. Malays J Math Sci 7(2):219–246
44. Rasappan S, Vaidyanathan S (2014) Global chaos synchronization of WINDMI and Coullet chaotic systems using adaptive backstepping control design. Kyungpook Math J 54(1):293–320
45. Rössler OE (1976) An equation for continuous chaos. Phys Lett A 57(5):397–398
46. Sampath S, Vaidyanathan S, Volos CK, Pham VT (2015) An eight-term novel four-scroll chaotic system with cubic nonlinearity and its circuit simulation. J Eng Sci Technol Rev 8(2): 1–6
47. Sarasu P, Sundarapandian V (2011) Active controller design for generalized projective synchronization of four-scroll chaotic systems. Int J Syst Signal Control Eng Appl 4(2):26–33
48. Sarasu P, Sundarapandian V (2011) The generalized projective synchronization of hyperchaotic Lorenz and hyperchaotic Qi systems via active control. Int J Soft Comput 6(5):216–223
49. Sarasu P, Sundarapandian V (2012) Adaptive controller design for the generalized projective synchronization of 4-scroll systems. Int J Syst Signal Control Eng Appl 5(2):21–30
50. Sarasu P, Sundarapandian V (2012) Generalized projective synchronization of three-scroll chaotic systems via adaptive control. Eur J Sci Res 72(4):504–522
51. Sarasu P, Sundarapandian V (2012) Generalized projective synchronization of two-scroll systems via adaptive control. Eur J Sci Res 72(4):146–156

52. Shirkhani N, Khanesar M, Teshnehlab M (2016) Indirect model reference fuzzy control of SISO fractional order nonlinear chaotic systems. Proc Comput Sci 102:309–316
53. Sprott JC (1994) Some simple chaotic flows. Phys Rev E 50(2):647–650
54. Sundarapandian V (2013) Adaptive control and synchronization design for the Lu-Xiao chaotic system. Lecture Notes Electr Eng 131:319–327
55. Sundarapandian V (2013) Analysis and anti-synchronization of a novel chaotic system via active and adaptive controllers. J Eng Sci Technol Rev 6(4):45–52
56. Sundarapandian V, Karthikeyan R (2011) Anti-synchronization of hyperchaotic Lorenz and hyperchaotic Chen systems by adaptive control. Int J Syst Signal Control Eng Appl 4(2):18–25
57. Sundarapandian V, Karthikeyan R (2012) Hybrid synchronization of hyperchaotic Lorenz and hyperchaotic Chen systems via active control. Int J Syst Signal Control Eng Appl 7(3):254–264
58. Sundarapandian V, Pehlivan I (2012) Analysis, control, synchronization, and circuit design of a novel chaotic system. Math Comput Model 55(7–8):1904–1915
59. Sundarapandian V, Sivaperumal S (2011) Sliding controller design of hybrid synchronization of four-wing chaotic systems. Int J Soft Comput 6(5):224–231
60. Suresh R, Sundarapandian V (2013) Global chaos synchronization of a family of n-scroll hyperchaotic Chua circuits using backstepping control with recursive feedback. Far East J Math Sci 73(1):73–95
61. Szmit Z, Warminski J (2016) Nonlinear dynamics of electro-mechanical system composed of two pendulums and rotating hub. Proc Eng 144:953–958
62. Tacha OI, Volos CK, Kyprianidis IM, Stouboulos IN, Vaidyanathan S, Pham VT (2016) Analysis, adaptive control and circuit simulation of a novel nonlinear finance system. Appl Math Comput 276:200–217
63. Tigan G, Opris D (2008) Analysis of a 3D chaotic system. Chaos Solitons Fractals 36:1315–1319
64. Utkin VI (1977) Variable structure systems with sliding modes. IEEE Trans Autom Control 22(2):212–222
65. Utkin VI (1993) Sliding mode control design principles and applications to electric drives. IEEE Trans Ind Electr 40(1):23–36
66. Vaidyanathan S (2011) Analysis and synchronization of the hyperchaotic Yujun systems via sliding mode control. Adv Intell Syst Comput 176:329–337
67. Vaidyanathan S (2012) Anti-synchronization of Sprott-L and Sprott-M chaotic systems via adaptive control. Int J Control Theory Appl 5(1):41–59
68. Vaidyanathan S (2012) Global chaos control of hyperchaotic Liu system via sliding control method. Int J Control Theory Appl 5(2):117–123
69. Vaidyanathan S (2012) Sliding mode control based global chaos control of Liu-Liu-Liu-Su chaotic system. Int J Control Theory Appl 5(1):15–20
70. Vaidyanathan S (2013) A new six-term 3-D chaotic system with an exponential nonlinearity. Far East J Math Sci 79(1):135–143
71. Vaidyanathan S (2013) Analysis and adaptive synchronization of two novel chaotic systems with hyperbolic sinusoidal and cosinusoidal nonlinearity and unknown parameters. J Eng Sci Technol Rev 6(4):53–65
72. Vaidyanathan S (2014) A new eight-term 3-D polynomial chaotic system with three quadratic nonlinearities. Far East J Math Sci 84(2):219–226
73. Vaidyanathan S (2014) Analysis and adaptive synchronization of eight-term 3-D polynomial chaotic systems with three quadratic nonlinearities. Eur Phys J: Spec Topics 223(8):1519–1529
74. Vaidyanathan S (2014) Analysis, control and synchronisation of a six-term novel chaotic system with three quadratic nonlinearities. Int J Model Identif Control 22(1):41–53
75. Vaidyanathan S (2014) Generalised projective synchronisation of novel 3-D chaotic systems with an exponential non-linearity via active and adaptive control. Int J Model Identif Control 22(3):207–217

76. Vaidyanathan S (2014) Global chaos synchronisation of identical Li-Wu chaotic systems via sliding mode control. Int J Model Identif Control 22(2):170–177
77. Vaidyanathan S (2015) A 3-D novel highly chaotic system with four quadratic nonlinearities, its adaptive control and anti-synchronization with unknown parameters. J Eng Sci Technol Rev 8(2):106–115
78. Vaidyanathan S (2015) A novel chemical chaotic reactor system and its adaptive control. Int J ChemTech Res 8(7):146–158
79. Vaidyanathan S (2015) A novel chemical chaotic reactor system and its output regulation via integral sliding mode control. Int J ChemTech Res 8(11):669–683
80. Vaidyanathan S (2015) Active control design for the anti-synchronization of Lotka-Volterra biological systems with four competitive species. Inte J PharmTech Res 8(7):58–70
81. Vaidyanathan S (2015) Adaptive control design for the anti-synchronization of novel 3-D chemical chaotic reactor systems. Int J ChemTech Res 8(11):654–668
82. Vaidyanathan S (2015) Adaptive control of a chemical chaotic reactor. Int J PharmTech Res 8(3):377–382
83. Vaidyanathan S (2015) Adaptive control of the FitzHugh-Nagumo chaotic neuron model. Int J PharmTech Res 8(6):117–127
84. Vaidyanathan S (2015) Adaptive synchronization of chemical chaotic reactors. Int J ChemTech Res 8(2):612–621
85. Vaidyanathan S (2015) Adaptive synchronization of generalized Lotka-Volterra three-species biological systems. Int J PharmTech Res 8(5):928–937
86. Vaidyanathan S (2015) Adaptive synchronization of novel 3-D chemical chaotic reactor systems. Int J ChemTech Res 8(7):159–171
87. Vaidyanathan S (2015) Analysis, properties and control of an eight-term 3-D chaotic system with an exponential nonlinearity. Int J Model Identif Control 23(2):164–172
88. Vaidyanathan S (2015) Anti-synchronization of brusselator chemical reaction systems via adaptive control. Int J ChemTech Res 8(6):759–768
89. Vaidyanathan S (2015) Anti-synchronization of brusselator chemical reaction systems via integral sliding mode control. Int J ChemTech Res 8(11):700–713
90. Vaidyanathan S (2015) Anti-synchronization of chemical chaotic reactors via adaptive control method. Int J ChemTech Res 8(8):73–85
91. Vaidyanathan S (2015) Anti-synchronization of the FitzHugh-Nagumo chaotic neuron models via adaptive control method. Int J PharmTech Res 8(7):71–83
92. Vaidyanathan S (2015) Chaos in neurons and synchronization of Birkhoff-Shaw strange chaotic attractors via adaptive control. Int J PharmTech Res 8(6):1–11
93. Vaidyanathan S (2015) Dynamics and control of brusselator chemical reaction. Int J ChemTech Res 8(6):740–749
94. Vaidyanathan S (2015) Dynamics and control of Tokamak system with symmetric and magnetically confined plasma. Int J ChemTech Res 8(6):795–802
95. Vaidyanathan S (2015) Global chaos synchronization of chemical chaotic reactors via novel sliding mode control method. Int J ChemTech Res 8(7):209–221
96. Vaidyanathan S (2015) Global chaos synchronization of Duffing double-well chaotic oscillators via integral sliding mode control. Int J ChemTech Res 8(11):141–151
97. Vaidyanathan S (2015) Global chaos synchronization of the forced Van der Pol chaotic oscillators via adaptive control method. Int J PharmTech Res 8(6):156–166
98. Vaidyanathan S (2015) Hybrid chaos synchronization of the FitzHugh-Nagumo chaotic neuron models via adaptive control method. Int J PharmTech Res 8(8):48–60
99. Vaidyanathan S (2015) Integral sliding mode control design for the global chaos synchronization of identical novel chemical chaotic reactor systems. Int J ChemTech Res 8(11):684–699
100. Vaidyanathan S (2015) Output regulation of the forced Van der Pol chaotic oscillator via adaptive control method. Int J PharmTech Res 8(6):106–116
101. Vaidyanathan S (2015) Sliding controller design for the global chaos synchronization of forced Van der Pol chaotic oscillators. Int J PharmTech Res 8(7):100–111

102. Vaidyanathan S (2015) Synchronization of Tokamak systems with symmetric and magnetically confined plasma via adaptive control. Int J ChemTech Res 8(6):818–827
103. Vaidyanathan S (2016) A novel 3-D conservative chaotic system with a sinusoidal nonlinearity and its adaptive control. Int J Control Theory Appl 9(1):115–132
104. Vaidyanathan S (2016) A novel 3-D jerk chaotic system with three quadratic nonlinearities and its adaptive control. Arch Control Sci 26(1):19–47
105. Vaidyanathan S (2016) A novel 3-D jerk chaotic system with two quadratic nonlinearities and its adaptive backstepping control. Int J Control Theory Appl 9(1):199–219
106. Vaidyanathan S (2016) Anti-synchronization of 3-cells cellular neural network attractors via integral sliding mode control. Int J PharmTech Res 9(1):193–205
107. Vaidyanathan S (2016) Generalized projective synchronization of Vaidyanathan chaotic system via active and adaptive control. In: Vaidyanathan S, Volos C (eds) Advances and applications in nonlinear control systems. Springer, Berlin, pp 97–116
108. Vaidyanathan S (2016) Global chaos control of the FitzHugh-Nagumo chaotic neuron model via integral sliding mode control. Int J PharmTech Res 9(4):413–425
109. Vaidyanathan S (2016) Global chaos control of the generalized Lotka-Volterra three-species system via integral sliding mode control. Int J PharmTech Res 9(4):399–412
110. Vaidyanathan S (2016) Global chaos regulation of a symmetric nonlinear gyro system via integral sliding mode control. Int J ChemTech Res 9(5):462–469
111. Vaidyanathan S (2016) Hybrid synchronization of the generalized Lotka-Volterra three-species biological systems via adaptive control. Int J PharmTech Res 9(1):179–192
112. Vaidyanathan S (2016) Mathematical analysis, adaptive control and synchronization of a ten-term novel three-scroll chaotic system with four quadratic nonlinearities. Int J Control Theory Appl 9(1):1–20
113. Vaidyanathan S, Azar AT (2015) Analysis, control and synchronization of a nine-term 3-D novel chaotic system. In: Azar AT, Vaidyanathan S (eds) Chaos modelling and control systems design. Studies in computational intelligence, vol 581. Springer, Germany, pp 19–38
114. Vaidyanathan S, Azar AT (2016) Takagi-Sugeno fuzzy logic controller for Liu-Chen four-scroll chaotic system. Int J Intell Eng Inf 4(2):135–150
115. Vaidyanathan S, Boulkroune A (2016) A novel 4-D hyperchaotic chemical reactor system and its adaptive control. In: Vaidyanathan S, Volos C (eds) Advances and applications in chaotic systems. Springer, Berlin, pp 447–469
116. Vaidyanathan S, Madhavan K (2013) Analysis, adaptive control and synchronization of a seven-term novel 3-D chaotic system. Int J Control Theory Appl 6(2):121–137
117. Vaidyanathan S, Pakiriswamy (2016) A five-term 3-D novel conservative chaotic system and its generalized projective synchronization via adaptive control method. Int J Control Theory Appl 9(1):61–78
118. Vaidyanathan S, Pakiriswamy S (2013) Generalized projective synchronization of six-term Sundarapandian chaotic systems by adaptive control. Int J Control Theory Appl 6(2):153–163
119. Vaidyanathan S, Pakiriswamy S (2015) A 3-D novel conservative chaotic system and its generalized projective synchronization via adaptive control. J Eng Sci Technol Rev 8(2):52–60
120. Vaidyanathan S, Pakiriswamy S (2016) Adaptive control and synchronization design of a seven-term novel chaotic system with a quartic nonlinearity. Int J Control Theory Appl 9(1):237–256
121. Vaidyanathan S, Rajagopal K (2011) Hybrid synchronization of hyperchaotic Wang-Chen and hyperchaotic Lorenz systems by active non-linear control. Int J Syst Signal Control Eng Appl 4(3):55–61
122. Vaidyanathan S, Rajagopal K (2012) Global chaos synchronization of hyperchaotic Pang and hyperchaotic Wang systems via adaptive control. Eur J Sci Res 7(1):28–37
123. Vaidyanathan S, Rajagopal K (2016) Adaptive control, synchronization and LabVIEW implementation of Rucklidge chaotic system for nonlinear double convection. Int J Control Theory Appl 9(1):175–197
124. Vaidyanathan S, Rajagopal K (2016) Analysis, control, synchronization and LabVIEW implementation of a seven-term novel chaotic system. Int J Control Theory Appl 9(1):151–174

125. Vaidyanathan S, Rasappan S (2014) Global chaos synchronization of n-scroll Chua circuit and Lur'e system using backstepping control design with recursive feedback. Arab J Sci Eng 39(4):3351–3364
126. Vaidyanathan S, Sampath S (2011) Global chaos synchronization of hyperchaotic Lorenz systems by sliding mode control. Commun Comput Inf Sci 205:156–164
127. Vaidyanathan S, Volos C (2015) Analysis and adaptive control of a novel 3-D conservative no-equilibrium chaotic system. Arch Control Sci 25(3):333–353
128. Vaidyanathan S, Volos C (2016) Advances and applications in chaotic systems. Springer, Berlin
129. Vaidyanathan S, Volos C (2016) Advances and applications in nonlinear control systems. Springer, Berlin
130. Vaidyanathan S, Volos C (2017) Advances in memristors, memristive devices and systems. Springer, Berlin
131. Vaidyanathan S, Volos C, Pham VT, Madhavan K, Idowu BA (2014) Adaptive backstepping control, synchronization and circuit simulation of a 3-D novel jerk chaotic system with two hyperbolic sinusoidal nonlinearities. Arch Control Sci 24(3):375–403
132. Vaidyanathan S, Azar AT, Rajagopal K, Alexander P (2015) Design and SPICE implementation of a 12-term novel hyperchaotic system and its synchronisation via active control. Int J Model Identif Control 23(3):267–277
133. Vaidyanathan S, Idowu BA, Azar AT (2015) Backstepping controller design for the global chaos synchronization of Sprott's jerk systems. In: Azar AT, Vaidyanathan S (eds) Chaos modeling and control systems design. Springer, Berlin, pp 39–58
134. Vaidyanathan S, Rajagopal K, Volos CK, Kyprianidis IM, Stouboulos IN (2015) Analysis, adaptive control and synchronization of a seven-term novel 3-D chaotic system with three quadratic nonlinearities and its digital implementation in LabVIEW. J Eng Sci Technol Rev 8(2):130–141
135. Vaidyanathan S, Volos CK, Kyprianidis IM, Stouboulos IN, Pham VT (2015) Analysis, adaptive control and anti-synchronization of a six-term novel jerk chaotic system with two exponential nonlinearities and its circuit simulation. J Eng Sci Technol Rev 8(2):24–36
136. Vaidyanathan S, Volos CK, Pham VT (2015) Analysis, adaptive control and adaptive synchronization of a nine-term novel 3-D chaotic system with four quadratic nonlinearities and its circuit simulation. J Eng Sci Technol Rev 8(2):181–191
137. Vaidyanathan S, Volos CK, Pham VT (2015) Global chaos control of a novel nine-term chaotic system via sliding mode control. In: Azar AT, Zhu Q (eds) Advances and applications in sliding mode control systems. Studies in computational intelligence, vol 576. Springer, Germany, pp 571–590
138. Vaidyanathan S, Volos CK, Rajagopal K, Kyprianidis IM, Stouboulos IN (2015) Adaptive backstepping controller design for the anti-synchronization of identical WINDMI chaotic systems with unknown parameters and its SPICE implementation. J Eng Sci Technol Rev 8(2):74–82
139. Vaidyanathan S, Madhavan K, Idowu BA (2016) Backstepping control design for the adaptive stabilization and synchronization of the Pandey jerk chaotic system with unknown parameters. Int J Control Theory Appl 9(1):299–319
140. Volos CK, Kyprianidis IM, Stouboulos IN, Tlelo-Cuautle E, Vaidyanathan S (2015) Memristor: a new concept in synchronization of coupled neuromorphic circuits. J Eng Sci Technol Rev 8(2):157–173
141. Volos CK, Pham VT, Vaidyanathan S, Kyprianidis IM, Stouboulos IN (2015) Synchronization phenomena in coupled Colpitts circuits. J Eng Sci Technol Rev 8(2):142–151
142. Volos CK, Pham VT, Vaidyanathan S, Kyprianidis IM, Stouboulos IN (2016) Synchronization phenomena in coupled hyperchaotic oscillators with hidden attractors using a nonlinear open loop controller. In: Vaidyanathan S, Volos C (eds) Advances and applications in chaotic systems. Springer, Berlin, pp 1–38
143. Volos CK, Pham VT, Vaidyanathan S, Kyprianidis IM, Stouboulos IN (2016) The case of bidirectionally coupled nonlinear circuits via a memristor. In: Vaidyanathan S, Volos C (eds) Advances and applications in nonlinear control systems. Springer, Berlin, pp 317–350

144. Volos CK, Prousalis D, Kyprianidis IM, Stouboulos I, Vaidyanathan S, Pham VT (2016) Synchronization and anti-synchronization of coupled Hindmarsh-Rose neuron models. Int J Control Theory Appl 9(1):101–114
145. Volos CK, Vaidyanathan S, Pham VT, Maaita JO, Giakoumis A, Kyprianidis IM, Stouboulos IN (2016) A novel design approach of a nonlinear resistor based on a memristor emulator. In: Azar AT, Vaidyanathan S (eds) Advances in chaos theory and intelligent control. Springer, Berlin, pp 3–34
146. Wang B, Zhong SM, Dong XC (2016) On the novel chaotic secure communication scheme design. Commun Nonlinear Sci Numer Simul 39:108–117
147. Wolf A, Swift JB, Swinney HL, Vastano JA (1985) Determining Lyapunov exponents from a time series. Physica D 16:285–317
148. Wu T, Sun W, Zhang X, Zhang S (2016) Concealment of time delay signature of chaotic output in a slave semiconductor laser with chaos laser injection. Opt Commun 381:174–179
149. Xu G, Liu F, Xiu C, Sun L, Liu C (2016) Optimization of hysteretic chaotic neural network based on fuzzy sliding mode control. Neurocomputing 189:72–79
150. Xu H, Tong X, Meng X (2016) An efficient chaos pseudo-random number generator applied to video encryption. Optik 127(20):9305–9319
151. Zhou W, Xu Y, Lu H, Pan L (2008) On dynamics analysis of a new chaotic attractor. Phys Lett A 372(36):5773–5777
152. Zhu C, Liu Y, Guo Y (2010) Theoretic and numerical study of a new chaotic system. Intell Inf Manag 2:104–109

Control and Synchronization of a Novel Hyperchaotic Two-Disk Dynamo System via Adaptive Integral Sliding Mode Control

Sundarapandian Vaidyanathan

Abstract Chaos and hyperchaos have important applications in physics, chemistry, biology, ecology, secure communications, cryptosystems and many scientific branches. Control and synchronization of chaotic and hyperchaotic systems are important research problems in chaos theory. Sliding mode control is an important method used to solve various problems in control systems engineering. In robust control systems, the sliding mode control is often adopted due to its inherent advantages of easy realization, fast response and good transient performance as well as insensitivity to parameter uncertainties and disturbance. This work proposes a novel 4-D hyperchaotic two-disk dynamo system without any equilibrium point. Thus, the proposed novel hyperchaotic two-disk dynamo system exhibits hidden attractors. Also, we show that the Lyapunov exponents of the hyperchaotic two-disk dynamo system are $L_1 = 0.0922, L_2 = 0.0743, L_3 = 0$ and $L_4 = -2.1665$. The Kaplan-Yorke dimension of the hyperchaotic two-disk dynamo system is derived as $D_{KY} = 3.0769$, which shows the high complexity of the system. Next, an adaptive integral sliding mode control scheme is proposed to globally stabilize all the trajectories of the hyperchaotic two-disk dynamo system. Furthermore, an adaptive integral sliding mode control scheme is proposed for the global hyperchaos synchronization of identical hyperchaotic two-disk dynamo systems. The adaptive control mechanism helps the control design by estimating the unknown parameters. Numerical simulations using MATLAB are shown to illustrate all the main results derived in this work.

Keywords Chaos · Chaotic systems · Chaos control · Integral sliding mode control · Adaptive control · Dynamo system

S. Vaidyanathan (✉)
Research and Development Centre, Vel Tech University, Avadi,
Chennai 600062, Tamil Nadu, India
e-mail: sundarcontrol@gmail.com

© Springer International Publishing AG 2017
S. Vaidyanathan and C.-H. Lien (eds.), *Applications of Sliding Mode Control in Science and Engineering*, Studies in Computational Intelligence 709,
DOI 10.1007/978-3-319-55598-0_11

235

1 Introduction

Chaos theory describes the quantitative study of unstable aperiodic dynamic behavior in deterministic nonlinear dynamical systems. For the motion of a dynamical system to be chaotic, the system variables should contain some nonlinear terms and the system must satisfy three properties: boundedness, infinite recurrence and sensitive dependence on initial conditions [3–5, 124–126].

A hyperchaotic system is defined as a chaotic system with at least two positive Lyapunov exponents [3, 4, 124]. Combined with one null exponent along the flow and one negative exponent to ensure the boundedness of the solution, the minimal dimension for a continuous-time hyperchaotic system is four.

Some classical examples of hyperchaotic systems are hyperchaotic Rössler system [41], hyperchaotic Lorenz system [19], hyperchaotic Chen system [18], hyperchaotic Lü system [9], etc. Some recent examples of hyperchaotic systems are hyperchaotic Dadras system [12], hyperchaotic Vaidyanathan systems [57, 64, 91, 95–101, 107, 109, 110, 112, 113, 115, 127, 129, 130, 132, 135, 137], hyperchaotic Sampath system [43], hyperchaotic Pham system [37], etc.

The problem of global control of a chaotic system is to device feedback control laws so that the closed-loop system is globally asymptotically stable. Global chaos control of various chaotic systems has been investigated via various methods in the control literature [3, 4, 124, 125].

The synchronization of chaotic systems deals with the problem of synchronizing the states of two chaotic systems called as *master* and *slave* systems asymptotically with time. The design goal of the complete synchronization problem is to use the output of the master system to control of the output of the slave system so that the outputs of the two systems are synchronized asymptotically with time.

Because of the *butterfly effect* [3], which causes the exponential divergence of the trajectories of two identical chaotic systems started with nearly the same initial conditions, synchronizing two chaotic systems is seemingly a challenging research problem in the chaos literature.

The synchronization of chaotic systems was first researched by Yamada and Fujisaka [17] with subsequent work by Pecora and Carroll [27, 29]. In the last few decades, several different methods have been devised for the synchronization of chaotic and hyperchaotic systems [3–5, 124–126].

Many new chaotic systems have been discovered in the recent years such as Zhou system [149], Zhu system [150], Li system [23], Sundarapandian systems [46, 47], Vaidyanathan systems [56, 58–62, 65, 75, 92–94, 108, 111, 116–119, 121, 123, 128, 131, 133, 134, 136], Pehlivan system [30], Sampath system [42], Tacha system [50], Pham systems [32, 35, 36, 39], Akgul system [2], etc.

Chaos theory has applications in several fields such as memristors [31, 33, 34, 38–40, 138, 141, 143], fuzzy logic [6, 44, 114, 147], communication systems [14, 15, 144], cryptosystems [8, 10], electromechanical systems [16, 49], lasers [7, 24, 146], encryption [25, 26, 148], electrical circuits [1, 2, 20, 120, 139], chemical reactions [66, 67, 69, 70, 72, 74, 76–78, 81, 83, 87, 115], oscillators [84, 85, 88,

89, 140], tokamak systems [82, 90], neurology [71, 79, 80, 86, 103, 142], ecology [68, 73, 104, 106], etc.

The adaptive control mechanism helps the control design by estimating the unknown parameters [3, 4, 124]. The sliding mode control approach is recognized as an efficient tool for designing robust controllers for linear or nonlinear control systems operating under uncertainty conditions [51, 52].

A major advantage of sliding mode control is low sensitivity to parameter variations in the plant and disturbances affecting the plant, which eliminates the necessity of exact modeling of the plant. In the sliding mode control, the control dynamics will have two sequential modes, viz. the reaching mode and the sliding mode. Basically, a sliding mode controller design consists of two parts: hyperplane design and controller design. A hyperplane is first designed via the pole-placement approach and a controller is then designed based on the sliding condition. The stability of the overall system is guaranteed by the sliding condition and by a stable hyperplane. Sliding mode control method is a popular method for the control and synchronization of chaotic systems [22, 28, 48, 53–55, 63, 102, 105, 122].

In this research work, we propose a novel 4-D hyperchaotic two-disk dynamo system without any equilibrium point. Thus, the proposed novel hyperchaotic two-disk dynamo system exhibits hidden attractors. Also, we show that the Lyapunov exponents of the hyperchaotic two-disk dynamo system are $L_1 = 0.0922$, $L_2 = 0.0743$, $L_3 = 0$ and $L_4 = -2.1665$. The Kaplan-Yorke dimension of the hyperchaotic two-disk dynamo system is derived as $D_{KY} = 3.0769$, which shows the high complexity of the system. We show that the novel hyperchaotic two-disk dynamo system has rotation symmetry about the x_3-axis. Thus, every non-trivial trajectory of the novel hyperchaotic two-disk dynamo system must have a twin trajectory. We also that the novel hyperchaotic two-disk dynamo system has no equilibrium point. This shows that the novel hyperchaotic two-disk dynamo system exhibits hidden attractors [13].

Next, an adaptive integral sliding mode control scheme is proposed to globally stabilize all the trajectories of the hyperchaotic two-disk dynamo system. Furthermore, an adaptive integral sliding mode control scheme is proposed for the global hyperchaos synchronization of identical hyperchaotic two-disk dynamo systems.

This work is organized as follows. Section 2 describes a 3-D chaotic two-disk dynamo system [11] and its qualitative properties. Section 3 describes a novel 4-D hyperchaotic two-disk dynamo system and describes its phase portraits. Section 4 details the properties of the novel 4-D hyperchaotic two-disk dynamo system. Section 5 contains new results on the adaptive integral sliding mode controller design for the global stabilization of the novel 4-D hyperchaotic two-disk dynamo system. Section 6 contains new results on the adaptive integral sliding mode controller design for the global synchronization of the novel identical 4-D hyperchaotic two-disk dynamo systems. Section 7 contains the conclusions of this work.

2 Two-Disk Dynamo System

In this section, we describe the two-disk dynamo chaotic system obtained by Cook [11].

A system of coupled dynamos can be described by the dynamics

$$
\begin{aligned}
\dot{x}_1 &= -a_1 x_1 + \omega_1 x_2 \\
\dot{x}_2 &= -a_2 x_2 + \omega_2 x_1 \\
\dot{\omega}_1 &= q_1 - \epsilon_1 \omega_1 - x_1 x_2 \\
\dot{\omega}_2 &= q_2 - \epsilon_2 \omega_2 - x_1 x_2
\end{aligned}
\tag{1}
$$

where x_1, x_2 represent the currents and ω_1, ω_2 are the angular velocities of the rotors of the two dynamos. Also, q_1, q_2 are the torques applied to the rotors and $a_1, a_2, \epsilon_1, \epsilon_2$ are positive constants representing the dissipative efforts.

After some simplifications, the fourth-order model (1) of the coupled dynamos can be represented as the third-order system

$$
\begin{aligned}
\dot{x}_1 &= -ax_1 + (x_3 + b)x_2 \\
\dot{x}_2 &= -ax_2 + (x_3 - b)x_1 \\
\dot{x}_3 &= 1 - x_1 x_2
\end{aligned}
\tag{2}
$$

where a and b are positive parameters.

Cook [11] observed that the two-disk dynamo system (2) is *chaotic* when the system parameters take the values

$$
a = 1, \quad b = 1.875
\tag{3}
$$

For numerical simulations, we take the initial state of the two-disk dynamo system (2) as

$$
x_1(0) = 1, \quad x_2(0) = 1, \quad x_3(0) = 1
\tag{4}
$$

For the parameter values (3) and the initial state (4), the Lyapunov exponents of the two-disk dynamo system (2) are calculated by Wolf's algorithm [145] as

$$
L_1 = 0.1877, \quad L_2 = 0, \quad L_3 = -2.1877
\tag{5}
$$

This shows that the two-disk dynamo system (2) is chaotic and dissipative.

Also, the Kaplan-Yorke dimension of the two-disk dynamo system (2) is determined as

$$
D_{KY} = 2 + \frac{L_1 + L_2}{|L_3|} = 2.0858,
\tag{6}
$$

which is fractional.

It is easy to see that the two-disk dynamo system (2) is invariant under the coordinates transformation

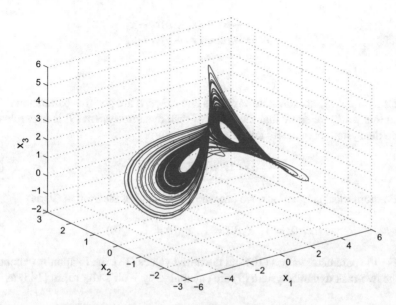

Fig. 1 3-D phase portrait of the two-disk dynamo chaotic system

$$(x_1, x_2, x_3) \mapsto (-x_1, -x_2, x_3) \tag{7}$$

Thus, the two-disk dynamo system (2) has rotation symmetry about the x_3-axis. Hence, it follows that any non-trivial trajectory of the two-disk dynamo system (2) must have a twin-trajectory.

For the parameter values (3), the two-disk dynamo system (2) has two equilibrium points given by

$$E_1 = \begin{bmatrix} 2.0000 \\ 0.5000 \\ 2.1250 \end{bmatrix}, \ E_2 = \begin{bmatrix} -2.0000 \\ -0.5000 \\ 2.1250 \end{bmatrix} \tag{8}$$

It is easy to verify that both equilibrium points E_1 and E_2 are marginally stable.

Figure 1 shows the 3-D phase portrait of the two-disk dynamo chaotic system (2). It is clear that the two-disk dynamo system exhibits a *two-scroll* chaotic attractor.

3 Hyperchaotic Two-Disk Dynamo System

In this section, we derive a new 4-D hyperchaotic two-disk dynamo system by adding a feedback control to the two-disk dynamo chaotic system (2).

Thus, our 4-D novel two-disk dynamo system is given by the dynamics

240 S. Vaidyanathan

$$
\begin{aligned}
\dot{x}_1 &= -ax_1 + (x_3 + b)x_2 - x_4 \\
\dot{x}_2 &= -ax_2 + (x_3 - b)x_1 - x_4 \\
\dot{x}_3 &= 1 - x_1 x_2 \\
\dot{x}_4 &= cx_2
\end{aligned}
\tag{9}
$$

where x_1, x_2, x_3, x_4 are the state variables and a, b, c, d are positive parameters.

In this work, we show that the 4-D two-disk dynamo system (9) is *hyperchaotic* when the system parameters take the values

$$
a = 1, \quad b = 2, \quad c = 0.7
\tag{10}
$$

For numerical simulations, we take the initial state of the system (9) as

$$
x_1(0) = 1, \quad x_2(0) = 1, \quad x_3(0) = 1, \quad x_4(0) = 1
\tag{11}
$$

For the parameter values (10) and the initial values (11), the Lyapunov exponents of the two-disk dynamo system (9) are calculated by Wolf's algorithm [145] as

$$
L_1 = 0.0922, \quad L_2 = 0.0743, \quad L_3 = 0, \quad L_4 = -2.1665
\tag{12}
$$

Since there are two positive Lyapunov exponents in the LE spectrum (12), it is immediate that the 4-D novel two-disk dynamo system (9) is hyperchaotic.

Also, the Kaplan-Yorke dimension of the new 4-D hyperchaotic two-disk dynamo system (9) is obtained as

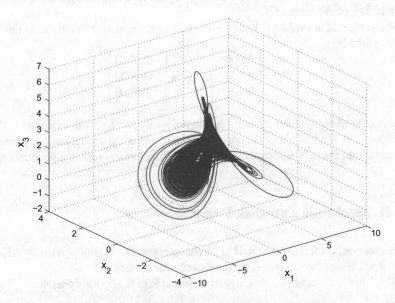

Fig. 2 3-D phase portrait of the new hyperchaotic system in (x_1, x_2, x_3) space

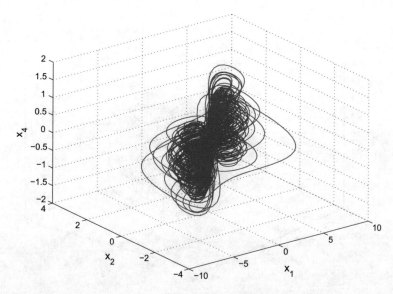

Fig. 3 3-D phase portrait of the new hyperchaotic system in (x_1, x_2, x_4) space

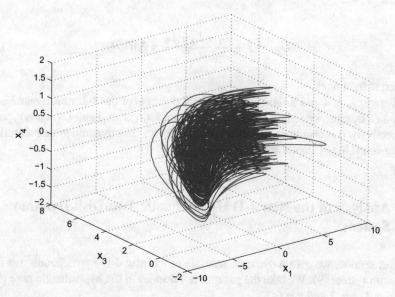

Fig. 4 3-D phase portrait of the new hyperchaotic system in (x_1, x_3, x_4) space

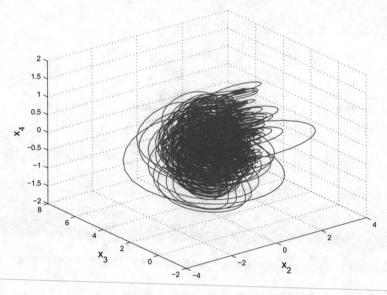

Fig. 5 3-D phase portrait of the new hyperchaotic system in (x_2, x_3, x_4) space

$$D_{KY} = 3 + \frac{L_1 + L_2 + L_3}{|L_4|} = 3.0769, \tag{13}$$

which is fractional.

Figures 2, 3, 4 and 5 show the 3-D phase portraits of the 4-D new hyperchaotic two-disk dynamo system in (x_1, x_2, x_3), (x_1, x_2, x_4), (x_1, x_3, x_4) and (x_2, x_3, x_4) spaces respectively. It is clear that the new hyperchaotic two-disk dynamo system describes a *two-scroll* hyperchaotic attractor.

4 Analysis of the New 4-D Hyperchaotic Two-Disk Dynamo System

In this section, we give a dynamic analysis of the new 4-D hyperchaotic two-disk dynamo system (9). We take the parameter values as in the hyperchaotic case (10), i.e. $a = 1$, $b = 2$ and $c = 0.7$.

4.1 Dissipativity

In vector notation, the new hyperchaotic two-disk dynamo system (9) can be expressed as

$$\dot{\mathbf{x}} = f(\mathbf{x}) = \begin{bmatrix} f_1(x_1, x_2, x_3, x_4) \\ f_2(x_1, x_2, x_3, x_4) \\ f_3(x_1, x_2, x_3, x_4) \\ f_4(x_1, x_2, x_3, x_4) \end{bmatrix}, \tag{14}$$

where

$$\begin{cases} f_1(x_1, x_2, x_3, x_4) = -ax_1 + (x_3 + b)x_2 - x_4 \\ f_2(x_1, x_2, x_3, x_4) = -ax_2 + (x_3 - b)x_1 - x_4 \\ f_3(x_1, x_2, x_3, x_4) = 1 - x_1 x_2 \\ f_4(x_1, x_2, x_3, x_4) = cx_2 \end{cases} \tag{15}$$

Let Ω be any region in $\mathbf{R}^4$ with a smooth boundary and also, $\Omega(t) = \Phi_t(\Omega)$, where Φ_t is the flow of f. Furthermore, let $V(t)$ denote the hypervolume of $\Omega(t)$.

By Liouville's theorem, we know that

$$\dot{V}(t) = \int_{\Omega(t)} (\nabla \cdot f) \, dx_1 \, dx_2 \, dx_3 \, dx_4 \tag{16}$$

The divergence of the hyperchaotic system (14) is found as:

$$\nabla \cdot f = \frac{\partial f_1}{\partial x_1} + \frac{\partial f_2}{\partial x_2} + \frac{\partial f_3}{\partial x_3} + \frac{\partial f_4}{\partial x_4} = -(a + a) = -2a < 0 \tag{17}$$

Inserting the value of $\nabla \cdot f$ from (17) into (16), we get

$$\dot{V}(t) = \int_{\Omega(t)} (-2a) \, dx_1 \, dx_2 \, dx_3 \, dx_4 = -2aV(t) \tag{18}$$

Integrating the first order linear differential equation (18), we get

$$V(t) = \exp(-2at)V(0) \tag{19}$$

Since $a > 0$, it follows from Eq. (19) that $V(t) \to 0$ exponentially as $t \to \infty$. This shows that the new hyperchaotic two-disk dynamo system (9) is dissipative. Hence, the system limit sets are ultimately confined into a specific limit set of zero hyper-volume, and the asymptotic motion of the new hyperchaotic system (9) settles onto a strange attractor of the system.

4.2 Equilibrium Points

We take the parameter values as in the hyperchaotic case (10).

The equilibrium points of the new hyperchaotic two-disk dynamo system (9) are obtained by solving the following system of equations.

$$-ax_1 + (x_3 + b)x_2 - x_4 = 0 \tag{20a}$$
$$-ax_2 + (x_3 - b)x_1 - x_4 = 0 \tag{20b}$$
$$1 - x_1 x_2 = 0 \tag{20c}$$
$$cx_2 = 0 \tag{20d}$$

From (20d), it is immediate that $x_2 = 0$. Substituting $x_2 = 0$ in (20c), we obtain a contradiction.

This shows that the system of equations (20) has no solution.

In other words, the new hyperchaotic two-disk dynamo system (9) has no equilibrium point. Hence, we deduce that the new hyperchaotic two-disk dynamo system (9) exhibits hidden attractors [13].

4.3 Rotation Symmetry About the x_3-axis

It is easy to see that the new 4-D hyperchaotic two-disk dynamo system (9) is invariant under the change of coordinates

$$(x_1, x_2, x_3, x_4) \mapsto (-x_1, -x_2, x_3, -x_4) \tag{21}$$

Since the transformation (21) persists for all values of the system parameters, it follows that the new 4-D hyperchaotic two-disk dynamo system (9) has rotation symmetry about the x_3-axis and that any non-trivial trajectory must have a twin trajectory.

4.4 Invariance

It is easy to see that the x_3-axis is invariant under the flow of the 4-D novel hyperchaotic system (9).

The invariant motion along the x_3-axis is characterized by the scalar dynamics

$$\dot{x}_3 = 1, \tag{22}$$

which is unstable.

4.5 Lyapunov Exponents and Kaplan-Yorke Dimension

We take the parameter values of the new hyperchaotic two-disk dynamo system (9) as in the hyperchaotic case (10), i.e.

$$a = 1, \quad b = 2, \quad c = 0.7 \tag{23}$$

We take the initial state of the new hyperchaotic two-disk dynamo system (9) as (11), i.e.

$$x_1(0) = 1, \quad x_2(0) = 1, \quad x_3(0) = 1, \quad x_4(0) = 1 \tag{24}$$

Then the Lyapunov exponents of the system (9) are numerically obtained using MATLAB as

$$L_1 = 0.0922, \quad L_2 = 0.0743, \quad L_3 = 0, \quad L_4 = -2.1665 \tag{25}$$

Since there are two positive Lyapunov exponents in (25), the new 4-D two-disk dynamo system (9) exhibits *hyperchaotic* behavior.

The maximal Lyapunov exponent (MLE) of the new hyperchaotic two-disk dynamo system (9) is obtained as $L_1 = 0.0922$.

Since $L_1 + L_2 + L_3 + L_4 = -2 < 0$, it follows that the new hyperchaotic two-disk dynamo system (9) is dissipative.

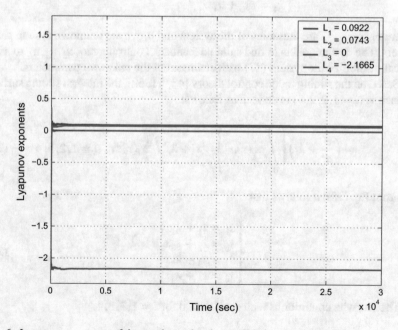

Fig. 6 Lyapunov exponents of the new hyperchaotic two-disk dynamo system

Also, the Kaplan-Yorke dimension of the new hyperchaotic two-disk dynamo system (9) is calculated as

$$D_{KY} = 3 + \frac{L_1 + L_2 + L_3}{|L_4|} = 3.0769 \tag{26}$$

Figure 6 shows the Lyapunov exponents of the new hyperchaotic two-disk dynamo system (9).

5   Global Hyperchaos Control of the New Hyperchaotic Two-Wing Dynamo System

In this section, we use adaptive integral sliding mode control for the global hyperchaos control of new hyperchaotic two-wing dynamo system with unknown system parameters. The adaptive control mechanism helps the control design by estimating the unknown parameters [3, 4, 124].

The controlled new hyperchaotic two-wing dynamo system is described by

$$\begin{cases} \dot{x}_1 = -ax_1 + (x_3 + b)x_2 - x_4 + u_1 \\ \dot{x}_2 = -ax_2 + (x_3 - b)x_1 - x_4 + u_2 \\ \dot{x}_3 = 1 - x_1 x_2 + u_3 \\ \dot{x}_4 = cx_2 + u_4 \end{cases} \tag{27}$$

where x_1, x_2, x_3, x_4 are the states of the system and a, b, c are unknown system parameters. The design goal is to find suitable feedback controllers u_1, u_2, u_3, u_4 so as to globally stabilize the system (27) with estimates of the unknown parameters.

Based on the sliding mode control theory [45, 51, 52], the integral sliding surface of each x_i ($i = 1, 2, 3, 4$) is defined as follows:

$$s_i = \left(\frac{d}{dt} + \lambda_i \right) \left(\int_0^t x_i(\tau) d\tau \right) = x_i + \lambda_i \int_0^t x_i(\tau) d\tau, \quad i = 1, 2, 3, 4 \tag{28}$$

From Eq. (28), it follows that

$$\begin{cases} \dot{s}_1 = \dot{x}_1 + \lambda_1 x_1 \\ \dot{s}_2 = \dot{x}_2 + \lambda_2 x_2 \\ \dot{s}_3 = \dot{x}_3 + \lambda_3 x_3 \\ \dot{s}_4 = \dot{x}_4 + \lambda_4 x_4 \end{cases} \tag{29}$$

The Hurwitz condition is realized if $\lambda_i > 0$ for $i = 1, 2, 3, 4$.

We consider the adaptive feedback control given by

$$
\begin{cases}
u_1 = \hat{a}(t)x_1 - (x_3 + \hat{b}(t))x_2 + x_4 - \lambda_1 x_1 - \eta_1 \operatorname{sgn}(s_1) - k_1 s_1 \\
u_2 = \hat{a}(t)x_2 - (x_3 - \hat{b}(t))x_1 + x_4 - \lambda_2 x_2 - \eta_2 \operatorname{sgn}(s_2) - k_2 s_2 \\
u_3 = -1 + x_1 x_2 - \lambda_3 x_3 - \eta_3 \operatorname{sgn}(s_3) - k_3 s_3 \\
u_4 = -\hat{c}(t)x_2 - \lambda_4 x_4 - \eta_4 \operatorname{sgn}(s_4) - k_4 s_4
\end{cases}
\tag{30}
$$

where $\eta_i > 0$ and $k_i > 0$ for $i = 1, 2, 3, 4$.

Substituting (30) into (27), we obtain the closed-loop control system given by

$$
\begin{cases}
\dot{x}_1 = -[a - \hat{a}(t)]x_1 - [b - \hat{b}(t)]x_2 - \lambda_1 x_1 - \eta_1 \operatorname{sgn}(s_1) - k_1 s_1 \\
\dot{x}_2 = -[a - \hat{a}(t)]x_2 - [b - \hat{b}(t)]x_1 - \lambda_2 x_2 - \eta_2 \operatorname{sgn}(s_2) - k_2 s_2 \\
\dot{x}_3 = -\lambda_3 x_3 - \eta_3 \operatorname{sgn}(s_3) - k_3 s_3 \\
\dot{x}_4 = [c - \hat{c}(t)]x_2 - \lambda_4 x_4 - \eta_4 \operatorname{sgn}(s_4) - k_4 s_4
\end{cases}
\tag{31}
$$

We define the parameter estimation errors as

$$
\begin{cases}
e_a(t) = a - \hat{a}(t) \\
e_b(t) = b - \hat{b}(t) \\
e_c(t) = c - \hat{c}(t)
\end{cases}
\tag{32}
$$

Using (32), we can simplify the closed-loop system (31) as

$$
\begin{cases}
\dot{x}_1 = -e_a x_1 - e_b x_2 - \lambda_1 x_1 - \eta_1 \operatorname{sgn}(s_1) - k_1 s_1 \\
\dot{x}_2 = -e_a x_2 - e_b x_1 - \lambda_2 x_2 - \eta_2 \operatorname{sgn}(s_2) - k_2 s_2 \\
\dot{x}_3 = -\lambda_3 x_3 - \eta_3 \operatorname{sgn}(s_3) - k_3 s_3 \\
\dot{x}_4 = e_c x_2 - \lambda_4 x_4 - \eta_4 \operatorname{sgn}(s_4) - k_4 s_4
\end{cases}
\tag{33}
$$

Differentiating (32) with respect to t, we get

$$
\begin{cases}
\dot{e}_a = -\dot{\hat{a}} \\
\dot{e}_b = -\dot{\hat{b}} \\
\dot{e}_c = -\dot{\hat{c}}
\end{cases}
\tag{34}
$$

Next, we state and prove the main result of this section.

Theorem 1 *The new hyperchaotic two-wing dynamo system (27) is rendered globally asymptotically stable for all initial conditions $\mathbf{x}(0) \in \mathbf{R}^4$ by the adaptive integral sliding mode control law (30) and the parameter update law*

$$\begin{cases} \dot{\hat{a}} = -s_1 x_1 - s_2 x_2 \\ \dot{\hat{b}} = -s_1 x_2 - s_2 x_1 \\ \dot{\hat{c}} = s_4 x_2 \end{cases} \tag{35}$$

where λ_i, η_i, k_i are positive constants for $i = 1, 2, 3, 4$.

Proof We consider the quadratic Lyapunov function defined by

$$V(s_1, s_2, s_3, s_4, e_a, e_b, e_c) = \frac{1}{2} \left(s_1^2 + s_2^2 + s_3^2 + s_4^2 \right) + \frac{1}{2} \left(e_a^2 + e_b^2 + e_c^2 \right) \tag{36}$$

Clearly, V is positive definite on $\mathbf{R}^7$.
Using (29), (33) and (34), the time-derivative of V is obtained as

$$\dot{V} = -\eta_1 |s_1| - k_1 s_1^2 - \eta_2 |s_2| - k_2 s_2^2 - \eta_3 |s_3| - k_3 s_3^2 - \eta_4 |s_4| - k_4 s_4^2 \\ + e_a \left(-s_1 x_1 - s_2 x_2 - \dot{\hat{a}} \right) + e_b \left(-s_1 x_2 - s_2 x_1 - \dot{\hat{b}} \right) + e_c \left(s_4 x_2 - \dot{\hat{c}} \right) \tag{37}$$

Using the parameter update law (35), we obtain

$$\dot{V} = -\eta_1 |s_1| - k_1 s_1^2 - \eta_2 |s_2| - k_2 s_2^2 - \eta_3 |s_3| - k_3 s_3^2 - \eta_4 |s_4| - k_4 s_4^2 \tag{38}$$

which shows that $\dot{V}$ is negative semi-definite on $\mathbf{R}^7$.

Hence, by Barbalat's lemma [21], it is immediate that $\mathbf{x}(t)$ is globally asymptotically stable for all values of $\mathbf{x}(0) \in \mathbf{R}^4$.

This completes the proof. ∎

For numerical simulations, we take the parameter values of the new hyperchaotic two-disk dynamo system (27) as in the hyperchaotic case (10), i.e. $a = 1$, $b = 2$ and $c = 0.7$.

We take the values of the control parameters as

$$k_i = 20, \quad \eta_i = 0.1, \quad \lambda_i = 50, \quad \text{where } i = 1, 2, 3, 4 \tag{39}$$

We take the estimates of the system parameters as

$$\hat{a}(0) = 10.5, \quad \hat{b}(0) = 7.3, \quad \hat{c}(0) = 8.2 \tag{40}$$

We take the initial state of the hyperchaotic two-disk dynamo system (27) as

$$x_1(0) = 5.9, \quad x_2(0) = 16.4, \quad x_3(0) = 12.7, \quad x_4(0) = 22.6 \tag{41}$$

Figure 7 shows the time-history of the controlled states x_1, x_2, x_3, x_4.

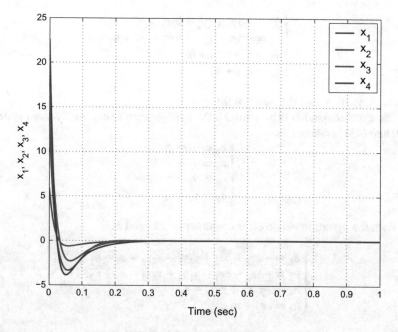

Fig. 7 Time-history of the controlled states x_1, x_2, x_3, x_4

6 Global Hyperchaos Synchronization of the New Hyperchaotic Two-Wing Dynamo Systems

In this section, we use adaptive integral sliding mode control for the global hyperchaos synchronization of new hyperchaotic two-wing dynamo systems with unknown system parameters. The adaptive control mechanism helps the control design by estimating the unknown parameters [3, 4, 124].

As the master system, we consider the new hyperchaotic two-wing dynamo system given by

$$\begin{cases} \dot{x}_1 = -ax_1 + (x_3 + b)x_2 - x_4 \\ \dot{x}_2 = -ax_2 + (x_3 - b)x_1 - x_4 \\ \dot{x}_3 = 1 - x_1 x_2 \\ \dot{x}_4 = cx_2 \end{cases} \tag{42}$$

where x_1, x_2, x_3, x_4 are the states of the system and a, b, c are unknown system parameters.

As the slave system, we consider the controlled new hyperchaotic two-wing dynamo system given by

$$\begin{cases} \dot{y}_1 = -ay_1 + (y_3 + b)y_2 - y_4 + u_1 \\ \dot{y}_2 = -ay_2 + (y_3 - b)y_1 - y_4 + u_2 \\ \dot{y}_3 = 1 - y_1 y_2 + u_3 \\ \dot{y}_4 = cy_2 + u_4 \end{cases} \tag{43}$$

where y_1, y_2, y_3, y_4 are the states of the system.

The synchronization error between the hyperchaotic two-wing dynamo systems (42) and (43) is defined as

$$\begin{cases} e_1 = y_1 - x_1 \\ e_2 = y_2 - x_2 \\ e_3 = y_3 - x_3 \\ e_4 = y_4 - x_4 \end{cases} \tag{44}$$

Then the synchronization error dynamics is obtained as

$$\begin{cases} \dot{e}_1 = -ae_1 + be_2 + y_2 y_3 - x_2 x_3 - e_4 + u_1 \\ \dot{e}_2 = -ae_2 - be_1 - y_1 y_3 + x_1 x_3 - e_4 + u_2 \\ \dot{e}_3 = -y_1 y_2 + x_1 x_2 + u_3 \\ \dot{e}_4 = ce_2 + u_4 \end{cases} \tag{45}$$

Based on the sliding mode control theory [45, 51, 52], the integral sliding surface of each e_i ($i = 1, 2, 3, 4$) is defined as follows:

$$s_i = \left(\frac{d}{dt} + \lambda_i \right) \left(\int_0^t e_i(\tau) d\tau \right) = e_i + \lambda_i \int_0^t e_i(\tau) d\tau, \quad i = 1, 2, 3, 4 \tag{46}$$

From Eq. (46), it follows that

$$\begin{cases} \dot{s}_1 = \dot{e}_1 + \lambda_1 e_1 \\ \dot{s}_2 = \dot{e}_2 + \lambda_2 e_2 \\ \dot{s}_3 = \dot{e}_3 + \lambda_3 e_3 \\ \dot{s}_4 = \dot{e}_4 + \lambda_4 e_4 \end{cases} \tag{47}$$

The Hurwitz condition is realized if $\lambda_i > 0$ for $i = 1, 2, 3, 4$.

We consider the adaptive feedback control given by

$$\begin{cases} u_1 = \hat{a}(t)e_1 - \hat{b}(t)e_2 - y_2 y_3 + x_2 x_3 + e_4 - \lambda_1 e_1 - \eta_1 \, \text{sgn}(s_1) - k_1 s_1 \\ u_2 = \hat{a}(t)e_2 + \hat{b}(t)e_1 + y_1 y_3 - x_1 x_3 + e_4 - \lambda_2 e_2 - \eta_2 \, \text{sgn}(s_2) - k_2 s_2 \\ u_3 = y_1 y_2 - x_1 x_2 - \lambda_3 e_3 - \eta_3 \, \text{sgn}(s_3) - k_3 s_3 \\ u_4 = -\hat{c}(t)e_2 - \lambda_4 e_4 - \eta_4 \, \text{sgn}(s_4) - k_4 s_4 \end{cases} \tag{48}$$

where $\eta_i > 0$ and $k_i > 0$ for $i = 1, 2, 3, 4$.

Substituting (48) into (45), we obtain the closed-loop error dynamics as

$$\begin{cases} \dot{e}_1 = -[a - \hat{a}(t)]e_1 - [b - \hat{b}(t)]e_2 - \lambda_1 e_1 - \eta_1 \, \mathrm{sgn}(s_1) - k_1 s_1 \\ \dot{e}_2 = -[a - \hat{a}(t)]e_2 - [b - \hat{b}(t)]e_1 - \lambda_2 e_2 - \eta_2 \, \mathrm{sgn}(s_2) - k_2 s_2 \\ \dot{e}_3 = -\lambda_3 e_3 - \eta_3 \, \mathrm{sgn}(s_3) - k_3 s_3 \\ \dot{e}_4 = [c - \hat{c}(t)]e_2 - \lambda_4 e_4 - \eta_4 \, \mathrm{sgn}(s_4) - k_4 s_4 \end{cases} \tag{49}$$

We define the parameter estimation errors as

$$\begin{cases} e_a(t) = a - \hat{a}(t) \\ e_b(t) = b - \hat{b}(t) \\ e_c(t) = c - \hat{c}(t) \end{cases} \tag{50}$$

Using (50), we can simplify the closed-loop system (49) as

$$\begin{cases} \dot{e}_1 = -e_a e_1 - e_b e_2 - \lambda_1 e_1 - \eta_1 \, \mathrm{sgn}(s_1) - k_1 s_1 \\ \dot{e}_2 = -e_a e_2 - e_b e_1 - \lambda_2 e_2 - \eta_2 \, \mathrm{sgn}(s_2) - k_2 s_2 \\ \dot{e}_3 = -\lambda_3 e_3 - \eta_3 \, \mathrm{sgn}(s_3) - k_3 s_3 \\ \dot{e}_4 = e_c e_2 - \lambda_4 e_4 - \eta_4 \, \mathrm{sgn}(s_4) - k_4 s_4 \end{cases} \tag{51}$$

Differentiating (50) with respect to t, we get

$$\begin{cases} \dot{e}_a = -\dot{\hat{a}} \\ \dot{e}_b = -\dot{\hat{b}} \\ \dot{e}_c = -\dot{\hat{c}} \end{cases} \tag{52}$$

Next, we state and prove the main result of this section.

Theorem 2 *The new hyperchaotic two-wing dynamo systems (42) and (43) are globally and asymptotically synchronized for all initial conditions* $\mathbf{x}(0), \mathbf{y}(0) \in \mathbf{R}^4$ *by the adaptive integral sliding mode control law (48) and the parameter update law*

$$\begin{cases} \dot{\hat{a}} = -s_1 e_1 - s_2 e_2 \\ \dot{\hat{b}} = -s_1 e_2 - s_2 e_1 \\ \dot{\hat{c}} = s_4 e_2 \end{cases} \tag{53}$$

where λ_i, η_i, k_i *are positive constants for* $i = 1, 2, 3, 4$.

Proof We consider the quadratic Lyapunov function defined by

$$V(s_1, s_2, s_3, s_4, e_a, e_b, e_c) = \frac{1}{2}\left(s_1^2 + s_2^2 + s_3^2 + s_4^2\right) + \frac{1}{2}\left(e_a^2 + e_b^2 + e_c^2\right) \tag{54}$$

Clearly, V is positive definite on $\mathbf{R}^7$.
Using (47), (51) and (52), the time-derivative of V is obtained as

$$\dot{V} = -\eta_1|s_1| - k_1 s_1^2 - \eta_2|s_2| - k_2 s_2^2 - \eta_3|s_3| - k_3 s_3^2 - \eta_4|s_4| - k_4 s_4^2$$
$$+ e_a \left(-s_1 e_1 - s_2 e_2 - \dot{\hat{a}}\right) + e_b \left(-s_1 e_2 - s_2 e_1 - \dot{\hat{b}}\right) + e_c \left(s_4 e_2 - \dot{\hat{c}}\right) \tag{55}$$

Using the parameter update law (53), we obtain

$$\dot{V} = -\eta_1|s_1| - k_1 s_1^2 - \eta_2|s_2| - k_2 s_2^2 - \eta_3|s_3| - k_3 s_3^2 - \eta_4|s_4| - k_4 s_4^2 \tag{56}$$

which shows that $\dot{V}$ is negative semi-definite on $\mathbf{R}^7$.

Hence, by Barbalat's lemma [21], it is immediate that $\mathbf{e}(t)$ is globally asymptotically stable for all values of $\mathbf{e}(0) \in \mathbf{R}^4$.

Hence, it follows that the new hyperchaotic two-wing dynamo systems (42) and (43) are globally and asymptotically synchronized for all initial conditions $\mathbf{x}(0), \mathbf{y}(0) \in \mathbf{R}^4$.

This completes the proof.   ∎

For numerical simulations, we take the parameter values of the new hyperchaotic two-disk dynamo systems (42) and (43) as in the hyperchaotic case (10), i.e.

$$a = 1, \quad b = 2, \quad c = 0.7 \tag{57}$$

We take the values of the control parameters as

$$k_i = 20, \quad \eta_i = 0.1, \quad \lambda_i = 50, \quad \text{where } i = 1, 2, 3, 4 \tag{58}$$

We take the estimates of the system parameters as

$$\hat{a}(0) = 8.2, \quad \hat{b}(0) = 12.4, \quad \hat{c}(0) = 6.5 \tag{59}$$

We take the initial state of the master system (42) as

$$x_1(0) = 15.6, \quad x_2(0) = 4.7, \quad x_3(0) = 7.3, \quad x_4(0) = 29.4 \tag{60}$$

We take the initial state of the slave system (43) as

$$y_1(0) = 4.2, \quad y_2(0) = 22.4, \quad y_3(0) = 19.1, \quad y_4(0) = 11.3 \tag{61}$$

Figures 8, 9, 10 and 11 show the complete synchronization of the new hyperchaotic systems (42) and (43). Figure 12 shows the time-history of the synchronization errors e_1, e_2, e_3, e_4.

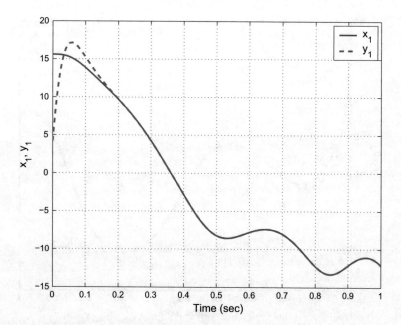

Fig. 8 Complete synchronization of the states x_1 and y_1

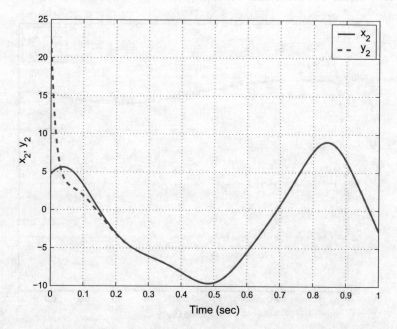

Fig. 9 Complete synchronization of the states x_2 and y_2

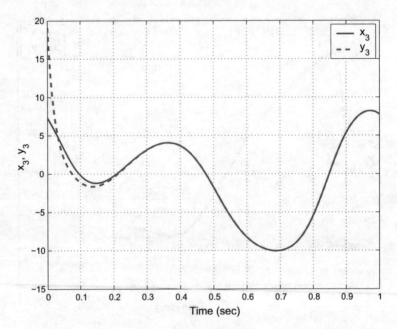

Fig. 10 Complete synchronization of the states x_3 and y_3

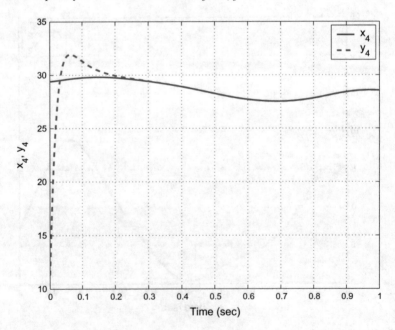

Fig. 11 Complete synchronization of the states x_4 and y_4

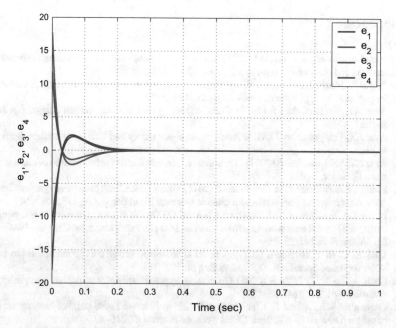

Fig. 12 Time-history of the synchronization errors e_1, e_2, e_3, e_4

7 Conclusions

Control and synchronization of chaotic and hyperchaotic systems are miscellaneous research problems in chaos theory. Sliding mode control is an important method used to solve various problems in control systems engineering. In robust control systems, the sliding mode control is often adopted due to its inherent advantages of easy realization, fast response and good transient performance as well as insensitivity to parameter uncertainties and disturbance. In this work, we described a new 4-D hyperchaotic two-disk dynamo system without any equilibrium point. Also, we showed that the Lyapunov exponents of the hyperchaotic two-disk dynamo system are $L_1 = 0.0922, L_2 = 0.0743, L_3 = 0$ and $L_4 = -2.1665$. The Kaplan-Yorke dimension of the hyperchaotic two-disk dynamo system has been derived as $D_{KY} = 3.0769$, which shows the high complexity of the system. Next, an adaptive integral sliding mode control scheme was derived to globally stabilize all the trajectories of the hyperchaotic two-disk dynamo system. Furthermore, an adaptive integral sliding mode control scheme was derived for the global hyperchaos synchronization of identical hyperchaotic two-disk dynamo systems. The adaptive control mechanism is useful in the control design by estimating the unknown parameters. Numerical simulations using MATLAB were exhibited to demonstrate all the main results derived in this work.

References

1. Akgul A, Hussain S, Pehlivan I (2016) A new three-dimensional chaotic system, its dynamical analysis and electronic circuit applications. Optik 127(18):7062–7071
2. Akgul A, Moroz I, Pehlivan I, Vaidyanathan S (2016) A new four-scroll chaotic attractor and its engineering applications. Optik 127(13):5491–5499
3. Azar AT, Vaidyanathan S (2015) Chaos modeling and control systems design. Springer, Berlin
4. Azar AT, Vaidyanathan S (2016) Advances in chaos theory and intelligent control. Springer, Berlin
5. Azar AT, Vaidyanathan S (2017) Fractional order control and synchronization of chaotic systems. Springer, Berlin
6. Boulkroune A, Bouzeriba A, Bouden T (2016) Fuzzy generalized projective synchronization of incommensurate fractional-order chaotic systems. Neurocomputing 173:606–614
7. Burov DA, Evstigneev NM, Magnitskii NA (2017) On the chaotic dynamics in two coupled partial differential equations for evolution of surface plasmon polaritons. Commun Nonlinear Sci Numer Simul 46:26–36
8. Chai X, Chen Y, Broyde L (2017) A novel chaos-based image encryption algorithm using DNA sequence operations. Opt Lasers Eng 88:197–213
9. Chen A, Lu J, Lü J, Yu S (2006) Generating hyperchaotic Lü attractor via state feedback control. Physica A 364:103–110
10. Chenaghlu MA, Jamali S, Khasmakhi NN (2016) A novel keyed parallel hashing scheme based on a new chaotic system. Chaos Solitons Fractals 87:216–225
11. Cook PA (1986) Nonlinear dynamical systems. Prentice-Hall, Englewood Cliffs, NJ
12. Dadras S, Momeni HR, Qi G, Lin Wang Z (2012) Four-wing hyperchaotic attractor generated from a new 4D system with one equilibrium and its fractional-order form. Nonlinear Dyn 67:1161–1173
13. Dudkowski D, Jafari S, Kapitaniak T, Kuznetsov NV, Leonov GA, Prasad A (2016) Hidden attractors in dynamical systems. Phys Rep 637:1–50
14. Fallahi K, Leung H (2010) A chaos secure communication scheme based on multiplication modulation. Commun Nonlinear Sci Numer Simul 15(2):368–383
15. Fontes RT, Eisencraft M (2016) A digital bandlimited chaos-based communication system. Commun Nonlinear Sci Numer Simul 37:374–385
16. Fotsa RT, Woafo P (2016) Chaos in a new bistable rotating electromechanical system. Chaos Solitons Fractals 93:48–57
17. Fujisaka H, Yamada T (1983) Stability theory of synchronized motion in coupled-oscillator systems. Progr Theor Phys 63:32–47
18. Gao T, Chen Z, Yuan Z, Chen G (2006) A hyperchaos generated from Chen's system. Int J Modern Phys C 17(4):471–478
19. Jia Q (2007) Hyperchaos generated from the Lorenz chaotic system and its control. Phys Lett A 366(3):217–222
20. Kacar S (2016) Analog circuit and microcontroller based RNG application of a new easy realizable 4D chaotic system. Optik 127(20):9551–9561
21. Khalil HK (2002) Nonlinear systems. Prentice Hall, New York
22. Lakhekar GV, Waghmare LM, Vaidyanathan S (2016) Diving autopilot design for underwater vehicles using an adaptive neuro-fuzzy sliding mode controller. In: Vaidyanathan S, Volos C (eds) Advances and applications in nonlinear control systems. Springer, Berlin, pp 477–503
23. Li D (2008) A three-scroll chaotic attractor. Phys Lett A 372(4):387–393
24. Liu H, Ren B, Zhao Q, Li N (2016) Characterizing the optical chaos in a special type of small networks of semiconductor lasers using permutation entropy. Opt Commun 359:79–84
25. Liu W, Sun K, Zhu C (2016) A fast image encryption algorithm based on chaotic map. Opt Lasers Eng 84:26–36
26. Liu X, Mei W, Du H (2016) Simultaneous image compression, fusion and encryption algorithm based on compressive sensing and chaos. Opt Commun 366:22–32

27. Pl M, Carroll TL (1991) Synchronizing chaotic circuits. IEEE Trans Circuits Syst 38:453–456
28. Moussaoui S, Boulkroune A, Vaidyanathan S (2016) Fuzzy adaptive sliding-mode control scheme for uncertain underactuated systems. In: Vaidyanathan S, Volos C (eds) Advances and applications in nonlinear control systems. Springer, Berlin, pp 351–367
29. Pecora LM, Carroll TL (1990) Synchronization in chaotic systems. Phys Rev Lett 64:821–824
30. Pehlivan I, Moroz IM, Vaidyanathan S (2014) Analysis, synchronization and circuit design of a novel butterfly attractor. J Sound Vib 333(20):5077–5096
31. Pham VT, Volos C, Jafari S, Wang X, Vaidyanathan S (2014) Hidden hyperchaotic attractor in a novel simple memristive neural network. Optoelectr Adv Mater Rapid Commun 8(11–12):1157–1163
32. Pham VT, Volos CK, Vaidyanathan S (2015) Multi-scroll chaotic oscillator based on a first-order delay differential equation. In: Azar AT, Vaidyanathan S (eds) Chaos modeling and control systems design. Studies in computational intelligence, vol 581. Springer, Germany, pp 59–72
33. Pham VT, Volos CK, Vaidyanathan S, Le TP, Vu VY (2015) A memristor-based hyperchaotic system with hidden attractors: Dynamics, synchronization and circuital emulating. J Eng Sci Technol Rev 8(2):205–214
34. Pham VT, Jafari S, Vaidyanathan S, Volos C, Wang X (2016) A novel memristive neural network with hidden attractors and its circuitry implementation. Sci Chin Technol Sci 59(3):358–363
35. Pham VT, Jafari S, Volos C, Giakoumis A, Vaidyanathan S, Kapitaniak T (2016) A chaotic system with equilibria located on the rounded square loop and its circuit implementation. IEEE Trans Circuits Syst II: Express Briefs 63(9):878–882
36. Pham VT, Jafari S, Volos C, Vaidyanathan S, Kapitaniak T (2016) A chaotic system with infinite equilibria located on a piecewise linear curve. Optik 127(20):9111–9117
37. Pham VT, Vaidyanathan S, Volos C, Jafari S, Kingni ST (2016) A no-equilibrium hyperchaotic system with a cubic nonlinear term. Optik 127(6):3259–3265
38. Pham VT, Vaidyanathan S, Volos CK, Hoang TM, Yem VV (2016) Dynamics, synchronization and SPICE implementation of a memristive system with hidden hyperchaotic attractor. In: Azar AT, Vaidyanathan S (eds) Advances in chaos theory and intelligent control. Springer, Berlin, pp 35–52
39. Pham VT, Vaidyanathan S, Volos CK, Jafari S, Kuznetsov NV, Hoang TM (2016) A novel memristive time-delay chaotic system without equilibrium points. Eur Phys J: Spec Topics 225(1):127–136
40. Pham VT, Vaidyanathan S, Volos CK, Jafari S, Wang X (2016) A chaotic hyperjerk system based on memristive device. In: Vaidyanathan S, Volos C (eds) Advances and applications in chaotic systems. Springer, Berlin, pp 39–58
41. Rössler O (1979) An equation for hyperchaos. Phys Lett A 71(2–3):155–157
42. Sampath S, Vaidyanathan S, Volos CK, Pham VT (2015) An eight-term novel four-scroll chaotic system with cubic nonlinearity and its circuit simulation. J Eng Sci Technol Rev 8(2):1–6
43. Sampath S, Vaidyanathan S, Pham VT (2016) A novel 4-D hyperchaotic system with three quadratic nonlinearities, its adaptive control and circuit simulation. Int J Control Theory Appl 9(1):339–356
44. Shirkhani N, Khanesar M, Teshnehlab M (2016) Indirect model reference fuzzy control of SISO fractional order nonlinear chaotic systems. Proc Comput Sci 102:309–316
45. Slotine J, Li W (1991) Applied nonlinear control. Prentice-Hall, Englewood Cliffs, NJ
46. Sundarapandian V (2013) Analysis and anti-synchronization of a novel chaotic system via active and adaptive controllers. J Eng Sci Technol Rev 6(4):45–52
47. Sundarapandian V, Pehlivan I (2012) Analysis, control, synchronization, and circuit design of a novel chaotic system. Math Comput Modell 55(7–8):1904–1915
48. Sundarapandian V, Sivaperumal S (2011) Sliding controller design of hybrid synchronization of four-wing chaotic systems. Int J Soft Comput 6(5):224–231

49. Szmit Z, Warminski J (2016) Nonlinear dynamics of electro-mechanical system composed of two pendulums and rotating hub. Proc Eng 144:953–958
50. Tacha OI, Volos CK, Kyprianidis IM, Stouboulos IN, Vaidyanathan S, Pham VT (2016) Analysis, adaptive control and circuit simulation of a novel nonlinear finance system. Appl Math Comput 276:200–217
51. Utkin VI (1977) Variable structure systems with sliding modes. IEEE Transactions on Automatic Control 22(2):212–222
52. Utkin VI (1993) Sliding mode control design principles and applications to electric drives. IEEE Trans Ind Electr 40(1):23–36
53. Vaidyanathan S (2011) Analysis and synchronization of the hyperchaotic Yujun systems via sliding mode control. Adv Intell Syst Comput 176:329–337
54. Vaidyanathan S (2012) Global chaos control of hyperchaotic Liu system via sliding control method. Int J Control Theory Appl 5(2):117–123
55. Vaidyanathan S (2012) Sliding mode control based global chaos control of Liu-Liu-Liu-Su chaotic system. Int J Control Theory Appl 5(1):15–20
56. Vaidyanathan S (2013) A new six-term 3-D chaotic system with an exponential nonlinearity. Far East J Math Sci 79(1):135–143
57. Vaidyanathan S (2013) A ten-term novel 4-D hyperchaotic system with three quadratic nonlinearities and its control. Int J Control Theory Appl 6(2):97–109
58. Vaidyanathan S (2013) Analysis and adaptive synchronization of two novel chaotic systems with hyperbolic sinusoidal and cosinusoidal nonlinearity and unknown parameters. J Eng Sci Technol Rev 6(4):53–65
59. Vaidyanathan S (2014) A new eight-term 3-D polynomial chaotic system with three quadratic nonlinearities. Far East J Math Sci 84(2):219–226
60. Vaidyanathan S (2014) Analysis and adaptive synchronization of eight-term 3-D polynomial chaotic systems with three quadratic nonlinearities. Eur Phys J: Spec Topics 223(8):1519–1529
61. Vaidyanathan S (2014) Analysis, control and synchronisation of a six-term novel chaotic system with three quadratic nonlinearities. Int J Model Identif Control 22(1):41–53
62. Vaidyanathan S (2014) Generalised projective synchronisation of novel 3-D chaotic systems with an exponential non-linearity via active and adaptive control. Int J Model Identif Control 22(3):207–217
63. Vaidyanathan S (2014) Global chaos synchronisation of identical Li-Wu chaotic systems via sliding mode control. Int J Model Identif Control 22(2):170–177
64. Vaidyanathan S (2014) Qualitative analysis and control of an eleven-term novel 4-D hyperchaotic system with two quadratic nonlinearities. Int J Control Theory Appl 7(1):35–47
65. Vaidyanathan S (2015) A 3-D novel highly chaotic system with four quadratic nonlinearities, its adaptive control and anti-synchronization with unknown parameters. J Eng Sci Technol Rev 8(2):106–115
66. Vaidyanathan S (2015) A novel chemical chaotic reactor system and its adaptive control. Int J ChemTech Res 8(7):146–158
67. Vaidyanathan S (2015) A novel chemical chaotic reactor system and its output regulation via integral sliding mode control. Int J ChemTech Res 8(11):669–683
68. Vaidyanathan S (2015) Active control design for the anti-synchronization of Lotka-Volterra biological systems with four competitive species. Int J PharmTech Res 8(7):58–70
69. Vaidyanathan S (2015) Adaptive control design for the anti-synchronization of novel 3-D chemical chaotic reactor systems. Int J ChemTech Res 8(11):654–668
70. Vaidyanathan S (2015) Adaptive control of a chemical chaotic reactor. Int J PharmTech Res 8(3):377–382
71. Vaidyanathan S (2015) Adaptive control of the FitzHugh-Nagumo chaotic neuron model. Int J PharmTech Res 8(6):117–127
72. Vaidyanathan S (2015) Adaptive synchronization of chemical chaotic reactors. Int J ChemTech Res 8(2):612–621

73. Vaidyanathan S (2015) Adaptive synchronization of generalized Lotka-Volterra three-species biological systems. Int J PharmTech Res 8(5):928–937
74. Vaidyanathan S (2015) Adaptive synchronization of novel 3-D chemical chaotic reactor systems. Int J ChemTech Res 8(7):159–171
75. Vaidyanathan S (2015) Analysis, properties and control of an eight-term 3-D chaotic system with an exponential nonlinearity. Int J Model Identif Control 23(2):164–172
76. Vaidyanathan S (2015) Anti-synchronization of brusselator chemical reaction systems via adaptive control. Int J ChemTech Res 8(6):759–768
77. Vaidyanathan S (2015) Anti-synchronization of brusselator chemical reaction systems via integral sliding mode control. Int J ChemTech Res 8(11):700–713
78. Vaidyanathan S (2015) Anti-synchronization of chemical chaotic reactors via adaptive control method. Int J ChemTech Res 8(8):73–85
79. Vaidyanathan S (2015) Anti-synchronization of the FitzHugh-Nagumo chaotic neuron models via adaptive control method. Int J PharmTech Res 8(7):71–83
80. Vaidyanathan S (2015) Chaos in neurons and synchronization of Birkhoff-Shaw strange chaotic attractors via adaptive control. Int J PharmTech Res 8(6):1–11
81. Vaidyanathan S (2015) Dynamics and control of brusselator chemical reaction. Int J ChemTech Res 8(6):740–749
82. Vaidyanathan S (2015) Dynamics and control of Tokamak system with symmetric and magnetically confined plasma. Int J ChemTech Res 8(6):795–802
83. Vaidyanathan S (2015) Global chaos synchronization of chemical chaotic reactors via novel sliding mode control method. Int J ChemTech Res 8(7):209–221
84. Vaidyanathan S (2015) Global chaos synchronization of Duffing double-well chaotic oscillators via integral sliding mode control. Int J ChemTech Res 8(11):141–151
85. Vaidyanathan S (2015) Global chaos synchronization of the forced Van der Pol chaotic oscillators via adaptive control method. Int J PharmTech Res 8(6):156–166
86. Vaidyanathan S (2015) Hybrid chaos synchronization of the FitzHugh-Nagumo chaotic neuron models via adaptive control method. Int J PharmTech Res 8(8):48–60
87. Vaidyanathan S (2015) Integral sliding mode control design for the global chaos synchronization of identical novel chemical chaotic reactor systems. Int J ChemTech Res 8(11):684–699
88. Vaidyanathan S (2015) Output regulation of the forced Van der Pol chaotic oscillator via adaptive control method. Int J PharmTech Res 8(6):106–116
89. Vaidyanathan S (2015) Sliding controller design for the global chaos synchronization of forced Van der Pol chaotic oscillators. Int J PharmTech Res 8(7):100–111
90. Vaidyanathan S (2015) Synchronization of Tokamak systems with symmetric and magnetically confined plasma via adaptive control. Int J ChemTech Res 8(6):818–827
91. Vaidyanathan S (2016) A no-equilibrium novel 4-D highly hyperchaotic system with four quadratic nonlinearities and its adaptive control. In: Vaidyanathan S, Volos C (eds) Advances and applications in nonlinear control systems. Springer, Berlin, pp 235–258
92. Vaidyanathan S (2016) A novel 3-D conservative chaotic system with a sinusoidal nonlinearity and its adaptive control. Int J Control Theory Appl 9(1):115–132
93. Vaidyanathan S (2016) A novel 3-D jerk chaotic system with three quadratic nonlinearities and its adaptive control. Arch Control Sci 26(1):19–47
94. Vaidyanathan S (2016) A novel 3-D jerk chaotic system with two quadratic nonlinearities and its adaptive backstepping control. Int J Control Theory Appl 9(1):199–219
95. Vaidyanathan S (2016) A novel 4-D hyperchaotic thermal convection system and its adaptive control. In: Azar AT, Vaidyanathan S (eds) Advances in chaos theory and intelligent control. Springer, Berlin, pp 75–100
96. Vaidyanathan S (2016) A novel 5-D hyperchaotic system with a line of equilibrium points and its adaptive control. In: Vaidyanathan S, Volos C (eds) Advances and applications in chaotic systems. Springer, Berlin, pp 471–494
97. Vaidyanathan S (2016) A novel highly hyperchaotic system and its adaptive control. In: Vaidyanathan S, Volos C (eds) Advances and applications in chaotic systems. Springer, Berlin, pp 513–535

98. Vaidyanathan S (2016) A novel hyperchaotic hyperjerk system with two nonlinearities, its analysis, adaptive control and synchronization via backstepping control method. Int J Control Theory Appl 9(1):257–278

99. Vaidyanathan S (2016) An eleven-term novel 4-D hyperchaotic system with three quadratic nonlinearities, analysis, control and synchronization via adaptive control method. Int J Control Theory Appl 9(1):21–43

100. Vaidyanathan S (2016) Analysis, adaptive control and synchronization of a novel 4-D hyperchaotic hyperjerk system via backstepping control method. Arch Control Sci 26(3):311–338

101. Vaidyanathan S (2016) Analysis, control and synchronization of a novel 4-D highly hyperchaotic system with hidden attractors. In: Azar AT, Vaidyanathan S (eds) Advances in chaos theory and intelligent control. Springer, Berlin, pp 529–552

102. Vaidyanathan S (2016) Anti-synchronization of 3-cells cellular neural network attractors via integral sliding mode control. Int J PharmTech Res 9(1):193–205

103. Vaidyanathan S (2016) Global chaos control of the FitzHugh-Nagumo chaotic neuron model via integral sliding mode control. Int J PharmTech Res 9(4):413–425

104. Vaidyanathan S (2016) Global chaos control of the generalized Lotka-Volterra three-species system via integral sliding mode control. Int J PharmTech Res 9(4):399–412

105. Vaidyanathan S (2016) Global chaos regulation of a symmetric nonlinear gyro system via integral sliding mode control. Int J ChemTech Res 9(5):462–469

106. Vaidyanathan S (2016) Hybrid synchronization of the generalized Lotka-Volterra three-species biological systems via adaptive control. Int J PharmTech Res 9(1):179–192

107. Vaidyanathan S (2016) Hyperchaos, adaptive control and synchronization of a novel 4-D hyperchaotic system with two quadratic nonlinearities. Arch Control Sci 26(4):471–495

108. Vaidyanathan S (2016) Mathematical analysis, adaptive Control and synchronization of a ten-term novel three-scroll chaotic system with four quadratic nonlinearities. Int J Control Theory Appl 9(1):1–20

109. Vaidyanathan S (2016) Qualitative analysis and properties of a novel 4-D hyperchaotic system with two quadratic nonlinearities and its adaptive control. In: Azar AT, Vaidyanathan S (eds) Advances in chaos theory and intelligent control. Springer, Berlin, pp 455–480

110. Vaidyanathan S, Azar AT (2015) Analysis and control of a 4-D novel hyperchaotic system. In: Azar AT, Vaidyanathan S (eds) Chaos modeling and control systems design. Studies in computational intelligence, vol 581. Springer, Germany, pp 3–17

111. Vaidyanathan S, Azar AT (2015) Analysis, control and synchronization of a nine-term 3-D novel chaotic system. In: Azar AT, Vaidyanathan S (eds) Chaos modelling and control systems Design. Studies in computational intelligence, vol 581. Springer, Germany, pp 19–38

112. Vaidyanathan S, Azar AT (2016) A novel 4-D four-wing chaotic system with four quadratic nonlinearities and its synchronization via adaptive control method. In: Azar AT, Vaidyanathan S (eds) Advances in chaos theory and intelligent control. Springer, Berlin, pp 203–224

113. Vaidyanathan S, Azar AT (2016) Qualitative study and adaptive control of a novel 4-D hyperchaotic system with three quadratic nonlinearities. In: Azar AT, Vaidyanathan S (eds) Advances in chaos theory and intelligent control. Springer, Berlin, pp 179–202

114. Vaidyanathan S, Azar AT (2016) Takagi-Sugeno fuzzy logic controller for Liu-Chen four-scroll chaotic system. Int J Intell Eng Inf 4(2):135–150

115. Vaidyanathan S, Boulkroune A (2016) A novel 4-D hyperchaotic chemical reactor system and its adaptive control. In: Vaidyanathan S, Volos C (eds) Advances and applications in chaotic systems. Springer, Berlin, pp 447–469

116. Vaidyanathan S, Madhavan K (2013) Analysis, adaptive control and synchronization of a seven-term novel 3-D chaotic system. Int J Control Theory Appl 6(2):121–137

117. Vaidyanathan S, Pakiriswamy (2016) A five-term 3-D novel conservative chaotic system and its generalized projective synchronization via adaptive control method. Int J Control Theory Appl 9(1):61–78

118. Vaidyanathan S, Pakiriswamy S (2015) A 3-D novel conservative chaotic system and its generalized projective synchronization via adaptive control. J Eng Sci Technol Rev 8(2):52–60

119. Vaidyanathan S, Pakiriswamy S (2016) Adaptive control and synchronization design of a seven-term novel chaotic system with a quartic nonlinearity. Int J Control Theory Appl 9(1):237–256

120. Vaidyanathan S, Rajagopal K (2016) Adaptive control, synchronization and LabVIEW implementation of Rucklidge chaotic system for nonlinear double convection. Int J Control Theory Appl 9(1):175–197

121. Vaidyanathan S, Rajagopal K (2016) Analysis, control, synchronization and LabVIEW implementation of a seven-term novel chaotic system. Int J Control Theory Appl 9(1):151–174

122. Vaidyanathan S, Sampath S (2011) Global chaos synchronization of hyperchaotic Lorenz systems by sliding mode control. Commun Comput Inf Sci 205:156–164

123. Vaidyanathan S, Volos C (2015) Analysis and adaptive control of a novel 3-D conservative no-equilibrium chaotic system. Arch Control Sci 25(3):333–353

124. Vaidyanathan S, Volos C (2016) Advances and applications in chaotic systems. Springer, Berlin

125. Vaidyanathan S, Volos C (2016) Advances and applications in nonlinear control systems. Springer, Berlin

126. Vaidyanathan S, Volos C (2017) Advances in memristors. Memristive devices and systems, Springer, Berlin

127. Vaidyanathan S, Volos C, Pham VT (2014) Hyperchaos, adpative control and synchronization of a novel 5-D hyperchaotic system with three positive Lyapunov exponents and its SPICE implementation. Arch Control Sci 24(4):409–446

128. Vaidyanathan S, Volos C, Pham VT, Madhavan K, Idowu BA (2014) Adaptive backstepping control, synchronization and circuit simulation of a 3-D novel jerk chaotic system with two hyperbolic sinusoidal nonlinearities. Arch Control Sci 24(3):375–403

129. Vaidyanathan S, Azar AT, Rajagopal K, Alexander P (2015) Design and SPICE implementation of a 12-term novel hyperchaotic system and its synchronisation via active control. Int J Modell Identif Control 23(3):267–277

130. Vaidyanathan S, Pham VT, Volos CK (2015) A 5-D hyperchaotic Rikitake dynamo system with hidden attractors. Eur Phys J: Spec Topics 224(8):1575–1592

131. Vaidyanathan S, Rajagopal K, Volos CK, Kyprianidis IM, Stouboulos IN (2015) Analysis, adaptive control and synchronization of a seven-term novel 3-D chaotic system with three quadratic nonlinearities and its digital implementation in LabVIEW. J Eng Sci Technol Rev 8(2):130–141

132. Vaidyanathan S, Volos C, Pham VT, Madhavan K (2015) Analysis, adaptive control and synchronization of a novel 4-D hyperchaotic hyperjerk system and its SPICE implementation. Arch Control Sci 25(1):135–158

133. Vaidyanathan S, Volos CK, Kyprianidis IM, Stouboulos IN, Pham VT (2015) Analysis, adaptive control and anti-synchronization of a six-term novel jerk chaotic system with two exponential nonlinearities and its circuit simulation. J Eng Sci Technol Rev 8(2):24–36

134. Vaidyanathan S, Volos CK, Pham VT (2015) Analysis, adaptive control and adaptive synchronization of a nine-term novel 3-D chaotic system with four quadratic nonlinearities and its circuit simulation. J Eng Sci Technol Rev 8(2):181–191

135. Vaidyanathan S, Volos CK, Pham VT (2015) Analysis, control, synchronization and SPICE implementation of a novel 4-D hyperchaotic Rikitake dynamo system without equilibrium. J Eng Sci Technol Rev 8(2):232–244

136. Vaidyanathan S, Volos CK, Pham VT (2015) Global chaos control of a novel nine-term chaotic system via sliding mode control. In: Azar AT, Zhu Q (eds) Advances and applications in sliding mode control systems. Studies in computational intelligence, vol 576. Springer, Germany, pp 571–590

137. Vaidyanathan S, Volos CK, Pham VT (2016) Hyperchaos, control, synchronization and circuit simulation of a novel 4-D hyperchaotic system with three quadratic nonlinearities. In: Azar AT, Vaidyanathan S (eds) Advances in chaos theory and intelligent control. Springer, Berlin, pp 297–325

138. Volos CK, Kyprianidis IM, Stouboulos IN, Tlelo-Cuautle E, Vaidyanathan S (2015) Memristor: a new concept in synchronization of coupled neuromorphic circuits. J Eng Sci Technol Rev 8(2):157–173
139. Volos CK, Pham VT, Vaidyanathan S, Kyprianidis IM, Stouboulos IN (2015) Synchronization phenomena in coupled Colpitts circuits. J Eng Sci Technol Rev 8(2):142–151
140. Volos CK, Pham VT, Vaidyanathan S, Kyprianidis IM, Stouboulos IN (2016) Synchronization phenomena in coupled hyperchaotic oscillators with hidden attractors using a nonlinear open loop controller. In: Vaidyanathan S, Volos C (eds) Advances and applications in chaotic systems. Springer, Berlin, pp 1–38
141. Volos CK, Pham VT, Vaidyanathan S, Kyprianidis IM, Stouboulos IN (2016) The case of bidirectionally coupled nonlinear circuits via a memristor. In: Vaidyanathan S, Volos C (eds) Advances and applications in nonlinear control systems. Springer, Berlin, pp 317–350
142. Volos CK, Prousalis D, Kyprianidis IM, Stouboulos I, Vaidyanathan S, Pham VT (2016) Synchronization and anti-synchronization of coupled Hindmarsh-Rose neuron models. International Journal of Control Theory and Applications 9(1):101–114
143. Volos CK, Vaidyanathan S, Pham VT, Maaita JO, Giakoumis A, Kyprianidis IM, Stouboulos IN (2016) A novel design approach of a nonlinear resistor based on a memristor emulator. In: Azar AT, Vaidyanathan S (eds) Advances in chaos theory and intelligent control. Springer, Berlin, pp 3–34
144. Wang B, Zhong SM, Dong XC (2016) On the novel chaotic secure communication scheme design. Commun Nonlinear Sci Numer Simul 39:108–117
145. Wolf A, Swift JB, Swinney HL, Vastano JA (1985) Determining Lyapunov exponents from a time series. Physica D 16:285–317
146. Wu T, Sun W, Zhang X, Zhang S (2016) Concealment of time delay signature of chaotic output in a slave semiconductor laser with chaos laser injection. Opt Commun 381:174–179
147. Xu G, Liu F, Xiu C, Sun L, Liu C (2016) Optimization of hysteretic chaotic neural network based on fuzzy sliding mode control. Neurocomputing 189:72–79
148. Xu H, Tong X, Meng X (2016) An efficient chaos pseudo-random number generator applied to video encryption. Optik 127(20):9305–9319
149. Zhou W, Xu Y, Lu H, Pan L (2008) On dynamics analysis of a new chaotic attractor. Phys Lett A 372(36):5773–5777
150. Zhu C, Liu Y, Guo Y (2010) Theoretic and numerical study of a new chaotic system. Intell Inf Manag 2:104–109

Adaptive Integral Sliding Mode Controller Design for the Control and Synchronization of a Rod-Type Plasma Torch Chaotic System

Sundarapandian Vaidyanathan

Abstract Chaos has important applications in physics, chemistry, biology, ecology, secure communications, cryptosystems and many scientific branches. Control and synchronization of chaotic systems are important research problems in chaos theory. Sliding mode control is an important method used to solve various problems in control systems engineering. In robust control systems, the sliding mode control is often adopted due to its inherent advantages of easy realization, fast response and good transient performance as well as insensitivity to parameter uncertainties and disturbance. In this work, we first describe the Ghorui jerk chaotic system (2000), which is an important model of a rod-type plasma torch chaotic system. This jerk system describes a strange chaotic attractor of a thermal arc plasma system based on triple convection theory. The phase portraits of the rod-type plasma torch chaotic system are displayed and the qualitative properties of the rod-type plasma torch chaotic system are discussed. We demonstrate that the rod-type plasma torch chaotic system has three unstable equilibrium points on the x_1-axis of $\mathbf{R}^3$. The Lyapunov exponents of the rod-type plasma torch chaotic system are obtained as $L_1 = 0.3781$, $L_2 = 0$ and $L_3 = -1.3781$. The Kaplan-Yorke dimension of the rod-type plasma torch chaotic system is derived as $D_{KY} = 2.2744$, which shows the complexity of the system. Next, an adaptive integral sliding mode control scheme is proposed to globally stabilize all the trajectories of the rod-type plasma torch chaotic system. Furthermore, an adaptive integral sliding mode control scheme is proposed for the global chaos synchronization of identical rod-type plasma torch chaotic systems. The adaptive control mechanism helps the control design by estimating the unknown parameters. Numerical simulations using MATLAB are shown to illustrate all the main results derived in this work.

Keywords Chaos · Chaotic systems · Jerk systems · Chaos control · Integral sliding mode control · Adaptive control · Plasma system

S. Vaidyanathan (✉)
Research and Development Centre, Vel Tech University,
Avadi, Chennai 600062, Tamil Nadu, India
e-mail: sundarcontrol@gmail.com

© Springer International Publishing AG 2017
S. Vaidyanathan and C.-H. Lien (eds.), *Applications of Sliding Mode Control in Science and Engineering*, Studies in Computational Intelligence 709,
DOI 10.1007/978-3-319-55598-0_12

1 Introduction

Chaos theory describes the quantitative study of unstable aperiodic dynamic behavior in deterministic nonlinear dynamical systems. For the motion of a dynamical system to be chaotic, the system variables should contain some nonlinear terms and the system must satisfy three properties: boundedness, infinite recurrence and sensitive dependence on initial conditions [3–5, 101–103].

The problem of global control of a chaotic system is to device feedback control laws so that the closed-loop system is globally asymptotically stable. Global chaos control of various chaotic systems has been investigated via various methods in the control literature [3, 4, 101, 102].

The synchronization of chaotic systems deals with the problem of synchronizing the states of two chaotic systems called as *master* and *slave* systems asymptotically with time. The design goal of the complete synchronization problem is to use the output of the master system to control of the output of the slave system so that the outputs of the two systems are synchronized asymptotically with time.

Because of the *butterfly effect* [3], which causes the exponential divergence of the trajectories of two identical chaotic systems started with nearly the same initial conditions, synchronizing two chaotic systems is seemingly a challenging research problem in the chaos literature.

The synchronization of chaotic systems was first researched by Yamada and Fujisaka [13] with subsequent work by Pecora and Carroll [23, 24]. In the last few decades, several different methods have been devised for the synchronization of chaotic and hyperchaotic systems [3–5, 101–103].

Many new chaotic systems have been discovered in the recent years such as Zhou system [120], Zhu system [121], Li system [18], Sundarapandian systems [38, 39], Vaidyanathan systems [48–53, 55, 65, 81–83, 89, 90, 93–96, 98, 100, 104–108], Pehlivan system [25], Sampath system [35], Tacha system [42], Pham systems [27, 30, 31, 33], Akgul system [2], etc.

Chaos theory has applications in several fields such as memristors [26, 28, 29, 32–34, 109, 112, 114], fuzzy logic [6, 36, 91, 118], communication systems [10, 11, 115], cryptosystems [8, 9], electromechanical systems [12, 41], lasers [7, 19, 117], encryption [20, 21, 119], electrical circuits [1, 2, 15, 97, 110], chemical reactions [56, 57, 59, 60, 62, 64, 66–68, 71, 73, 77, 92], oscillators [74, 75, 78, 79, 111], tokamak systems [72, 80], neurology [61, 69, 70, 76, 85, 113], ecology [58, 63, 86, 88], etc.

The adaptive control mechanism helps the control design by estimating the unknown parameters [3, 4, 101]. The sliding mode control approach is recognized as an efficient tool for designing robust controllers for linear or nonlinear control systems operating under uncertainty conditions [43, 44].

A major advantage of sliding mode control is low sensitivity to parameter variations in the plant and disturbances affecting the plant, which eliminates the necessity of exact modeling of the plant. In the sliding mode control, the control dynamics will have two sequential modes, viz. the reaching mode and the sliding mode. Basically, a sliding mode controller design consists of two parts: hyperplane design and controller design. A hyperplane is first designed via the pole-placement approach and a controller is then designed based on the sliding condition. The stability of the overall system is guaranteed by the sliding condition and by a stable hyperplane. Sliding mode control method is a popular method for the control and synchronization of chaotic systems [17, 22, 40, 45–47, 54, 84, 87, 99].

In the recent decades, there is some good interest in finding jerk chaotic systems, which are described by the third-order ordinary differential equation

$$\dddot{x} = f(x, \dot{x}, \ddot{x}) \tag{1}$$

The differential equation (1) is called a *jerk system*, because the third-order time derive in mechanical systems is called *jerk*.

By defining phase variables $x_1 = x$, $x_2 = \dot{x}$ and $x_3 = \ddot{x}$, the jerk differential equation (1) can be expressed as a 3-D system given by

$$\begin{cases} \dot{x}_1 = x_2 \\ \dot{x}_2 = x_3 \\ \dot{x}_3 = f(x_1, x_2, x_3) \end{cases} \tag{2}$$

Thermal plasma technology is of great importance in research and industry. In industry, thermal plasma technology is employed in the manufacture of novel materials, eliminating poisonous waste and in enabling secure and effective production. The efficiency of plasma technology in modern industry is affected mainly by the instruments used to produce plasma such as plasma fluctuations. In the recent decades, a diagrammatic plasma torch has been proposed for the study of fluctuations in practical tests. Especially, it was shown that the inherent variations in plasma instruments can exhibit chaotic dynamical behavior [14].

In this research work, we study Ghorui's rod-type plasma torch chaotic system [14], which is based on triple convection theory. First, we display the phase portraits of the Ghouri's rod-type plasma torch chaotic system. Next, we show that the rod-type plasma torch chaotic system has three unstable equilibrium points on the x_1-axis of $\mathbf{R}^3$. The Lyapunov exponents of the rod-type plasma torch chaotic system are obtained as $L_1 = 0.3781$, $L_2 = 0$ and $L_3 = -1.3781$. The Kaplan-Yorke dimension of the rod-type plasma toch chaotic system is derived as $D_{KY} = 2.2744$.

Next, an adaptive integral sliding mode control scheme is proposed to globally stabilize all the trajectories of the rod-type plasma torch chaotic system. Furthermore, an adaptive integral sliding mode control scheme is proposed for the global chaos synchronization of identical rod-type plasma torch chaotic systems.

This work is organized as follows. Section 2 describes a 3-D rod-type plasma torch chaotic system [14] and its phase portraits. Section 3 details the qualitative properties of the rod-type plasma torch chaotic system. Section 4 contains new results on the adaptive integral sliding mode controller design for the global stabilization of the rod-type plasma torch chaotic system. Section 5 contains new results on the adaptive integral sliding mode controller design for the global synchronization of the rod-type plasma torch chaotic systems.

2   Rod-Type Plasma Torch Chaotic System

In [14], the following differential equation is modeled for a thermal arc plasma based on triple convection theory:

$$\dddot{F} + \Omega_2 \ddot{F} + \Omega_1 \dot{F} + \Omega_0 F = \pm F^3 \tag{3}$$

Thermo-physical parameters such as the plasma torch tool, flow speed of plasma gas and arc current determine the parameters of Eq. (3). For the study of dynamical behavior of the plasma torch, the coefficients in Eq. (3) are considered as in [14].

Thus, we rewrite the Ghorui's thermal arc plasma equation (3) as

$$\dddot{F} + \ddot{F} + b\dot{F} - aF = -F^3 \tag{4}$$

where a and b are constant, positive, parameters.

The state-space model of the rod-type plasma torch chaotic differential equation (4) can be described as follows.

$$\begin{cases} \dot{x}_1 = x_2 \\ \dot{x}_2 = x_3 \\ \dot{x}_3 = ax_1 - bx_2 - x_3 - x_1^3 \end{cases} \tag{5}$$

where x_1, x_2, x_3 are the states and a, b are constant, positive parameters.

In [14], it was observed that the rod-type plasma torch system (5) is *chaotic* when the system parameters take the values

$$a = 130, \quad b = 50 \tag{6}$$

For numerical simulations, we take the initial state of the rod-type plasma torch system (5) as

$$x_1(0) = 0.1, \quad x_2(0) = 0.1, \quad x_3(0) = 0.1 \tag{7}$$

For the parameter values (6) and the initial state (7), the Lyapunov exponents of the rod-type plasma torch system (5) are calculated by Wolf's algorithm [116] as

$$L_1 = 0.3781, \quad L_2 = 0, \quad L_3 = -1.3781 \tag{8}$$

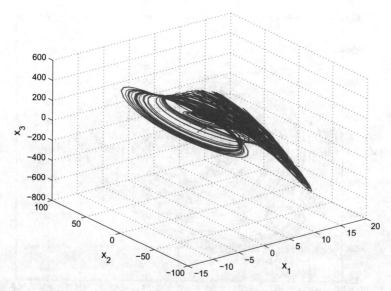

Fig. 1 3-D phase portrait of the rod-type plasma torch chaotic system

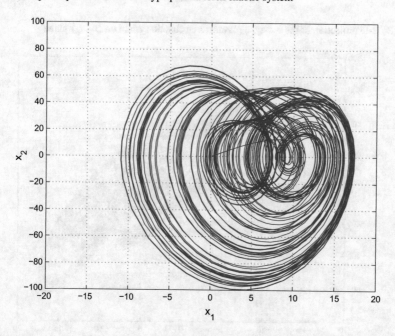

Fig. 2 2-D projection of the rod-type plasma torch chaotic system on (x_1, x_2)-plane

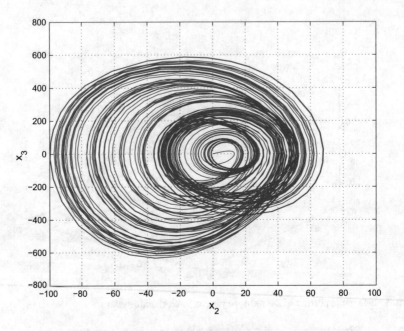

Fig. 3 2-D projection of the rod-type plasma torch chaotic system on (x_2, x_3)-plane

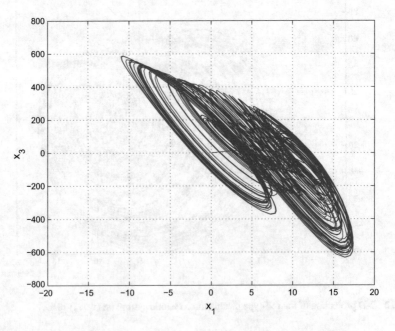

Fig. 4 2-D projection of the rod-type plasma torch chaotic system on (x_1, x_3)-plane

This shows that the rod-type plasma torch system (5) is chaotic and dissipative. Also, the Kaplan-Yorke dimension of the rod-type plasma torch chaotic system (5) is determined as

$$D_{KY} = 2 + \frac{L_1 + L_2}{|L_3|} = 2.2744 \tag{9}$$

Figure 1 shows the 3-D phase portrait of the rod-type plasma torch chaotic system (5). Figures 2, 3 and 4 show the 2-D projections of the rod-type plasma torch chaotic system (5).

3 Analysis of the 3-D Rod-Type Plasma Torch Chaotic System

3.1 Dissipativity

In vector notation, the rod-type plasma torch system (5) can be expressed as

$$\dot{\mathbf{x}} = f(\mathbf{x}) = \begin{bmatrix} f_1(x_1, x_2, x_3) \\ f_2(x_1, x_2, x_3) \\ f_3(x_1, x_2, x_3) \end{bmatrix}, \tag{10}$$

where

$$\begin{cases} f_1(x_1, x_2, x_3) = x_2 \\ f_2(x_1, x_2, x_3) = x_3 \\ f_3(x_1, x_2, x_3) = ax_1 - bx_2 - x_3 - x_1^3 \end{cases} \tag{11}$$

Let Ω be any region in $\mathbf{R}^3$ with a smooth boundary and also, $\Omega(t) = \Phi_t(\Omega)$, where Φ_t is the flow of f. Furthermore, let $V(t)$ denote the volume of $\Omega(t)$ (Fig. 5).

By Liouville's theorem, we know that

$$\dot{V}(t) = \int_{\Omega(t)} (\nabla \cdot f) \, dx_1 \, dx_2 \, dx_3 \tag{12}$$

The divergence of the rod-type plasma torch system (10) is found as:

$$\nabla \cdot f = \frac{\partial f_1}{\partial x_1} + \frac{\partial f_2}{\partial x_2} + \frac{\partial f_3}{\partial x_3} = -1 < 0 \tag{13}$$

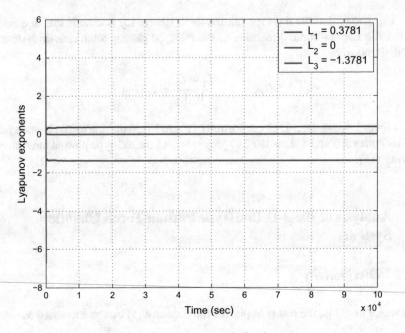

Fig. 5 Lyapunov exponents of the rod-type plasma torch chaotic system

Inserting the value of $\nabla \cdot f$ from (13) into (12), we get

$$\dot{V}(t) = \int\limits_{\Omega(t)} (-1) \, dx_1 \, dx_2 \, dx_3 = -V(t) \tag{14}$$

Integrating the first order linear differential equation (14), we get

$$V(t) = \exp(-t)V(0) \tag{15}$$

It is clear from Eq. (15) that $V(t) \rightarrow 0$ exponentially as $t \rightarrow \infty$. This shows that the rod-type plasma torch chaotic system (5) is dissipative. Hence, the system limit sets are ultimately confined into a specific limit set of zero volume, and the asymptotic motion of the rod-type plasma torch chaotic system (5) settles onto a strange attractor of the system.

3.2 Equilibrium Points

The equilibrium points of the rod-type plasma torch chaotic system (5) are obtained by solving the equations

$$\begin{cases} f_1(x_1,x_2,x_3) = x_2 & = 0 \\ f_2(x_1,x_2,x_3) = x_3 & = 0 \\ f_3(x_1,x_2,x_3) = ax_1 - bx_2 - x_3 - x_1^3 = 0 \end{cases} \quad (16)$$

We take the parameter values as in the chaotic case, viz. $a = 130$ and $b = 50$.

Solving the Eq. (16), we get three equilibrium points of the rod-type plasma torch chaotic system (5) as

$$E_0 = \begin{bmatrix} 0 \\ 0 \\ 0 \end{bmatrix}, \quad E_1 = \begin{bmatrix} \sqrt{130} \\ 0 \\ 0 \end{bmatrix}, \quad E_2 = \begin{bmatrix} -\sqrt{130} \\ 0 \\ 0 \end{bmatrix} \quad (17)$$

To test the stability type of the equilibrium points, we calculate the Jacobian matrix of the rod-type plasma torch chaotic system (5) at any point x:

$$J(x) = \begin{bmatrix} 0 & 1 & 0 \\ 0 & 0 & 1 \\ 130 - 3x_1^2 & -50 & -1 \end{bmatrix} \quad (18)$$

We find that

$$J_0 \overset{\Delta}{=} J(E_0) = \begin{bmatrix} 0 & 1 & 0 \\ 0 & 0 & 1 \\ 130 & -50 & -1 \end{bmatrix} \quad (19)$$

The matrix J_0 has the eigenvalues

$$\lambda_1 = 2.2650, \quad \lambda_{2,3} = -1.6325 \pm 7.3980\,i \quad (20)$$

This shows that the equilibrium point E_0 is a saddle-focus, which is unstable. Next, we find that

$$J_1 \overset{\Delta}{=} J(E_1) = \begin{bmatrix} 0 & 1 & 0 \\ 0 & 0 & 1 \\ -260 & -50 & -1 \end{bmatrix} \quad (21)$$

The matrix J_1 has the eigenvalues

$$\lambda_1 = -4.1312, \quad \lambda_{2,3} = 1.5656 \pm 7.7772\,i \quad (22)$$

This shows that the equilibrium point E_1 is a saddle-focus, which is unstable. We also find that

$$J_2 \overset{\Delta}{=} J(E_2) = \begin{bmatrix} 0 & 1 & 0 \\ 0 & 0 & 1 \\ -260 & -50 & -1 \end{bmatrix} \quad (23)$$

The matrix J_2 has the eigenvalues

$$\lambda_1 = -4.1312, \quad \lambda_{2,3} = 1.5656 \pm 7.7772\,i \tag{24}$$

This shows that the equilibrium point E_2 is a saddle-focus, which is unstable.

Thus, the rod-type plasma torch chaotic system (5) has three unstable equilibrium points on the x_1-axis.

3.3   Invariance

The rod-type plasma torch chaotic system (5) is invariant under the coordinates transformation

$$(x_1, x_2, x_3) \mapsto (-x_1, -x_2, -x_3) \tag{25}$$

This shows that the rod-type plasma torch chaotic system (5) has point-reflection symmetry about the origin and every non-trivial trajectory of the system (5) must have a twin trajectory.

3.4   Lyapunov Exponents and Kaplan-Yorke Dimension

We take the parameter values of the rod-type plasma torch system (5) as $a = 130$ and $b = 50$. We take the initial state of the rod-type plasma torch system (5) as given in (7).

Then the Lyapunov exponents of the rod-type plasma torch system (5) are numerically obtained using MATLAB as

$$L_1 = 0.3781, \quad L_2 = 0, \quad L_3 = -1.3781 \tag{26}$$

Thus, the maximal Lyapunov exponent (MLE) of the rod-type plasma torch system (5) is positive, which shows that the system (5) has a chaotic behavior.

Since $L_1 + L_2 + L_3 = -1 < 0$, it follows that the rod-type plasma torch chaotic system (5) is dissipative.

Also, the Kaplan-Yorke dimension of the rod-type plasma torch chaotic system (5) is derived as

$$D_{KY} = 2 + \frac{L_1 + L_2}{|L_3|} = 2.2744 \tag{27}$$

4 Global Chaos Control of the Rod-Type Plasma Torch Chaotic System

In this section, we apply adaptive integral sliding mode control for the global chaos control of the rod-type plasma torch chaotic system with unknown system parameters. The adaptive control mechanism helps the control design by estimating the unknown parameters [3, 4, 101].

The controlled rod-type plasma torch chaotic system is described by

$$
\begin{cases}
\dot{x}_1 = x_2 + u_1 \\
\dot{x}_2 = x_3 + u_2 \\
\dot{x}_3 = ax_1 - bx_2 - x_3 - x_1^3 + u_3
\end{cases}
\tag{28}
$$

where x_1, x_2, x_3 are the states of the system and a, b are unknown system parameters. The design goal is to find suitable feedback controllers u_1, u_2, u_3 so as to globally stabilize the system (28) with estimates of the unknown parameters.

Based on the sliding mode control theory [37, 43, 44], the integral sliding surface of each x_i $(i = 1, 2, 3)$ is defined as follows:

$$
s_i = \left(\frac{d}{dt} + \lambda_i \right) \left(\int_0^t x_i(\tau) d\tau \right) = x_i + \lambda_i \int_0^t x_i(\tau) d\tau, \quad i = 1, 2, 3
\tag{29}
$$

From Eq. (29), it follows that

$$
\begin{cases}
\dot{s}_1 = \dot{x}_1 + \lambda_1 x_1 \\
\dot{s}_2 = \dot{x}_2 + \lambda_2 x_2 \\
\dot{s}_3 = \dot{x}_3 + \lambda_3 x_3
\end{cases}
\tag{30}
$$

The Hurwitz condition is realized if $\lambda_i > 0$ for $i = 1, 2, 3$.
We consider the adaptive feedback control given by

$$
\begin{cases}
u_1 = -x_2 - \lambda_1 x_1 - \eta_1 \, \text{sgn}(s_1) - k_1 s_1 \\
u_2 = -x_3 - \lambda_2 x_2 - \eta_2 \, \text{sgn}(s_2) - k_2 s_2 \\
u_3 = -\hat{a}(t)x_1 + \hat{b}(t)x_2 + x_3 + x_1^3 - \lambda_3 x_3 - \eta_3 \, \text{sgn}(s_3) - k_3 s_3
\end{cases}
\tag{31}
$$

where $\eta_i > 0$ and $k_i > 0$ for $i = 1, 2, 3$.
Substituting (31) into (28), we obtain the closed-loop control system given by

$$\begin{cases} \dot{x}_1 = -\lambda_1 x_1 - \eta_1 \,\mathrm{sgn}(s_1) - k_1 s_1 \\ \dot{x}_2 = -\lambda_2 x_2 - \eta_2 \,\mathrm{sgn}(s_2) - k_2 s_2 \\ \dot{x}_3 = [a - \hat{a}(t)]x_1 - [b - \hat{b}(t)]x_2 - \lambda_3 x_3 - \eta_3 \,\mathrm{sgn}(s_3) - k_3 s_3 \end{cases} \tag{32}$$

We define the parameter estimation errors as

$$\begin{cases} e_a(t) = a - \hat{a}(t) \\ e_b(t) = b - \hat{b}(t) \end{cases} \tag{33}$$

Using (33), we can simplify the closed-loop system (32) as

$$\begin{cases} \dot{x}_1 = -\lambda_1 x_1 - \eta_1 \,\mathrm{sgn}(s_1) - k_1 s_1 \\ \dot{x}_2 = -\lambda_2 x_2 - \eta_2 \,\mathrm{sgn}(s_2) - k_2 s_2 \\ \dot{x}_3 = e_a x_1 - e_b x_2 - \lambda_3 x_3 - \eta_3 \,\mathrm{sgn}(s_3) - k_3 s_3 \end{cases} \tag{34}$$

Differentiating (33) with respect to t, we get

$$\begin{cases} \dot{e}_a = -\dot{\hat{a}} \\ \dot{e}_b = -\dot{\hat{b}} \end{cases} \tag{35}$$

Next, we state and prove the main result of this section.

Theorem 1 *The rod-type plasma torch chaotic system (28) is rendered globally asymptotically stable for all initial conditions $\mathbf{x}(0) \in \mathbf{R}^3$ by the adaptive integral sliding mode control law (31) and the parameter update law*

$$\begin{cases} \dot{\hat{a}} = s_3 x_1 \\ \dot{\hat{b}} = -s_3 x_2 \end{cases} \tag{36}$$

where λ_i, η_i, k_i are positive constants for $i = 1, 2, 3$.

Proof We consider the quadratic Lyapunov function defined by

$$V(s_1, s_2, s_3, e_a, e_b) = \frac{1}{2}\left(s_1^2 + s_2^2 + s_3^2\right) + \frac{1}{2}\left(e_a^2 + e_b^2\right) \tag{37}$$

Clearly, V is positive definite on $\mathbf{R}^5$.
Using (30), (34) and (35), the time-derivative of V is obtained as

$$\begin{aligned} \dot{V} = {}& s_1(-\eta_1 \,\mathrm{sgn}(s_1) - k_1 s_1) + s_2(-\eta_2 \,\mathrm{sgn}(s_2) - k_1 s_2) \\ & + s_3 \left[e_a x_1 - e_b x_2 - \eta_3 \,\mathrm{sgn}(s_3) - k_3 s_3\right] - e_a \dot{\hat{a}} - e_b \dot{\hat{b}} \end{aligned} \tag{38}$$

i.e.

$$\dot{V} = -\eta_1|s_1| - k_1s_1^2 - \eta_2|s_2| - k_2s_2^2 - \eta_3|s_3| - k_3s_3^2 \\ + e_a\left(s_3x_1 - \dot{\hat{a}}\right) + e_b\left(-s_3x_2 - \dot{\hat{b}}\right)$$

(39)

Using the parameter update law (36), we obtain

$$\dot{V} = -\eta_1|s_1| - k_1s_1^2 - \eta_2|s_2| - k_2s_2^2 - \eta_3|s_3| - k_3s_3^2$$

(40)

which shows that $\dot{V}$ is negative semi-definite on $\mathbf{R}^5$.

Hence, by Barbalat's lemma [16], it is immediate that $\mathbf{x}(t)$ is globally asymptotically stable for all values of $\mathbf{x}(0) \in \mathbf{R}^3$.

This completes the proof. ∎

For numerical simulations, we take the parameter values of the new rod-type plasma torch chaotic system (28) as in the chaotic case (6), i.e.

$$a = 130, \quad b = 50$$

(41)

We take the values of the control parameters as

$$k_i = 20, \quad \eta_i = 0.1, \quad \lambda_i = 50, \quad \text{where} \quad i = 1, 2, 3$$

(42)

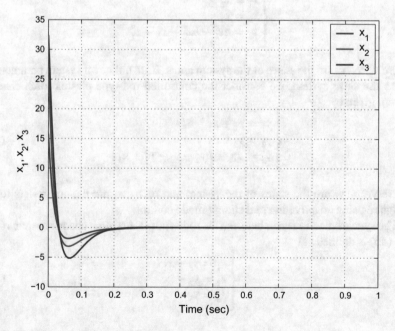

Fig. 6 Time-history of the controlled states x_1, x_2, x_3

We take the estimates of the system parameters as

$$\hat{a}(0) = 24.3, \quad \hat{b}(0) = 17.4 \tag{43}$$

We take the initial state of the system (28) as

$$x_1(0) = 15.8, \quad x_2(0) = 26.9, \quad x_3(0) = 32.4 \tag{44}$$

Figure 6 shows the time-history of the controlled states x_1, x_2, x_3.

5   Global Chaos Synchronization of the Rod-Type Plasma Torch Chaotic Systems

In this section, we use adaptive integral sliding mode control for the global chaos synchronization of the rod-type plasma torch chaotic systems with unknown system parameters. The adaptive control mechanism helps the control design by estimating the unknown parameters [3, 4, 101].

As the master system, we consider the rod-type plasma torch chaotic system given by

$$\begin{cases} \dot{x}_1 = x_2 \\ \dot{x}_2 = x_3 \\ \dot{x}_3 = ax_1 - bx_2 - x_3 - x_1^3 \end{cases} \tag{45}$$

where x_1, x_2, x_3 are the states of the system and a, b are unknown system parameters.

As the slave system, we consider the controlled rod-type plasma torch chaotic system given by

$$\begin{cases} \dot{y}_1 = y_2 + u_1 \\ \dot{y}_2 = y_3 + u_2 \\ \dot{y}_3 = ay_1 - by_2 - y_3 - y_1^3 + u_3 \end{cases} \tag{46}$$

where y_1, y_2, y_3 are the states of the system and u_1, u_2, u_3 are the controllers to be designed using adaptive integral sliding mode control.

The synchronization error between the rod-type plasma torch chaotic systems (45) and (46) is defined as

$$\begin{cases} e_1 = y_1 - x_1 \\ e_2 = y_2 - x_2 \\ e_3 = y_3 - x_3 \end{cases} \tag{47}$$

Then the synchronization error dynamics is obtained as

$$\begin{cases} \dot{e}_1 = e_2 + u_1 \\ \dot{e}_2 = e_3 + u_2 \\ \dot{e}_3 = ae_1 - be_2 - e_3 - y_1^3 + x_1^3 + u_3 \end{cases} \tag{48}$$

Based on the sliding mode control theory [37, 43, 44], the integral sliding surface of each e_i ($i = 1, 2, 3$) is defined as follows:

$$s_i = \left(\frac{d}{dt} + \lambda_i \right) \left(\int_0^t e_i(\tau) d\tau \right) = e_i + \lambda_i \int_0^t e_i(\tau) d\tau, \quad i = 1, 2, 3 \tag{49}$$

From Eq. (49), it follows that

$$\begin{cases} \dot{s}_1 = \dot{e}_1 + \lambda_1 e_1 \\ \dot{s}_2 = \dot{e}_2 + \lambda_2 e_2 \\ \dot{s}_3 = \dot{e}_3 + \lambda_3 e_3 \end{cases} \tag{50}$$

The Hurwitz condition is realized if $\lambda_i > 0$ for $i = 1, 2, 3$.
We consider the adaptive feedback control given by

$$\begin{cases} u_1 = -e_2 - \lambda_1 e_1 - \eta_1 \operatorname{sgn}(s_1) - k_1 s_1 \\ u_2 = -e_3 - \lambda_2 e_2 - \eta_2 \operatorname{sgn}(s_2) - k_2 s_2 \\ u_3 = -\hat{a}(t) e_1 + \hat{b}(t) e_2 + e_3 + y_1^3 - x_1^3 - \lambda_3 e_3 - \eta_3 \operatorname{sgn}(s_3) - k_3 s_3 \end{cases} \tag{51}$$

where $\eta_i > 0$ and $k_i > 0$ for $i = 1, 2, 3$.
Substituting (51) into (48), we obtain the closed-loop error dynamics as

$$\begin{cases} \dot{e}_1 = -\lambda_1 e_1 - \eta_1 \operatorname{sgn}(s_1) - k_1 s_1 \\ \dot{e}_2 = -\lambda_2 e_2 - \eta_2 \operatorname{sgn}(s_2) - k_2 s_2 \\ \dot{e}_3 = [a - \hat{a}(t)] e_1 - [b - \hat{b}(t)] e_2 - \lambda_3 e_3 - \eta_3 \operatorname{sgn}(s_3) - k_3 s_3 \end{cases} \tag{52}$$

We define the parameter estimation errors as

$$\begin{cases} e_a(t) = a - \hat{a}(t) \\ e_b(t) = b - \hat{b}(t) \end{cases} \tag{53}$$

Using (53), we can simplify the closed-loop system (52) as

$$
\begin{cases}
\dot{e}_1 = -\lambda_1 e_1 - \eta_1 \, \text{sgn}(s_1) - k_1 s_1 \\
\dot{e}_2 = -\lambda_2 e_2 - \eta_2 \, \text{sgn}(s_2) - k_2 s_2 \\
\dot{e}_3 = e_a e_1 - e_b e_2 - \lambda_3 e_3 - \eta_3 \, \text{sgn}(s_3) - k_3 s_3
\end{cases}
\tag{54}
$$

Differentiating (53) with respect to t, we get

$$
\begin{cases}
\dot{e}_a = -\dot{\hat{a}} \\
\dot{e}_b = -\dot{\hat{b}} \\
\dot{e}_c = -\dot{\hat{c}}
\end{cases}
\tag{55}
$$

Next, we state and prove the main result of this section.

Theorem 2 *The rod-type plasma torch chaotic systems (45) and (46) are globally and asymptotically synchronized for all initial conditions $\mathbf{x}(0), \mathbf{y}(0) \in \mathbf{R}^3$ by the adaptive integral sliding mode control law (51) and the parameter update law*

$$
\begin{cases}
\dot{\hat{a}} = s_3 e_1 \\
\dot{\hat{b}} = -s_3 e_2
\end{cases}
\tag{56}
$$

where λ_i, η_i, k_i are positive constants for $i = 1, 2, 3$.

Proof We consider the quadratic Lyapunov function defined by

$$
V(s_1, s_2, s_3, e_a, e_b) = \frac{1}{2} \left(s_1^2 + s_2^2 + s_3^2 \right) + \frac{1}{2} \left(e_a^2 + e_b^2 \right)
\tag{57}
$$

Clearly, V is positive definite on $\mathbf{R}^5$.
Using (50), (54) and (55), the time-derivative of V is obtained as

$$
\dot{V} = s_1(-\eta_1 \, \text{sgn}(s_1) - k_1 s_1) + s_2(-\eta_2 \, \text{sgn}(s_2) - k_1 s_2) \\
+ s_3 \left[e_a e_1 - e_b e_2 - \eta_3 \, \text{sgn}(s_3) - k_3 s_3 \right] - e_a \dot{\hat{a}} - e_b \dot{\hat{b}}
\tag{58}
$$

i.e.

$$
\dot{V} = -\eta_1 |s_1| - k_1 s_1^2 - \eta_2 |s_2| - k_2 s_2^2 - \eta_3 |s_3| - k_3 s_3^2 \\
+ e_a \left(s_3 e_1 - \dot{\hat{a}} \right) + e_b \left(-s_3 e_2 - \dot{\hat{b}} \right)
\tag{59}
$$

Using the parameter update law (56), we obtain

$$
\dot{V} = -\eta_1 |s_1| - k_1 s_1^2 - \eta_2 |s_2| - k_2 s_2^2 - \eta_3 |s_3| - k_3 s_3^2
\tag{60}
$$

which shows that $\dot{V}$ is negative semi-definite on $\mathbf{R}^5$.

Hence, by Barbalat's lemma [16], it is immediate that $\mathbf{e}(t)$ is globally asymptotically stable for all values of $\mathbf{e}(0) \in \mathbf{R}^3$.

This completes the proof.   ∎

For numerical simulations, we take the parameter values of the rod-type plasma torch chaotic systems (45) and (46) as in the chaotic case (6) i.e. $a = 130$ and $b = 50$. We take the values of the control parameters as

$$k_i = 20, \quad \eta_i = 0.1, \quad \lambda_i = 50, \quad \text{where} \quad i = 1, 2, 3 \tag{61}$$

We take the estimates of the system parameters as

$$\hat{a}(0) = 18.2, \quad \hat{b}(0) = 22.4 \tag{62}$$

We take the initial state of the master system (45) as

$$x_1(0) = 5.4, \quad x_2(0) = 14.7, \quad x_3(0) = 7.3 \tag{63}$$

We take the initial state of the slave system (46) as

$$y_1(0) = 14.9, \quad y_2(0) = 9.3, \quad y_3(0) = 12.1 \tag{64}$$

Figures 7, 8 and 9 show the complete synchronization of the rod-type plasma torch chaotic systems (45) and (46). Figure 10 shows the time-history of the synchronization errors e_1, e_2, e_3.

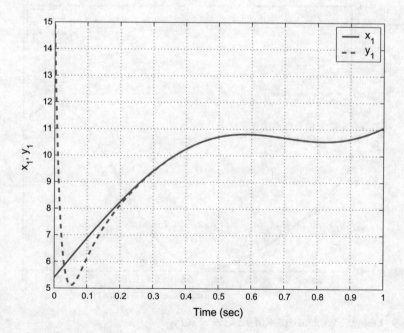

Fig. 7 Complete synchronization of the states x_1 and y_1

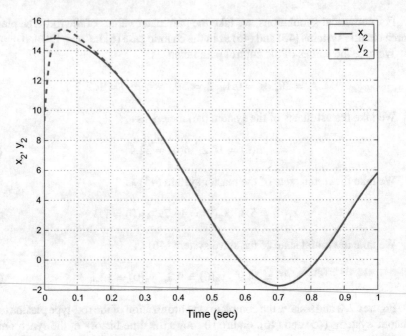

Fig. 8 Complete synchronization of the states x_2 and y_2

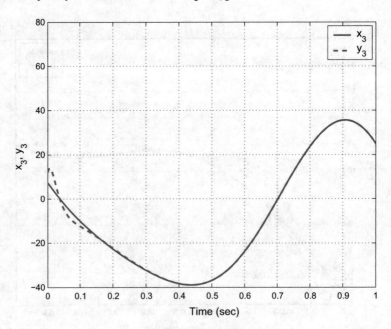

Fig. 9 Complete synchronization of the states x_3 and y_3

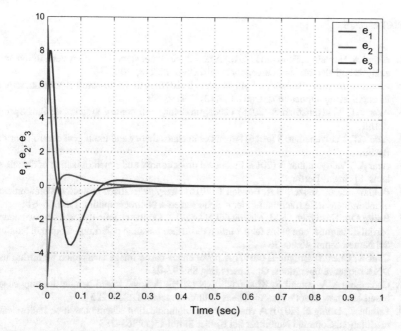

Fig. 10 Time-history of the synchronization errors e_1, e_2, e_3

6 Conclusions

In this work, we first described the Ghorui jerk chaotic system (2000). Ghorui jerk system is an important model of a rod-type plasma torch chaotic system, which describes a strange chaotic attractor of a thermal arc plasma system based on triple convection theory. The phase portraits of the rod-type plasma torch chaotic system were depicted and the qualitative properties of the rod-type plasma torch chaotic system were also discussed. We showed that the rod-type plasma torch chaotic system has three unstable equilibrium points on the x_1-axis of $\mathbf{R}^3$. The Lyapunov exponents of the rod-type plasma torch chaotic system have been obtained as $L_1 = 0.3781$, $L_2 = 0$ and $L_3 = -1.3781$. The Kaplan-Yorke dimension of the rod-type plasma toch chaotic system has been deduced as $D_{KY} = 2.2744$. Next, an adaptive integral sliding mode control scheme was proposed to globally stabilize all the trajectories of the rod-type plasma torch chaotic system. Furthermore, an adaptive integral sliding mode control scheme was proposed for the global chaos synchronization of identical rod-type plasma torch chaotic systems. The adaptive control results were established using Lyapunov stability theory. Numerical simulations using MATLAB were shown to demonstrate all the main results derived in this work.

References

1. Akgul A, Hussain S, Pehlivan I (2016) A new three-dimensional chaotic system, its dynamical analysis and electronic circuit applications. Optik 127(18):7062–7071
2. Akgul A, Moroz I, Pehlivan I, Vaidyanathan S (2016) A new four-scroll chaotic attractor and its engineering applications. Optik 127(13):5491–5499
3. Azar AT, Vaidyanathan S (2015) Chaos modeling and control systems design. Springer, Berlin
4. Azar AT, Vaidyanathan S (2016) Advances in chaos theory and intelligent control. Springer, Berlin
5. Azar AT, Vaidyanathan S (2017) Fractional order control and synchronization of chaotic systems. Springer, Berlin
6. Boulkroune A, Bouzeriba A, Bouden T (2016) Fuzzy generalized projective synchronization of incommensurate fractional-order chaotic systems. Neurocomputing 173:606–614
7. Burov DA, Evstigneev NM, Magnitskii NA (2017) On the chaotic dynamics in two coupled partial differential equations for evolution of surface plasmon polaritons. Commun Nonlinear Sci Numer Simul 46:26–36
8. Chai X, Chen Y, Broyde L (2017) A novel chaos-based image encryption algorithm using DNA sequence operations. Opt Lasers Eng 88:197–213
9. Chenaghlu MA, Jamali S, Khasmakhi NN (2016) A novel keyed parallel hashing scheme based on a new chaotic system. Chaos Solitons Fractals 87:216–225
10. Fallahi K, Leung H (2010) A chaos secure communication scheme based on multiplication modulation. Commun Nonlinear Sci Numer Simul 15(2):368–383
11. Fontes RT, Eisencraft M (2016) A digital bandlimited chaos-based communication system. Commun Nonlinear Sci Numer Simul 37:374–385
12. Fotsa RT, Woafo P (2016) Chaos in a new bistable rotating electromechanical system. Chaos Solitons Fractals 93:48–57
13. Fujisaka H, Yamada T (1983) Stability theory of synchronized motion in coupled-oscillator systems. Progr Theor Phys 63:32–47
14. Ghorui S, Sahasrabudhe SN, Muryt PSS, Das AK, Venkatramani N (2000) Experimental evidence of chaotic behavior in atmosphere pressure arc discharge. IEEE Trans Plasma Sci 28(1):253–260
15. Kacar S (2016) Analog circuit and microcontroller based RNG application of a new easy realizable 4D chaotic system. Optik 127(20):9551–9561
16. Khalil HK (2002) Nonlinear systems. Prentice Hall, New York
17. Lakhekar GV, Waghmare LM, Vaidyanathan S (2016) Diving autopilot design for underwater vehicles using an adaptive neuro-fuzzy sliding mode controller. In: Vaidyanathan S, Volos C (eds) Advances and applications in nonlinear control systems. Springer, Berlin, pp 477–503
18. Li D (2008) A three-scroll chaotic attractor. Phys Lett A 372(4):387–393
19. Liu H, Ren B, Zhao Q, Li N (2016) Characterizing the optical chaos in a special type of small networks of semiconductor lasers using permutation entropy. Optics Commun 359:79–84
20. Liu W, Sun K, Zhu C (2016) A fast image encryption algorithm based on chaotic map. Opt Lasers Eng 84:26–36
21. Liu X, Mei W, Du H (2016) Simultaneous image compression, fusion and encryption algorithm based on compressive sensing and chaos. Optics Commun 366:22–32
22. Moussaoui S, Boulkroune A, Vaidyanathan S (2016) Fuzzy adaptive sliding-mode control scheme for uncertain underactuated systems. In: Vaidyanathan S, Volos C (eds) Advances and applications in nonlinear control systems. Springer, Berlin, pp 351–367
23. Pecora LM, Carroll TL (1990) Synchronization in chaotic systems. Phys Rev Lett 64:821–824
24. Pecora LM, Carroll TL (1991) Synchronizing chaotic circuits. IEEE Trans Circuits Syst 38:453–456
25. Pehlivan I, Moroz IM, Vaidyanathan S (2014) Analysis, synchronization and circuit design of a novel butterfly attractor. J Sound Vib 333(20):5077–5096

26. Pham VT, Volos C, Jafari S, Wang X, Vaidyanathan S (2014) Hidden hyperchaotic attractor in a novel simple memristive neural network. Optoelectr Adv Mater Rapid Commun 8(11–12):1157–1163
27. Pham VT, Volos CK, Vaidyanathan S (2015) Multi-scroll chaotic oscillator based on a first-order delay differential equation. In: Azar AT, Vaidyanathan S (eds) Chaos modeling and control systems design. Studies in computational intelligence, vol 581. Springer, Germany, pp 59–72
28. Pham VT, Volos CK, Vaidyanathan S, Le TP, Vu VY (2015) A memristor-based hyperchaotic system with hidden attractors: dynamics, synchronization and circuital emulating. J Eng Sci Technol Rev 8(2):205–214
29. Pham VT, Jafari S, Vaidyanathan S, Volos C, Wang X (2016) A novel memristive neural network with hidden attractors and its circuitry implementation. Sci Chin Technol Sci 59(3):358–363
30. Pham VT, Jafari S, Volos C, Giakoumis A, Vaidyanathan S, Kapitaniak T (2016) A chaotic system with equilibria located on the rounded square loop and its circuit implementation. IEEE Trans Circuits Syst II: Express Briefs 63(9):878–882
31. Pham VT, Jafari S, Volos C, Vaidyanathan S, Kapitaniak T (2016) A chaotic system with infinite equilibria located on a piecewise linear curve. Optik 127(20):9111–9117
32. Pham VT, Vaidyanathan S, Volos CK, Hoang TM, Yem VV (2016) Dynamics, synchronization and SPICE implementation of a memristive system with hidden hyperchaotic attractor. In: Azar AT, Vaidyanathan S (eds) Advances in chaos theory and intelligent control. Springer, Berlin, pp 35–52
33. Pham VT, Vaidyanathan S, Volos CK, Jafari S, Kuznetsov NV, Hoang TM (2016) A novel memristive time-delay chaotic system without equilibrium points. Eur Phys J: Spec Topics 225(1):127–136
34. Pham VT, Vaidyanathan S, Volos CK, Jafari S, Wang X (2016) A chaotic hyperjerk system based on memristive device. In: Vaidyanathan S, Volos C (eds) Advances and applications in chaotic systems. Springer, Berlin, pp 39–58
35. Sampath S, Vaidyanathan S, Volos CK, Pham VT (2015) An eight-term novel four-scroll chaotic system with cubic nonlinearity and its circuit simulation. J Eng Sci Technol Rev 8(2):1–6
36. Shirkhani N, Khanesar M, Teshnehlab M (2016) Indirect model reference fuzzy control of SISO fractional order nonlinear chaotic systems. Proc Comput Sci 102:309–316
37. Slotine J, Li W (1991) Applied nonlinear control. Prentice-Hall, Englewood Cliffs, NJ
38. Sundarapandian V (2013) Analysis and anti-synchronization of a novel chaotic system via active and adaptive controllers. J Eng Sci Technol Rev 6(4):45–52
39. Sundarapandian V, Pehlivan I (2012) Analysis, control, synchronization, and circuit design of a novel chaotic system. Math Comput Model 55(7–8):1904–1915
40. Sundarapandian V, Sivaperumal S (2011) Sliding controller design of hybrid synchronization of four-wing chaotic systems. Int J Soft Comput 6(5):224–231
41. Szmit Z, Warminski J (2016) Nonlinear dynamics of electro-mechanical system composed of two pendulums and rotating hub. Proc Eng 144:953–958
42. Tacha OI, Volos CK, Kyprianidis IM, Stouboulos IN, Vaidyanathan S, Pham VT (2016) Analysis, adaptive control and circuit simulation of a novel nonlinear finance system. Appl Math Comput 276:200–217
43. Utkin VI (1977) Variable structure systems with sliding modes. IEEE Trans Autom Control 22(2):212–222
44. Utkin VI (1993) Sliding mode control design principles and applications to electric drives. IEEE Trans Ind Electr 40(1):23–36
45. Vaidyanathan S (2011) Analysis and synchronization of the hyperchaotic Yujun systems via sliding mode control. Adv Intell Syst Comput 176:329–337
46. Vaidyanathan S (2012) Global chaos control of hyperchaotic Liu system via sliding control method. Int J Control Theory Appl 5(2):117–123

47. Vaidyanathan S (2012) Sliding mode control based global chaos control of Liu-Liu-Liu-Su chaotic system. Int J Control Theory Appl 5(1):15–20
48. Vaidyanathan S (2013) A new six-term 3-D chaotic system with an exponential nonlinearity. Far East J Math Sc 79(1):135–143
49. Vaidyanathan S (2013) Analysis and adaptive synchronization of two novel chaotic systems with hyperbolic sinusoidal and cosinusoidal nonlinearity and unknown parameters. J Eng Sci Technol Rev 6(4):53–65
50. Vaidyanathan S (2014) A new eight-term 3-D polynomial chaotic system with three quadratic nonlinearities. Far East J Math Sci 84(2):219–226
51. Vaidyanathan S (2014) Analysis and adaptive synchronization of eight-term 3-D polynomial chaotic systems with three quadratic nonlinearities. Eur Phys J: Spec Topics 223(8):1519–1529
52. Vaidyanathan S (2014) Analysis, control and synchronisation of a six-term novel chaotic system with three quadratic nonlinearities. Int J Model Identif Control 22(1):41–53
53. Vaidyanathan S (2014) Generalised projective synchronisation of novel 3-D chaotic systems with an exponential non-linearity via active and adaptive control. Int J Model Identif Control 22(3):207–217
54. Vaidyanathan S (2014) Global chaos synchronisation of identical Li-Wu chaotic systems via sliding mode control. Int J Model Identif Control 22(2):170–177
55. Vaidyanathan S (2015) A 3-D novel highly chaotic system with four quadratic nonlinearities, its adaptive control and anti-synchronization with unknown parameters. J Eng Sci Technol Rev 8(2):106–115
56. Vaidyanathan S (2015) A novel chemical chaotic reactor system and its adaptive control. Int J ChemTech Res 8(7):146–158
57. Vaidyanathan S (2015) A novel chemical chaotic reactor system and its output regulation via integral sliding mode control. Int J ChemTech Res 8(11):669–683
58. Vaidyanathan S (2015) Active control design for the anti-synchronization of Lotka-Volterra biological systems with four competitive species. Int J PharmTech Res 8(7):58–70
59. Vaidyanathan S (2015) Adaptive control design for the anti-synchronization of novel 3-D chemical chaotic reactor systems. Int J ChemTech Res 8(11):654–668
60. Vaidyanathan S (2015) Adaptive control of a chemical chaotic reactor. Int J PharmTech Res 8(3):377–382
61. Vaidyanathan S (2015) Adaptive control of the FitzHugh-Nagumo chaotic neuron model. Int J PharmTech Res 8(6):117–127
62. Vaidyanathan S (2015) Adaptive synchronization of chemical chaotic reactors. Int J ChemTech Res 8(2):612–621
63. Vaidyanathan S (2015) Adaptive synchronization of generalized Lotka-Volterra three-species biological systems. Int J PharmTech Res 8(5):928–937
64. Vaidyanathan S (2015) Adaptive synchronization of novel 3-D chemical chaotic reactor systems. Int J ChemTech Res 8(7):159–171
65. Vaidyanathan S (2015) Analysis, properties and control of an eight-term 3-D chaotic system with an exponential nonlinearity. Int J Model Identif Control 23(2):164–172
66. Vaidyanathan S (2015) Anti-synchronization of brusselator chemical reaction systems via adaptive control. Int J ChemTech Res 8(6):759–768
67. Vaidyanathan S (2015) Anti-synchronization of brusselator chemical reaction systems via integral sliding mode control. Int J ChemTech Res 8(11):700–713
68. Vaidyanathan S (2015) Anti-synchronization of chemical chaotic reactors via adaptive control method. Int J ChemTech Res 8(8):73–85
69. Vaidyanathan S (2015) Anti-synchronization of the FitzHugh-Nagumo chaotic neuron models via adaptive control method. Int J PharmTech Res 8(7):71–83
70. Vaidyanathan S (2015) Chaos in neurons and synchronization of Birkhoff-Shaw strange chaotic attractors via adaptive control. Int J PharmTech Res 8(6):1–11
71. Vaidyanathan S (2015) Dynamics and control of brusselator chemical reaction. Int J ChemTech Res 8(6):740–749

72. Vaidyanathan S (2015) Dynamics and control of Tokamak system with symmetric and magnetically confined plasma. Int J ChemTech Res 8(6):795–802
73. Vaidyanathan S (2015) Global chaos synchronization of chemical chaotic reactors via novel sliding mode control method. Int J ChemTech Res 8(7):209–221
74. Vaidyanathan S (2015) Global chaos synchronization of Duffing double-well chaotic oscillators via integral sliding mode control. Int J ChemTech Res 8(11):141–151
75. Vaidyanathan S (2015) Global chaos synchronization of the forced Van der Pol chaotic oscillators via adaptive control method. Int J PharmTech Res 8(6):156–166
76. Vaidyanathan S (2015) Hybrid chaos synchronization of the FitzHugh-Nagumo chaotic neuron models via adaptive control method. Int J PharmTech Res 8(8):48–60
77. Vaidyanathan S (2015) Integral sliding mode control design for the global chaos synchronization of identical novel chemical chaotic reactor systems. Int J ChemTech Res 8(11):684–699
78. Vaidyanathan S (2015) Output regulation of the forced Van der Pol chaotic oscillator via adaptive control method. Int J PharmTech Res 8(6):106–116
79. Vaidyanathan S (2015) Sliding controller design for the global chaos synchronization of forced Van der Pol chaotic oscillators. Int J PharmTech Res 8(7):100–111
80. Vaidyanathan S (2015) Synchronization of Tokamak systems with symmetric and magnetically confined plasma via adaptive control. Int J ChemTech Res 8(6):818–827
81. Vaidyanathan S (2016) A novel 3-D conservative chaotic system with a sinusoidal nonlinearity and its adaptive control. Int J Control Theory Appl 9(1):115–132
82. Vaidyanathan S (2016) A novel 3-D jerk chaotic system with three quadratic nonlinearities and its adaptive control. Arch Control Sci 26(1):19–47
83. Vaidyanathan S (2016) A novel 3-D jerk chaotic system with two quadratic nonlinearities and its adaptive backstepping control. Int J Control Theory Appl 9(1):199–219
84. Vaidyanathan S (2016) Anti-synchronization of 3-cells cellular neural network attractors via integral sliding mode control. Int J PharmTech Res 9(1):193–205
85. Vaidyanathan S (2016) Global chaos control of the FitzHugh-Nagumo chaotic neuron model via integral sliding mode control. Int J PharmTech Res 9(4):413–425
86. Vaidyanathan S (2016) Global chaos control of the generalized Lotka-Volterra three-species system via integral sliding mode control. Int J PharmTech Res 9(4):399–412
87. Vaidyanathan S (2016) Global chaos regulation of a symmetric nonlinear gyro system via integral sliding mode control. Int J ChemTech Res 9(5):462–469
88. Vaidyanathan S (2016) Hybrid synchronization of the generalized Lotka-Volterra three-species biological systems via adaptive control. Int J PharmTech Res 9(1):179–192
89. Vaidyanathan S (2016) Mathematical analysis, adaptive Control and synchronization of a ten-term novel three-scroll chaotic system with four quadratic nonlinearities. Int J Control Theory Appl 9(1):1–20
90. Vaidyanathan S, Azar AT (2015) Analysis, control and synchronization of a nine-term 3-D novel chaotic system. In: Azar AT, Vaidyanathan S (eds) Chaos modelling and control systems design. Studies in computational intelligence, vol 581. Springer, Germany, pp 19–38
91. Vaidyanathan S, Azar AT (2016) Takagi-Sugeno fuzzy logic controller for Liu-Chen four-scroll chaotic system. Int J Intell Eng Inf 4(2):135–150
92. Vaidyanathan S, Boulkroune A (2016) A novel 4-D hyperchaotic chemical reactor system and its adaptive control. In: Vaidyanathan S, Volos C (eds) Advances and applications in chaotic systems. Springer, Berlin, pp 447–469
93. Vaidyanathan S, Madhavan K (2013) Analysis, adaptive control and synchronization of a seven-term novel 3-D chaotic system. Int J Control Theory Appl 6(2):121–137
94. Vaidyanathan S, Pakiriswamy (2016) A five-term 3-D novel conservative chaotic system and its generalized projective synchronization via adaptive control method. Int J Control Theory Appl 9(1):61–78
95. Vaidyanathan S, Pakiriswamy S (2015) A 3-D novel conservative chaotic system and its generalized projective synchronization via adaptive control. J Eng Sci Technol Rev 8(2):52–60
96. Vaidyanathan S, Pakiriswamy S (2016) Adaptive control and synchronization design of a seven-term novel chaotic system with a quartic nonlinearity. Int J Control Theory Appl 9(1):237–256

97. Vaidyanathan S, Rajagopal K (2016) Adaptive control, synchronization and LabVIEW implementation of Rucklidge chaotic system for nonlinear double convection. Int J Control Theory Appl 9(1):175–197
98. Vaidyanathan S, Rajagopal K (2016) Analysis, control, synchronization and LabVIEW implementation of a seven-term novel chaotic system. Int J Control Theory Appl 9(1):151–174
99. Vaidyanathan S, Sampath S (2011) Global chaos synchronization of hyperchaotic Lorenz systems by sliding mode control. Commun Comput Inf Sci 205:156–164
100. Vaidyanathan S, Volos C (2015) Analysis and adaptive control of a novel 3-D conservative no-equilibrium chaotic system. Arch Control Sci 25(3):333–353
101. Vaidyanathan S, Volos C (2016a) Advances and Applications in Chaotic Systems. Springer, Berlin, Germany
102. Vaidyanathan S, Volos C (2016b) Advances and Applications in Nonlinear Control Systems. Springer, Berlin, Germany
103. Vaidyanathan S, Volos C (2017) Advances in Memristors. Memristive Devices and Systems, Springer, Berlin, Germany
104. Vaidyanathan S, Volos C, Pham VT, Madhavan K, Idowu BA (2014) Adaptive backstepping control, synchronization and circuit simulation of a 3-D novel jerk chaotic system with two hyperbolic sinusoidal nonlinearities. Arch Control Sci 24(3):375–403
105. Vaidyanathan S, Rajagopal K, Volos CK, Kyprianidis IM, Stouboulos IN (2015) Analysis, adaptive control and synchronization of a seven-term novel 3-D chaotic system with three quadratic nonlinearities and its digital implementation in LabVIEW. J Eng Sci Technol Rev 8(2):130–141
106. Vaidyanathan S, Volos CK, Kyprianidis IM, Stouboulos IN, Pham VT (2015) Analysis, adaptive control and anti-synchronization of a six-term novel jerk chaotic system with two exponential nonlinearities and its circuit simulation. Journal of Engineering Science and Technology Review 8(2):24–36
107. Vaidyanathan S, Volos CK, Pham VT (2015) Analysis, adaptive control and adaptive synchronization of a nine-term novel 3-D chaotic system with four quadratic nonlinearities and its circuit simulation. J Eng Sci Technol Rev 8(2):181–191
108. Vaidyanathan S, Volos CK, Pham VT (2015) Global chaos control of a novel nine-term chaotic system via sliding mode control. In: Azar AT, Zhu Q (eds) Advances and applications in sliding mode control systems. Studies in computational intelligence, vol 576. Springer, Germany, pp 571–590
109. Volos CK, Kyprianidis IM, Stouboulos IN, Tlelo-Cuautle E, Vaidyanathan S (2015) Memristor: a new concept in synchronization of coupled neuromorphic circuits. J Eng Sci Technol Rev 8(2):157–173
110. Volos CK, Pham VT, Vaidyanathan S, Kyprianidis IM, Stouboulos IN (2015) Synchronization phenomena in coupled Colpitts circuits. J Eng Sci Technol Rev 8(2):142–151
111. Volos CK, Pham VT, Vaidyanathan S, Kyprianidis IM, Stouboulos IN (2016) Synchronization phenomena in coupled hyperchaotic oscillators with hidden attractors using a nonlinear open loop controller. In: Vaidyanathan S, Volos C (eds) Advances and applications in chaotic systems. Springer, Berlin, Germany, pp 1–38
112. Volos CK, Pham VT, Vaidyanathan S, Kyprianidis IM, Stouboulos IN (2016) The case of bidirectionally coupled nonlinear circuits via a memristor. In: Vaidyanathan S, Volos C (eds) Advances and applications in nonlinear control systems. Springer, Berlin, pp 317–350
113. Volos CK, Prousalis D, Kyprianidis IM, Stouboulos I, Vaidyanathan S, Pham VT (2016) Synchronization and anti-synchronization of coupled Hindmarsh-Rose neuron models. Int J Control Theory Appl 9(1):101–114
114. Volos CK, Vaidyanathan S, Pham VT, Maaita JO, Giakoumis A, Kyprianidis IM, Stouboulos IN (2016) A novel design approach of a nonlinear resistor based on a memristor emulator. In: Azar AT, Vaidyanathan S (eds) Advances in chaos theory and intelligent control. Springer, Berlin, pp 3–34
115. Wang B, Zhong SM, Dong XC (2016) On the novel chaotic secure communication scheme design. Commun Nonlinear Sci Numer Simul 39:108–117

116. Wolf A, Swift JB, Swinney HL, Vastano JA (1985) Determining Lyapunov exponents from a time series. Phys D 16:285–317
117. Wu T, Sun W, Zhang X, Zhang S (2016) Concealment of time delay signature of chaotic output in a slave semiconductor laser with chaos laser injection. Opt Commun 381:174–179
118. Xu G, Liu F, Xiu C, Sun L, Liu C (2016) Optimization of hysteretic chaotic neural network based on fuzzy sliding mode control. Neurocomputing 189:72–79
119. Xu H, Tong X, Meng X (2016) An efficient chaos pseudo-random number generator applied to video encryption. Optik 127(20):9305–9319
120. Zhou W, Xu Y, Lu H, Pan L (2008) On dynamics analysis of a new chaotic attractor. Phys Lett A 372(36):5773–5777
121. Zhu C, Liu Y, Guo Y (2010) Theoretic and numerical study of a new chaotic system. Intell Inf Management 2:104–109

Adaptive Integral Sliding Mode Controller Design for the Regulation and Synchronization of a Novel Hyperchaotic Finance System with a Stable Equilibrium

Sundarapandian Vaidyanathan

Abstract Chaos and hyperchaos have important applications in finance, physics, chemistry, biology, ecology, secure communications, cryptosystems and many scientific branches. Control and synchronization of chaotic and hyperchaotic systems are important research problems in chaos theory. Sliding mode control is an important method used to solve various problems in control systems engineering. In robust control systems, the sliding mode control is often adopted due to its inherent advantages of easy realization, fast response and good transient performance as well as insensitivity to parameter uncertainties and disturbance. This work proposes a novel 4-D hyperchaotic finance system. We show that the hyperchaotic finance system has a unique equilibrium point, which is locally exponentially stable. Also, we show that the Lyapunov exponents of the hyperchaotic finance system are $L_1 = 0.0365$, $L_2 = 0.0172$, $L_3 = 0$ and $L_4 = -0.8727$. The Kaplan-Yorke dimension of the hyperchaotic finance system is derived as $D_{KY} = 3.0615$, which shows the high complexity of the system. Next, an adaptive integral sliding mode control scheme is proposed for the global regulation of all the trajectories of the hyperchaotic finance system. Furthermore, an adaptive integral sliding mode control scheme is proposed for the global hyperchaos synchronization of identical hyperchaotic finance systems. The adaptive control mechanism helps the control design by estimating the unknown parameters. Numerical simulations using MATLAB are shown to illustrate all the main results derived in this work.

Keywords Chaos · Hyperchaos · Chaotic systems · Control · Synchronization · Integral sliding mode control · Adaptive control · Finance system

S. Vaidyanathan (✉)
Research and Development Centre, Vel Tech University, Avadi, Chennai 600062,
Tamil Nadu, India
e-mail: sundarcontrol@gmail.com

© Springer International Publishing AG 2017
S. Vaidyanathan and C.-H. Lien (eds.), *Applications of Sliding Mode Control in Science and Engineering*, Studies in Computational Intelligence 709,
DOI 10.1007/978-3-319-55598-0_13

1 Introduction

Chaos theory describes the quantitative study of unstable aperiodic dynamic behavior in deterministic nonlinear dynamical systems. For the motion of a dynamical system to be chaotic, the system variables should contain some nonlinear terms and the system must satisfy three properties: boundedness, infinite recurrence and sensitive dependence on initial conditions [3–5, 123–125].

A hyperchaotic system is defined as a chaotic system with at least two positive Lyapunov exponents [3, 4, 123]. Combined with one null exponent along the flow and one negative exponent to ensure the boundedness of the solution, the minimal dimension for a continuous-time hyperchaotic system is four.

Some classical examples of hyperchaotic systems are hyperchaotic Rössler system [40], hyperchaotic Lorenz system [18], hyperchaotic Chen system [16], hyperchaotic Lü system [9], etc. Some recent examples of hyperchaotic systems are hyperchaotic Dadras system [11], hyperchaotic Vaidyanathan systems [56, 63, 90, 94–100, 106, 108, 109, 111, 112, 114, 126, 128, 129, 131, 134, 136], hyperchaotic Sampath system [42], hyperchaotic Pham system [36], etc.

The problem of global control of a chaotic system is to device feedback control laws so that the closed-loop system is globally asymptotically stable. Global chaos control of various chaotic systems has been investigated via various methods in the control literature [3, 4, 123, 124].

The synchronization of chaotic systems deals with the problem of synchronizing the states of two chaotic systems called as *master* and *slave* systems asymptotically with time. The design goal of the complete synchronization problem is to use the output of the master system to control of the output of the slave system so that the outputs of the two systems are synchronized asymptotically with time.

Because of the *butterfly effect* [3], which causes the exponential divergence of the trajectories of two identical chaotic systems started with nearly the same initial conditions, synchronizing two chaotic systems is seemingly a challenging research problem in the chaos literature.

The synchronization of chaotic systems was first researched by Yamada and Fujisaka [15] with subsequent work by Pecora and Carroll [27, 28]. In the last few decades, several different methods have been devised for the synchronization of chaotic and hyperchaotic systems [3–5, 123–125].

Many new chaotic systems have been discovered in the recent years such as Zhou system [148], Zhu system [149], Li system [22], Sundarapandian systems [45, 46], Vaidyanathan systems [55, 57–61, 64, 74, 91–93, 107, 110, 115–118, 120, 122, 127, 130, 132, 133, 135], Pehlivan system [29], Sampath system [41], Tacha system [49], Pham systems [31, 34, 35, 38], Akgul system [2], etc.

Chaos theory has applications in several fields such as memristors [30, 32, 33, 37–39, 137, 140, 142], fuzzy logic [6, 43, 113, 146], communication systems [12, 13, 143], cryptosystems [8, 10], electromechanical systems [14, 48], lasers [7, 23, 145], encryption [24, 25, 147], electrical circuits [1, 2, 19, 119, 138], chemical reactions [65, 66, 68, 69, 71, 73, 75–77, 80, 82, 86, 114], oscillators [83, 84, 87,

88, 139], tokamak systems [81, 89], neurology [70, 78, 79, 85, 102, 141], ecology [67, 72, 103, 105], etc.

The adaptive control mechanism helps the control design by estimating the unknown parameters [3, 4, 123]. The sliding mode control approach is recognized as an efficient tool for designing robust controllers for linear or nonlinear control systems operating under uncertainty conditions [50, 51].

A major advantage of sliding mode control is low sensitivity to parameter variations in the plant and disturbances affecting the plant, which eliminates the necessity of exact modeling of the plant. In the sliding mode control, the control dynamics will have two sequential modes, viz. the reaching mode and the sliding mode. Basically, a sliding mode controller design consists of two parts: hyperplane design and controller design. A hyperplane is first designed via the pole-placement approach and a controller is then designed based on the sliding condition. The stability of the overall system is guaranteed by the sliding condition and by a stable hyperplane. Sliding mode control method is a popular method for the control and synchronization of chaotic systems [21, 26, 47, 52–54, 62, 101, 104, 121].

Recently, Huang and Li obtained a 3-D nonlinear finance system undergoing chaotic behavior [17]. This finance model is a 3-D nonlinear system consisting of interest rate, investment demand and price exponent as three states. Huang and Li exhibited chaotic behavior of the 3-D finance system for a set of values of the system parameters.

In this work, we introduce a feedback control to the 3-D nonlinear finance system and obtain a new 4-D nonlinear finance system undergoing hyperchaotic behavior. We show that the novel hyperchaotic finance system has a unique equilibrium point, which is locally exponentially stable.

Also, we show that the Lyapunov exponents of the hyperchaotic finance system are $L_1 = 0.0365, L_2 = 0.0172, L_3 = 0$ and $L_4 = -0.8727$. The Kaplan-Yorke dimension of the hyperchaotic finance system is derived as $D_{KY} = 3.0615$, which shows the high complexity of the system.

Next, an adaptive integral sliding mode control scheme is proposed for the regulation of all the trajectories of the hyperchaotic finance system. Furthermore, an adaptive integral sliding mode control scheme is proposed for the global hyperchaos synchronization of identical hyperchaotic finance systems.

This work is organized as follows. Section 2 describes a 3-D chaotic finance system [17] and its qualitative properties. Section 3 describes a novel 4-D hyperchaotic finance system and describes its phase portraits. Section 4 details the properties of the novel 4-D hyperchaotic finance system. Section 5 contains new results on the adaptive integral sliding mode controller design for the global regulation of the novel 4-D hyperchaotic finance system. Section 6 contains new results on the adaptive integral sliding mode controller design for the global synchronization of the novel identical 4-D hyperchaotic finance systems. Section 7 contains the conclusions of this work.

2   Chaotic Finance System

In this section, we describe the 3-D chaotic finance system obtained by Huang and Li [17].

The chaotic finance system [17] is given by the 3-D dynamics

$$\begin{cases} \dot{x}_1 = x_3 + (x_2 - a)x_1 \\ \dot{x}_2 = 1 - bx_2 - x_1^2 \\ \dot{x}_3 = -x_1 - cx_3 \end{cases} \tag{1}$$

where x_1 is the interest rate, x_2 is the investment demand and x_3 is the price exponent. Also, a is the saving, b is the per investment cost and c is the elasticity demand of the commercials. It is clear that the parameters a, b and c are positive constants.

In [17], it was shown that the finance system (1) exhibits chaotic behavior when the parameters take the values

$$a = 0.9, \quad b = 0.2, \quad c = 1.2 \tag{2}$$

For numerical simulations, we take the initial state of the finance system (1) as

$$x_1(0) = 1, \quad x_2(0) = 1, \quad x_3(0) = 1 \tag{3}$$

For the parameter values (2) and the initial state (3), the Lyapunov exponents of the nonlinear finance system (1) are calculated by Wolf's algorithm [144] as

$$L_1 = 0.0797, \quad L_2 = 0, \quad L_3 = -0.6964 \tag{4}$$

This shows that the nonlinear finance system (1) is chaotic and dissipative.

Also, the Kaplan-Yorke dimension of the nonlinear finance system (1) is determined as

$$D_{KY} = 2 + \frac{L_1 + L_2}{|L_3|} = 2.1144, \tag{5}$$

which is fractional.

The nonlinear finance system (1) is invariant under the coordinates transformation

$$(x_1, x_2, x_3) \mapsto (-x_1, x_2, -x_3) \tag{6}$$

This shows that the nonlinear finance system (1) has rotation symmetry about the y-axis. Thus, every non-trivial trajectory of the nonlinear finance system (1) must have a twin trajectory.

For the parameter values (2), the nonlinear finance system (1) has three equilibrium points given by

$$E_1 = \begin{bmatrix} 0 \\ 0.5 \\ 0 \end{bmatrix}, \quad E_2 = \begin{bmatrix} 0.8033 \\ 1.7333 \\ -0.6736 \end{bmatrix}, \quad E_3 = \begin{bmatrix} -0.8033 \\ 1.7333 \\ 0.6736 \end{bmatrix} \tag{7}$$

In vector notation, we can write the nonlinear finance system (1) as

$$\dot{\mathbf{x}} = f(\mathbf{x}) \tag{8}$$

It is easy to verify that $J_1 = Df(E_1)$ has the eigenvalues

$$\lambda_1 = -0.2, \quad \lambda_{2,3} = -0.8000 \pm 0.9165i \tag{9}$$

which shows that E_1 is locally exponentially stable.

Since the nonlinear finance system (1) has rotation symmetry about the x_2-axis, the equilibrium points E_2 and E_3 have the same stability type.

It is easy to verify that $J_2 = Df(E_2)$ has the eigenvalues

$$\lambda_1 = -0.9182, \quad \lambda_{2,3} = 0.1757 \pm 1.2868i \tag{10}$$

which shows that E_2 is a saddle-focus point.

It can be easily checked that $J_3 = Df(E_3)$ has the eigenvalues

$$\lambda_1 = -0.9182, \quad \lambda_{2,3} = 0.1757 \pm 1.2868i \tag{11}$$

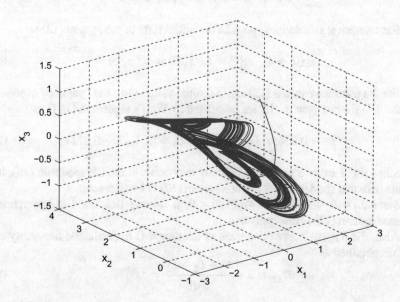

Fig. 1 3-D phase portrait of the finance chaotic system

which shows that E_3 is also a saddle-focus point.

Thus, E_2 and E_3 are both unstable equilibrium points.

Figure 1 shows the 3-D phase portrait of the finance chaotic system (1).

3 Hyperchaotic Finance System

In this section, we derive a new 4-D hyperchaotic finance system by adding a feed-back control to the finance chaotic system (1).

Thus, our novel 4-D finance system is given by the dynamics

$$\begin{cases} \dot{x}_1 = x_3 + (x_2 - a)x_1 - x_4 \\ \dot{x}_2 = 1 - bx_2 - x_1^2 \\ \dot{x}_3 = -x_1 - cx_3 \\ \dot{x}_4 = dx_1 \end{cases} \tag{12}$$

where x_1, x_2, x_3, x_4 are the state variables and a, b, c, d are positive parameters.

In (12), x_1 is the interest rate, x_2 is the investment demand, x_3 is the price exponent and x_4 is the average profit margin.

In this work, we show that the 4-D finance system (12) is *hyperchaotic* when the system parameters take the values

$$a = 1, \quad b = 0.2, \quad c = 1.3, \quad d = 0.1 \tag{13}$$

For numerical simulations, we take the initial state of the system (12) as

$$x_1(0) = 1, \quad x_2(0) = 1, \quad x_3(0) = 1, \quad x_4(0) = 1 \tag{14}$$

For the parameter values (13) and the initial values (14), the Lyapunov exponents of the 4-D finance system (12) are calculated by Wolf's algorithm [144] as

$$L_1 = 0.0365, \quad L_2 = 0.0172, \quad L_3 = 0, \quad L_4 = -0.8727 \tag{15}$$

Since there are two positive Lyapunov exponents in the LE spectrum (15), it is immediate that the 4-D novel finance system (12) is hyperchaotic.

Since $L_1 + L_2 + L_3 + L_4 = -0.8190 < 0$, it follows that the 4-D hyperchaotic finance system (12) is dissipative.

Also, the Kaplan-Yorke dimension of the new 4-D hyperchaotic finance system (12) is obtained as

$$D_{KY} = 3 + \frac{L_1 + L_2 + L_3}{|L_4|} = 3.0674, \tag{16}$$

which shows the complexity of the system.

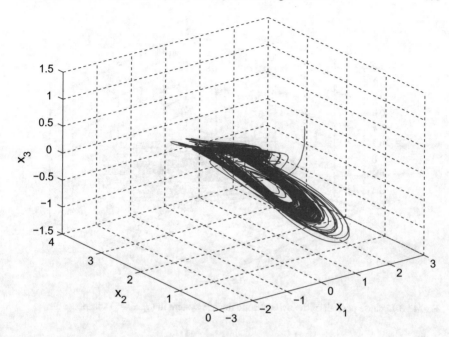

Fig. 2 3-D phase portrait of the hyperchaotic finance system in (x_1, x_2, x_3) space

Figures 2, 3, 4 and 5 show the 3-D phase portraits of the 4-D new hyperchaotic finance system in (x_1, x_2, x_3), (x_1, x_2, x_4), (x_1, x_3, x_4) and (x_2, x_3, x_4) spaces respectively.

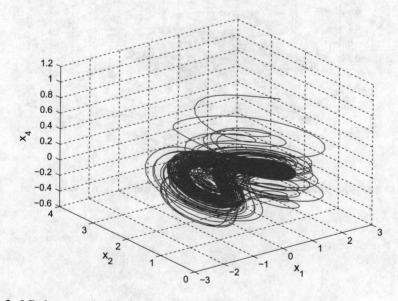

Fig. 3 3-D phase portrait of the hyperchaotic finance system in (x_1, x_2, x_4) space

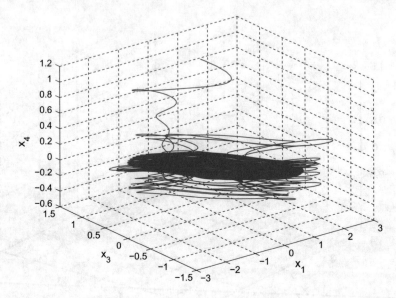

Fig. 4 3-D phase portrait of the hyperchaotic finance system in (x_1, x_3, x_4) space

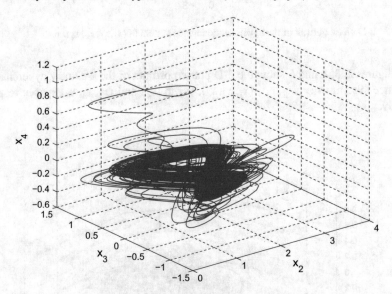

Fig. 5 3-D phase portrait of the hyperchaotic finance system in (x_2, x_3, x_4) space

4 Analysis of the New 4-D Hyperchaotic Finance System

4.1 Equilibrium Points

We take the parameter values of the 4-D finance system (12) as in the hyperchaotic case (13), i.e. $a = 1$, $b = 0.2$, $c = 1.3$ and $d = 0.1$.

The equilibrium points of the new hyperchaotic finance system (12) are obtained by solving the following system of equations.

$$x_3 + (x_2 - a)x_1 - x_4 = 0 \tag{17a}$$

$$1 - bx_2 - x_1^2 = 0 \tag{17b}$$

$$-x_1 - cx_3 = 0 \tag{17c}$$

$$dx_1 = 0 \tag{17d}$$

From (17d), it is immediate that $x_1 = 0$. Substituting $x_1 = 0$ in (17c), we obtain $x_3 = 0$. Substituting $x_1 = 0$ and $x_3 = 0$ in (17a), we get $x_4 = 0$. Substituting $x_1 = 0$ in (17b), we get $x_2 = 1/b = 0.2$.

This shows that the system of equations (17) has the unique solution

$$E_1 = \begin{bmatrix} 0 \\ 0.5 \\ 0 \\ 0 \end{bmatrix} \tag{18}$$

In vector notation, we can write the 4-D nonlinear finance system (12) as

$$\dot{\mathbf{x}} = f(\mathbf{x}) \tag{19}$$

It is easy to verify that $J_1 = Df(E_1)$ has the eigenvalues

$$\lambda_1 = -0.2, \quad \lambda_2 = -0.0807, \quad \lambda_{3,4} = -0.8597 \pm 0.9340i \tag{20}$$

which shows that E_1 is locally exponentially stable.

4.2 Rotation Symmetry About the x_2-Axis

It is easy to see that the new 4-D hyperchaotic finance system (12) is invariant under the change of coordinates

$$(x_1, x_2, x_3, x_4) \mapsto (-x_1, x_2, -x_3, -x_4) \tag{21}$$

Since the transformation (21) persists for all values of the system parameters, it follows that the new 4-D hyperchaotic finance system (12) has rotation symmetry about the x_2-axis and that any non-trivial trajectory of the system must have a twin trajectory.

4.3 Invariance

It is easy to see that the x_2-axis is invariant under the flow of the 4-D novel hyperchaotic system (12).

The invariant motion along the x_2-axis is characterized by the scalar dynamics

$$\dot{x}_2 = 1 - bx_2, \tag{22}$$

which is unstable.

4.4 Lyapunov Exponents and Kaplan-Yorke Dimension

We take the parameter values of the new hyperchaotic finance system (12) as in the hyperchaotic case (13), i.e.

$$a = 1, \quad b = 0.2, \quad c = 1.3, \quad d = 0.1 \tag{23}$$

We take the initial state of the new hyperchaotic finance system (12) as (14), i.e.

$$x_1(0) = 1, \quad x_2(0) = 1, \quad x_3(0) = 1, \quad x_4(0) = 1 \tag{24}$$

Then the Lyapunov exponents of the 4-D finance system (12) are numerically obtained using MATLAB as

$$L_1 = 0.0365, \quad L_2 = 0.0172, \quad L_3 = 0, \quad L_4 = -0.8727 \tag{25}$$

Since there are two positive Lyapunov exponents in the above equation, the new 4-D hyperchaotic finance system (12) is a hyperchaotic system.

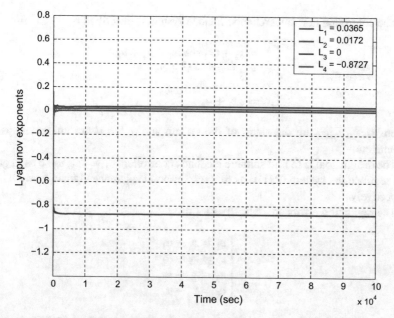

Fig. 6 Lyapunov exponents of the new hyperchaotic finance system

Since $L_1 + L_2 + L_3 + L_4 = -0.8190 < 0$, it follows that the new hyperchaotic finance system (12) is dissipative.

Also, the Kaplan-Yorke dimension of the new hyperchaotic finance system (12) is calculated as

$$D_{KY} = 3 + \frac{L_1 + L_2 + L_3}{|L_4|} = 3.0615 \tag{26}$$

which shows the high complexity of the hyperchaotic finance system.

Figure 6 shows the Lyapunov exponents of the new hyperchaotic finance system (12).

5 Global Hyperchaos Regulation of the New Hyperchaotic Finance System

In this section, we use adaptive integral sliding mode control for the global hyperchaos regulation of new hyperchaotic finance system with unknown system parameters. The adaptive control mechanism helps the control design by estimating the unknown parameters [3, 4, 123].

The controlled new hyperchaotic finance system is described by

$$
\begin{cases}
\dot{x}_1 = x_3 + (x_2 - a)x_1 - x_4 + u_1 \\
\dot{x}_2 = 1 - bx_2 - x_1^2 + u_2 \\
\dot{x}_3 = -x_1 - cx_3 + u_3 \\
\dot{x}_4 = dx_1 + u_4
\end{cases}
\tag{27}
$$

where x_1, x_2, x_3, x_4 are the states of the system and a, b, c, d are unknown system parameters.

The design goal is to find suitable feedback controllers u_1, u_2, u_3, u_4 so as to globally regulate the system (27) so as to track set-point reference signals $\alpha_1, \alpha_2, \alpha_3, \alpha_4$, respectively.

The regulation errors are defined as follows:

$$
\begin{cases}
e_1 = x_1 - \alpha_1 \\
e_2 = x_2 - \alpha_2 \\
e_3 = x_3 - \alpha_3 \\
e_4 = x_4 - \alpha_4
\end{cases}
\tag{28}
$$

where $\alpha_i (i = 1, 2, 3, 4)$ are constant reference signals (*set-point controls*).

The regulation error dynamics is obtained as

$$
\begin{cases}
\dot{e}_1 = e_3 + \alpha_3 + (e_2 + \alpha_2 - a)(e_1 + \alpha_1) - e_4 - \alpha_4 + u_1 \\
\dot{e}_2 = 1 - b(e_2 + \alpha_2) - (e_1 + \alpha_1)^2 + u_2 \\
\dot{e}_3 = -e_1 - \alpha_1 - c(e_3 + \alpha_3) + u_3 \\
\dot{e}_4 = d(e_1 + \alpha_1) + u_4
\end{cases}
\tag{29}
$$

Based on the sliding mode control theory [44, 50, 51], the integral sliding surface of each e_i ($i = 1, 2, 3, 4$) is defined as follows:

$$
s_i = \left(\frac{d}{dt} + \lambda_i \right) \left(\int_0^t e_i(\tau) d\tau \right) = e_i + \lambda_i \int_0^t e_i(\tau) d\tau, \quad i = 1, 2, 3, 4
\tag{30}
$$

From Eq. (30), it follows that

$$
\begin{cases}
\dot{s}_1 = \dot{e}_1 + \lambda_1 e_1 \\
\dot{s}_2 = \dot{e}_2 + \lambda_2 e_2 \\
\dot{s}_3 = \dot{e}_3 + \lambda_3 e_3 \\
\dot{s}_4 = \dot{e}_4 + \lambda_4 e_4
\end{cases}
\tag{31}
$$

The Hurwitz condition is realized if $\lambda_i > 0$ for $i = 1, 2, 3, 4$.

We consider the adaptive feedback control given by

$$
\begin{cases}
u_1 = -e_3 - \alpha_3 - [e_2 + \alpha_2 - \hat{a}(t)](e_1 + \alpha_1) + e_4 + \alpha_4 \\
\qquad - \lambda_1 e_1 - \eta_1 \operatorname{sgn}(s_1) - k_1 s_1 \\
u_2 = -1 + \hat{b}(t)(e_2 + \alpha_2) + (e_1 + \alpha_1)^2 - \lambda_2 e_2 - \eta_2 \operatorname{sgn}(s_2) - k_2 s_2 \\
u_3 = e_1 + \alpha_1 + \hat{c}(t)(e_3 + \alpha_3) - \lambda_3 e_3 - \eta_3 \operatorname{sgn}(s_3) - k_3 s_3 \\
u_4 = -\hat{d}(t)(e_1 + \alpha_1) - \lambda_4 e_4 - \eta_4 \operatorname{sgn}(s_4) - k_4 s_4
\end{cases}
\tag{32}
$$

where $\eta_i > 0$ and $k_i > 0$ for $i = 1, 2, 3, 4$.

Substituting (32) into (27), we obtain the closed-loop control system given by

$$
\begin{cases}
\dot{e}_1 = -[a - \hat{a}(t)](e_1 + \alpha_1) - \lambda_1 e_1 - \eta_1 \operatorname{sgn}(s_1) - k_1 s_1 \\
\dot{e}_2 = -[b - \hat{b}(t)](e_2 + \alpha_2) - \lambda_2 e_2 - \eta_2 \operatorname{sgn}(s_2) - k_2 s_2 \\
\dot{e}_3 = -[c - \hat{c}(t)](e_3 + \alpha_3) - \lambda_3 e_3 - \eta_3 \operatorname{sgn}(s_3) - k_3 s_3 \\
\dot{e}_4 = [d - \hat{d}(t)](e_1 + \alpha_1) - \lambda_4 e_4 - \eta_4 \operatorname{sgn}(s_4) - k_4 s_4
\end{cases}
\tag{33}
$$

We define the parameter estimation errors as

$$
\begin{cases}
e_a(t) = a - \hat{a}(t) \\
e_b(t) = b - \hat{b}(t) \\
e_c(t) = c - \hat{c}(t) \\
e_d(t) = d - \hat{d}(t)
\end{cases}
\tag{34}
$$

Using (34), we can simplify the closed-loop system (33) as

$$
\begin{cases}
\dot{e}_1 = -e_a(e_1 + \alpha_1) - \lambda_1 e_1 - \eta_1 \operatorname{sgn}(s_1) - k_1 s_1 \\
\dot{e}_2 = -e_b(e_2 + \alpha_2) - \lambda_2 e_2 - \eta_2 \operatorname{sgn}(s_2) - k_2 s_2 \\
\dot{e}_3 = -e_c(e_3 + \alpha_3) - \lambda_3 e_3 - \eta_3 \operatorname{sgn}(s_3) - k_3 s_3 \\
\dot{e}_4 = e_d(e_1 + \alpha_1) - \lambda_4 e_4 - \eta_4 \operatorname{sgn}(s_4) - k_4 s_4
\end{cases}
\tag{35}
$$

Differentiating (34) with respect to t, we get

$$
\begin{cases}
\dot{e}_a = -\dot{\hat{a}} \\
\dot{e}_b = -\dot{\hat{b}} \\
\dot{e}_c = -\dot{\hat{c}} \\
\dot{e}_d = -\dot{\hat{d}}
\end{cases}
\tag{36}
$$

Next, we state and prove the main result of this section.

Theorem 1 *The new hyperchaotic two-wing dynamo system (27) is globally regulated to track the set-point controls $\alpha_i, (i = 1, 2, 3, 4)$ for all initial conditions $\mathbf{x}(0) \in \mathbf{R}^4$ by the adaptive integral sliding mode control law (32) and the parameter update law*

$$\begin{cases} \dot{\hat{a}} = -s_1(e_1 + \alpha_1) \\ \dot{\hat{b}} = -s_2(e_2 + \alpha_2) \\ \dot{\hat{c}} = -s_3(e_3 + \alpha_3) \\ \dot{\hat{d}} = s_4(e_1 + \alpha_1) \end{cases} \qquad (37)$$

where λ_i, η_i, k_i are positive constants for $i = 1, 2, 3, 4$.

Proof We consider the quadratic Lyapunov function defined by

$$V(s_1, s_2, s_3, s_4, e_a, e_b, e_c, e_d) = \frac{1}{2}\left(s_1^2 + s_2^2 + s_3^2 + s_4^2\right) + \frac{1}{2}\left(e_a^2 + e_b^2 + e_c^2 + e_d^2\right) \quad (38)$$

Clearly, V is positive definite on $\mathbf{R}^8$.

Using (31), (35) and (36), the time-derivative of V is obtained as

$$\begin{aligned} \dot{V} &= s_1[-e_a(e_1 + \alpha_1) - \eta_1 \operatorname{sgn}(s_1) - k_1 s_1] - e_a \dot{\hat{a}} \\ &\quad + s_2[-e_b(e_2 + \alpha_2) - \eta_2 \operatorname{sgn}(s_2) - k_2 s_2] - e_b \dot{\hat{b}} \\ &\quad + s_3[-e_c(e_3 + \alpha_3) - \eta_3 \operatorname{sgn}(s_3) - k_3 s_3] - e_c \dot{\hat{c}} \\ &\quad + s_4[e_d(e_1 + \alpha_1) - \eta_4 \operatorname{sgn}(s_4) - k_4 s_4] - e_d \dot{\hat{d}} \end{aligned} \qquad (39)$$

i.e.

$$\begin{aligned} \dot{V} &= -\eta_1 |s_1| - k_1 s_1^2 - \eta_2 |s_2| - k_2 s_2^2 - \eta_3 |s_3| - k_3 s_3^2 - \eta_4 |s_4| - k_4 s_4^2 \\ &\quad + e_a \left[-s_1(e_1 + \alpha_1) - \dot{\hat{a}}\right] + e_b \left[-s_2(e_2 + \alpha_2) - \dot{\hat{b}}\right] \\ &\quad + e_c \left[-s_3(e_3 + \alpha_3) - \dot{\hat{c}}\right] + e_d \left[s_4(e_1 + \alpha_1) - \dot{\hat{d}}\right] \end{aligned} \qquad (40)$$

Using the parameter update law (37), we obtain

$$\dot{V} = -\eta_1 |s_1| - k_1 s_1^2 - \eta_2 |s_2| - k_2 s_2^2 - \eta_3 |s_3| - k_3 s_3^2 - \eta_4 |s_4| - k_4 s_4^2 \qquad (41)$$

which shows that $\dot{V}$ is negative semi-definite on $\mathbf{R}^7$.

Hence, by Barbalat's lemma [20], it is immediate that the regulation error $\mathbf{e}(t)$ is globally asymptotically stable for all values of $\mathbf{e}(0) \in \mathbf{R}^4$.

This completes the proof. ∎

For numerical simulations, we take the parameter values of the new hyperchaotic finance system (27) as in the hyperchaotic case (13), i.e. $a = 1$, $b = 0.2$, $c = 1.3$ and $d = 0.1$. We take the values of the control parameters as $k_i = 20$, $\eta_i = 0.1$ and $\lambda_i = 50$, where $i = 1, 2, 3, 4$.

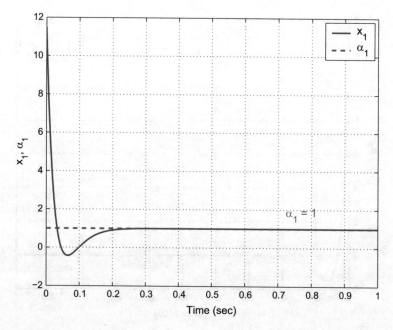

Fig. 7 Output regulation of the state x_1

We take the set-point controls as

$$\alpha_1 = 1, \quad \alpha_2 = 2, \quad \alpha_3 = 3, \quad \alpha_4 = 4 \tag{42}$$

We take the estimates of the system parameters as

$$\hat{a}(0) = 11.5, \quad \hat{b}(0) = 8.3, \quad \hat{c}(0) = 10.2, \quad \hat{d}(0) = 5.4 \tag{43}$$

We take the initial state of the hyperchaotic finance system (27) as

$$x_1(0) = 7.2, \quad x_2(0) = 12.3, \quad x_3(0) = 18.1, \quad x_4(0) = 10.6 \tag{44}$$

Figures 7, 8, 9 and 10 show the tracking of the states x_1, x_2, x_3, x_4 to the set-point controls $\alpha_1, \alpha_2, \alpha_3, \alpha_4$, respectively.

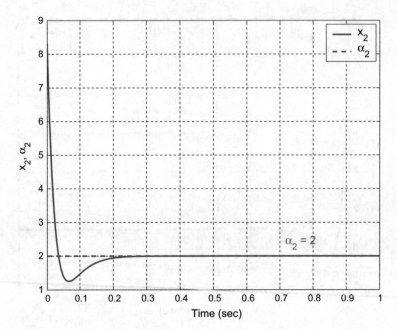

Fig. 8 Output regulation of the state x_2

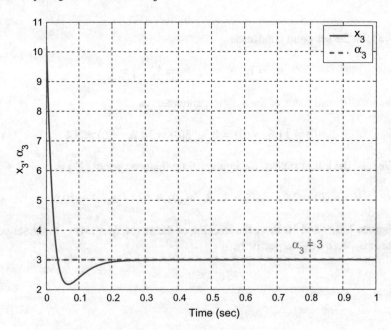

Fig. 9 Output regulation of the state x_3

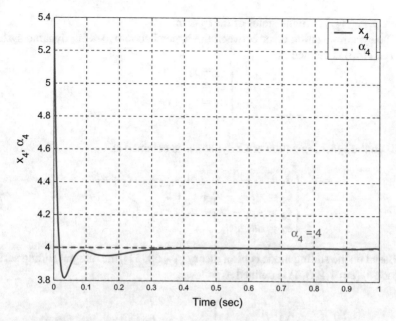

Fig. 10 Output regulation of the state x_4

6 Global Hyperchaos Synchronization of the New Hyperchaotic Finance Systems

In this section, we use adaptive integral sliding mode control for the global hyperchaos synchronization of the new hyperchaotic finance systems with unknown system parameters. The adaptive control mechanism helps the control design by estimating the unknown parameters [3, 4, 123].

As the master system, we consider the new hyperchaotic finance system given by

$$\begin{cases} \dot{x}_1 = x_3 + (x_2 - a)x_1 - x_4 \\ \dot{x}_2 = 1 - bx_2 - x_1^2 \\ \dot{x}_3 = -x_1 - cx_3 \\ \dot{x}_4 = dx_1 \end{cases} \tag{45}$$

where x_1, x_2, x_3, x_4 are the states of the system and a, b, c, d are unknown system parameters.

As the slave system, we consider the controlled new hyperchaotic finance system given by

$$\begin{cases} \dot{y}_1 = y_3 + (y_2 - a)y_1 - y_4 + u_1 \\ \dot{y}_2 = 1 - by_2 - y_1^2 + u_2 \\ \dot{y}_3 = -y_1 - cy_3 + u_3 \\ \dot{y}_4 = dy_1 + u_4 \end{cases} \tag{46}$$

where y_1, y_2, y_3, y_4 are the states of the system.

The synchronization error between the hyperchaotic two-wing dynamo systems (45) and (46) is defined as

$$\begin{cases} e_1 = y_1 - x_1 \\ e_2 = y_2 - x_2 \\ e_3 = y_3 - x_3 \\ e_4 = y_4 - x_4 \end{cases} \tag{47}$$

Then the synchronization error dynamics is obtained as

$$\begin{cases} \dot{e}_1 = e_3 - ae_1 - e_4 + y_1y_2 - x_1x_2 + u_1 \\ \dot{e}_2 = -be_2 - y_1^2 + x_1^2 + u_2 \\ \dot{e}_3 = -e_1 - ce_3 + u_3 \\ \dot{e}_4 = de_1 + u_4 \end{cases} \tag{48}$$

Based on the sliding mode control theory [44, 50, 51], the integral sliding surface of each e_i $(i = 1, 2, 3, 4)$ is defined as follows:

$$s_i = \left(\frac{d}{dt} + \lambda_i\right)\left(\int_0^t e_i(\tau)d\tau\right) = e_i + \lambda_i \int_0^t e_i(\tau)d\tau, \quad i = 1, 2, 3, 4 \tag{49}$$

From Eq. (49), it follows that

$$\begin{cases} \dot{s}_1 = \dot{e}_1 + \lambda_1 e_1 \\ \dot{s}_2 = \dot{e}_2 + \lambda_2 e_2 \\ \dot{s}_3 = \dot{e}_3 + \lambda_3 e_3 \\ \dot{s}_4 = \dot{e}_4 + \lambda_4 e_4 \end{cases} \tag{50}$$

The Hurwitz condition is realized if $\lambda_i > 0$ for $i = 1, 2, 3, 4$.

We consider the adaptive feedback control given by

$$\begin{cases} u_1 = -e_3 + \hat{a}(t)e_1 + e_4 - y_1y_2 + x_1x_2 - \lambda_1 e_1 - \eta_1 \operatorname{sgn}(s_1) - k_1 s_1 \\ u_2 = \hat{b}(t)e_2 + y_1^2 - x_1^2 - \lambda_2 e_2 - \eta_2 \operatorname{sgn}(s_2) - k_2 s_2 \\ u_3 = e_1 + \hat{c}(t)e_3 - \lambda_3 e_3 - \eta_3 \operatorname{sgn}(s_3) - k_3 s_3 \\ u_4 = -\hat{d}(t)e_1 - \lambda_4 e_4 - \eta_4 \operatorname{sgn}(s_4) - k_4 s_4 \end{cases} \tag{51}$$

where $\eta_i > 0$ and $k_i > 0$ for $i = 1, 2, 3, 4$.

Substituting (51) into (48), we obtain the closed-loop error dynamics as

$$\begin{cases} \dot{e}_1 = -[a - \hat{a}(t)]e_1 - \lambda_1 e_1 - \eta_1 \operatorname{sgn}(s_1) - k_1 s_1 \\ \dot{e}_2 = -[b - \hat{b}(t)]e_2 - \lambda_2 e_2 - \eta_2 \operatorname{sgn}(s_2) - k_2 s_2 \\ \dot{e}_3 = -[c - \hat{c}(t)]e_3 - \lambda_3 e_3 - \eta_3 \operatorname{sgn}(s_3) - k_3 s_3 \\ \dot{e}_4 = [d - \hat{d}(t)]e_1 - \lambda_4 e_4 - \eta_4 \operatorname{sgn}(s_4) - k_4 s_4 \end{cases} \tag{52}$$

We define the parameter estimation errors as

$$
\begin{cases}
e_a(t) = a - \hat{a}(t) \\
e_b(t) = b - \hat{b}(t) \\
e_c(t) = c - \hat{c}(t) \\
e_d(t) = d - \hat{d}(t)
\end{cases}
\tag{53}
$$

Using (53), we can simplify the closed-loop system (52) as

$$
\begin{cases}
\dot{e}_1 = -e_a e_1 - \lambda_1 e_1 - \eta_1 \operatorname{sgn}(s_1) - k_1 s_1 \\
\dot{e}_2 = -e_b e_2 - \lambda_2 e_2 - \eta_2 \operatorname{sgn}(s_2) - k_2 s_2 \\
\dot{e}_3 = -e_c e_3 - \lambda_3 e_3 - \eta_3 \operatorname{sgn}(s_3) - k_3 s_3 \\
\dot{e}_4 = e_d e_1 - \lambda_4 e_4 - \eta_4 \operatorname{sgn}(s_4) - k_4 s_4
\end{cases}
\tag{54}
$$

Differentiating (53) with respect to t, we get

$$
\begin{cases}
\dot{e}_a = -\dot{\hat{a}} \\
\dot{e}_b = -\dot{\hat{b}} \\
\dot{e}_c = -\dot{\hat{c}} \\
\dot{e}_d = -\dot{\hat{d}}
\end{cases}
\tag{55}
$$

Next, we state and prove the main result of this section.

Theorem 2 *The new hyperchaotic two-wing dynamo systems (45) and (46) are globally and asymptotically synchronized for all initial conditions $\mathbf{x}(0), \mathbf{y}(0) \in \mathbf{R}^4$ by the adaptive integral sliding mode control law (51) and the parameter update law*

$$
\begin{cases}
\dot{\hat{a}} = -s_1 e_1 \\
\dot{\hat{b}} = -s_2 e_2 \\
\dot{\hat{c}} = -s_3 e_3 \\
\dot{\hat{d}} = s_4 e_1
\end{cases}
\tag{56}
$$

where λ_i, η_i, k_i are positive constants for $i = 1, 2, 3, 4$.

Proof We consider the quadratic Lyapunov function defined by

$$
V(s_1, s_2, s_3, s_4, e_a, e_b, e_c, e_d) = \frac{1}{2}\left(s_1^2 + s_2^2 + s_3^2 + s_4^2\right) + \frac{1}{2}\left(e_a^2 + e_b^2 + e_c^2 + e_d^2\right) \tag{57}
$$

Clearly, V is positive definite on $\mathbf{R}^8$.
Using (50), (54) and (55), the time-derivative of V is obtained as

$$\dot{V} = s_1[-e_a e_1 - \eta_1 \operatorname{sgn}(s_1) - k_1 s_1] - e_a \dot{\hat{a}}$$
$$+s_2[-e_b e_2 - \eta_2 \operatorname{sgn}(s_2) - k_2 s_2] - e_b \dot{\hat{b}}$$
$$+s_3[-e_c e_3 - \eta_3 \operatorname{sgn}(s_3) - k_3 s_3] - e_c \dot{\hat{c}} \qquad (58)$$
$$+s_4[e_d e_1 - \eta_4 \operatorname{sgn}(s_4) - k_4 s_4] - e_d \dot{\hat{d}}$$

i.e.

$$\dot{V} = -\eta_1 |s_1| - k_1 s_1^2 - \eta_2 |s_2| - k_2 s_2^2 - \eta_3 |s_3| - k_3 s_3^2 - \eta_4 |s_4| - k_4 s_4^2$$
$$+e_a \left(-s_1 e_1 - \dot{\hat{a}} \right) + e_b \left(-s_2 e_2 - \dot{\hat{b}} \right) \qquad (59)$$
$$+e_c \left(-s_3 e_3 - \dot{\hat{c}} \right) + e_d \left(s_4 e_1 - \dot{\hat{d}} \right)$$

Using the parameter update law (56), we obtain

$$\dot{V} = -\eta_1 |s_1| - k_1 s_1^2 - \eta_2 |s_2| - k_2 s_2^2 - \eta_3 |s_3| - k_3 s_3^2 - \eta_4 |s_4| - k_4 s_4^2 \qquad (60)$$

which shows that $\dot{V}$ is negative semi-definite on $\mathbf{R}^8$.

Hence, by Barbalat's lemma [20], it is immediate that $\mathbf{e}(t)$ is globally asymptotically stable for all values of $\mathbf{e}(0) \in \mathbf{R}^4$.

Hence, it follows that the new hyperchaotic two-wing dynamo systems (45) and (46) are globally and asymptotically synchronized for all initial conditions $\mathbf{x}(0), \mathbf{y}(0) \in \mathbf{R}^4$.

This completes the proof. ∎

For numerical simulations, we take the parameter values of the new hyperchaotic two-disk dynamo systems (45) and (46) as in the hyperchaotic case (13), i.e.

$$a = 1, \quad b = 0.2, \quad c = 1.3, \quad d = 0.1 \qquad (61)$$

We take the values of the control parameters as

$$k_i = 20, \quad \eta_i = 0.1, \quad \lambda_i = 50, \quad \text{where } i = 1, 2, 3, 4 \qquad (62)$$

We take the estimates of the system parameters as

$$\hat{a}(0) = 14.2, \quad \hat{b}(0) = 15.4, \quad \hat{c}(0) = 7.5, \quad \hat{d}(0) = 15.2 \qquad (63)$$

We take the initial state of the master system (45) as

$$x_1(0) = 24.6, \quad x_2(0) = 14.7, \quad x_3(0) = 27.3, \quad x_4(0) = 19.4 \qquad (64)$$

We take the initial state of the slave system (46) as

$$y_1(0) = 12.5, \quad y_2(0) = 22.4, \quad y_3(0) = 4.1, \quad y_4(0) = 25.2 \qquad (65)$$

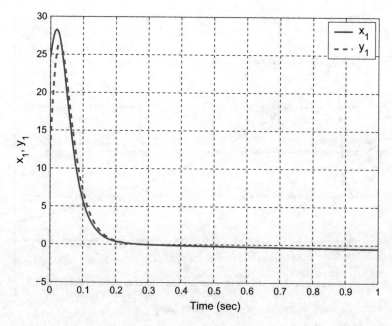

Fig. 11 Complete synchronization of the states x_1 and y_1

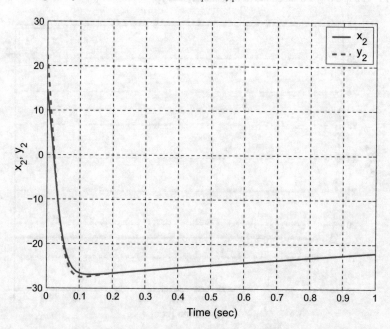

Fig. 12 Complete synchronization of the states x_2 and y_2

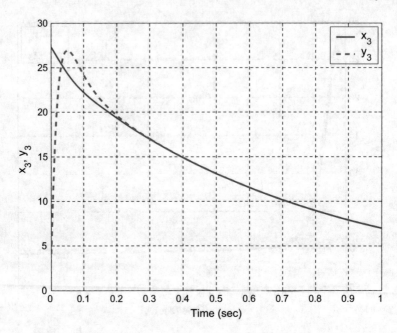

Fig. 13 Complete synchronization of the states x_3 and y_3

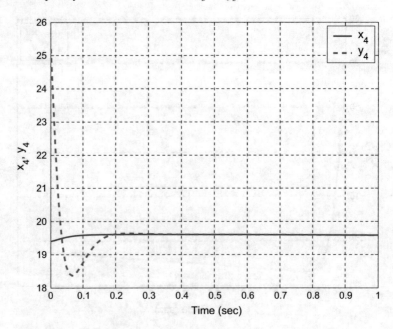

Fig. 14 Complete synchronization of the states x_4 and y_4

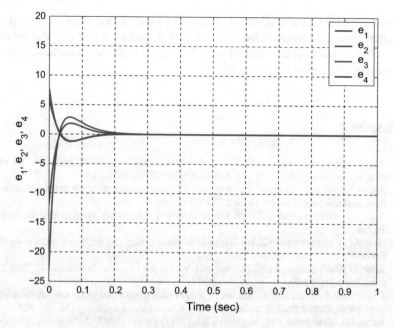

Fig. 15 Time-history of the synchronization errors e_1, e_2, e_3, e_4

Figures 11, 12, 13 and 14 show the complete synchronization of the new hyper-chaotic finance systems (45) and (46). Figure 15 shows the time-history of the synchronization errors e_1, e_2, e_3, e_4.

7 Conclusions

Control and synchronization of chaotic and hyperchaotic systems are miscellaneous research problems in chaos theory. Sliding mode control is an important method used to solve various problems in control systems engineering. In robust control systems, the sliding mode control is often adopted due to its inherent advantages of easy realization, fast response and good transient performance as well as insensitivity to parameter uncertainties and disturbance. In this work, we described a novel 4-D hyperchaotic finance system. We showed that the hyperchaotic finance system has a unique equilibrium point, which is locally exponentially stable. Also, we established that the Lyapunov exponents of the hyperchaotic finance system are $L_1 = 0.0365$, $L_2 = 0.0172$, $L_3 = 0$ and $L_4 = -0.8727$. The Kaplan-Yorke dimension of the hyperchaotic finance system has been derived as $D_{KY} = 3.0615$, which shows the high complexity of the system. Next, an adaptive integral sliding mode control scheme was derived for the global regulation of all the trajectories of the hyperchaotic finance system. Also, an adaptive integral sliding mode control scheme was proposed for the global hyperchaos synchronization of identical hyperchaotic finance systems. The adaptive control mechanism helps the control design by estimating the unknown parameters. Numerical simulations using MATLAB were shown to demonstrate all the main results derived in this work.

References

1. Akgul A, Hussain S, Pehlivan I (2016) A new three-dimensional chaotic system, its dynamical analysis and electronic circuit applications. Optik 127(18):7062–7071
2. Akgul A, Moroz I, Pehlivan I, Vaidyanathan S (2016) A new four-scroll chaotic attractor and its engineering applications. Optik 127(13):5491–5499
3. Azar AT, Vaidyanathan S (2015) Chaos modeling and control systems design. Springer, Berlin
4. Azar AT, Vaidyanathan S (2016) Advances in chaos theory and intelligent control. Springer, Berlin
5. Azar AT, Vaidyanathan S (2017) Fractional order control and synchronization of chaotic systems. Springer, Berlin
6. Boulkroune A, Bouzeriba A, Bouden T (2016) Fuzzy generalized projective synchronization of incommensurate fractional-order chaotic systems. Neurocomputing 173:606–614
7. Burov DA, Evstigneev NM, Magnitskii NA (2017) On the chaotic dynamics in two coupled partial differential equations for evolution of surface plasmon polaritons. Commun Nonlinear Sci Numer Simul 46:26–36
8. Chai X, Chen Y, Broyde L (2017) A novel chaos-based image encryption algorithm using DNA sequence operations. Opt Lasers Eng 88:197–213
9. Chen A, Lu J, Lü J, Yu S (2006) Generating hyperchaotic Lü attractor via state feedback control. Phys A 364:103–110
10. Chenaghlu MA, Jamali S, Khasmakhi NN (2016) A novel keyed parallel hashing scheme based on a new chaotic system. Chaos Solitons Fractals 87:216–225
11. Dadras S, Momeni HR, Qi G, lin Wang Z, (2012) Four-wing hyperchaotic attractor generated from a new 4D system with one equilibrium and its fractional-order form. Nonlinear Dyn 67:1161–1173
12. Fallahi K, Leung H (2010) A chaos secure communication scheme based on multiplication modulation. Commun Nonlinear Sci Numer Simul 15(2):368–383
13. Fontes RT, Eisencraft M (2016) A digital bandlimited chaos-based communication system. Commun Nonlinear Sci Numer Simul 37:374–385
14. Fotsa RT, Woafo P (2016) Chaos in a new bistable rotating electromechanical system. Chaos Solitons Fractals 93:48–57
15. Fujisaka H, Yamada T (1983) Stability theory of synchronized motion in coupled-oscillator systems. Progress Theoret Phys 63:32–47
16. Gao T, Chen Z, Yuan Z, Chen G (2006) A hyperchaos generated from Chen's system. Int J Mod Phys C 17(4):471–478
17. Huang D, Li H (1993) Theory and method of the nonlinear economics. Publishing House of Sichuan University, Chengdu
18. Jia Q (2007) Hyperchaos generated from the Lorenz chaotic system and its control. Phys Lett A 366(3):217–222
19. Kacar S (2016) Analog circuit and microcontroller based RNG application of a new easy realizable 4D chaotic system. Optik 127(20):9551–9561
20. Khalil HK (2002) Nonlinear systems. Prentice Hall, New York
21. Lakhekar GV, Waghmare LM, Vaidyanathan S (2016) Diving autopilot design for underwater vehicles using an adaptive neuro-fuzzy sliding mode controller. In: Vaidyanathan S, Volos C

(eds) Advances and applications in nonlinear control systems. Springer, Germany, pp 477–503

22. Li D (2008) A three-scroll chaotic attractor. Phys Lett A 372(4):387–393
23. Liu H, Ren B, Zhao Q, Li N (2016) Characterizing the optical chaos in a special type of small networks of semiconductor lasers using permutation entropy. Opt Commun 359:79–84
24. Liu W, Sun K, Zhu C (2016) A fast image encryption algorithm based on chaotic map. Opt Lasers Eng 84:26–36
25. Liu X, Mei W, Du H (2016) Simultaneous image compression, fusion and encryption algorithm based on compressive sensing and chaos. Opt Commun 366:22–32
26. Moussaoui S, Boulkroune A, Vaidyanathan S (2016) Fuzzy adaptive sliding-mode control scheme for uncertain underactuated systems. In: Vaidyanathan S, Volos C (eds) Advances and applications in nonlinear control systems. Springer, Germany, pp 351–367
27. Pecora LM, Carroll TL (1991) Synchronizing chaotic circuits. IEEE Trans Circuits Syst 38:453–456
28. Pecora LM, Carroll TL (1990) Synchronization in chaotic systems. Phys Rev Lett 64:821–824
29. Pehlivan I, Moroz IM, Vaidyanathan S (2014) Analysis, synchronization and circuit design of a novel butterfly attractor. J Sound Vib 333(20):5077–5096
30. Pham VT, Volos C, Jafari S, Wang X, Vaidyanathan S (2014) Hidden hyperchaotic attractor in a novel simple memristive neural network. Optoelectron Adv Mater Rapid Commun 8(11–12):1157–1163
31. Pham VT, Volos CK, Vaidyanathan S (2015) Multi-scroll chaotic oscillator based on a first-order delay differential equation. In: Azar AT, Vaidyanathan S (eds) Chaos modeling and control systems design. Studies in computational intelligence, vol 581. Springer, Germany, pp 59–72
32. Pham VT, Volos CK, Vaidyanathan S, Le TP, Vu VY (2015) A memristor-based hyperchaotic system with hidden attractors: dynamics, synchronization and circuital emulating. J Eng Sci Technol Rev 8(2):205–214
33. Pham VT, Jafari S, Vaidyanathan S, Volos C, Wang X (2016) A novel memristive neural network with hidden attractors and its circuitry implementation. Sci China Technol Sci 59(3):358–363
34. Pham VT, Jafari S, Volos C, Giakoumis A, Vaidyanathan S, Kapitaniak T (2016) A chaotic system with equilibria located on the rounded square loop and its circuit implementation. IEEE Trans Circuits Syst II: Express Briefs 63(9):878–882
35. Pham VT, Jafari S, Volos C, Vaidyanathan S, Kapitaniak T (2016) A chaotic system with infinite equilibria located on a piecewise linear curve. Optik 127(20):9111–9117
36. Pham VT, Vaidyanathan S, Volos C, Jafari S, Kingni ST (2016) A no-equilibrium hyper-chaotic system with a cubic nonlinear term. Optik 127(6):3259–3265
37. Pham VT, Vaidyanathan S, Volos CK, Hoang TM, Yem VV (2016) Dynamics, synchronization and SPICE implementation of a memristive system with hidden hyperchaotic attractor. In: Azar AT, Vaidyanathan S (eds) Advances in chaos theory and intelligent control. Springer, Berlin, pp 35–52
38. Pham VT, Vaidyanathan S, Volos CK, Jafari S, Kuznetsov NV, Hoang TM (2016) A novel memristive time-delay chaotic system without equilibrium points. Eur Phys J Spec Top 225(1):127–136
39. Pham VT, Vaidyanathan S, Volos CK, Jafari S, Wang X (2016) A chaotic hyperjerk system based on memristive device. In: Vaidyanathan S, Volos C (eds) Advances and applications in chaotic systems. Springer, Berlin, pp 39–58
40. Rössler O (1979) An equation for hyperchaos. Phys Lett A 71(2–3):155–157
41. Sampath S, Vaidyanathan S, Volos CK, Pham VT (2015) An eight-term novel four-scroll chaotic system with cubic nonlinearity and its circuit simulation. J Eng Sci Technol Rev 8(2):1–6
42. Sampath S, Vaidyanathan S, Pham VT (2016) A novel 4-D hyperchaotic system with three quadratic nonlinearities, its adaptive control and circuit simulation. Int J Control Theory Appl 9(1):339–356

43. Shirkhani N, Khanesar M, Teshnehlab M (2016) Indirect model reference fuzzy control of SISO fractional order nonlinear chaotic systems. Procedia Comput Sci 102:309–316
44. Slotine J, Li W (1991) Applied nonlinear control. Prentice-Hall, Englewood Cliffs
45. Sundarapandian V (2013) Analysis and anti-synchronization of a novel chaotic system via active and adaptive controllers. J Eng Sci Technol Rev 6(4):45–52
46. Sundarapandian V, Pehlivan I (2012) Analysis, control, synchronization, and circuit design of a novel chaotic system. Math Comput Model 55(7–8):1904–1915
47. Sundarapandian V, Sivaperumal S (2011) Sliding controller design of hybrid synchronization of four-wing chaotic systems. Int J Soft Comput 6(5):224–231
48. Szmit Z, Warminski J (2016) Nonlinear dynamics of electro-mechanical system composed of two pendulums and rotating hub. Procedia Eng 144:953–958
49. Tacha OI, Volos CK, Kyprianidis IM, Stouboulos IN, Vaidyanathan S, Pham VT (2016) Analysis, adaptive control and circuit simulation of a novel nonlinear finance system. Appl Math Comput 276:200–217
50. Utkin VI (1977) Variable structure systems with sliding modes. IEEE Trans Autom Control 22(2):212–222
51. Utkin VI (1993) Sliding mode control design principles and applications to electric drives. IEEE Trans Industr Electron 40(1):23–36
52. Vaidyanathan S (2011) Analysis and synchronization of the hyperchaotic Yujun systems via sliding mode control. Adv Intell Syst Comput 176:329–337
53. Vaidyanathan S (2012) Global chaos control of hyperchaotic Liu system via sliding control method. Int J Control Theory Appl 5(2):117–123
54. Vaidyanathan S (2012) Sliding mode control based global chaos control of Liu-Liu-Liu-Su chaotic system. Int J Control Theory Appl 5(1):15–20
55. Vaidyanathan S (2013) A new six-term 3-D chaotic system with an exponential nonlinearity. Far East J Math Sci 79(1):135–143
56. Vaidyanathan S (2013) A ten-term novel 4-D hyperchaotic system with three quadratic non-linearities and its control. Int J Control Theory Appl 6(2):97–109
57. Vaidyanathan S (2013) Analysis and adaptive synchronization of two novel chaotic systems with hyperbolic sinusoidal and cosinusoidal nonlinearity and unknown parameters. J Eng Sci Technol Rev 6(4):53–65
58. Vaidyanathan S (2014) A new eight-term 3-D polynomial chaotic system with three quadratic nonlinearities. Far East J Math Sci 84(2):219–226
59. Vaidyanathan S (2014) Analysis and adaptive synchronization of eight-term 3-D polynomial chaotic systems with three quadratic nonlinearities. Eur Phys J Spec Top 223(8):1519–1529
60. Vaidyanathan S (2014) Analysis, control and synchronisation of a six-term novel chaotic system with three quadratic nonlinearities. Int J Model Ident Control 22(1):41–53
61. Vaidyanathan S (2014) Generalised projective synchronisation of novel 3-D chaotic systems with an exponential non-linearity via active and adaptive control. Int J Model Ident Control 22(3):207–217
62. Vaidyanathan S (2014) Global chaos synchronisation of identical Li-Wu chaotic systems via sliding mode control. Int J Model Ident Control 22(2):170–177
63. Vaidyanathan S (2014) Qualitative analysis and control of an eleven-term novel 4-D hyperchaotic system with two quadratic nonlinearities. Int J Control Theory Appl 7(1):35–47
64. Vaidyanathan S (2015) A 3-D novel highly chaotic system with four quadratic nonlinearities, its adaptive control and anti-synchronization with unknown parameters. J Eng Sci Technol Rev 8(2):106–115
65. Vaidyanathan S (2015) A novel chemical chaotic reactor system and its adaptive control. Int J ChemTech Res 8(7):146–158
66. Vaidyanathan S (2015) A novel chemical chaotic reactor system and its output regulation via integral sliding mode control. Int J ChemTech Res 8(11):669–683
67. Vaidyanathan S (2015) Active control design for the anti-synchronization of Lotka-Volterra biological systems with four competitive species. Int J PharmTech Res 8(7):58–70

68. Vaidyanathan S (2015) Adaptive control design for the anti-synchronization of novel 3-D chemical chaotic reactor systems. Int J ChemTech Res 8(11):654–668
69. Vaidyanathan S (2015) Adaptive control of a chemical chaotic reactor. Int J PharmTech Res 8(3):377–382
70. Vaidyanathan S (2015) Adaptive control of the FitzHugh-Nagumo chaotic neuron model. Int J PharmTech Res 8(6):117–127
71. Vaidyanathan S (2015) Adaptive synchronization of chemical chaotic reactors. Int J ChemTech Res 8(2):612–621
72. Vaidyanathan S (2015) Adaptive synchronization of generalized Lotka-Volterra three-species biological systems. Int J PharmTech Res 8(5):928–937
73. Vaidyanathan S (2015) Adaptive synchronization of novel 3-D chemical chaotic reactor systems. Int J ChemTech Res 8(7):159–171
74. Vaidyanathan S (2015) Analysis, properties and control of an eight-term 3-D chaotic system with an exponential nonlinearity. Int J Model Ident Control 23(2):164–172
75. Vaidyanathan S (2015) Anti-synchronization of brusselator chemical reaction systems via adaptive control. Int J ChemTech Res 8(6):759–768
76. Vaidyanathan S (2015) Anti-synchronization of brusselator chemical reaction systems via integral sliding mode control. Int J ChemTech Res 8(11):700–713
77. Vaidyanathan S (2015) Anti-synchronization of chemical chaotic reactors via adaptive control method. Int J ChemTech Res 8(8):73–85
78. Vaidyanathan S (2015) Anti-synchronization of the FitzHugh-Nagumo chaotic neuron models via adaptive control method. Int J PharmTech Res 8(7):71–83
79. Vaidyanathan S (2015) Chaos in neurons and synchronization of Birkhoff-Shaw strange chaotic attractors via adaptive control. Int J PharmTech Res 8(6):1–11
80. Vaidyanathan S (2015) Dynamics and control of brusselator chemical reaction. Int J ChemTech Res 8(6):740–749
81. Vaidyanathan S (2015) Dynamics and control of Tokamak system with symmetric and magnetically confined plasma. Int J ChemTech Res 8(6):795–802
82. Vaidyanathan S (2015) Global chaos synchronization of chemical chaotic reactors via novel sliding mode control method. Int J ChemTech Res 8(7):209–221
83. Vaidyanathan S (2015) Global chaos synchronization of Duffing double-well chaotic oscillators via integral sliding mode control. Int J ChemTech Res 8(11):141–151
84. Vaidyanathan S (2015) Global chaos synchronization of the forced Van der Pol chaotic oscillators via adaptive control method. Int J PharmTech Res 8(6):156–166
85. Vaidyanathan S (2015) Hybrid chaos synchronization of the FitzHugh-Nagumo chaotic neuron models via adaptive control method. Int J PharmTech Res 8(8):48–60
86. Vaidyanathan S (2015) Integral sliding mode control design for the global chaos synchronization of identical novel chemical chaotic reactor systems. Int J ChemTech Res 8(11):684–699
87. Vaidyanathan S (2015) Output regulation of the forced Van der Pol chaotic oscillator via adaptive control method. Int J PharmTech Res 8(6):106–116
88. Vaidyanathan S (2015) Sliding controller design for the global chaos synchronization of forced Van der Pol chaotic oscillators. Int J PharmTech Res 8(7):100–111
89. Vaidyanathan S (2015) Synchronization of Tokamak systems with symmetric and magnetically confined plasma via adaptive control. Int J ChemTech Res 8(6):818–827
90. Vaidyanathan S (2016) A no-equilibrium novel 4-D highly hyperchaotic system with four quadratic nonlinearities and its adaptive control. In: Vaidyanathan S, Volos C (eds) Advances and applications in nonlinear control systems. Springer, Berlin, pp 235–258
91. Vaidyanathan S (2016) A novel 3-D conservative chaotic system with a sinusoidal nonlinearity and its adaptive control. Int J Control Theory Appl 9(1):115–132
92. Vaidyanathan S (2016) A novel 3-D jerk chaotic system with three quadratic nonlinearities and its adaptive control. Arch Control Sci 26(1):19–47
93. Vaidyanathan S (2016) A novel 3-D jerk chaotic system with two quadratic nonlinearities and its adaptive backstepping control. Int J Control Theory Appl 9(1):199–219

94. Vaidyanathan S (2016) A novel 4-D hyperchaotic thermal convection system and its adaptive control. In: Azar AT, Vaidyanathan S (eds) Advances in chaos theory and intelligent control. Springer, Berlin, pp 75–100
95. Vaidyanathan S (2016) A novel 5-D hyperchaotic system with a line of equilibrium points and its adaptive control. In: Vaidyanathan S, Volos C (eds) Advances and applications in chaotic systems. Springer, Berlin, pp 471–494
96. Vaidyanathan S (2016) A novel highly hyperchaotic system and its adaptive control. In: Vaidyanathan S, Volos C (eds) Advances and applications in chaotic systems. Springer, Berlin, pp 513–535
97. Vaidyanathan S (2016) A novel hyperchaotic hyperjerk system with two nonlinearities, its analysis, adaptive control and synchronization via backstepping control method. Int J Control Theory Appl 9(1):257–278
98. Vaidyanathan S (2016) An eleven-term novel 4-D hyperchaotic system with three quadratic nonlinearities, analysis, control and synchronization via adaptive control method. Int J Control Theory Appl 9(1):21–43
99. Vaidyanathan S (2016) Analysis, adaptive control and synchronization of a novel 4-D hyperchaotic hyperjerk system via backstepping control method. Arch Control Sci 26(3):311–338
100. Vaidyanathan S (2016) Analysis, control and synchronization of a novel 4-D highly hyperchaotic system with hidden attractors. In: Azar AT, Vaidyanathan S (eds) Advances in chaos theory and intelligent control. Springer, Berlin, pp 529–552
101. Vaidyanathan S (2016) Anti-synchronization of 3-cells cellular neural network attractors via integral sliding mode control. Int J PharmTech Res 9(1):193–205
102. Vaidyanathan S (2016) Global chaos control of the FitzHugh-Nagumo chaotic neuron model via integral sliding mode control. Int J PharmTech Res 9(4):413–425
103. Vaidyanathan S (2016) Global chaos control of the generalized Lotka-Volterra three-species system via integral sliding mode control. Int J PharmTech Res 9(4):399–412
104. Vaidyanathan S (2016) Global chaos regulation of a symmetric nonlinear gyro system via integral sliding mode control. Int J ChemTech Res 9(5):462–469
105. Vaidyanathan S (2016) Hybrid synchronization of the generalized Lotka-Volterra three-species biological systems via adaptive control. Int J PharmTech Res 9(1):179–192
106. Vaidyanathan S (2016) Hyperchaos, adaptive control and synchronization of a novel 4-D hyperchaotic system with two quadratic nonlinearities. Arch Control Sci 26(4):471–495
107. Vaidyanathan S (2016) Mathematical analysis, adaptive control and synchronization of a ten-term novel three-scroll chaotic system with four quadratic nonlinearities. Int J Control Theory Appl 9(1):1–20
108. Vaidyanathan S (2016) Qualitative analysis and properties of a novel 4-D hyperchaotic system with two quadratic nonlinearities and its adaptive control. In: Azar AT, Vaidyanathan S (eds) Advances in chaos theory and intelligent control. Springer, Berlin, pp 455–480
109. Vaidyanathan S, Azar AT (2015) Analysis and control of a 4-D novel hyperchaotic system. In: Azar AT, Vaidyanathan S (eds) Chaos modeling and control systems design. Studies in computational intelligence, vol 581. Springer, Germany, pp 3–17
110. Vaidyanathan S, Azar AT (2015) Analysis, control and synchronization of a nine-term 3-D novel chaotic system. In: Azar AT, Vaidyanathan S (eds) Chaos modelling and control systems design. Studies in computational intelligence, vol 581. Springer, Germany, pp 19–38
111. Vaidyanathan S, Azar AT (2016) A novel 4-D four-wing chaotic system with four quadratic nonlinearities and its synchronization via adaptive control method. In: Azar AT, Vaidyanathan S (eds) Advances in chaos theory and intelligent control. Springer, Berlin, pp 203–224
112. Vaidyanathan S, Azar AT (2016) Qualitative study and adaptive control of a novel 4-D hyperchaotic system with three quadratic nonlinearities. In: Azar AT, Vaidyanathan S (eds) Advances in chaos theory and intelligent control. Springer, Berlin, pp 179–202
113. Vaidyanathan S, Azar AT (2016) Takagi-Sugeno fuzzy logic controller for Liu-Chen four-scroll chaotic system. Int J Intell Eng Inform 4(2):135–150
114. Vaidyanathan S, Boulkroune A (2016) A novel 4-D hyperchaotic chemical reactor system and its adaptive control. In: Vaidyanathan S, Volos C (eds) Advances and applications in chaotic systems. Springer, Berlin, pp 447–469

115. Vaidyanathan S, Madhavan K (2013) Analysis, adaptive control and synchronization of a seven-term novel 3-D chaotic system. Int J Control Theory Appl 6(2):121–137
116. Vaidyanathan S, Pakiriswamy S (2016) A five-term 3-D novel conservative chaotic system and its generalized projective synchronization via adaptive control method. Int J Control Theory Appl 9(1):61–78
117. Vaidyanathan S, Pakiriswamy S (2015) A 3-D novel conservative chaotic system and its generalized projective synchronization via adaptive control. J Eng Sci Technol Rev 8(2):52–60
118. Vaidyanathan S, Pakiriswamy S (2016) Adaptive control and synchronization design of a seven-term novel chaotic system with a quartic nonlinearity. Int J Control Theory Appl 9(1):237–256
119. Vaidyanathan S, Rajagopal K (2016) Adaptive control, synchronization and LabVIEW implementation of Rucklidge chaotic system for nonlinear double convection. Int J Control Theory Appl 9(1):175–197
120. Vaidyanathan S, Rajagopal K (2016) Analysis, control, synchronization and LabVIEW implementation of a seven-term novel chaotic system. Int J Control Theory Appl 9(1):151–174
121. Vaidyanathan S, Sampath S (2011) Global chaos synchronization of hyperchaotic Lorenz systems by sliding mode control. Commun Comput Inf Sci 205:156–164
122. Vaidyanathan S, Volos C (2015) Analysis and adaptive control of a novel 3-D conservative no-equilibrium chaotic system. Arch Control Sci 25(3):333–353
123. Vaidyanathan S, Volos C (2016) Advances and applications in chaotic systems. Springer, Berlin
124. Vaidyanathan S, Volos C (2016) Advances and applications in nonlinear control systems. Springer, Berlin
125. Vaidyanathan S, Volos C (2017) Advances in memristors, memristive devices and systems. Springer, Berlin
126. Vaidyanathan S, Volos C, Pham VT (2014) Hyperchaos, adpative control and synchronization of a novel 5-D hyperchaotic system with three positive Lyapunov exponents and its SPICE implementation. Arch Control Sci 24(4):409–446
127. Vaidyanathan S, Volos C, Pham VT, Madhavan K, Idowu BA (2014) Adaptive backstepping control, synchronization and circuit simulation of a 3-D novel jerk chaotic system with two hyperbolic sinusoidal nonlinearities. Arch Control Sci 24(3):375–403
128. Vaidyanathan S, Azar AT, Rajagopal K, Alexander P (2015) Design and SPICE implementation of a 12-term novel hyperchaotic system and its synchronisation via active control. Int J Model Ident Control 23(3):267–277
129. Vaidyanathan S, Pham VT, Volos CK (2015) A 5-D hyperchaotic Rikitake dynamo system with hidden attractors. Eur Phys J Spec Top 224(8):1575–1592
130. Vaidyanathan S, Rajagopal K, Volos CK, Kyprianidis IM, Stouboulos IN (2015) Analysis, adaptive control and synchronization of a seven-term novel 3-D chaotic system with three quadratic nonlinearities and its digital implementation in LabVIEW. J Eng Sci Technol Rev 8(2):130–141
131. Vaidyanathan S, Volos C, Pham VT, Madhavan K (2015) Analysis, adaptive control and synchronization of a novel 4-D hyperchaotic hyperjerk system and its SPICE implementation. Arch Control Sci 25(1):135–158
132. Vaidyanathan S, Volos CK, Kyprianidis IM, Stouboulos IN, Pham VT (2015) Analysis, adaptive control and anti-synchronization of a six-term novel jerk chaotic system with two exponential nonlinearities and its circuit simulation. J Eng Sci Technol Rev 8(2):24–36
133. Vaidyanathan S, Volos CK, Pham VT (2015) Analysis, adaptive control and adaptive synchronization of a nine-term novel 3-D chaotic system with four quadratic nonlinearities and its circuit simulation. J Eng Sci Technol Rev 8(2):181–191
134. Vaidyanathan S, Volos CK, Pham VT (2015) Analysis, control, synchronization and SPICE implementation of a novel 4-D hyperchaotic Rikitake dynamo system without equilibrium. J Eng Sci Technol Rev 8(2):232–244
135. Vaidyanathan S, Volos CK, Pham VT (2015) Global chaos control of a novel nine-term chaotic system via sliding mode control. In: Azar AT, Zhu Q (eds) Advances and applications

in sliding mode control systems. Studies in computational intelligence, vol 576. Springer, Germany, pp 571–590

136. Vaidyanathan S, Volos CK, Pham VT (2016) Hyperchaos, control, synchronization and circuit simulation of a novel 4-D hyperchaotic system with three quadratic nonlinearities. In: Azar AT, Vaidyanathan S (eds) Advances in chaos theory and intelligent control. Springer, Berlin, pp 297–325

137. Volos CK, Kyprianidis IM, Stouboulos IN, Tlelo-Cuautle E, Vaidyanathan S (2015) Memristor: a new concept in synchronization of coupled neuromorphic circuits. J Eng Sci Technol Rev 8(2):157–173

138. Volos CK, Pham VT, Vaidyanathan S, Kyprianidis IM, Stouboulos IN (2015) Synchronization phenomena in coupled Colpitts circuits. J Eng Sci Technol Rev 8(2):142–151

139. Volos CK, Pham VT, Vaidyanathan S, Kyprianidis IM, Stouboulos IN (2016) Synchronization phenomena in coupled hyperchaotic oscillators with hidden attractors using a nonlinear open loop controller. In: Vaidyanathan S, Volos C (eds) Advances and applications in chaotic systems. Springer, Berlin, pp 1–38

140. Volos CK, Pham VT, Vaidyanathan S, Kyprianidis IM, Stouboulos IN (2016) The case of bidirectionally coupled nonlinear circuits via a memristor. In: Vaidyanathan S, Volos C (eds) Advances and applications in nonlinear control systems. Springer, Berlin, pp 317–350

141. Volos CK, Prousalis D, Kyprianidis IM, Stouboulos I, Vaidyanathan S, Pham VT (2016) Synchronization and anti-synchronization of coupled Hindmarsh-Rose neuron models. Int J Control Theory Appl 9(1):101–114

142. Volos CK, Vaidyanathan S, Pham VT, Maaita JO, Giakoumis A, Kyprianidis IM, Stouboulos IN (2016) A novel design approach of a nonlinear resistor based on a memristor emulator. In: Azar AT, Vaidyanathan S (eds) Advances in chaos theory and intelligent control. Springer, Berlin, pp 3–34

143. Wang B, Zhong SM, Dong XC (2016) On the novel chaotic secure communication scheme design. Commun Nonlinear Sci Numer Simul 39:108–117

144. Wolf A, Swift JB, Swinney HL, Vastano JA (1985) Determining Lyapunov exponents from a time series. Phys D 16:285–317

145. Wu T, Sun W, Zhang X, Zhang S (2016) Concealment of time delay signature of chaotic output in a slave semiconductor laser with chaos laser injection. Opt Commun 381:174–179

146. Xu G, Liu F, Xiu C, Sun L, Liu C (2016) Optimization of hysteretic chaotic neural network based on fuzzy sliding mode control. Neurocomputing 189:72–79

147. Xu H, Tong X, Meng X (2016) An efficient chaos pseudo-random number generator applied to video encryption. Optik 127(20):9305–9319

148. Zhou W, Xu Y, Lu H, Pan L (2008) On dynamics analysis of a new chaotic attractor. Phys Lett A 372(36):5773–5777

149. Zhu C, Liu Y, Guo Y (2010) Theoretic and numerical study of a new chaotic system. Intell Inf Manage 2:104–109

Adaptive Integral Sliding Mode Controller Design for the Control of a Novel 6-D Coupled Double Convection Hyperchaotic System

Sundarapandian Vaidyanathan

Abstract Chaos and hyperchaos have important applications in finance, physics, chemistry, biology, ecology, secure communications, cryptosystems and many scientific branches. Control and synchronization of chaotic and hyperchaotic systems are important research problems in chaos theory. Sliding mode control is an important method used to solve various problems in control systems engineering. In robust control systems, the sliding mode control is often adopted due to its inherent advantages of easy realization, fast response and good transient performance as well as insensitivity to parameter uncertainties and disturbance. In the recent decades, there is great interest shown in the finding of chaotic motion and oscillations in nonlinear dynamical systems arising in science and engineering. Rucklidge chaotic system (1992) is a famous 3-D chaotic system for nonlinear double convection in fluid mechanics. By bidirectional coupling of Rucklidge chaotic systems, we obtain a novel 6-D coupled double convection hyperchaotic system. We describe the phase portraits and qualitative properties of the novel 6-D hyperchaotic system. Also, we show that the Lyapunov exponents of the novel hyperchaotic system are $L_1 = 0.5065$, $L_2 = 0.1917$, $L_3 = 0$, $L_4 = -0.0180$, $L_5 = -2.6885$ and $L_6 = -3.1917$. The Kaplan-Yorke dimension of the 6-D hyperchaotic system is derived as $D_{KY} = 4.2530$, which shows the high complexity of the system. Next, an adaptive integral sliding mode control scheme is proposed for the global control of all the trajectories of the 6-D hyperchaotic system. The adaptive control mechanism helps the control design by estimating the unknown parameters. Numerical simulations using MATLAB are shown to illustrate all the main results derived in this work.

Keywords Chaos · Hyperchaos · Chaotic systems · Control · Integral sliding mode control · Adaptive control · Double convection

S. Vaidyanathan (✉)
Research and Development Centre, Vel Tech University, Avadi,
Chennai 600062, Tamil Nadu, India
e-mail: sundarcontrol@gmail.com

© Springer International Publishing AG 2017
S. Vaidyanathan and C.-H. Lien (eds.), *Applications of Sliding Mode Control
in Science and Engineering*, Studies in Computational Intelligence 709,
DOI 10.1007/978-3-319-55598-0_14

319

1 Introduction

Chaos theory describes the quantitative study of unstable aperiodic dynamic behavior in deterministic nonlinear dynamical systems. For the motion of a dynamical system to be chaotic, the system variables should contain some nonlinear terms and the system must satisfy three properties: boundedness, infinite recurrence and sensitive dependence on initial conditions [3–5, 122–124].

A hyperchaotic system is defined as a chaotic system with at least two positive Lyapunov exponents [3, 4, 122]. Combined with one null exponent along the flow and one negative exponent to ensure the boundedness of the solution, the minimal dimension for a continuous-time hyperchaotic system is four.

Some classical examples of hyperchaotic systems are hyperchaotic Rössler system [37], hyperchaotic Lorenz system [17], hyperchaotic Chen system [15], hyperchaotic Lü system [9], etc. Some recent examples of hyperchaotic systems are hyperchaotic Dadras system [11], hyperchaotic Vaidyanathan systems [55, 62, 89, 93–99, 105, 107, 108, 110, 111, 113, 125, 127, 128, 130, 133, 135], hyperchaotic Sampath system [40], hyperchaotic Pham system [33], etc.

The problem of global control of a chaotic system is to device feedback control laws so that the closed-loop system is globally asymptotically stable. Global chaos control of various chaotic systems has been investigated via various methods in the control literature [3, 4, 122, 123].

Many new chaotic systems have been discovered in the recent years such as Zhou system [147], Zhu system [148], Li system [21], Sundarapandian systems [44, 45], Vaidyanathan systems [54, 56–60, 63, 73, 90–92, 106, 109, 114–117, 119, 121, 126, 129, 131, 132, 134], Pehlivan system [26], Sampath system [39], Tacha system [48], Pham systems [28, 31, 32, 35], Akgul system [2], etc.

Chaos theory has applications in several fields such as memristors [27, 29, 30, 34–36, 136, 139, 141], fuzzy logic [6, 42, 112, 145], communication systems [12, 13, 142], cryptosystems [8, 10], electromechanical systems [14, 47], lasers [7, 22, 144], encryption [23, 24, 146], electrical circuits [1, 2, 18, 118, 137], chemical reactions [64, 65, 67, 68, 70, 72, 74–76, 79, 81, 85, 113], oscillators [82, 83, 86, 87, 138], tokamak systems [80, 88], neurology [69, 77, 78, 84, 101, 140], ecology [66, 71, 102, 104], etc.

The adaptive control mechanism helps the control design by estimating the unknown parameters [3, 4, 122]. The sliding mode control approach is recognized as an efficient tool for designing robust controllers for linear or nonlinear control systems operating under uncertainty conditions [49, 50].

A major advantage of sliding mode control is low sensitivity to parameter variations in the plant and disturbances affecting the plant, which eliminates the necessity of exact modeling of the plant. In the sliding mode control, the control dynamics will have two sequential modes, viz. the reaching mode and the sliding mode. Basically, a sliding mode controller design consists of two parts: hyperplane design and controller design. A hyperplane is first designed via the pole-placement approach and a controller is then designed based on the sliding condition. The stability of the overall system is guaranteed by the sliding condition and by a stable hyperplane. Sliding mode control method is a popular method for the control and synchronization of chaotic systems [20, 25, 46, 51–53, 61, 100, 103, 120].

In the recent decades, there is great interest shown in the finding of chaotic motion and oscillations in nonlinear dynamical systems arising in science and engineering. Rucklidge chaotic system (1992) is a famous 3-D chaotic system for nonlinear double convection in fluid mechanics [38].

Bidirectional coupling has important applications in chaos theory and secure communications [16, 41]. By bidirectional coupling of Rucklidge chaotic systems, we obtain a novel 6-D coupled double convection hyperchaotic system. We describe the phase portraits and qualitative properties of the novel 6-D hyperchaotic system.

Next, we derive the qualitative properties of the 6-D hyperchaotic system. Also, we show that the Lyapunov exponents of the novel hyperchaotic system are $L_1 = 0.5065$, $L_2 = 0.1917$, $L_3 = 0$, $L_4 = -0.0180$, $L_5 = -2.6885$ and $L_6 = -3.1917$. The Kaplan-Yorke dimension of the 6-D hyperchaotic system is derived as $D_{KY} = 4.2530$, which shows the high complexity of the system.

Next, an adaptive integral sliding mode control scheme is proposed for the global control of all the trajectories of the 6-D hyperchaotic system. The adaptive control mechanism helps the control design by estimating the unknown parameters. Numerical simulations using MATLAB are shown to illustrate all the main results derived in this work.

This work is organized as follows. Section 2 describes a 3-D Rucklidge chaotic system for double convection and its phase portrait. In this section, we also discuss the properties of the Rucklidge chaotic system. Section 3 describes a novel 6-D hyperchaotic system obtained by bidirectional coupling of Rucklidge double convection chaotic systems, and describes its phase portraits. Section 4 details the qualitative properties of the novel 6-D hyperchaotic system. We determine the Lyapunov exponents and Kaplan-Yorke dimension of the novel 6-D hyperchaotic system. Section 5 contains new results on the adaptive integral sliding mode controller design for the global control of the novel 6-D hyperchaotic system. Section 6 contains the conclusions of this work.

2   Rucklidge Chaotic System

In this section, we describe the 3-D double convection chaotic system obtained by Rucklidge [38] and discuss the properties of the Rucklidge chaotic attractor.

In fluid mechanics modeling, cases of two-dimensional convection in a horizontal layer of Boussineq fluid with lateral constraints were considered by Rucklidge [38]. When the convection takes place in a fluid layer rotating uniformly about a vertical axis and in the limit of tall thin rolls, convection in an imposed vertical magnetic field and convection in a rotating fluid layer were both modeled by a third-order set of ordinary differential equations in the Rucklidge system [38].

The double convection Rucklidge chaotic system [38] is described by the 3-D model

$$
\begin{cases}
\dot{x}_1 = -ax_1 + bx_2 - x_2x_3 \\
\dot{x}_2 = x_1 \\
\dot{x}_3 = -x_3 + x_2^2
\end{cases}
\tag{1}
$$

where x_1, x_2, x_3 are states and a, b are positive parameters.

In [38], it was shown that the double convection system (1) exhibits chaotic behavior when the parameters take the values

$$
a = 2, \quad b = 6.7
\tag{2}
$$

For numerical simulations, we take the initial state of the finance system (1) as

$$
x_1(0) = 0.1, \quad x_2(0) = 0.1, \quad x_3(0) = 0.1
\tag{3}
$$

For the parameter values (2) and the initial state (3), the Lyapunov exponents of the nonlinear double convection system (1) are calculated by Wolf's algorithm [143] as

$$
L_1 = 0.1907, \quad L_2 = 0, \quad L_3 = -3.1907
\tag{4}
$$

This shows that the Rucklidge chaotic system (1) is chaotic and dissipative.

Also, the Kaplan-Yorke dimension of the Rucklidge chaotic system (1) is determined as

$$
D_{KY} = 2 + \frac{L_1 + L_2}{|L_3|} = 2.0598,
\tag{5}
$$

which is fractional.

Rucklidge chaotic system (1) is invariant under the coordinates transformation

$$
(-x_1, -x_2, x_3) \mapsto (x_1, x_2, x_3)
\tag{6}
$$

This shows that the Rucklidge chaotic system (1) has rotation symmetry about the x_3-axis. Thus, every non-trivial trajectory of the Rucklidge chaotic system (1) must have a twin trajectory.

For the parameter values (2), the Rucklidge chaotic system (1) has three equilibrium points given by

$$
E_0 = \begin{bmatrix} 0 \\ 0 \\ 0 \end{bmatrix}, \quad E_1 = \begin{bmatrix} 0 \\ 2.5884 \\ 6.7 \end{bmatrix}, \quad E_2 = \begin{bmatrix} 0 \\ -2.5884 \\ 6.7 \end{bmatrix}
\tag{7}
$$

It is easy to verify that E_0 is a saddle point and E_1, E_2 are saddle-foci. Thus, all the three equilibrium points of the Rucklidge chaotic system (1) are unstable.

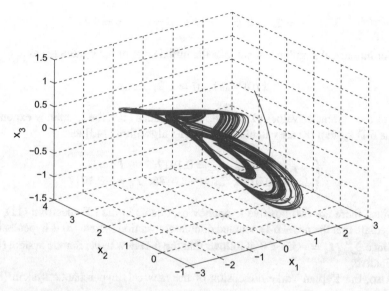

Fig. 1 3-D phase portrait of the Rucklidge chaotic system

Figure 1 shows the 3-D phase portrait of the Rucklidge chaotic system (1).

3 A Novel 6-D Coupled Double Convection Hyperchaotic System

In this section, we derive a new 6-D coupled double convection hyperchaotic system by bidirectional coupling of Rucklidge chaotic systems discussed in Sect. 2.

After coupling, we obtain the novel system given by the 6-D dynamics

$$
\begin{cases}
\dot{x}_1 = -ax_1 + bx_2 - x_2x_3 + p(x_1 - x_4) \\
\dot{x}_2 = x_1 \\
\dot{x}_3 = -x_3 + x_2^2 \\
\dot{x}_4 = -cx_4 + dx_5 - x_5x_6 + p(x_4 - x_1) \\
\dot{x}_5 = x_4 \\
\dot{x}_6 = -x_6 + x_5^2
\end{cases}
\tag{8}
$$

where $x_i, (i = 1, 2, \ldots, 6)$ are the state variables and a, b, c, d, p are positive parameters. In (8), p represents the *coupling strength*.

In this work, we show that the 6-D coupled system (8) is *hyperchaotic* when the system parameters take the values

$$a = 2, \quad b = 6.7, \quad c = 2, \quad d = 6.7, \quad p = 0.4 \tag{9}$$

For numerical simulations, we take the initial state of the system (8) as

$$x_i(0) = 0.1, \quad (i = 1, 2, \dots, 6) \tag{10}$$

For the parameter values (9) and the initial values (10), the Lyapunov exponents of the 6-D system (8) are calculated by Wolf's algorithm [143] as

$$\begin{cases} L_1 = 0.5065, & L_2 = 0.1917, & L_3 = 0, \\ L_4 = -0.0180, & L_5 = -2.6885, & L_6 = -3.1917 \end{cases} \tag{11}$$

Since there are two positive Lyapunov exponents in the LE spectrum (11), it is immediate that the novel 6-D coupled double-convection system (8) is hyperchaotic.

Since $\sum_{i=1}^{6} L_i = -5.2 < 0$, it follows that the 6-D new hyperchaotic system (8) is dissipative.

Also, the Kaplan-Yorke dimension of the new 6-D hyperchaotic system (8) is obtained as

$$D_{KY} = 4 + \frac{L_1 + L_2 + L_3 + L_4}{|L_5|} = 4.2530, \tag{12}$$

which shows the high complexity of the 6-D hyperchaotic system.

Figures 2, 3, 4 and 5 show the 3-D phase portraits of the 6-D new hyperchaotic system in (x_1, x_2, x_3), (x_2, x_3, x_4), (x_3, x_4, x_5) and (x_4, x_5, x_6) spaces respectively.

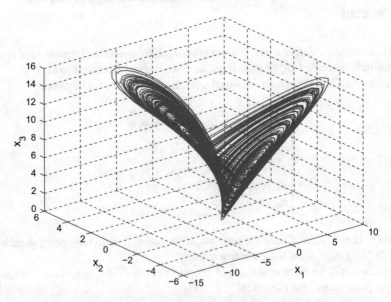

Fig. 2 3-D phase portrait of the new 6-D hyperchaotic system in (x_1, x_2, x_3) space

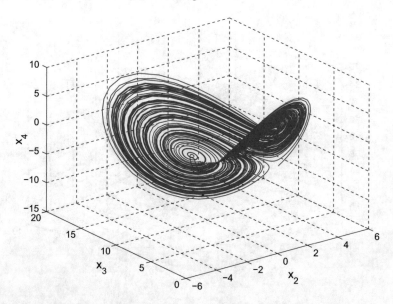

Fig. 3 3-D phase portrait of the new 6-D hyperchaotic system in (x_2, x_3, x_4) space

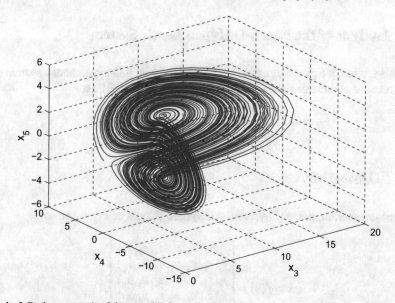

Fig. 4 3-D phase portrait of the new 6-D hyperchaotic system in (x_3, x_4, x_5) space

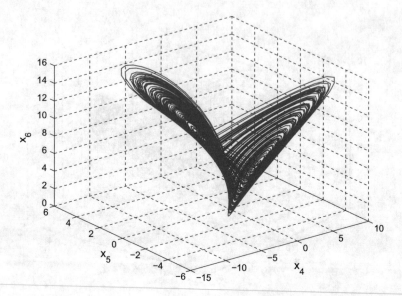

Fig. 5 3-D phase portrait of the new 6-D hyperchaotic system in (x_4, x_5, x_6) space

4 Analysis of the New 6-D Hyperchaotic System

In this section, we give a dynamic analysis of the new 6-D hyperchaotic system (8). We take the parameter values as in the hyperchaotic case (9), i.e.

$$a = 2, \quad b = 6.7, \quad c = 2, \quad d = 6.7, \quad p = 0.4 \tag{13}$$

4.1 Dissipativity

In vector notation, the new hyperchaotic system (8) can be expressed as

$$\dot{\mathbf{x}} = f(\mathbf{x}) = \begin{bmatrix} f_1(x_1, x_2, x_3, x_4, x_5, x_6) \\ f_2(x_1, x_2, x_3, x_4, x_5, x_6) \\ f_3(x_1, x_2, x_3, x_4, x_5, x_6) \\ f_4(x_1, x_2, x_3, x_4, x_5, x_6) \\ f_5(x_1, x_2, x_3, x_4, x_5, x_6) \\ f_6(x_1, x_2, x_3, x_4, x_5, x_6) \end{bmatrix}, \tag{14}$$

where

$$
\begin{cases}
f_1(x_1, x_2, x_3, x_4, x_5, x_6) = -ax_1 + bx_2 - x_2x_3 + p(x_1 - x_4) \\
f_2(x_1, x_2, x_3, x_4, x_5, x_6) = x_1 \\
f_3(x_1, x_2, x_3, x_4, x_5, x_6) = -x_3 + x_2^2 \\
f_4(x_1, x_2, x_3, x_4, x_5, x_6) = -cx_4 + dx_5 - x_5x_6 + p(x_4 - x_1) \\
f_5(x_1, x_2, x_3, x_4, x_5, x_6) = x_4 \\
f_6(x_1, x_2, x_3, x_4, x_5, x_6) = -x_6 + x_5^2
\end{cases}
\tag{15}
$$

Let Ω be any region in $\mathbf{R}^6$ with a smooth boundary and also, $\Omega(t) = \Phi_t(\Omega)$, where Φ_t is the flow of f. Furthermore, let $V(t)$ denote the hypervolume of $\Omega(t)$.

By Liouville's theorem, we know that

$$
\dot{V}(t) = \int_{\Omega(t)} (\nabla \cdot f) \, dx_1 \, dx_2 \, dx_3 \, dx_4
\tag{16}
$$

The divergence of the hyperchaotic system (14) is found as:

$$
\nabla \cdot f = \frac{\partial f_1}{\partial x_1} + \frac{\partial f_2}{\partial x_2} + \frac{\partial f_3}{\partial x_3} + \frac{\partial f_4}{\partial x_4} + \frac{\partial f_5}{\partial x_5} + \frac{\partial f_6}{\partial x_6} = -2(a - p + 1) = -\mu
\tag{17}
$$

where $\mu = 2(a - p + 1) = 5.2 > 0$.

Inserting the value of $\nabla \cdot f$ from (17) into (16), we get

$$
\dot{V}(t) = \int_{\Omega(t)} (-\mu) \, dx_1 \, dx_2 \, dx_3 \, dx_4 \, dx_5 \, dx_6 = -\mu V(t)
\tag{18}
$$

Integrating the first order linear differential equation (18), we get

$$
V(t) = \exp(-\mu t) V(0)
\tag{19}
$$

Since $\mu > 0$, it follows from Eq. (19) that $V(t) \to 0$ exponentially as $t \to \infty$. This shows that the new 6-D hyperchaotic system (8) is dissipative. Hence, the system limit sets are ultimately confined into a specific limit set of zero hypervolume, and the asymptotic motion of the new hyperchaotic system (8) settles onto a strange attractor of the system.

4.2 Equilibrium Points

We take the parameter values of the new 6-D hyperchaotic system (8) as in the hyperchaotic case (9), i.e. $a = 2$, $b = 6.7$, $c = 2$, $d = 6.7$ and $p = 0.4$.

The equilibrium points of the new 6-D hyperchaotic system (8) are obtained by solving the following system of equations.

$$
\begin{cases}
-ax_1 + bx_2 - x_2x_3 + p(x_1 - x_4) = 0 \\
x_1 = 0 \\
-x_3 + x_2^2 = 0 \\
-cx_4 + dx_5 - x_5x_6 + p(x_4 - x_1) = 0 \\
x_4 = 0 \\
-x_6 + x_5^2 = 0
\end{cases}
\tag{20}
$$

Simplifying the system (20), we get the following equations

$$
x_1 = 0, \quad x_4 = 0, \quad x_2(b - x_3) = 0, \quad x_5(d - x_6) = 0, \quad x_3 = x_2^2, \quad x_6 = x_5^2
\tag{21}
$$

Solving the Eq. (21), we find that the 6-D system has a set of nine equilibrium points given by $E_0, E_1, \ldots, E_8$, where

$$
E_0 = \begin{bmatrix} 0 \\ 0 \\ 0 \\ 0 \\ 0 \\ 0 \end{bmatrix}, \quad
E_1 = \begin{bmatrix} 0 \\ 0 \\ 0 \\ 0 \\ 2.5884 \\ 6.7 \end{bmatrix}, \quad
E_2 = \begin{bmatrix} 0 \\ 0 \\ 0 \\ 0 \\ -2.5884 \\ 6.7 \end{bmatrix}
\tag{22}
$$

$$
E_4 = \begin{bmatrix} 0 \\ 2.5884 \\ 6.7 \\ 0 \\ 0 \\ 0 \end{bmatrix}, \quad
E_5 = \begin{bmatrix} 0 \\ -2.5884 \\ 6.7 \\ 0 \\ 0 \\ 0 \end{bmatrix}, \quad
E_6 = \begin{bmatrix} 0 \\ 2.5884 \\ 6.7 \\ 0 \\ 2.5884 \\ 6.7 \end{bmatrix}
\tag{23}
$$

and

$$
E_7 = \begin{bmatrix} 0 \\ 2.5884 \\ 6.7 \\ 0 \\ -2.5884 \\ 6.7 \end{bmatrix}, \quad
E_8 = \begin{bmatrix} 0 \\ -2.5884 \\ 6.7 \\ 0 \\ 2.5884 \\ 6.7 \end{bmatrix}, \quad
E_9 = \begin{bmatrix} 0 \\ -2.5884 \\ 6.7 \\ 0 \\ -2.5884 \\ 6.7 \end{bmatrix}
\tag{24}
$$

It is easy to verify that E_0 is a saddle point and the other equilibrium points are all saddle-foci. Thus, all the nine equilibrium points of the new 6-D hyperchaotic system are unstable.

4.3 Rotation Symmetry About the (x_3, x_6)-Plane

It is easy to see that the new 6-D hyperchaotic system (8) is invariant under the change of coordinates

$$(x_1, x_2, x_3, x_4) \mapsto (-x_1, -x_2, x_3, -x_4, -x_5, x_6) \tag{25}$$

Since the transformation (25) persists for all values of the system parameters, it follows that the new 6-D hyperchaotic system (8) has rotation symmetry about the (x_3, x_6)-plane and that any non-trivial trajectory of the system must have a twin trajectory.

4.4 Invariance

It is easy to see that the x_3-axis and x_6-axis are invariant under the flow of the 4-D novel hyperchaotic system (8).

The invariant motion along the x_3-axis is characterized by the scalar dynamics

$$\dot{x}_3 = -x_3, \tag{26}$$

which is globally exponentially stable.

The invariant motion along the x_6-axis is characterized by the scalar dynamics

$$\dot{x}_6 = -x_6, \tag{27}$$

which is also globally exponentially stable.

4.5 Lyapunov Exponents and Kaplan-Yorke Dimension

We take the parameter values of the new 6-D hyperchaotic system (8) as in the hyperchaotic case (9), i.e.

$$a = 1, \quad b = 0.2, \quad c = 1.3, \quad d = 0.1 \tag{28}$$

We take the initial state of the new hyperchaotic finance system (8) as (10), i.e.

$$x_i(0) = 0.1, \quad (i = 1, 2, \ldots, 6) \tag{29}$$

Then the Lyapunov exponents of the 6-D hyperchaotic system (8) are numerically obtained using MATLAB as

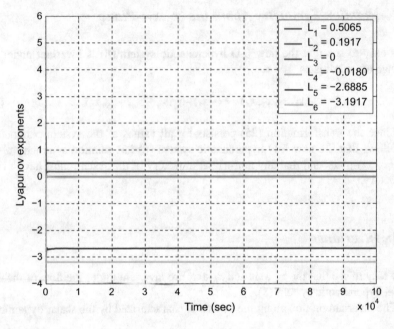

Fig. 6 Lyapunov exponents of the new 6-D hyperchaotic system

$$\begin{cases} L_1 = 0.5065, & L_2 = 0.1917, & L_3 = 0, \\ L_4 = -0.0180, & L_5 = -2.6885, & L_6 = -3.1917 \end{cases} \tag{30}$$

Since there are two positive Lyapunov exponents in the LE spectrum (30), it is immediate that the novel 6-D coupled double-convection system (8) is hyperchaotic.

Since $\sum_{i=1}^{6} L_i = -5.2 < 0$, it follows that the 6-D new hyperchaotic system (8) is dissipative.

Also, the Kaplan-Yorke dimension of the new 6-D hyperchaotic system (8) is obtained as

$$D_{KY} = 4 + \frac{L_1 + L_2 + L_3 + L_4}{|L_5|} = 4.2530, \tag{31}$$

which shows the high complexity of the 6-D hyperchaotic system.

Figure 6 shows the Lyapunov exponents of the new 6-D hyperchaotic system (8).

5 Global Hyperchaos Control of the New 6-D Hyperchaotic System

In this section, we use adaptive integral sliding mode control for the global hyperchaos control of the new 6-D hyperchaotic system with unknown system parameters.

The adaptive control mechanism helps the control design by estimating the unknown parameters [3, 4, 122].

The controlled new 6-D hyperchaotic system is described by

$$
\begin{cases}
\dot{x}_1 = -ax_1 + bx_2 - x_2x_3 + p(x_1 - x_4) + u_1 \\
\dot{x}_2 = x_1 + u_2 \\
\dot{x}_3 = -x_3 + x_2^2 + u_3 \\
\dot{x}_4 = -cx_4 + dx_5 - x_5x_6 + p(x_4 - x_1) + u_4 \\
\dot{x}_5 = x_4 + u_5 \\
\dot{x}_6 = -x_6 + x_5^2 + u_6
\end{cases}
\tag{32}
$$

where $x_i, (i = 1, 2, \ldots, 6)$ are the states of the system and a, b, c, d, p are unknown system parameters.

The design goal is to find suitable feedback controllers $u_i, (i = 1, 2, \ldots, 6)$ so as to globally stabilize the system (32) for all initial conditions $\mathbf{x}(0) \in \mathbf{R}^6$.

Based on the sliding mode control theory [43, 49, 50], the integral sliding surface of each e_i $(i = 1, 2, \ldots, 6)$ is defined as follows:

$$
s_i = \left(\frac{d}{dt} + \lambda_i \right) \left(\int_0^t x_i(\tau)d\tau \right) = x_i + \lambda_i \int_0^t x_i(\tau)d\tau, \quad i = 1, 2, \ldots, 6
\tag{33}
$$

From Eq. (33), it follows that

$$
\begin{cases}
\dot{s}_1 = \dot{x}_1 + \lambda_1 x_1 \\
\dot{s}_2 = \dot{x}_2 + \lambda_2 x_2 \\
\dot{s}_3 = \dot{x}_3 + \lambda_3 x_3 \\
\dot{s}_4 = \dot{x}_4 + \lambda_4 x_4 \\
\dot{s}_5 = \dot{x}_5 + \lambda_5 x_5 \\
\dot{s}_6 = \dot{x}_6 + \lambda_6 x_6
\end{cases}
\tag{34}
$$

The Hurwitz condition is realized if $\lambda_i > 0$ for $i = 1, 2, \ldots, 6$.
We consider the adaptive feedback control given by

$$
\begin{cases}
u_1 = \hat{a}(t)x_1 - \hat{b}(t)x_2 + x_2x_3 - \hat{p}(t)(x_1 - x_4) - \lambda_1 x_1 - \eta_1 \operatorname{sgn}(s_1) - k_1 s_1 \\
u_2 = -x_1 - \lambda_2 x_2 - \eta_2 \operatorname{sgn}(s_2) - k_2 s_2 \\
u_3 = x_3 - x_2^2 - \lambda_3 x_3 - \eta_3 \operatorname{sgn}(s_3) - k_3 s_3 \\
u_4 = \hat{c}(t)x_4 - \hat{d}(t)x_5 + x_5x_6 - \hat{p}(t)(x_4 - x_1) - \lambda_4 x_4 - \eta_4 \operatorname{sgn}(s_4) - k_4 s_4 \\
u_5 = -x_4 - \lambda_5 x_5 - \eta_5 \operatorname{sgn}(s_5) - k_5 s_5 \\
u_6 = x_6 - x_5^2 - \lambda_6 x_6 - \eta_6 \operatorname{sgn}(s_6) - k_6 s_6
\end{cases}
\tag{35}
$$

where $\eta_i > 0$ and $k_i > 0$ for $i = 1, \ldots, 6$.

Substituting (35) into (32), we obtain the closed-loop control system given by

$$
\begin{cases}
\dot{x}_1 = -[a - \hat{a}(t)]x_1 + [b - \hat{b}(t)]x_2 + [p - \hat{p}(t)](x_1 - x_4) \\
\qquad -\lambda_1 x_1 - \eta_1 \operatorname{sgn}(s_1) - k_1 s_1 \\
\dot{x}_2 = -\lambda_2 x_2 - \eta_2 \operatorname{sgn}(s_2) - k_2 s_2 \\
\dot{x}_3 = -\lambda_3 x_3 - \eta_3 \operatorname{sgn}(s_3) - k_3 s_3 \\
\dot{x}_4 = -[c - \hat{c}(t)]x_4 + [d - \hat{d}(t)]x_5 + [p - \hat{p}(t)](x_4 - x_1) \\
\qquad -\lambda_4 x_4 - \eta_4 \operatorname{sgn}(s_4) - k_4 s_4 \\
\dot{x}_5 = -\lambda_5 x_5 - \eta_5 \operatorname{sgn}(s_5) - k_5 s_5 \\
\dot{x}_6 = -\lambda_6 x_2 - \eta_6 \operatorname{sgn}(s_6) - k_6 s_6
\end{cases}
\tag{36}
$$

We define the parameter estimation errors as

$$
\begin{cases}
e_a(t) = a - \hat{a}(t) \\
e_b(t) = b - \hat{b}(t) \\
e_c(t) = c - \hat{c}(t) \\
e_d(t) = d - \hat{d}(t) \\
e_p(t) = p - \hat{p}(t)
\end{cases}
\tag{37}
$$

Using (37), we can simplify the closed-loop system (36) as

$$
\begin{cases}
\dot{x}_1 = -e_a x_1 + e_b x_2 + e_p(x_1 - x_4) - \lambda_1 x_1 - \eta_1 \operatorname{sgn}(s_1) - k_1 s_1 \\
\dot{x}_2 = -\lambda_2 x_2 - \eta_2 \operatorname{sgn}(s_2) - k_2 s_2 \\
\dot{x}_3 = -\lambda_3 x_3 - \eta_3 \operatorname{sgn}(s_3) - k_3 s_3 \\
\dot{x}_4 = -e_c x_4 + e_d x_5 + e_p(x_4 - x_1) - \lambda_4 x_4 - \eta_4 \operatorname{sgn}(s_4) - k_4 s_4 \\
\dot{x}_5 = -\lambda_5 x_5 - \eta_5 \operatorname{sgn}(s_5) - k_5 s_5 \\
\dot{x}_6 = -\lambda_6 x_2 - \eta_6 \operatorname{sgn}(s_6) - k_6 s_6
\end{cases}
\tag{38}
$$

Differentiating (37) with respect to t, we get

$$
\begin{cases}
\dot{e}_a = -\dot{\hat{a}} \\
\dot{e}_b = -\dot{\hat{b}} \\
\dot{e}_c = -\dot{\hat{c}} \\
\dot{e}_d = -\dot{\hat{d}} \\
\dot{e}_p = -\dot{\hat{p}}
\end{cases}
\tag{39}
$$

Next, we state and prove the main result of this section.

Theorem 1 *The new 6-D hyperchaotic system (32) is globally stabilized for all initial conditions* $\mathbf{x}(0) \in \mathbf{R}^6$ *by the adaptive integral sliding mode control law (35) and the parameter update law*

$$\begin{cases} \dot{\hat{a}} = -s_1 x_1 \\ \dot{\hat{b}} = s_1 x_2 \\ \dot{\hat{c}} = -s_4 x_4 \\ \dot{\hat{d}} = s_4 x_5 \\ \dot{\hat{p}} = (s_1 - s_4)(x_1 - x_4) \end{cases} \tag{40}$$

where λ_i, η_i, k_i are positive constants for $i = 1, 2, \ldots, 6$.

Proof We consider the quadratic Lyapunov function defined by

$$V(s_1, s_2, s_3, s_4, s_5, s_6, e_a, e_b, e_c, e_d, e_p) = \frac{1}{2} \sum_{i=1}^{6} s_i^2 + \frac{1}{2} \left(e_a^2 + e_b^2 + e_c^2 + e_d^2 + e_p^2 \right) \tag{41}$$

Clearly, V is positive definite on $\mathbf{R}^{11}$.

Using (34), (38) and (39), the time-derivative of V is obtained as

$$\begin{aligned} \dot{V} &= s_1[-e_a x_1 + e_b x_2 + e_p(x_1 - x_4) - \eta_1 \operatorname{sgn}(s_1) - k_1 s_1] - e_a \dot{\hat{a}} \\ &+ s_2[-\eta_2 \operatorname{sgn}(s_2) - k_2 s_2] + s_3[-\eta_3 \operatorname{sgn}(s_3) - k_3 s_3] - e_b \dot{\hat{b}} \\ &+ s_4[-e_c x_4 + e_d x_5 + e_p(x_4 - x_1) - \eta_4 \operatorname{sgn}(s_4) - k_4 s_4] - e_c \dot{\hat{c}} \\ &+ s_5[-\eta_5 \operatorname{sgn}(s_5) - k_5 s_5] + s_6[-\eta_6 \operatorname{sgn}(s_6) - k_6 s_6] - e_d \dot{\hat{d}} - e_p \dot{\hat{p}} \end{aligned} \tag{42}$$

i.e.

$$\begin{aligned} \dot{V} &= -\eta_1 |s_1| - k_1 s_1^2 - \eta_2 |s_2| - k_2 s_2^2 - \eta_3 |s_3| - k_3 s_3^2 - \eta_4 |s_4| - k_4 s_4^2 \\ &-\eta_5 |s_5| - k_5 s_5^2 - \eta_6 |s_6| - k_6 s_6^2 \\ &+e_a \left[-s_1 x_1 - \dot{\hat{a}} \right] + e_b \left[s_1 x_2 - \dot{\hat{b}} \right] \\ &+e_c \left[-s_4 x_4 - \dot{\hat{c}} \right] \\ &+e_d \left[s_4 x_5 - \dot{\hat{d}} \right] + e_p[(s_1 - s_4)(x_1 - x_4) - \dot{\hat{p}}] \end{aligned} \tag{43}$$

Using the parameter update law (40), we obtain

$$\dot{V} = -\sum_{i=1}^{6} \eta_i |s_i| - \sum_{i=1}^{6} k_i s_i^2 \tag{44}$$

which shows that $\dot{V}$ is negative semi-definite on $\mathbf{R}^{11}$.

Hence, by Barbalat's lemma [19], it is immediate that the state vector $\mathbf{x}(t)$ is globally asymptotically stable for all values of $\mathbf{x}(0) \in \mathbf{R}^6$.

This completes the proof. ∎

For numerical simulations, we take the parameter values of the new 6-D hyperchaotic system (32) as in the hyperchaotic case (9), i.e.

$$a = 2, \quad b = 6.7, \quad c = 2, \quad d = 6.7, \quad p = 0.4 \tag{45}$$

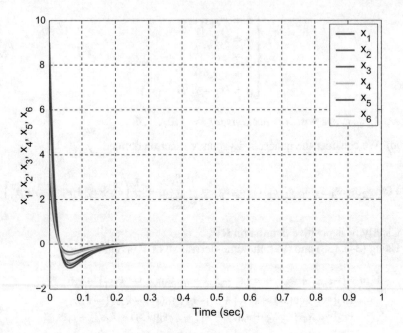

Fig. 7 Time-history of the controlled states $x_1, x_2, x_3, x_4, x_5, x_6$

We take the values of the control parameters as $k_i = 20$, $\eta_i = 0.1$ and $\lambda_i = 50$, where $i = 1, 2 \ldots, 6$.

We take the estimates of the system parameters as

$$\hat{a}(0) = 12.5, \quad \hat{b}(0) = 18.3, \quad \hat{c}(0) = 22.4, \quad \hat{d}(0) = 15.3, \quad \hat{p}(0) = 7.4 \qquad (46)$$

We take the initial state of the 6-D hyperchaotic system (32) as

$$x_1(0) = 7, \quad x_2(0) = 9, \quad x_3(0) = 3, \quad x_4(0) = 6, \quad x_5(0) = 8, \quad x_6(0) = 4 \qquad (47)$$

Figure 7 shows the time-history of the controlled states x_i, $(i = 1, 2, \ldots, 6)$.

6 Conclusions

By bidirectional coupling of Rucklidge chaotic systems (1992), a novel 6-D coupled double convection hyperchaotic system was derived in this work. We described the phase portraits and qualitative properties of the novel 6-D hyperchaotic system. Next, an adaptive integral sliding mode control scheme was designed for the global control of all the trajectories of the 6-D hyperchaotic system. MATLAB simulations were depicted to illustrate all the main results derived in this work.

References

1. Akgul A, Hussain S, Pehlivan I (2016) A new three-dimensional chaotic system, its dynamical analysis and electronic circuit applications. Optik 127(18):7062–7071
2. Akgul A, Moroz I, Pehlivan I, Vaidyanathan S (2016) A new four-scroll chaotic attractor and its engineering applications. Optik 127(13):5491–5499
3. Azar AT, Vaidyanathan S (2015) Chaos modeling and control systems design. Springer, Berlin, Germany
4. Azar AT, Vaidyanathan S (2016) Advances in chaos theory and intelligent control. Springer, Berlin, Germany
5. Azar AT, Vaidyanathan S (2017) Fractional order control and synchronization of chaotic systems. Springer, Berlin, Germany
6. Boulkroune A, Bouzeriba A, Bouden T (2016) Fuzzy generalized projective synchronization of incommensurate fractional-order chaotic systems. Neurocomputing 173:606–614
7. Burov DA, Evstigneev NM, Magnitskii NA (2017) On the chaotic dynamics in two coupled partial differential equations for evolution of surface plasmon polaritons. Commun Nonlinear Sci Numer Simul 46:26–36
8. Chai X, Chen Y, Broyde L (2017) A novel chaos-based image encryption algorithm using DNA sequence operations. Opt Lasers Eng 88:197–213
9. Chen A, Lu J, Lü J, Yu S (2006) Generating hyperchaotic Lü attractor via state feedback control. Physica A 364:103–110
10. Chenaghlu MA, Jamali S, Khasmakhi NN (2016) A novel keyed parallel hashing scheme based on a new chaotic system. Chaos Solitons Fractals 87:216–225
11. Dadras S, Momeni HR, Qi G, Lin Wang Z (2012) Four-wing hyperchaotic attractor generated from a new 4D system with one equilibrium and its fractional-order form. Nonlinear Dyn 67:1161–1173
12. Fallahi K, Leung H (2010) A chaos secure communication scheme based on multiplication modulation. Commun Nonlinear Sci Numer Simul 15(2):368–383
13. Fontes RT, Eisencraft M (2016) A digital bandlimited chaos-based communication system. Commun Nonlinear Sci Numer Simul 37:374–385
14. Fotsa RT, Woafo P (2016) Chaos in a new bistable rotating electromechanical system. Chaos Solitons Fractals 93:48–57
15. Gao T, Chen Z, Yuan Z, Chen G (2006) A hyperchaos generated from Chen's system. Int J Modern Phys C 17(4):471–478
16. Hu J, Ma J, Lin J (2010) Chaos synchronization and communication of mutual coupling lasers ring based on incoherent injection. Optik 121(24):2227–2229
17. Jia Q (2007) Hyperchaos generated from the Lorenz chaotic system and its control. Phys Lett A 366(3):217–222
18. Kacar S (2016) Analog circuit and microcontroller based RNG application of a new easy realizable 4D chaotic system. Optik 127(20):9551–9561
19. Khalil HK (2002) Nonlinear Syst. Prentice Hall, New York, USA
20. Lakhekar GV, Waghmare LM, Vaidyanathan S (2016) Diving autopilot design for underwater vehicles using an adaptive neuro-fuzzy sliding mode controller. In: Vaidyanathan S, Volos C (eds) Advances and applications in nonlinear control systems. Springer, Berlin, Germany, pp 477–503
21. Li D (2008) A three-scroll chaotic attractor. Phys Lett A 372(4):387–393
22. Liu H, Ren B, Zhao Q, Li N (2016) Characterizing the optical chaos in a special type of small networks of semiconductor lasers using permutation entropy. Opt Commun 359:79–84
23. Liu W, Sun K, Zhu C (2016) A fast image encryption algorithm based on chaotic map. Opt Lasers Eng 84:26–36
24. Liu X, Mei W, Du H (2016) Simultaneous image compression, fusion and encryption algorithm based on compressive sensing and chaos. Opt Commun 366:22–32

25. Moussaoui S, Boulkroune A, Vaidyanathan S (2016) Fuzzy adaptive sliding-mode control scheme for uncertain underactuated systems. In: Vaidyanathan S, Volos C (eds) Advances and applications in nonlinear control systems. Springer, Berlin, Germany, pp 351–367
26. Pehlivan I, Moroz IM, Vaidyanathan S (2014) Analysis, synchronization and circuit design of a novel butterfly attractor. J Sound Vib 333(20):5077–5096
27. Pham VT, Volos C, Jafari S, Wang X, Vaidyanathan S (2014) Hidden hyperchaotic attractor in a novel simple memristive neural network. Optoelectron Adv Mater Rapid Commun 8(11–12):1157–1163
28. Pham VT, Volos CK, Vaidyanathan S (2015) Multi-scroll chaotic oscillator based on a first-order delay differential equation. In: Azar AT, Vaidyanathan S (eds) Chaos modeling and control systems design. Studies in computational intelligence, vol 581. Springer, Germany, pp 59–72
29. Pham VT, Volos CK, Vaidyanathan S, Le TP, Vu VY (2015) A memristor-based hyperchaotic system with hidden attractors: dynamics, synchronization and circuital emulating. J Eng Sci Technol Rev 8(2):205–214
30. Pham VT, Jafari S, Vaidyanathan S, Volos C, Wang X (2016) A novel memristive neural network with hidden attractors and its circuitry implementation. Sci China Technol Sci 59(3):358–363
31. Pham VT, Jafari S, Volos C, Giakoumis A, Vaidyanathan S, Kapitaniak T (2016) A chaotic system with equilibria located on the rounded square loop and its circuit implementation. IEEE Trans Circuits Syst II: Express Briefs 63(9):878–882
32. Pham VT, Jafari S, Volos C, Vaidyanathan S, Kapitaniak T (2016) A chaotic system with infinite equilibria located on a piecewise linear curve. Optik 127(20):9111–9117
33. Pham VT, Vaidyanathan S, Volos C, Jafari S, Kingni ST (2016) A no-equilibrium hyper-chaotic system with a cubic nonlinear term. Optik 127(6):3259–3265
34. Pham VT, Vaidyanathan S, Volos CK, Hoang TM, Yem VV (2016) Dynamics, synchronization and SPICE implementation of a memristive system with hidden hyperchaotic attractor. In: Azar AT, Vaidyanathan S (eds) Advances in chaos theory and intelligent control. Springer, Berlin, Germany, pp 35–52
35. Pham VT, Vaidyanathan S, Volos CK, Jafari S, Kuznetsov NV, Hoang TM (2016) A novel memristive time-delay chaotic system without equilibrium points. Eur Phys J: Spec Top 225(1):127–136
36. Pham VT, Vaidyanathan S, Volos CK, Jafari S, Wang X (2016) A chaotic hyperjerk system based on memristive device. In: Vaidyanathan S, Volos C (eds) Advances and applications in chaotic systems. Springer, Berlin, Germany, pp 39–58
37. Rössler O (1979) An equation for hyperchaos. Phys Lett A 71(2–3):155–157
38. Rucklidge AM (1992) Chaos in models of double convection. J Fluid Mech 237:209–229
39. Sampath S, Vaidyanathan S, Volos CK, Pham VT (2015) An eight-term novel four-scroll chaotic system with cubic nonlinearity and its circuit simulation. J Eng Sci Technol Rev 8(2):1–6
40. Sampath S, Vaidyanathan S, Pham VT (2016) A novel 4-D hyperchaotic system with three quadratic nonlinearities, its adaptive control and circuit simulation. Int J Control Theory Appl 9(1):339–356
41. Shahverdiev E, Bayramov P, Qocayeva M (2013) Inverse chaos synchronization between bidi-rectionally coupled variable multiple time delay systems. Optik 124(18):3427–3429
42. Shirkhani N, Khanesar M, Teshnehlab M (2016) Indirect model reference fuzzy control of SISO fractional order nonlinear chaotic systems. Proc Comput Sci 102:309–316
43. Slotine J, Li W (1991) Applied nonlinear control. Prentice-Hall, Englewood Cliffs, NJ, USA
44. Sundarapandian V (2013) Analysis and anti-synchronization of a novel chaotic system via active and adaptive controllers. J Eng Sci Technol Rev 6(4):45–52
45. Sundarapandian V, Pehlivan I (2012) Analysis, control, synchronization, and circuit design of a novel chaotic system. Math Comput Modell 55(7–8):1904–1915
46. Sundarapandian V, Sivaperumal S (2011) Sliding controller design of hybrid synchronization of four-wing Chaotic systems. Int J Soft Comput 6(5):224–231

47. Szmit Z, Warminski J (2016) Nonlinear dynamics of electro-mechanical system composed of two pendulums and rotating hub. Proc Eng 144:953–958
48. Tacha OI, Volos CK, Kyprianidis IM, Stouboulos IN, Vaidyanathan S, Pham VT (2016) Analysis, adaptive control and circuit simulation of a novel nonlinear finance system. Appl Math Comput 276:200–217
49. Utkin VI (1977) Variable structure systems with sliding modes. IEEE Trans Autom Control 22(2):212–222
50. Utkin VI (1993) Sliding mode control design principles and applications to electric drives. IEEE Trans Ind Electron 40(1):23–36
51. Vaidyanathan S (2011) Analysis and synchronization of the hyperchaotic Yujun systems via sliding mode control. Adv Intell Syst Comput 176:329–337
52. Vaidyanathan S (2012) Global chaos control of hyperchaotic Liu system via sliding control method. Int J Control Theory Appl 5(2):117–123
53. Vaidyanathan S (2012) Sliding mode control based global chaos control of Liu-Liu-Liu-Su chaotic system. Int J Control Theory Appl 5(1):15–20
54. Vaidyanathan S (2013) A new six-term 3-D chaotic system with an exponential nonlinearity. Far East J Math Sci 79(1):135–143
55. Vaidyanathan S (2013) A ten-term novel 4-D hyperchaotic system with three quadratic non-linearities and its control. Int J Control Theory Appl 6(2):97–109
56. Vaidyanathan S (2013) Analysis and adaptive synchronization of two novel chaotic systems with hyperbolic sinusoidal and cosinusoidal nonlinearity and unknown parameters. J Eng Sci Technol Rev 6(4):53–65
57. Vaidyanathan S (2014) A new eight-term 3-D polynomial chaotic system with three quadratic nonlinearities. Far East J Math Sci 84(2):219–226
58. Vaidyanathan S (2014) Analysis and adaptive synchronization of eight-term 3-D polynomial chaotic systems with three quadratic nonlinearities. Eur Phys J: Spec Top 223(8):1519–1529
59. Vaidyanathan S (2014) Analysis, control and synchronisation of a six-term novel chaotic system with three quadratic nonlinearities. Int J Modell Ident Control 22(1):41–53
60. Vaidyanathan S (2014) Generalised projective synchronisation of novel 3-D chaotic systems with an exponential non-linearity via active and adaptive control. Int J Modell Ident Control 22(3):207–217
61. Vaidyanathan S (2014) Global chaos synchronisation of identical Li-Wu chaotic systems via sliding mode control. Int J Modell Ident Control 22(2):170–177
62. Vaidyanathan S (2014) Qualitative analysis and control of an eleven-term novel 4-D hyper-chaotic system with two quadratic nonlinearities. Int J Control Theory Appl 7(1):35–47
63. Vaidyanathan S (2015) A 3-D novel highly chaotic system with four quadratic nonlinearities, its adaptive control and anti-synchronization with unknown parameters. J Eng Sci Technol Rev 8(2):106–115
64. Vaidyanathan S (2015) A novel chemical chaotic reactor system and its adaptive control. Int J ChemTech Res 8(7):146–158
65. Vaidyanathan S (2015) A novel chemical chaotic reactor system and its output regulation via integral sliding mode control. Int J ChemTech Res 8(11):669–683
66. Vaidyanathan S (2015) Active control design for the anti-synchronization of Lotka-Volterra biological systems with four competitive species. Int J PharmTech Res 8(7):58–70
67. Vaidyanathan S (2015) Adaptive control design for the anti-synchronization of novel 3-D chemical chaotic reactor systems. Int J ChemTech Res 8(11):654–668
68. Vaidyanathan S (2015) Adaptive control of a chemical chaotic reactor. Int J PharmTech Res 8(3):377–382
69. Vaidyanathan S (2015) Adaptive control of the FitzHugh-Nagumo chaotic neuron model. Int J PharmTech Res 8(6):117–127
70. Vaidyanathan S (2015) Adaptive synchronization of chemical chaotic reactors. Int J ChemTech Res 8(2):612–621
71. Vaidyanathan S (2015) Adaptive synchronization of generalized Lotka-Volterra three-species biological systems. Int J PharmTech Res 8(5):928–937

72. Vaidyanathan S (2015) Adaptive synchronization of novel 3-D chemical chaotic reactor systems. Int J ChemTech Res 8(7):159–171
73. Vaidyanathan S (2015) Analysis, properties and control of an eight-term 3-D chaotic system with an exponential nonlinearity. Int J Modell Ident Control 23(2):164–172
74. Vaidyanathan S (2015) Anti-synchronization of brusselator chemical reaction systems via adaptive control. Int J ChemTech Res 8(6):759–768
75. Vaidyanathan S (2015) Anti-synchronization of brusselator chemical reaction systems via integral sliding mode control. Int J ChemTech Res 8(11):700–713
76. Vaidyanathan S (2015) Anti-synchronization of chemical chaotic reactors via adaptive control method. Int J ChemTech Res 8(8):73–85
77. Vaidyanathan S (2015) Anti-synchronization of the FitzHugh-Nagumo chaotic neuron models via adaptive control method. Int J PharmTech Res 8(7):71–83
78. Vaidyanathan S (2015) Chaos in neurons and synchronization of Birkhoff-Shaw strange chaotic attractors via adaptive control. Int J PharmTech Res 8(6):1–11
79. Vaidyanathan S (2015) Dynamics and control of brusselator chemical reaction. Int J ChemTech Res 8(6):740–749
80. Vaidyanathan S (2015) Dynamics and control of Tokamak system with symmetric and magnetically confined plasma. Int J ChemTech Res 8(6):795–802
81. Vaidyanathan S (2015) Global chaos synchronization of chemical chaotic reactors via novel sliding mode control method. Int J ChemTech Res 8(7):209–221
82. Vaidyanathan S (2015) Global chaos synchronization of Duffing double-well chaotic oscillators via integral sliding mode control. Int J ChemTech Res 8(11):141–151
83. Vaidyanathan S (2015) Global chaos synchronization of the forced Van der Pol chaotic oscillators via adaptive control method. Int J PharmTech Res 8(6):156–166
84. Vaidyanathan S (2015) Hybrid chaos synchronization of the FitzHugh-Nagumo chaotic neuron models via adaptive control method. Int J PharmTech Res 8(8):48–60
85. Vaidyanathan S (2015) Integral sliding mode control design for the global chaos synchronization of identical novel chemical chaotic reactor systems. Int J ChemTech Res 8(11):684–699
86. Vaidyanathan S (2015) Output regulation of the forced Van der Pol chaotic oscillator via adaptive control method. Int J PharmTech Res 8(6):106–116
87. Vaidyanathan S (2015) Sliding controller design for the global chaos synchronization of forced Van der Pol chaotic oscillators. Int J PharmTech Res 8(7):100–111
88. Vaidyanathan S (2015) Synchronization of Tokamak systems with symmetric and magnetically confined plasma via adaptive control. Int J ChemTech Res 8(6):818–827
89. Vaidyanathan S (2016) A no-equilibrium novel 4-D highly hyperchaotic system with four quadratic nonlinearities and its adaptive control. In: Vaidyanathan S, Volos C (eds) Advances and applications in nonlinear control systems. Springer, Berlin, Germany, pp 235–258
90. Vaidyanathan S (2016) A novel 3-D conservative chaotic system with a sinusoidal nonlinearity and its adaptive control. Int J Control Theory Appl 9(1):115–132
91. Vaidyanathan S (2016) A novel 3-D jerk chaotic system with three quadratic nonlinearities and its adaptive control. Arch Control Sci 26(1):19–47
92. Vaidyanathan S (2016) A novel 3-D jerk chaotic system with two quadratic nonlinearities and its adaptive backstepping control. Int J Control Theory Appl 9(1):199–219
93. Vaidyanathan S (2016) A novel 4-D hyperchaotic thermal convection system and its adaptive control. In: Azar AT, Vaidyanathan S (eds) Advances in chaos theory and intelligent control. Springer, Berlin, Germany, pp 75–100
94. Vaidyanathan S (2016) A novel 5-D hyperchaotic system with a line of equilibrium points and its adaptive control. In: Vaidyanathan S, Volos C (eds) Advances and applications in chaotic systems. Springer, Berlin, Germany, pp 471–494
95. Vaidyanathan S (2016) A novel highly hyperchaotic system and its adaptive control. In: Vaidyanathan S, Volos C (eds) Advances and applications in chaotic systems. Springer, Berlin, Germany, pp 513–535
96. Vaidyanathan S (2016) A novel hyperchaotic hyperjerk system with two nonlinearities, its analysis, adaptive control and synchronization via backstepping control method. Int J Control Theory Appl 9(1):257–278

97. Vaidyanathan S (2016) An eleven-term novel 4-D hyperchaotic system with three quadratic nonlinearities, analysis, control and synchronization via adaptive control method. Int J Control Theory Appl 9(1):21–43
98. Vaidyanathan S (2016) Analysis, adaptive control and synchronization of a novel 4-D hyperchaotic hyperjerk system via backstepping control method. Arch Control Sci 26(3):311–338
99. Vaidyanathan S (2016) Analysis, control and synchronization of a novel 4-D highly hyperchaotic system with hidden attractors. In: Azar AT, Vaidyanathan S (eds) Advances in chaos theory and intelligent control. Springer, Berlin, Germany, pp 529–552
100. Vaidyanathan S (2016) Anti-synchronization of 3-cells cellular neural network attractors via integral sliding mode control. Int J PharmTech Res 9(1):193–205
101. Vaidyanathan S (2016) Global chaos control of the FitzHugh-Nagumo chaotic neuron model via integral sliding mode control. Int J PharmTech Res 9(4):413–425
102. Vaidyanathan S (2016) Global chaos control of the generalized Lotka-Volterra three-species system via integral sliding mode control. Int J PharmTech Res 9(4):399–412
103. Vaidyanathan S (2016) Global chaos regulation of a symmetric nonlinear gyro system via integral sliding mode control. Int J ChemTech Res 9(5):462–469
104. Vaidyanathan S (2016) Hybrid synchronization of the generalized Lotka-Volterra three-species biological systems via adaptive control. Int J PharmTech Res 9(1):179–192
105. Vaidyanathan S (2016) Hyperchaos, adaptive control and synchronization of a novel 4-D hyperchaotic system with two quadratic nonlinearities. Arch Control Sci 26(4):471–495
106. Vaidyanathan S (2016) Mathematical analysis, adaptive control and synchronization of a ten-term novel three-scroll chaotic system with four quadratic nonlinearities. Int J Control Theory Appl 9(1):1–20
107. Vaidyanathan S (2016) Qualitative analysis and properties of a novel 4-D hyperchaotic system with two quadratic nonlinearities and its adaptive control. In: Azar AT, Vaidyanathan S (eds) Advances in chaos theory and intelligent control. Springer, Berlin, Germany, pp 455–480
108. Vaidyanathan S, Azar AT (2015) Analysis and control of a 4-D novel hyperchaotic system. In: Azar AT, Vaidyanathan S (eds) Chaos modeling and control systems design. Studies in computational intelligence, vol 581. Springer, Germany, pp 3–17
109. Vaidyanathan S, Azar AT (2015) Analysis, control and synchronization of a nine-term 3-D novel chaotic system. In: Azar AT, Vaidyanathan S (eds) Chaos modelling and control systems design. Studies in computational intelligence, vol 581. Springer, Germany, pp 19–38
110. Vaidyanathan S, Azar AT (2016) A novel 4-D four-wing chaotic system with four quadratic nonlinearities and its synchronization via adaptive control method. In: Azar AT, Vaidyanathan S (eds) Advances in chaos theory and intelligent control. Springer, Berlin, Germany, pp 203–224
111. Vaidyanathan S, Azar AT (2016) Qualitative study and adaptive control of a novel 4-D hyperchaotic system with three quadratic nonlinearities. In: Azar AT, Vaidyanathan S (eds) Advances in chaos theory and intelligent control. Springer, Berlin, Germany, pp 179–202
112. Vaidyanathan S, Azar AT (2016) Takagi-Sugeno fuzzy logic controller for Liu-Chen four-scroll chaotic system. Int J Intell Eng Inform 4(2):135–150
113. Vaidyanathan S, Boulkroune A (2016) A novel 4-D hyperchaotic chemical reactor system and its adaptive control. In: Vaidyanathan S, Volos C (eds) Advances and applications in chaotic systems. Springer, Berlin, Germany, pp 447–469
114. Vaidyanathan S, Madhavan K (2013) Analysis, adaptive control and synchronization of a seven-term novel 3-D chaotic system. Int J Control Theory Appl 6(2):121–137
115. Vaidyanathan S, Pakiriswamy S (2016) A five-term 3-D novel conservative chaotic system and its generalized projective synchronization via adaptive control method. Int J Control Theory Appl 9(1):61–78
116. Vaidyanathan S, Pakiriswamy S (2015) A 3-D novel conservative chaotic system and its generalized projective synchronization via adaptive control. J Eng Sci Technol Rev 8(2):52–60
117. Vaidyanathan S, Pakiriswamy S (2016) Adaptive control and synchronization design of a seven-term novel chaotic system with a quartic nonlinearity. Int J Control Theory Appl 9(1):237–256

118. Vaidyanathan S, Rajagopal K (2016) Adaptive control, synchronization and LabVIEW implementation of Rucklidge chaotic system for nonlinear double convection. Int J Control Theory Appl 9(1):175–197

119. Vaidyanathan S, Rajagopal K (2016) Analysis, control, synchronization and LabVIEW implementation of a seven-term novel chaotic system. Int J Control Theory Appl 9(1):151–174

120. Vaidyanathan S, Sampath S (2011) Global chaos synchronization of hyperchaotic Lorenz systems by sliding mode control. Commun Comput Inf Sci 205:156–164

121. Vaidyanathan S, Volos C (2015) Analysis and adaptive control of a novel 3-D conservative no-equilibrium chaotic system. Arch Control Sci 25(3):333–353

122. Vaidyanathan S, Volos C (2016) Advances and applications in chaotic systems. Springer, Berlin, Germany

123. Vaidyanathan S, Volos C (2016) Advances and applications in nonlinear control systems. Springer, Berlin, Germany

124. Vaidyanathan S, Volos C (2017) Advances in memristors, memristive devices and systems. Springer, Berlin, Germany

125. Vaidyanathan S, Volos C, Pham VT (2014) Hyperchaos, adpative control and synchronization of a novel 5-D hyperchaotic system with three positive Lyapunov exponents and its SPICE implementation. Arch Control Sci 24(4):409–446

126. Vaidyanathan S, Volos C, Pham VT, Madhavan K, Idowu BA (2014) Adaptive backstepping control, synchronization and circuit simulation of a 3-D novel jerk chaotic system with two hyperbolic sinusoidal nonlinearities. Arch Control Sci 24(3):375–403

127. Vaidyanathan S, Azar AT, Rajagopal K, Alexander P (2015) Design and SPICE implementation of a 12-term novel hyperchaotic system and its synchronisation via active control. Int J Modell Ident Control 23(3):267–277

128. Vaidyanathan S, Pham VT, Volos CK (2015) A 5-D hyperchaotic Rikitake dynamo system with hidden attractors. Eur Phys J: Spec Top 224(8):1575–1592

129. Vaidyanathan S, Rajagopal K, Volos CK, Kyprianidis IM, Stouboulos IN (2015) Analysis, adaptive control and synchronization of a seven-term novel 3-D chaotic system with three quadratic nonlinearities and its digital implementation in LabVIEW. J Eng Sci Technol Rev 8(2):130–141

130. Vaidyanathan S, Volos C, Pham VT, Madhavan K (2015) Analysis, adaptive control and synchronization of a novel 4-D hyperchaotic hyperjerk system and its SPICE implementation. Arch Control Sci 25(1):135–158

131. Vaidyanathan S, Volos CK, Kyprianidis IM, Stouboulos IN, Pham VT (2015) Analysis, adaptive control and anti-synchronization of a six-term novel jerk chaotic system with two exponential nonlinearities and its circuit simulation. J Eng Sci Technol Rev 8(2):24–36

132. Vaidyanathan S, Volos CK, Pham VT (2015) Analysis, adaptive control and adaptive synchronization of a nine-term novel 3-D chaotic system with four quadratic nonlinearities and its circuit simulation. J Eng Sci Technol Rev 8(2):181–191

133. Vaidyanathan S, Volos CK, Pham VT (2015) Analysis, control, synchronization and SPICE implementation of a novel 4-D hyperchaotic Rikitake dynamo system without equilibrium. J Eng Sci Technol Rev 8(2):232–244

134. Vaidyanathan S, Volos CK, Pham VT (2015) Global chaos control of a novel nine-term chaotic system via sliding mode control. In: Azar AT, Zhu Q (eds) Advances and applications in sliding mode control systems. Studies in computational intelligence, vol 576. Springer, Germany, pp 571–590

135. Vaidyanathan S, Volos CK, Pham VT (2016) Hyperchaos, control, synchronization and circuit simulation of a novel 4-D hyperchaotic system with three quadratic nonlinearities. In: Azar AT, Vaidyanathan S (eds) Advances in chaos theory and intelligent control. Springer, Berlin, Germany, pp 297–325

136. Volos CK, Kyprianidis IM, Stouboulos IN, Tlelo-Cuautle E, Vaidyanathan S (2015) Memristor: a new concept in synchronization of coupled neuromorphic circuits. J Eng Sci Technol Rev 8(2):157–173

137. Volos CK, Pham VT, Vaidyanathan S, Kyprianidis IM, Stouboulos IN (2015) Synchronization phenomena in coupled Colpitts circuits. J Eng Sci Technol Rev 8(2):142–151
138. Volos CK, Pham VT, Vaidyanathan S, Kyprianidis IM, Stouboulos IN (2016) Synchronization phenomena in coupled hyperchaotic oscillators with hidden attractors using a nonlinear open loop controller. In: Vaidyanathan S, Volos C (eds) Advances and applications in chaotic systems. Springer, Berlin, Germany, pp 1–38
139. Volos CK, Pham VT, Vaidyanathan S, Kyprianidis IM, Stouboulos IN (2016) The case of bidirectionally coupled nonlinear circuits via a memristor. In: Vaidyanathan S, Volos C (eds) Advances and applications in nonlinear control systems. Springer, Berlin, Germany, pp 317–350
140. Volos CK, Prousalis D, Kyprianidis IM, Stouboulos I, Vaidyanathan S, Pham VT (2016) Synchronization and anti-synchronization of coupled Hindmarsh-Rose neuron models. Int J Control Theory Appl 9(1):101–114
141. Volos CK, Vaidyanathan S, Pham VT, Maaita JO, Giakoumis A, Kyprianidis IM, Stouboulos IN (2016) A novel design approach of a nonlinear resistor based on a memristor emulator. In: Azar AT, Vaidyanathan S (eds) Advances in chaos theory and intelligent control. Springer, Berlin, Germany, pp 3–34
142. Wang B, Zhong SM, Dong XC (2016) On the novel chaotic secure communication scheme design. Commun Nonlinear Sci Numer Simul 39:108–117
143. Wolf A, Swift JB, Swinney HL, Vastano JA (1985) Determining Lyapunov exponents from a time series. Physica D 16:285–317
144. Wu T, Sun W, Zhang X, Zhang S (2016) Concealment of time delay signature of chaotic output in a slave semiconductor laser with chaos laser injection. Opt Commun 381:174–179
145. Xu G, Liu F, Xiu C, Sun L, Liu C (2016) Optimization of hysteretic chaotic neural network based on fuzzy sliding mode control. Neurocomputing 189:72–79
146. Xu H, Tong X, Meng X (2016) An efficient chaos pseudo-random number generator applied to video encryption. Optik 127(20):9305–9319
147. Zhou W, Xu Y, Lu H, Pan L (2008) On dynamics analysis of a new chaotic attractor. Phys Lett A 372(36):5773–5777
148. Zhu C, Liu Y, Guo Y (2010) Theoretic and numerical study of a new chaotic system. Intell Inf Manage 2:104–109

A Memristor-Based Hyperchaotic System with Hidden Attractor and Its Sliding Mode Control

Sundarapandian Vaidyanathan

Abstract Memristor-based systems and their potential applications, in which memristor is both a nonlinear element and a memory element, have been received significant attention in the control literature. In this work, we study a memristor-based hyperchaotic system with hidden attractors. First, we study the dynamic properties of the memristor-based hyperchaotic system such as equilibria, Lyapunov exponents, Kaplan-Yorke dimension, etc. We obtain the Lyapunov exponents of the memristor-based system as $L_1 = 0.2205$, $L_2 = 0.0305$, $L_3 = 0$ and $L_4 = -10.7862$. Since there are two positive Lyapunov exponents, the memristor-based system is hyperchaotic. Also, the Kaplan-Yorke fractional dimension of the memristor-based hyperchaotic system is obtained as $D_{KY} = 3.0233$, which shows the high complexity of the system. We show that the memristor-based hyperchaotic system has no equilibrium point, which shows that the system has a hidden attractor. Control and synchronization of chaotic and hyperchaotic systems are important research problems in chaos theory. Sliding mode control is an important method used to solve various problems in control systems engineering. In robust control systems, the sliding mode control is often adopted due to its inherent advantages of easy realization, fast response and good transient performance as well as insensitivity to parameter uncertainties and disturbance. Next, using integral sliding mode control, we design adaptive control and synchronization schemes for the memristor-based hyperchaotic system. The main adaptive control and synchronization results are established using Lyapunov stability theory. MATLAB simulations are shown to illustrate all the main results of this work.

Keywords Memristor · Hidden attractor · Chaos · Hyperchaos · Adaptive control · Synchronization · Sliding mode control

---

S. Vaidyanathan (✉)
Research and Development Centre, Vel Tech University, Avadi, Chennai
600062, Tamil Nadu, India
e-mail: sundarcontrol@gmail.com

© Springer International Publishing AG 2017
S. Vaidyanathan and C.-H. Lien (eds.), *Applications of Sliding Mode Control
in Science and Engineering*, Studies in Computational Intelligence 709,
DOI 10.1007/978-3-319-55598-0_15

343

1   Introduction

Chua's circuit [38], the Cellular Neural Networks (CNNs) [17] and the memristor
[15] are three attractive inventions of Prof. Leon O. Chua and these inventions are
widely regarded as the major breakthroughs in the literature of the nonlinear control
systems. Chua's circuit has been applied in various areas in engineering [4, 16, 21,
35, 73]. Cellular Neural Networks have been applied in various areas such as chaos
[119], secure communications [150], cryptosystem [14], etc. The studies on memristor [2, 3, 25, 58, 59, 148, 153] have received significant attention only recently after
the realization of a solid-state thin film two-terminal memristor at Hewlett-Packard
Laboratories [62].

Memristor was proposed by L.O. Chua as the fourth basic circuit element besides
the three conventional ones (resistor, inductor and capacitor) [74].

Memristor depicts the relationship between two fundamental circuit variables,
viz. the charge (q) and the flux (φ). Hence, there are two kinds of memristors:
(1) *charge-controlled memristor*, and (2) *flux-controlled memristor*.

A charge-controlled memristor is described by

$$v_M = M(q)i_M \tag{1}$$

where v_M is the *voltage* across the memristor and i_M is the *current* through the memristor. Here, the *memristance* (M) is defined by

$$M(q) = \frac{d\varphi(q)}{dq} \tag{2}$$

A flux-controlled memristor is given by

$$i_M = W(\varphi)v_M \tag{3}$$

where $W(\varphi)$ is the *memductance*, which is defined by

$$W(q) = \frac{dq(\varphi)}{d\varphi} \tag{4}$$

By generalizing the original definition of a memristor [15, 74], a *memristive system* is defined as

$$\begin{cases} \dot{\mathbf{x}} = f(\mathbf{x}, u, t) \\ y = g(\mathbf{x}, u, t)u \end{cases} \tag{5}$$

where $\mathbf{x}$ is the *state*, u is the *input* and y is the *output* of the system (5). We assume
that the function f is a continuously differentiable, n-dimensional vector field and g
is a continuous scalar function.

The intrinsic nonlinear characteristic of memristor has applications in implementing chaotic systems with complex dynamics as well as special features [23, 42]. For

example, a simple memristor-based chaotic system including only three elements (an inductor, a capacitor and a memristor) was introduced in [41]. Also, a system containing an HP memristor model and triangular wave sequence can generate multiscroll chaotic attractors [33]. Moreover, a four-dimensional hyperchaotic memristive system with a line equilibrium was presented by Li [34].

Chaos theory deals with the qualitative study of chaotic dynamical systems and their applications in science and engineering. A dynamical system is called *chaotic* if it satisfies the three properties: boundedness, infinite recurrence and sensitive dependence on initial conditions [6].

Some classical paradigms of 3-D chaotic systems in the literature are Lorenz system [36], Rössler system [51], ACT system [5], Sprott systems [61], Chen system [12], Lü system [37], Cai system [10], Tigan system [75], etc.

Many new chaotic systems have been discovered in the recent years such as Zhou system [154], Zhu system [155], Li system [31], Wei-Yang system [151], Sundarapandian systems [65, 70], Vaidyanathan systems [85, 86, 88–91, 94, 105, 106, 122, 123, 125, 133, 137, 139, 141, 143, 144], Pehlivan system [44], Sampath system [52], etc.

Chaos theory has many applications in science and engineering such as chemical systems [95, 99, 101, 103, 107, 111–113], biological systems [93, 96–98, 100, 102, 104, 108–110, 114–118], memristors [1, 45, 147], etc.

The study of control of a chaotic system investigates feedback control methods that globally or locally asymptotically stabilize or regulate the outputs of a chaotic system. Many methods have been designed for control and regulation of chaotic systems such as active control [63, 64, 80], adaptive control [136, 142, 145], backstepping control [32, 149], sliding mode control [82, 84], etc.

Synchronization of chaotic systems is a phenomenon that occurs when two or more chaotic systems are coupled or when a chaotic system drives another chaotic system. Because of the butterfly effect which causes exponential divergence of the trajectories of two identical chaotic systems started with nearly the same initial conditions, the synchronization of chaotic systems is a challenging research problem in the chaos literature [6–8, 134, 135].

Pecora and Carroll pioneered the research on synchronization of chaotic systems with their seminal papers [11, 43]. The active control method [26, 53, 54, 69, 79, 83, 126, 127, 130] is typically used when the system parameters are available for measurement. Adaptive control method [55–57, 66–68, 81, 87, 121, 124, 128, 129, 136, 140] is typically used when some or all the system parameters are not available for measurement and estimates for the uncertain parameters of the systems.

Sampled-data feedback control method [22, 152] and time-delay feedback control method [13, 24] are also used for synchronization of chaotic systems. Backstepping control method [46–50, 72, 131, 138, 146] is also applied for the synchronization of chaotic systems. Backstepping control is a recursive method for stabilizing the origin of a control system in strict-feedback form [27]. In this research work, we apply backstepping control method for the adaptive control and synchronization of the novel hyperjerk system.

According to a new classification of chaotic dynamics [19, 28, 30], there are two kinds of chaotic attractors: (1) *self-cited attractors*, and (2) *hidden attractors*. The classical attractors of Lorenz, Rössler, Chua, Chen, and other widely-known attractors are those excited from unstable equilibria [6–8, 134, 135]. From the computational point of view, this allows one to use numerical method, in which after transient process a trajectory, started from a point of unstable manifold in the neighborhood of equilibrium, reaches an attractor and identifies it. However there are attractors of another type: hidden attractors, a basin of attraction of which does not contain neighborhoods of equilibria. Hidden attractors cannot be reached by trajectory from neighborhoods of equilibria.

In this work, we derive a new memristor-based hyperchaotic system without any equilibrium point. Thus, the new memristor-based hyperchaotic system has a hidden attractor. First, we study the dynamic properties of the memristor-based hyperchaotic system such as equilibria, Lyapunov exponents, Kaplan-Yorke dimension, etc. We obtain the Lyapunov exponents of the memristor-based system as $L_1 = 0.2205$, $L_2 = 0.0305$, $L_3 = 0$ and $L_4 = -10.7862$. Since there are two positive Lyapunov exponents, the memristor-based system is hyperchaotic. Also, the Kaplan-Yorke fractional dimension of the memristor-based hyperchaotic system is obtained as $D_{KY} = 3.0233$, which shows the high complexity of the system.

Next, we use sliding mode control for the adaptive control and synchronization of the new memristor-based hyperchaotic system. A major advantage of sliding mode control is low sensitivity to parameter variations in the plant and disturbances affecting the plant, which eliminates the necessity of exact modeling of the plant. In the sliding mode control, the control dynamics will have two sequential modes, viz. the reaching mode and the sliding mode. Basically, a sliding mode controller design consists of two parts: hyperplane design and controller design. A hyperplane is first designed via the pole-placement approach and a controller is then designed based on the sliding condition. The stability of the overall system is guaranteed by the sliding condition and by a stable hyperplane. Sliding mode control method is a popular method for the control and synchronization of chaotic systems [29, 39, 71, 78, 82, 84, 92, 120, 132].

This work is organized as follows. Section 2 describes the model of the memristor-based system. Section 3 describes the qualitative properties of the new memristor-based hyperchaotic system. Section 4 derives new results for the global control of the memristor-based hyperchaotic system via adaptive sliding mode control. Section 5 discusses new results for the global hyperchaos synchronization of the memristor-based hyperchaotic systems via adaptive sliding mode control. Finally, Sect. 6 concludes this work with a summary of the main results.

2 Model of Memristor-Based System

This work describes a flux-controlled memristor. For this construction, we use the following memductance function [9, 20, 40]

$$W(\varphi) = 1 + 6\varphi^2 \tag{6}$$

Based on this memristor, a four-dimensional system is introduced as follows:

$$\begin{cases} \dot{x} = -a(x+y) - 5yz \\ \dot{y} = -6x + 6xz + byW(\varphi) + c \\ \dot{z} = -z - 5xy \\ \dot{\varphi} = x \end{cases} \tag{7}$$

where a, b, c are real parameters and $W(\varphi)$ is the memductance as defined in (6).
When $c = 0$, the memristor-based system (7) has the line equilibrium

$$E_\varphi = \begin{bmatrix} 0 \\ 0 \\ 0 \\ \varphi \end{bmatrix} \tag{8}$$

Also, when $c = 0$, the system (7) is hyperchaotic for different values of the parameters.

If we choose the parameter values as

$$a = 10, \quad b = 0.1, \quad c = 0 \tag{9}$$

then the memristor-based system (7) is hyperchaotic.

For numerical simulations, we take the initial values of the system (7) as

$$x(0) = 0.3, \quad y(0) = 0.3, \quad z(0) = 0.3, \quad w(0) = 0.3 \tag{10}$$

For the parameter values (9) and the initial values (10), the Lyapunov exponents of the system (7) are numerically determined as

$$L_1 = 0.2192, \quad L_2 = 0.0256, \quad L_3 = 0, \quad L_4 = -10.7844 \tag{11}$$

Figure 1 shows the Lyapunov exponents of the memristor-based hyperchaotic system (7).

Since there are two positive Lyapunov exponents in the LE spectrum (11), it is immediate that the memristor-based system (7) is hyperchaotic.

Also, the Kaplan-Yorke dimension of the system (7) is given by

$$D_{KY} = 3 + \frac{L_1 + L_2 + L_3}{|L_4|} = 3.0227, \tag{12}$$

which shows the high complexity of the system.

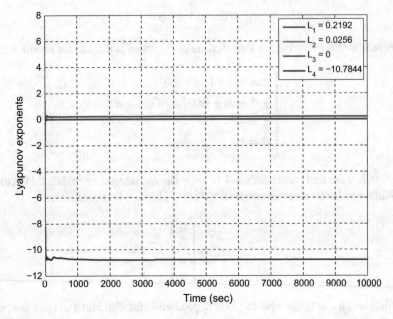

Fig. 1 Lyapunov exponents of the memristor-based system for $(a, b, c) = (10, 0.1, 0)$

Figures 2 and 3 show the 3-D projections of the memristor-based hyperchaotic system (7) in (x, y, z) and (y, z, φ) spaces, respectively.

3 Dynamics of the Memristor-Based System

We consider the memristor-based system (7) when $c \neq 0$.

We find the equilibrium points of the system (7) by solving the following system of equations

$$-a(x + y) - 5yz \quad = \quad 0 \qquad (13a)$$
$$-6x + 6xz + byW(\varphi) + c \quad = \quad 0 \qquad (13b)$$
$$-z - 5xy \quad = \quad 0 \qquad (13c)$$
$$x \quad = \quad 0 \qquad (13d)$$

From (13d), it is immediate that $x = 0$.

Substituting $x = 0$ in (13c), it is immediate that $z = 0$.

Substituting $x = 0$ and $z = 0$ in (13a), it follows that $y = 0$.

Thus, Eq. (13b) reduces to $c = 0$, which is a contradiction.

Hence, there is no equilibrium for the memristor-based system (7) when $c \neq 0$.

Next, we take the parameters of the memristor-based system (7) as

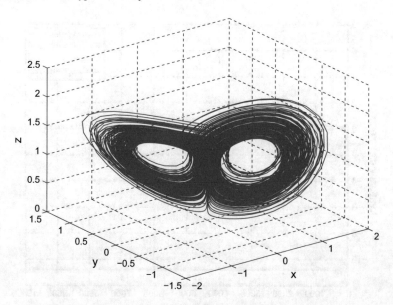

Fig. 2 3-D plot of the memristor-based system in (x, y, z) space for $(a, b, c) = (10, 0.1, 0)$

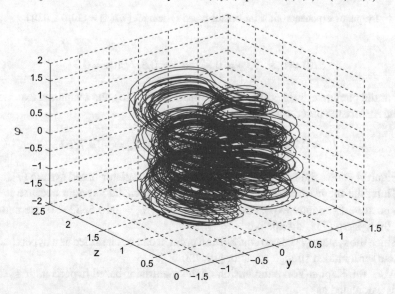

Fig. 3 3-D plot of the memristor-based system in (y, z, w) space for $(a, b, c) = (10, 0.1, 0)$

$$a = 10, \quad b = 0.1, \quad c = 0.01 \tag{14}$$

We choose the initial conditions of the system (7) as

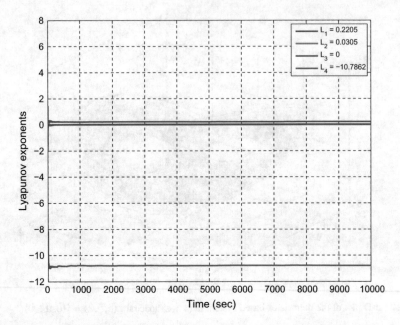

Fig. 4 Lyapunov exponents of the memristor-based system for $(a, b, c) = (10, 0.1, 0.01)$

$$x(0) = 0.3, \quad y(0) = 0.3, \quad z(0) = 0.3, \quad \varphi(0) = 0.3 \tag{15}$$

For the parameter values (14) and the initial values (15), the Lyapunov exponents of the memristor-based system (7) are obtained as

$$L_1 = 0.2205, \quad L_2 = 0.0305, \quad L_3 = 0, \quad L_4 = -10.7862 \tag{16}$$

Figure 4 shows the Lyapunov exponents of the memristor-based system (7).

Thus, the memristor-based system (7) is a hyperchaotic system because it has two positive Lyapunov exponents [7]. Since $\sum_{i=1}^{4} L_i = -10.5352 < 0$, we deduce that the system (7) is dissipative.

Since the system (7) has no equilibrium point, it can be classified as a hyperchaotic system with hidden strange attractor [19, 28, 30].

Also, the Kaplan-Yorke dimension of the memristor-based hyperchaotic system (7) is calculated as

$$D_{KY} = 3 + \frac{L_1 + L_2 + L_3}{|L_4|} = 3.0233, \tag{17}$$

which shows the high complexity of the system.

Figures 5, 6, 7 and 8 show the 3-D projections of the memristor-based hyperchaotic system in (x, y, z), (x, y, φ), (x, z, φ) and (y, z, φ) spaces, respectively.

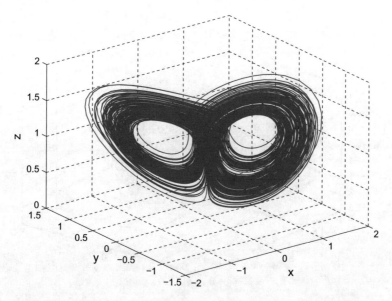

Fig. 5 3-D plot of the memristor-based system in (x, y, z) space for $(a, b, c) = (10, 0.1, 0.01)$

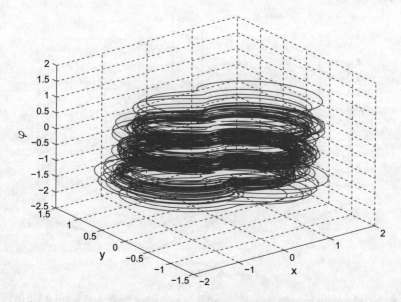

Fig. 6 3-D plot of the memristor-based system in (x, y, φ) space for $(a, b, c) = (10, 0.1, 0.01)$

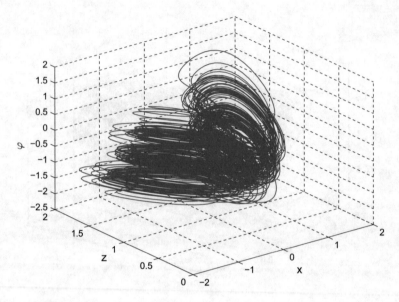

Fig. 7 3-D plot of the memristor-based system in (x, z, φ) space for $(a, b, c) = (10, 0.1, 0.01)$

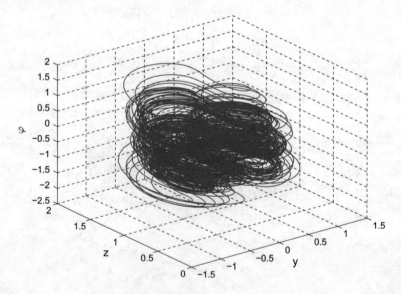

Fig. 8 3-D plot of the memristor-based system in (y, z, φ) space for $(a, b, c) = (10, 0.1, 0.01)$

4 Global Hyperchaos Control of the New Memristor-Based Hyperchaotic System

In this section, we use adaptive integral sliding mode control for the global hyperchaos control of the new memristor-based hyperchaotic system with unknown system parameters. The adaptive control mechanism helps the control design by estimating the unknown parameters [6–8, 134, 135].

The controlled new memristor-based hyperchaotic system is described by

$$\begin{cases} \dot{x} = -a(x+y) - 5yz + u_x \\ \dot{y} = -6x + 6xz + byW(\varphi) + c + u_y \\ \dot{z} = -z - 5xy + u_z \\ \dot{\varphi} = x + u_\varphi \end{cases} \tag{18}$$

where x, y, z, φ are the states of the system and a, b, c are unknown system parameters. As noted in Sect. 2, the memductance function is taken as

$$W(\varphi) = 1 + 6\varphi^2 \tag{19}$$

The design goal is to find suitable feedback controllers u_x, u_y, u_z, u_w so as to globally stabilize the system (18) with estimates of the unknown parameters.

Based on the sliding mode control theory [60, 76, 77], the integral sliding surface of each state is defined as follows:

$$\begin{cases} s_x = \left(\frac{d}{dt} + \lambda_x\right)\left(\int_0^t x(\tau)d\tau\right) = x + \lambda_x \int_0^t x(\tau)d\tau \\ s_y = \left(\frac{d}{dt} + \lambda_y\right)\left(\int_0^t y(\tau)d\tau\right) = y + \lambda_y \int_0^t y(\tau)d\tau \\ s_z = \left(\frac{d}{dt} + \lambda_z\right)\left(\int_0^t z(\tau)d\tau\right) = z + \lambda_z \int_0^t z(\tau)d\tau \\ s_\varphi = \left(\frac{d}{dt} + \lambda_\varphi\right)\left(\int_0^t \varphi(\tau)d\tau\right) = \varphi + \lambda_\varphi \int_0^t \varphi(\tau)d\tau \end{cases} \tag{20}$$

From Eq. (20), it follows that

$$\begin{cases} \dot{s}_x = \dot{x} + \lambda_x x \\ \dot{s}_y = \dot{y} + \lambda_y y \\ \dot{s}_z = \dot{z} + \lambda_z z \\ \dot{s}_\varphi = \dot{\varphi} + \lambda_\varphi \varphi \end{cases} \tag{21}$$

The Hurwitz condition is realized if $\lambda_x > 0$, $\lambda_y > 0$, $\lambda_z > 0$ and $\lambda_\varphi > 0$.
We consider the adaptive feedback control given by

$$
\begin{cases}
u_x = \hat{a}(t)(x + y) + 5yz - \lambda_x x - \eta_x \operatorname{sgn}(s_x) - k_x s_x \\
u_y = 6x - 6xz - \hat{b}(t)yW(\varphi) - \hat{c}(t) - \lambda_y y - \eta_y \operatorname{sgn}(s_y) - k_y s_y \\
u_z = z + 5xy - \lambda_z z - \eta_z \operatorname{sgn}(s_z) - k_z s_z \\
u_\varphi = -x - \lambda_\varphi \varphi - \eta_\varphi \operatorname{sgn}(s_\varphi) - k_\varphi s_\varphi
\end{cases}
\tag{22}
$$

where $\eta_x, \eta_y, \eta_z, \eta_\varphi, k_x, k_y, k_z, k_\varphi$ are positive constants.

Substituting (22) into (18), we obtain the closed-loop control system given by

$$
\begin{cases}
\dot{x} = -[a - \hat{a}(t)](x + y) - \lambda_x x - \eta_x \operatorname{sgn}(s_x) - k_x s_x \\
\dot{y} = [b - \hat{b}(t)]yW(\varphi) + [c - \hat{c}(t)] - \lambda_y y - \eta_y \operatorname{sgn}(s_y) - k_y s_y \\
\dot{z} = -\lambda_z z - \eta_z \operatorname{sgn}(s_z) - k_z s_z \\
\dot{\varphi} = -\lambda_\varphi \varphi - \eta_\varphi \operatorname{sgn}(s_\varphi) - k_\varphi s_\varphi
\end{cases}
\tag{23}
$$

We define the parameter estimation errors as

$$
\begin{cases}
e_a(t) = a - \hat{a}(t) \\
e_b(t) = b - \hat{b}(t) \\
e_c(t) = c - \hat{c}(t)
\end{cases}
\tag{24}
$$

Using (24), we can simplify the closed-loop system (23) as

$$
\begin{cases}
\dot{x} = -e_a(x + y) - \lambda_x x - \eta_x \operatorname{sgn}(s_x) - k_x s_x \\
\dot{y} = e_b yW(\varphi) + e_c - \lambda_y y - \eta_y \operatorname{sgn}(s_y) - k_y s_y \\
\dot{z} = -\lambda_z z - \eta_z \operatorname{sgn}(s_z) - k_z s_z \\
\dot{\varphi} = -\lambda_\varphi \varphi - \eta_\varphi \operatorname{sgn}(s_\varphi) - k_\varphi s_\varphi
\end{cases}
\tag{25}
$$

Differentiating (24) with respect to t, we get

$$
\begin{cases}
\dot{e}_a = -\dot{\hat{a}} \\
\dot{e}_b = -\dot{\hat{b}} \\
\dot{e}_c = -\dot{\hat{c}}
\end{cases}
\tag{26}
$$

Next, we state and prove the main result of this section.

Theorem 1 *The new memristor-based hyperchaotic system (18) is rendered globally asymptotically stable for all initial conditions $\mathbf{X}(0) \in \mathbf{R}^4$ by the adaptive integral sliding mode control law (22) and the parameter update law*

$$\begin{cases} \dot{a} = -s_x(x+y) \\ \dot{b} = s_y y W(\varphi) \\ \dot{c} = s_y \end{cases} \tag{27}$$

where $\lambda_x, \lambda_y, \lambda_z, \lambda_\varphi, \eta_x, \eta_y, \eta_z, \eta_\varphi, k_x, k_y, k_z, k_\varphi$ are positive constants.

Proof We consider the quadratic Lyapunov function defined by

$$V(s_x, s_y, s_z, s_\varphi, e_a, e_b, e_c) = \frac{1}{2}\left(s_x^2 + s_y^2 + s_z^2 + s_\varphi^2\right) + \frac{1}{2}\left(e_a^2 + e_b^2 + e_c^2\right) \tag{28}$$

Clearly, V is positive definite on $\mathbf{R}^7$.

Using (21), (25) and (26), the time-derivative of V is obtained as

$$\begin{aligned} \dot{V} = s_x &\left[-e_a(x+y) - \eta_x \operatorname{sgn}(s_x) - k_x s_x\right] - e_a \dot{a} \\ &+ s_y \left[e_b y W(\varphi) + e_c - \eta_y \operatorname{sgn}(s_y) - k_y s_y\right] - e_b \dot{b} - e_c \dot{c} \\ &+ s_z \left[-\eta_z \operatorname{sgn}(s_z) - k_z s_z\right] + s_\varphi \left[-\eta_\varphi \operatorname{sgn}(s_\varphi) - k_\varphi s_\varphi\right] \end{aligned} \tag{29}$$

i.e.

$$\begin{aligned} \dot{V} = &-\eta_x|s_x| - k_x s_x^2 - \eta_y|s_y| - k_y s_y^2 - \eta_z|s_z| - k_z s_z^2 - \eta_\varphi|s_\varphi| - k_\varphi s_\varphi^2 \\ &+ e_a\left[-s_x(x+y) - \dot{\hat{a}}\right] + e_b\left[s_y y W(\varphi) - \dot{\hat{b}}\right] + e_c\left[s_y - \dot{\hat{c}}\right] \end{aligned} \tag{30}$$

Using the parameter update law (27), we obtain

$$\dot{V} = -\eta_x|s_x| - k_x s_x^2 - \eta_y|s_y| - k_y s_y^2 - \eta_z|s_z| - k_z s_z^2 - \eta_\varphi|s_\varphi| - k_\varphi s_\varphi^2 \tag{31}$$

which shows that $\dot{V}$ is negative semi-definite on $\mathbf{R}^7$.

Hence, by Barbalat's lemma [27], it is immediate that $\mathbf{x}(t)$ is globally asymptotically stable for all values of $\mathbf{x}(0) \in \mathbf{R}^4$.

This completes the proof. ∎

For numerical simulations, we take the parameter values of the new memristor-based hyperchaotic system (18) as in the hyperchaotic case (14), i.e. $a = 10, b = 0.1$ and $c = 0.01$.

We take the values of the control parameters as $k_x = 20, k_y = 20, k_z = 20, k_\varphi = 20, \eta_x = 0.1, \eta_y = 0.1, \eta_z = 0.1, \eta_\varphi = 0.1, \lambda_x = 50, \lambda_y = 50, \lambda_z = 50$ and $\lambda_\varphi = 50$.

We take the estimates of the system parameters as

$$\hat{a}(0) = 12.6, \quad \hat{b}(0) = 17.3, \quad \hat{c}(0) = 15.4 \tag{32}$$

We take the initial state of the hyperchaotic system (18) as

$$x(0) = 4.9, \quad y(0) = 10.5, \quad z(0) = 5.7, \quad \varphi(0) = 5.2 \tag{33}$$

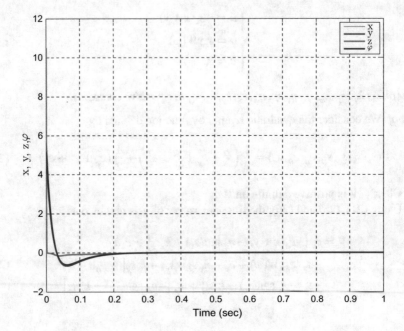

Fig. 9 Time-history of the controlled states x, y, z, φ

Figure 9 shows the time-history of the controlled states x, y, z, φ.

5 Global Hyperchaos Synchronization of the New Memristor-Based Hyperchaotic Systems

In this section, we use adaptive integral sliding mode control for the global hyperchaos synchronization of the new memristor-based hyperchaotic systems with unknown system parameters. The adaptive control mechanism helps the control design by estimating the unknown parameters [6–8, 134, 135].

As the master system, we consider the new memristor-based hyperchaotic system given by

$$\begin{cases} \dot{x}_1 = -a(x_1 + y_1) - 5y_1z_1 \\ \dot{y}_1 = -6x_1 + 6x_1z_1 + by_1W(\varphi_1) + c \\ \dot{z}_1 = -z_1 - 5x_1y_1 \\ \dot{\varphi}_1 = x_1 \end{cases} \tag{34}$$

where x_1, y_1, z_1, φ_1 are the states of the system and a, b, c are unknown system parameters.

As the slave system, we consider the new memristor-based hyperchaotic system given by

$$\begin{cases} \dot{x}_2 = -a(x_2 + y_2) - 5y_2z_2 + u_x \\ \dot{y}_2 = -6x_2 + 6x_2z_2 + by_2W(\varphi_2) + c + u_y \\ \dot{z}_2 = -z_2 - 5x_2y_2 + u_z \\ \dot{\varphi}_2 = x_2 + u_\varphi \end{cases} \tag{35}$$

where x_2, y_2, z_2, φ_2 are the states of the system and u_x, u_y, u_z, u_φ are the feedback controllers to be defined.

In (34) and (35), it is noted that

$$W(\varphi_1) = 1 + 6\varphi_1^2 \quad \text{and} \quad W(\varphi_2) = 1 + 6\varphi_2^2 \tag{36}$$

The synchronization error between the hyperchaotic systems (34) and (35) is defined as

$$\begin{cases} e_x = x_2 - x_1 \\ e_y = y_2 - y_1 \\ e_z = z_2 - z_1 \\ e_\varphi = \varphi_2 - \varphi_1 \end{cases} \tag{37}$$

Then the synchronization error dynamics is obtained as

$$\begin{cases} \dot{e}_x = -a(e_x + e_y) - 5(y_2z_2 - y_1z_1) + u_x \\ \dot{e}_y = -6e_x + 6(x_2z_2 - x_1z_1) + b[y_2W(\varphi_2) - y_1W(\varphi_1)] + u_y \\ \dot{e}_z = -e_z - 5(x_2y_2 - x_1y_1) + u_z \\ \dot{e}_\varphi = e_x + u_\varphi \end{cases} \tag{38}$$

Based on the sliding mode control theory [60, 76, 77], the integral sliding surface of each error variable is defined as follows:

$$\begin{cases} s_x = \left(\frac{d}{dt} + \lambda_x\right)\left(\int_0^t e_x(\tau)d\tau\right) = e_x + \lambda_x \int_0^t e_x(\tau)d\tau \\ s_y = \left(\frac{d}{dt} + \lambda_y\right)\left(\int_0^t e_y(\tau)d\tau\right) = e_y + \lambda_y \int_0^t e_y(\tau)d\tau \\ s_z = \left(\frac{d}{dt} + \lambda_z\right)\left(\int_0^t e_z(\tau)d\tau\right) = e_z + \lambda_z \int_0^t e_z(\tau)d\tau \\ s_\varphi = \left(\frac{d}{dt} + \lambda_\varphi\right)\left(\int_0^t e_\varphi(\tau)d\tau\right) = e_\varphi + \lambda_\varphi \int_0^t e_\varphi(\tau)d\tau \end{cases} \tag{39}$$

From Eq. (39), it follows that

$$\begin{cases} \dot{s}_x = \dot{e}_x + \lambda_x e_x \\ \dot{s}_y = \dot{e}_y + \lambda_y e_y \\ \dot{s}_z = \dot{e}_z + \lambda_z e_z \\ \dot{s}_\varphi = \dot{e}_\varphi + \lambda_\varphi e_\varphi \end{cases} \tag{40}$$

The Hurwitz condition is realized if $\lambda_x > 0, \lambda_y > 0, \lambda_z > 0$ and $\lambda_\varphi > 0$.
We consider the adaptive feedback control given by

$$\begin{cases} u_x = \hat{a}(t)(e_x + e_y) + 5(y_2 z_2 - y_1 z_1) - \lambda_x e_x - \eta_x \, \text{sgn}(s_x) - k_x s_x \\ u_y = 6e_x - 6(x_2 z_2 - x_1 z_1) - \hat{b}(t)[y_2 W(\varphi_2) - y_1 W(\varphi_1)] \\ \qquad - \lambda_y e_y - \eta_y \, \text{sgn}(s_y) - k_y s_y \\ u_z = e_z + 5(x_2 y_2 - x_1 y_1) - \lambda_z e_z - \eta_z \, \text{sgn}(s_z) - k_z s_z \\ u_\varphi = -e_x - \lambda_\varphi e_\varphi - \eta_\varphi \, \text{sgn}(s_\varphi) - k_\varphi s_\varphi \end{cases} \tag{41}$$

where $\eta_x, \eta_y, \eta_z, \eta_\varphi, k_x, k_y, k_z, k_\varphi$ are positive constants.

Substituting (41) into (38), we obtain the closed-loop error dynamics as

$$\begin{cases} \dot{e}_x = -[a - \hat{a}(t)](e_x + e_y) - \lambda_x e_x - \eta_x \, \text{sgn}(s_x) - k_x s_x \\ \dot{e}_y = [b - \hat{b}(t)][y_2 W(\varphi_2) - y_1 W(\varphi_1)] - \lambda_y e_y - \eta_y \, \text{sgn}(s_y) - k_y s_y \\ \dot{e}_z = -\lambda_z e_z - \eta_z \, \text{sgn}(s_z) - k_z s_z \\ \dot{e}_\varphi = -\lambda_\varphi e_\varphi - \eta_\varphi \, \text{sgn}(s_\varphi) - k_\varphi s_\varphi \end{cases} \tag{42}$$

We define the parameter estimation errors as

$$\begin{cases} e_a(t) = a - \hat{a}(t) \\ e_b(t) = b - \hat{b}(t) \end{cases} \tag{43}$$

Using (43), we can simplify the closed-loop system (42) as

$$\begin{cases} \dot{e}_x = -e_a(e_x + e_y) - \lambda_x e_x - \eta_x \, \text{sgn}(s_x) - k_x s_x \\ \dot{e}_y = e_b[y_2 W(\varphi_2) - y_1 W(\varphi_1)] - \lambda_y e_y - \eta_y \, \text{sgn}(s_y) - k_y s_y \\ \dot{e}_z = -\lambda_z e_z - \eta_z \, \text{sgn}(s_z) - k_z s_z \\ \dot{e}_\varphi = -\lambda_\varphi e_\varphi - \eta_\varphi \, \text{sgn}(s_\varphi) - k_\varphi s_\varphi \end{cases} \tag{44}$$

Differentiating (43) with respect to t, we get

$$\begin{cases} \dot{e}_a = -\dot{\hat{a}} \\ \dot{e}_b = -\dot{\hat{b}} \end{cases} \tag{45}$$

Theorem 2 *The new memristor-based hyperchaotic systems (34) and (35) are globally and asymptotically synchronized for all initial conditions by the adaptive integral sliding mode control law (41) and the parameter update law*

$$\begin{cases} \dot{\hat{a}} = -s_x(e_x + e_y) \\ \dot{\hat{b}} = s_y[y_2 W(\varphi_2) - y_1 W(\varphi_1)] \end{cases} \tag{46}$$

where $\lambda_x, \lambda_y, \lambda_z, \lambda_\varphi, \eta_x, \eta_y, \eta_z, \eta_\varphi, k_x, k_y, k_z, k_\varphi$ are positive constants.

Proof We consider the quadratic Lyapunov function defined by

$$V(s_x, s_y, s_z, s_\varphi, e_a, e_b) = \frac{1}{2}\left(s_x^2 + s_y^2 + s_z^2 + s_\varphi^2\right) + \frac{1}{2}\left(e_a^2 + e_b^2\right) \tag{47}$$

Clearly, V is positive definite on $\mathbf{R}^6$.
Using (40), (44) and (45), the time-derivative of V is obtained as

$$\begin{aligned} \dot{V} &= s_x\left[-e_a(e_x + e_y) - \eta_x \,\mathrm{sgn}(s_x) - k_x s_x\right] - e_a\dot{\hat{a}} \\ &\quad + s_y\left[e_b(y_2 W(\varphi_2) - y_1 W(\varphi_1)) - \eta_y \,\mathrm{sgn}(s_y) - k_y s_y\right] - e_b\dot{\hat{b}} \\ &\quad + s_z\left[-\eta_z \,\mathrm{sgn}(s_z) - k_z s_z\right] + s_\varphi\left[-\eta_\varphi \,\mathrm{sgn}(s_\varphi) - k_\varphi s_\varphi\right] \end{aligned} \tag{48}$$

i.e.

$$\begin{aligned} \dot{V} &= -\eta_x|s_x| - k_x s_x^2 - \eta_y|s_y| - k_y s_y^2 - \eta_z|s_z| - k_z s_z^2 - \eta_\varphi|s_\varphi| - k_\varphi s_\varphi^2 \\ &\quad + e_a\left[-s_x(e_x + e_y) - \dot{\hat{a}}\right] + e_b\left[s_y(y_2 W(\varphi_2) - y_1 W(\varphi_1)) - \dot{\hat{b}}\right] \end{aligned} \tag{49}$$

Using the parameter update law (46), we obtain

$$\dot{V} = -\eta_x|s_x| - k_x s_x^2 - \eta_y|s_y| - k_y s_y^2 - \eta_z|s_z| - k_z s_z^2 - \eta_\varphi|s_\varphi| - k_\varphi s_\varphi^2 \tag{50}$$

which shows that $\dot{V}$ is negative semi-definite on $\mathbf{R}^6$.
Hence, by Barbalat's lemma [27], it is immediate that $\mathbf{e}(t)$ is globally asymptotically stable for all values of $\mathbf{e}(0) \in \mathbf{R}^4$, where $\mathbf{e} = \left[e_x, e_y, e_z, e_\varphi\right]^T$.
This completes the proof. ∎

For numerical simulations, we take the parameter values of the new memristor-based systems (34) and (35) as in the hyperchaotic case (14), i.e. $a = 10, b = 0.1$ and $c = 0.01$.

We take the values of the control parameters as $k_x = 20$, $k_y = 20$, $k_z = 20$, $k_\varphi = 20$, $\eta_x = 0.1$, $\eta_y = 0.1$, $\eta_z = 0.1$, $\eta_\varphi = 0.1$, $\lambda_x = 50$, $\lambda_y = 50$, $\lambda_z = 50$ and $\lambda_\varphi = 50$. We take the estimates of the system parameters as

$$\hat{a}(0) = 15.2, \quad \hat{b}(0) = 12.4 \tag{51}$$

We take the initial state of the master system (34) as

$$x_1(0) = 4.1, \quad y_1(0) = 1.4, \quad z_1(0) = 2.7, \quad \varphi_1(0) = 3.8 \tag{52}$$

We take the initial state of the slave system (35) as

$$x_2(0) = 2.9, \quad y_2(0) = 0.5, \quad z_2(0) = 3.1, \quad \varphi_2(0) = 5.2 \tag{53}$$

Figures 10, 11, 12 and 13 show the complete synchronization of the new hyperchaotic systems (34) and (35). Figure 14 shows the time-history of the synchronization errors e_x, e_y, e_z, e_φ.

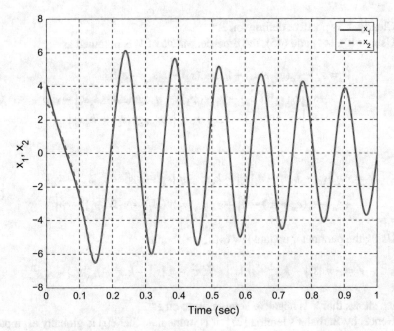

Fig. 10 Complete synchronization of the states x_1 and x_2

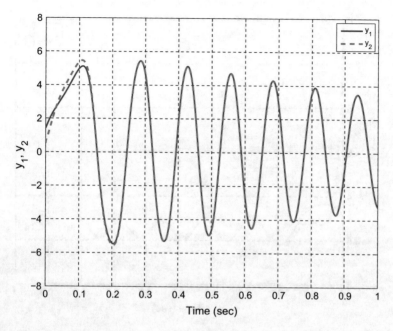

Fig. 11 Complete synchronization of the states y_1 and y_2

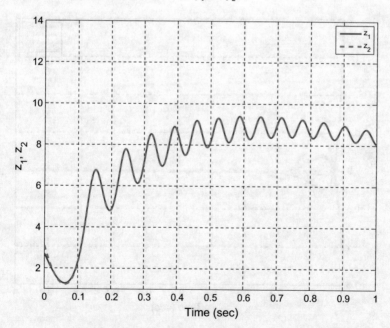

Fig. 12 Complete synchronization of the states z_1 and z_2

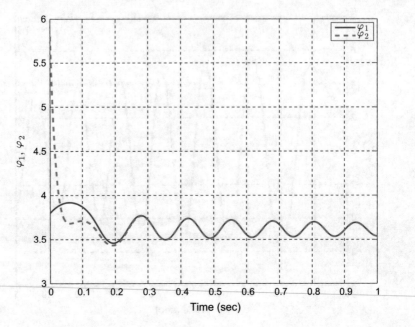

Fig. 13 Complete synchronization of the states φ_1 and φ_2

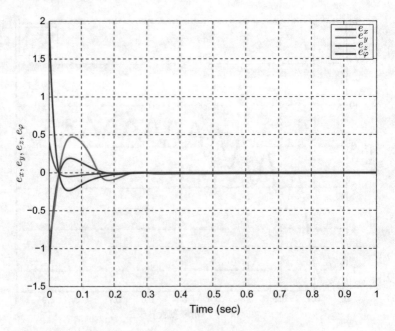

Fig. 14 Time-history of the synchronization errors e_x, e_y, e_z, e_φ

6 Conclusions

In this work, a memristor-based hyperchaotic system has been studied. This system displays rich dynamical behavior as confirmed by qualitative properties and numerical simulations. Moreover, the possibility of adaptive control and synchronization schemes of memristor-based hyperchaotic systems have been designed via adaptive sliding mode control method and Lyapunov stability theory. MATLAB simulations are shown to illustrate the phase portraits, adaptive control and synchronization results for the memristor-based hyperchaotic system. It is worth noting that the presence of the memristor creates some special and unusual features. For example, such memristor-based systems can exhibit hyperchaos although it has no equilibrium points. Also, it is well-known that hyperchaotic system, which is characterized by more than one positive Lyapunov exponent, exhibits a higher level of complexity than a conventional chaotic system. Hence, we can apply this memristor-based hyperchaotic system in practical applications like encryption, cryptosystems, neural networks and secure communications.

References

1. Abdurrahman A, Jiang H, Teng Z (2015) Finite-time synchronization for memristor-based neural networks with time-varying delays. Neural Netw 69:20–28
2. Adhikari SP, Yang C, Kim H, Chua LO (2012) Memristor bridge synapse-based neural network and its learning. IEEE Trans Neural Netw Learn Syst 23:1426–1435
3. Adhikari SP, Sad MP, Kim H, Chua LO (2013) Three fingerprints of memristor. IEEE Trans Circ Syst I Reg Papers 60(11):3008–3021
4. Albuquerque HA, Rubinger RM, Rech PC (2008) Self-similar structures in a 2D parameter-space of an inductorless Chua's circuit. Phys Lett A 372:4793–4798
5. Arneodo A, Coullet P, Tresser C (1981) Possible new strange attractors with spiral structure. Commun Math Phys 79(4):573–576
6. Azar AT, Vaidyanathan S (2015) Chaos modeling and control systems design, vol 581. Springer, Germany
7. Azar AT, Vaidyanathan S (2016) Advances in chaos theory and intelligent control. Springer, Berlin, Germany
8. Azar AT, Vaidyanathan S, Ouannas A (2017) Fractional order control and synchronization of chaotic systems. Springer, Berlin, Germany
9. Bao BC, Liu Z, Xu BP (2010) Dynamical analysis of memristor chaotic oscillator. Acta Phys Sin 59(6):3785–3793
10. Cai G, Tan Z (2007) Chaos synchronization of a new chaotic system via nonlinear control. J Uncertain Syst 1(3):235–240
11. Carroll TL, Pecora LM (1991) Synchronizing chaotic circuits. IEEE Trans Circ Syst 38(4):453–456
12. Chen G, Ueta T (1999) Yet another chaotic attractor. Int J Bifurc Chaos 9(7):1465–1466
13. Chen WH, Wei D, Lu X (2014) Global exponential synchronization of nonlinear time-delay Lur'e systems via delayed impulsive control. Commun Nonlinear Sci Numer Simul 19(9):3298–3312
14. Cheng CJ, Cheng CB (2013) An asymmetric image cryptosystem based on the adaptive synchronization of an uncertain unified chaotic system and a cellular neural network. Commun Nonlinear Sci Numer Simul 18(10):2825–2837

15. Chua LO (1971) Memristor-the missing circuit element. IEEE Trans Circ Theor 18(5):507–519
16. Chua LO (1994) Chua's circuit: an overview ten years later. J Circ Syst Comput 04:117–159
17. Chua LO, Yang L (1988) Cellular neural networks: applications. IEEE Trans Circ Syst 35:1273–1290
18. Chua LO, Yang L (1988) Cellular neural networks: theory. IEEE Trans Circ Syst 35:1257–1272
19. Dudkowski D, Jafari S, Kapitaniaka T, Kuznetsov NV, Leonov GA, Prasad A (2016) Hidden attractors in dynamical systems. Phys Rep 637:1–50
20. Fitch AL, Yu DS, Iu HHC, Sreeram V (2012) Hyperchaos in a memristor-based modified canonical Chua's circuit. Int J Bifurc Chaos 22(6):1250,133
21. Fortuna L, Frasca M, Xibilia MG (2009) Chua's circuit implementations: yesterday, today and tomorrow. World Scientific, Singapore
22. Gan Q, Liang Y (2012) Synchronization of chaotic neural networks with time delay in the leakage term and parametric uncertainties based on sampled-data control. J Frankl Inst 349(6):1955–1971
23. Itoh M, Chua LO (2008) Memristor oscillators. Int J Bifurc Chaos 18(11):3183–3206
24. Jiang GP, Zheng WX, Chen G (2004) Global chaos synchronization with channel time-delay. Chaos, Solitons & Fractals 20(2):267–275
25. Joglekar YN, Wolf SJ (2009) The elusive memristor: properties of basic electrical circuits. Eur J Phys 30(4):661–675
26. Karthikeyan R, Sundarapandian V (2014) Hybrid chaos synchronization of four-scroll systems via active control. J Electr Eng 65(2):97–103
27. Khalil HK (2001) Nonlinear Syst, 3rd edn. Prentice Hall, New Jersey, USA
28. Kuznetsov NV, Leonov GA (2014) Hidden attractors in dynamical systems: systems with no equilibria, multistability and coexisting attractors. IFAC Proc Vol 47(3):5445–5454
29. Lakhekar GV, Waghmare LM, Vaidyanathan S (2016) Diving autopilot design for underwater vehicles using an adaptive neuro-fuzzy sliding mode controller. In: Vaidyanathan S, Volos C (eds) Advances and applications in nonlinear control systems. Springer, Berlin, Germany, pp 477–503
30. Leonov GA, Kuznetsov NV, Vagaitsev VI (2011) Localization of hidden Chua's attractors. Phys Lett A 375(23):2230–2233
31. Li D (2008) A three-scroll chaotic attractor. Phys Lett A 372(4):387–393
32. Li GH, Zhou SP, Yang K (2007) Controlling chaos in Colpitts oscillator. Chaos Solitons Fractals 33:582–587
33. Li H, Wang L, Duan S (2014) A memristor-mased scroll chaotic system—design, analysis and circuit implementation. Int J Bifurc Chaos 24(07):1450,099
34. Li Q, Hu S, Tang S, Zeng G (2013) Hyperchaos and horseshoe in a 4D memristive system with a line of equilibria and its implementation. Int J Circ Theor Appl 42(11):1172–1188
35. Liu L, Wu X, Hu H (2004) Estimating system parameters of Chua's circuit from synchronizing signal. Phys Lett A 324(1):36–41
36. Lorenz EN (1963) Deterministic periodic flow. J Atmos Sci 20(2):130–141
37. Lü J, Chen G (2002) A new chaotic attractor coined. Int J Bifurc Chaos 12(3):659–661
38. Matsumoto T (1984) A chaotic attractor from Chua's circuit. IEEE Trans Circ Syst 31:1055–1058
39. Moussaoui S, Boulkroune A, Vaidyanathan S (2016) Fuzzy adaptive sliding-mode control scheme for uncertain underactuated systems. In: Vaidyanathan S, Volos C (eds) Advances and applications in nonlinear control systems. Springer, Berlin, Berlin, pp 351–367
40. Muthuswamy B (2010) Implementing memristor based chaotic circuits. Int J Bifurc Chaos 20(5):1335–1350
41. Muthuswamy B, Chua LO (2010) Simplest chaotic circuit. Int J Bifurc Chaos 20(5):1567–1580
42. Muthuswamy B, Kokate P (2009) Memristor based chaotic circuits. IETE Tech Rev 26(6):417–429

43. Pecora LM, Carroll TL (1990) Synchronization in chaotic systems. Phys Rev Lett 64(8):821–824
44. Pehlivan I, Moroz IM, Vaidyanathan S (2014) Analysis, synchronization and circuit design of a novel butterfly attractor. J Sound Vib 333(20):5077–5096
45. Pham VT, Volos CK, Vaidyanathan S, Le TP, Vu VY (2015) A memristor-based hyperchaotic system with hidden attractors: dynamics, synchronization and circuital emulating. J Eng Sci Technol Rev 8(2):205–214
46. Rasappan S, Vaidyanathan S (2012) Global chaos synchronization of WINDMI and Coullet chaotic systems by backstepping control. Far East J Math Sci 67(2):265–287
47. Rasappan S, Vaidyanathan S (2012) Hybrid synchronization of n-scroll Chua and Lur'e chaotic systems via backstepping control with novel feedback. Arch Control Sci 22(3):343–365
48. Rasappan S, Vaidyanathan S (2012) Synchronization of hyperchaotic Liu system via backstepping control with recursive feedback. Commun Comput Inf Sci 305:212–221
49. Rasappan S, Vaidyanathan S (2013) Hybrid synchronization of n-scroll chaotic Chua circuits using adaptive backstepping control design with recursive feedback. Malays J Math Sci 7(2):219–246
50. Rasappan S, Vaidyanathan S (2014) Global chaos synchronization of WINDMI and Coullet chaotic systems using adaptive backstepping control design. Kyungpook Math J 54(1):293–320
51. Rössler OE (1976) An equation for continuous chaos. Phys Lett A 57(5):397–398
52. Sampath S, Vaidyanathan S, Volos CK, Pham VT (2015) An eight-term novel four-scroll chaotic system with cubic nonlinearity and its circuit simulation. J Eng Sci Technol Rev 8(2):1–6
53. Sarasu P, Sundarapandian V (2011) Active controller design for the generalized projective synchronization of four-scroll chaotic systems. Int J Syst Signal Control Eng Appl 4(2):26–33
54. Sarasu P, Sundarapandian V (2011) The generalized projective synchronization of hyperchaotic Lorenz and hyperchaotic Qi systems via active control. Int J Soft Comput 6(5):216–223
55. Sarasu P, Sundarapandian V (2012) Adaptive controller design for the generalized projective synchronization of 4-scroll systems. Int J Syst Signal Control Eng Appl 5(2):21–30
56. Sarasu P, Sundarapandian V (2012) Generalized projective synchronization of three-scroll chaotic systems via adaptive control. Eur J Sci Res 72(4):504–522
57. Sarasu P, Sundarapandian V (2012) Generalized projective synchronization of two-scroll systems via adaptive control. Int J Soft Comput 7(4):146–156
58. Shang Y, Fei W, Yu H (2012) Analysis and modeling of internal state variables for dynamic effects of nonvolatile memory devices. IEEE Trans Circ Syst I Reg Pap 59:1906–1918
59. Shin S, Kim K, Kang SM (2011) Memristor applications for programmable analog ICs. IEEE Trans Nanotechnol 410:266–274
60. Slotine J, Li W (1991) Applied nonlinear control. Prentice-Hall, Englewood Cliffs, NJ, USA
61. Sprott JC (1994) Some simple chaotic flows. Phys Rev E 50(2):647–650
62. Strukov D, Snider G, Stewart G, Williams R (2008) The missing memristor found. Nature 453:80–83
63. Sundarapandian V (2010) Output regulation of the Lorenz attractor. Far East J Math Sci 42(2):289–299
64. Sundarapandian V (2011) Output regulation of the Arneodo-Coullet chaotic system. Commun Comput Inf Sci 133:98–107
65. Sundarapandian V (2013) Analysis and anti-synchronization of a novel chaotic system via active and adaptive controllers. J Eng Sci Technol Rev 6(4):45–52
66. Sundarapandian V, Karthikeyan R (2011) Anti-synchronization of hyperchaotic Lorenz and hyperchaotic Chen systems by adaptive control. Int J Syst Signal Control Eng Appl 4(2):18–25

67. Sundarapandian V, Karthikeyan R (2011) Anti-synchronization of Lü and Pan chaotic systems by adaptive nonlinear control. Eur J Sci Res 64(1):94–106
68. Sundarapandian V, Karthikeyan R (2012) Adaptive anti-synchronization of uncertain Tigan and Li systems. J Eng Appl Sci 7(1):45–52
69. Sundarapandian V, Karthikeyan R (2012) Hybrid synchronization of hyperchaotic Lorenz and hyperchaotic Chen systems via active control. J Eng Appl Sci 7(3):254–264
70. Sundarapandian V, Pehlivan I (2012) Analysis, control, synchronization, and circuit design of a novel chaotic system. Math Comput Model 55(7–8):1904–1915
71. Sundarapandian V, Sivaperumal S (2011) Sliding controller design of hybrid synchronization of four-wing Chaotic systems. Int J Soft Comput 6(5):224–231
72. Suresh R, Sundarapandian V (2013) Global chaos synchronization of a family of n-scroll hyperchaotic Chua circuits using backstepping control with recursive feedback. Far East J Math Sci 73(1):73–95
73. Tang F, Wang L (2005) An adaptive active control for the modified Chua's circuit. Phys Lett A 346:342–346
74. Tetzlaff R (2014) Memristors and memristive systems. Springer, Berlin, Germany
75. Tigan G, Opris D (2008) Analysis of a 3D chaotic system. Chaos, Solitons Fractals 36:1315–1319
76. Utkin VI (1977) Variable structure systems with sliding modes. IEEE Trans Autom Control 22(2):212–222
77. Utkin VI (1993) Sliding mode control design principles and applications to electric drives. IEEE Trans Ind Electr 40(1):23–36
78. Vaidyanathan S (2011) Analysis and synchronization of the hyperchaotic Yujun systems via sliding mode control. Adv Intell Syst Comput 176:329–337
79. Vaidyanathan S (2011) Hybrid chaos synchronization of Liu and Lü systems by active nonlinear control. Commun Comput Inf Sci 204:1–10
80. Vaidyanathan S (2011) Output regulation of the unified chaotic system. Commun Comput Inf Sci 204:84–93
81. Vaidyanathan S (2012) Anti-synchronization of Sprott-L and Sprott-M chaotic systems via adaptive control. Int J Control Theor Appl 5(1):41–59
82. Vaidyanathan S (2012) Global chaos control of hyperchaotic Liu system via sliding control method. Int J Control Theor Appl 5(2):117–123
83. Vaidyanathan S (2012) Output regulation of the Liu chaotic system. Appl Mech Mater 110–116:3982–3989
84. Vaidyanathan S (2012) Sliding mode control based global chaos control of Liu-Liu-Liu-Su chaotic system. Int J Control Theor Appl 5(1):15–20
85. Vaidyanathan S (2013) A new six-term 3-D chaotic system with an exponential nonlinearity. Far East J Math Sci 79(1):135–143
86. Vaidyanathan S (2013) Analysis and adaptive synchronization of two novel chaotic systems with hyperbolic sinusoidal and cosinusoidal nonlinearity and unknown parameters. J Eng Sci Technol Rev 6(4):53–65
87. Vaidyanathan S (2013) Analysis, control and synchronization of hyperchaotic Zhou system via adaptive control. Adv Intell Syst Comput 177:1–10
88. Vaidyanathan S (2014) A new eight-term 3-D polynomial chaotic system with three quadratic nonlinearities. Far East J Math Sci 84(2):219–226
89. Vaidyanathan S (2014) Analysis and adaptive synchronization of eight-term 3-D polynomial chaotic systems with three quadratic nonlinearities. Eur Phys J Spec Top 223(8):1519–1529
90. Vaidyanathan S (2014) Analysis, control and synchronisation of a six-term novel chaotic system with three quadratic nonlinearities. Int J Model Identif Control 22(1):41–53
91. Vaidyanathan S (2014) Generalized projective synchronisation of novel 3-D chaotic systems with an exponential non-linearity via active and adaptive control. Int J Model Identif Control 22(3):207–217
92. Vaidyanathan S (2014) Global chaos synchronisation of identical Li-Wu chaotic systems via sliding mode control. Int J Model Identif Control 22(2):170–177

93. Vaidyanathan S (2015) 3-cells cellular neural network (CNN) attractor and its adaptive biological control. Int J PharmTech Res 8(4):632–640
94. Vaidyanathan S (2015) A 3-D novel highly chaotic system with four quadratic nonlinearities, its adaptive control and anti-synchronization with unknown parameters. J Eng Sci Technol Rev 8(2):106–115
95. Vaidyanathan S (2015) A novel chemical chaotic reactor system and its adaptive control. Int J ChemTech Res 8(7):146–158
96. Vaidyanathan S (2015) Adaptive backstepping control of enzymes-substrates system with ferroelectric behaviour in brain waves. Int J PharmTech Res 8(2):256–261
97. Vaidyanathan S (2015) Adaptive biological control of generalized Lotka-Volterra three-species biological system. Int J PharmTech Res 8(4):622–631
98. Vaidyanathan S (2015) Adaptive chaotic synchronization of enzymes-substrates system with ferroelectric behaviour in brain waves. Int J PharmTech Res 8(5):964–973
99. Vaidyanathan S (2015) Adaptive control of a chemical chaotic reactor. Int J PharmTech Res 8(3):377–382
100. Vaidyanathan S (2015) Adaptive control of the FitzHugh-Nagumo chaotic neuron model. Int J PharmTech Res 8(6):117–127
101. Vaidyanathan S (2015) Adaptive synchronization of chemical chaotic reactors. Int J ChemTech Res 8(2):612–621
102. Vaidyanathan S (2015) Adaptive synchronization of generalized Lotka-Volterra three-species biological systems. Int J PharmTech Res 8(5):928–937
103. Vaidyanathan S (2015) Adaptive synchronization of novel 3-D chemical chaotic reactor systems. Int J ChemTech Res 8(7):159–171
104. Vaidyanathan S (2015) Adaptive synchronization of the identical FitzHugh-Nagumo chaotic neuron models. Int J PharmTech Res 8(6):167–177
105. Vaidyanathan S (2015) Analysis, control and synchronization of a 3-D novel jerk chaotic system with two quadratic nonlinearities. Kyungpook Math J 55:563–586
106. Vaidyanathan S (2015) Analysis, properties and control of an eight-term 3-D chaotic system with an exponential nonlinearity. Int J Model Identif Control 23(2):164–172
107. Vaidyanathan S (2015) Anti-synchronization of brusselator chemical reaction systems via adaptive control. Int J ChemTech Res 8(6):759–768
108. Vaidyanathan S (2015) Chaos in neurons and adaptive control of Birkhoff-Shaw strange chaotic attractor. Int J PharmTech Res 8(5):956–963
109. Vaidyanathan S (2015) Chaos in neurons and synchronization of Birkhoff-Shaw strange chaotic attractors via adaptive control. Int J PharmTech Res 8(6):1–11
110. Vaidyanathan S (2015) Coleman-Gomatam logarithmic competitive biology models and their ecological monitoring. Int J PharmTech Res 8(6):94–105
111. Vaidyanathan S (2015) Dynamics and control of brusselator chemical reaction. Int J ChemTech Res 8(6):740–749
112. Vaidyanathan S (2015) Dynamics and control of tokamak system with symmetric and magnetically confined plasma. Int J ChemTech Res 8(6):795–803
113. Vaidyanathan S (2015) Global chaos synchronization of chemical chaotic reactors via novel sliding mode control method. Int J ChemTech Res 8(7):209–221
114. Vaidyanathan S (2015) Global chaos synchronization of the forced Van der Pol chaotic oscillators via adaptive control method. Int J PharmTech Res 8(6):156–166
115. Vaidyanathan S (2015) Global chaos synchronization of the Lotka-Volterra biological systems with four competitive species via active control. Int J PharmTech Res 8(6):206–217
116. Vaidyanathan S (2015) Lotka-Volterra population biology models with negative feedback and their ecological monitoring. Int J PharmTech Res 8(5):974–981
117. Vaidyanathan S (2015) Lotka-Volterra two species competitive biology models and their ecological monitoring. Int J PharmTech Res 8(6):32–44
118. Vaidyanathan S (2015) Output regulation of the forced Van der Pol chaotic oscillator via adaptive control method. Int J PharmTech Res 8(6):106–116

119. Vaidyanathan S (2016) Anti-synchronization of 3-cells Cellular Neural Network attractors via integral sliding mode control. Int J PharmTech Res 9(1):193–205
120. Vaidyanathan S (2016) Global chaos regulation of a symmetric nonlinear gyro system via integral sliding mode control. Int J ChemTech Res 9(5):462–469
121. Vaidyanathan S, Azar AT (2015) Analysis and control of a 4-D novel hyperchaotic system. In: Azar AT, Vaidyanathan S (eds) Chaos modeling and control systems design. Studies in computational intelligence, vol 581. Springer, Germany, pp 19–38
122. Vaidyanathan S, Azar AT (2015) Analysis, control and synchronization of a nine-term 3-D novel chaotic system. In: Azar AT, Vaidyanathan S (eds) Chaos modelling and control systems design. Studies in computational intelligence, vol 581. Springer, Germany, pp 19–38
123. Vaidyanathan S, Madhavan K (2013) Analysis, adaptive control and synchronization of a seven-term novel 3-D chaotic system. Int J Control Theor Appl 6(2):121–137
124. Vaidyanathan S, Pakiriswamy S (2013) Generalized projective synchronization of six-term Sundarapandian chaotic systems by adaptive control. Int J Control Theor Appl 6(2):153–163
125. Vaidyanathan S, Pakiriswamy S (2015) A 3-D novel conservative chaotic system and its generalized projective synchronization via adaptive control. J Eng Sci Technol Rev 8(2):52–60
126. Vaidyanathan S, Rajagopal K (2011) Anti-synchronization of Li and T chaotic systems by active nonlinear control. Commun Comput Inf Sci 198:175–184
127. Vaidyanathan S, Rajagopal K (2011) Global chaos synchronization of hyperchaotic Pang and Wang systems by active nonlinear control. Commun Comput Inf Sci 204:84–93
128. Vaidyanathan S, Rajagopal K (2011) Global chaos synchronization of Lü and Pan systems by adaptive nonlinear control. Commun Comput Inf Sci 205:193–202
129. Vaidyanathan S, Rajagopal K (2012) Global chaos synchronization of hyperchaotic Pang and hyperchaotic Wang systems via adaptive control. Int J Soft Comput 7(1):28–37
130. Vaidyanathan S, Rasappan S (2011) Global chaos synchronization of hyperchaotic Bao and Xu systems by active nonlinear control. Commun Comput Inf Sci 198:10–17
131. Vaidyanathan S, Rasappan S (2014) Global chaos synchronization of n-scroll Chua circuit and Lur'e system using backstepping control design with recursive feedback. Arab J Sci Eng 39(4):3351–3364
132. Vaidyanathan S, Sampath S (2011) Global chaos synchronization of hyperchaotic Lorenz systems by sliding mode control. Commun Comput Inf Sci 205:156–164
133. Vaidyanathan S, Volos C (2015) Analysis and adaptive control of a novel 3-D conservative no-equilibrium chaotic system. Arch Control Sci 25(3):333–353
134. Vaidyanathan S, Volos C (2016) Advances and applications in chaotic systems. Springer, Berlin, Germany
135. Vaidyanathan S, Volos C (2016) Advances and applications in nonlinear control systems. Springer, Berlin, Germany
136. Vaidyanathan S, Volos C, Pham VT (2014) Hyperchaos, adaptive control and synchronization of a novel 5-D hyperchaotic system with three positive Lyapunov exponents and its SPICE implementation. Arch Control Sci 24(4):409–446
137. Vaidyanathan S, Volos C, Pham VT, Madhavan K, Idowu BA (2014) Adaptive backstepping control, synchronization and circuit simulation of a 3-D novel jerk chaotic system with two hyperbolic sinusoidal nonlinearities. Arch Control Sci 24(3):375–403
138. Vaidyanathan S, Idowu BA, Azar AT (2015) Backstepping controller design for the global chaos synchronization of Sprott's jerk systems. Stud Comput Intell 581:39–58
139. Vaidyanathan S, Rajagopal K, Volos CK, Kyprianidis IM, Stouboulos IN (2015) Analysis, adaptive control and synchronization of a seven-term novel 3-D chaotic system with three quadratic nonlinearities and its digital implementation in LabVIEW. J Eng Sci Technol Rev 8(2):130–141
140. Vaidyanathan S, Volos C, Pham VT, Madhavan K (2015) Analysis, adaptive control and synchronization of a novel 4-D hyperchaotic hyperjerk system and its SPICE implementation. Arch Control Sci 25(1):5–28
141. Vaidyanathan S, Volos CK, Kyprianidis IM, Stouboulos IN, Pham VT (2015) Analysis, adaptive control and anti-synchronization of a six-term novel jerk chaotic system with two exponential nonlinearities and its circuit simulation. J Eng Sci Technol Rev 8(2):24–36

142. Vaidyanathan S, Volos CK, Madhavan K (2015) Analysis, control, synchronization and SPICE implementation of a novel 4-D hyperchaotic Rikitake dynamo System without equilibrium. J Eng Sci Technol Rev 8(2):232–244
143. Vaidyanathan S, Volos CK, Pham VT (2015) Analysis, adaptive control and adaptive synchronization of a nine-term novel 3-D chaotic system with four quadratic nonlinearities and its circuit simulation. J Eng Sci Technol Rev 8(2):181–191
144. Vaidyanathan S, Volos CK, Pham VT (2015) Global chaos control of a novel nine-term chaotic system via sliding mode control. In: Azar AT, Zhu Q (eds) Advances and applications in sliding mode control systems. Studies in computational intelligence, vol 576. Springer, Germany, pp 571–590
145. Vaidyanathan S, Volos CK, Pham VT, Madhavan K (2015) Analysis, adaptive control and synchronization of a novel 4-D hyperchaotic hyperjerk system and its SPICE implementation. Arch Control Sci 25(1):135–158
146. Vaidyanathan S, Volos CK, Rajagopal K, Kyprianidis IM, Stouboulos IN (2015) Adaptive backstepping controller design for the anti-synchronization of identical WINDMI chaotic systems with unknown parameters and its SPICE implementation. J Eng Sci Technol Rev 8(2):74–82
147. Volos CK, Kyprianidis IM, Stouboulos IN, Tlelo-Cuautle E, Vaidyanathan S (2015) Memristor: a new concept in synchronization of coupled neuromorphic circuits. J Eng Sci Technol Rev 8(2):157–173
148. Wang L, Zhang C, Chen L, Lai J, Tong J (2012) A novel memristor-based rSRAM structure for multiple-bit upsets immunity. IEICE Electron Express 9:861–867
149. Wang X, Ge C (2008) Controlling and tracking of Newton-Leipnik system via backstepping design. Int J Nonlinear Sci 5(2):133–139
150. Wang X, Xu B, Luo C (2012) An asynchronous communication system based on the hyperchaotic system of 6th-order cellular neural network. Opt Commun 285(24):5401–5405
151. Wei Z, Yang Q (2010) Anti-control of Hopf bifurcation in the new chaotic system with two stable node-foci. Appl Math Comput 217(1):422–429
152. Xiao X, Zhou L, Zhang Z (2014) Synchronization of chaotic Lur'e systems with quantized sampled-data controller. Commun Nonlinear Sci Numer Simul 19(6):2039–2047
153. Yang JJ, Strukov DB, Stewart DR (2013) Memristive devices for computing. Nat Nanotechnol 8:13–24
154. Zhou W, Xu Y, Lu H, Pan L (2008) On dynamics analysis of a new chaotic attractor. Phys Lett A 372(36):5773–5777
155. Zhu C, Liu Y, Guo Y (2010) Theoretic and numerical study of a new chaotic system. Intell Inf Manag 2:104–109

Adaptive Integral Sliding Mode Control of a Chemical Chaotic Reactor System

Sundarapandian Vaidyanathan

Abstract Chaos in nonlinear dynamics occurs widely in physics, chemistry, biology, ecology, secure communications, cryptosystems and many scientific branches. Chaotic systems have important applications in science and engineering. In this work, we discuss the dynamics and qualitative properties of the 3-D chemical chaotic reactor obtained by Huang and Yang (2005). The phase portraits of the chemical reactor system are depicted and the qualitative properties of the chemical system are discussed. The Lyapunov exponents of the chemical chaotic reactor system are obtained as $L_1 = 0.2001, L_2 = 0$ and $L_3 = -10.8295$. Also, the Kaplan-Yorke dimension of the chemical chaotic reactor system is obtained as $D_{KY} = 2.0185$, which shows the complexity of the system. Since the sum of the Lyapunov exponents is negative, the chemical chaotic reactor system is dissipative. We show that the chemical chaotic reactor system has three unstable equilibrium points and a stable equilibrium point. Control and synchronization of chaotic systems are important research problems in chaos theory. Sliding mode control is an important method used to solve various problems in control systems engineering. In robust control systems, the sliding mode control is often adopted due to its inherent advantages of easy realization, fast response and good transient performance as well as insensitivity to parameter uncertainties and disturbance. Next, using integral sliding mode control, we design adaptive control and synchronization schemes for the chemical chaotic reactor system. The main adaptive control and synchronization results are established using Lyapunov stability theory. MATLAB simulations are shown to illustrate all the main results of this work.

Keywords Chaos · Chaotic systems · Chemical reactor · Sliding mode control · Adaptive control · Synchronization

S. Vaidyanathan (✉)
Research and Development Centre, Vel Tech University, Avadi, Chennai
600062, Tamil Nadu, India
e-mail: sundarcontrol@gmail.com

© Springer International Publishing AG 2017
S. Vaidyanathan and C.-H. Lien (eds.), *Applications of Sliding Mode Control
in Science and Engineering*, Studies in Computational Intelligence 709,
DOI 10.1007/978-3-319-55598-0_16

371

1 Introduction

Chaos theory describes the quantitative study of unstable aperiodic dynamic behaviour in deterministic nonlinear dynamical systems. For the motion of a dynamical system to be chaotic, the system variables should contain some nonlinear terms and the system must satisfy three properties: boundedness, infinite recurrence and sensitive dependence on initial conditions [3–5, 92, 93].

Some classical paradigms of 3-D chaotic systems in the literature are Lorenz system [13], Rössler system [20], ACT system [2], Sprott systems [26], Chen system [7], Lü system [14], Cai system [6], Tigan system [36], etc.

Many new chaotic systems have been discovered in the recent years such as Zhou system [103], Zhu system [104], Li system [12], Wei-Yang system [102], Sundara-pandian systems [29, 33], Vaidyanathan systems [46, 47, 49–52, 55, 66, 67, 81, 82, 84, 91, 94, 96, 98–100], Pehlivan system [16], Sampath system [21], etc.

Chaos theory has applications in several fields of science and engineering such as chemical reactors [56, 60, 62, 64, 68, 72–74], biological systems [54, 57–59, 61, 63, 65, 69–71, 75–79], memristors [1, 17, 101], etc.

The control of a chaotic system aims to stabilize or regulate the system with the help of a feedback control. There are many methods available for controlling a chaotic system such as active control [27, 40, 41], adaptive control [28, 42, 48], sliding mode control [44, 45], backstepping control [95], etc.

Major works on synchronization of chaotic systems deal with the complete synchronization of a pair of chaotic systems called the *master* and *slave* systems. The design goal of the complete synchronization is to apply the output of the master system to control the slave system so that the output of the slave system tracks the output of the master system asymptotically with time.

There are many methods available for chaos synchronization such as active control [9, 22, 23, 85, 87], adaptive control [24, 30–32, 43, 83, 86], sliding mode control [34, 53, 90, 97], backstepping control [18, 19, 35, 88], etc.

Recently, Huang and Yang derived a chemical reactor model by considering reactor dynamics with five steps [8]. The Lyapunov exponents of the chemical chaotic reactor system [8] are obtained as $L_1 = 0.2001$, $L_2 = 0$ and $L_3 = -10.8295$. Also, the Kaplan-Yorke dimension of the chemical chaotic reactor system is obtained as $D_{KY} = 2.0185$, which shows the complexity of the system.

Since the sum of the Lyapunov exponents is negative, the chemical chaotic reactor system is dissipative. We show that the chemical chaotic reactor system has three unstable equilibrium points and a stable equilibrium point.

Next, we use sliding mode control for the adaptive control and synchronization of the chemical reactor chaotic system. A major advantage of sliding mode control is low sensitivity to parameter variations in the plant and disturbances affecting the plant, which eliminates the necessity of exact modeling of the plant.

In the sliding mode control, the control dynamics will have two sequential modes, viz. the reaching mode and the sliding mode. Basically, a sliding mode controller design consists of two parts: hyperplane design and controller design. A hyperplane is first designed via the pole-placement approach and a controller is then designed based on the sliding condition. The stability of the overall system is guaranteed by the sliding condition and by a stable hyperplane. Sliding mode control method is a popular method for the control and synchronization of chaotic systems [11, 15, 34, 39, 44, 45, 53, 80, 89].

This work is organized as follows. Section 2 describes the dynamics and qualitative properties of the chemical chaotic reactor system. In Sect. 3, we design an adaptive integral sliding mode controller to globally stabilize the chemical chaotic reactor system with unknown parameters. In Sect. 4, we design an adaptive integral sliding mode controller to achieve global chaos synchronization of the identical chemical chaotic reactor systems with unknown parameters. Section 5 contains the conclusions of this work.

2   Chemical Chaotic Reactor System

The well-stirred chemical reactor dynamics model of Huang-Yang [8] consist of the following five steps given below.

$$A_1 + X \underset{k_1}{\overset{k_{-1}}{\rightleftharpoons}} 2X \tag{1a}$$

$$X + Y \overset{k_2}{\longrightarrow} 2Y \tag{1b}$$

$$A_5 + Y \overset{k_3}{\longrightarrow} A_2 \tag{1c}$$

$$X + Z \overset{k_4}{\longrightarrow} A_3 \tag{1d}$$

$$A_5 + Z \underset{k_5}{\overset{k_{-5}}{\rightleftharpoons}} 2Z \tag{1e}$$

Equations (1a) and (1e) indicate reversible steps, while Eqs. (1c)–(1e) indicate non-reversible steps of the Huang-Yang chemical reactor [8].

In (1), A_1, A_4, A_5 are initiators and A_2, A_3 are products. The intermediates whose dynamics are followed are X, Y and Z.

Assuming an ideal mixture and a well-stirred reactor, the macroscopic rate equations for the Huang-Yang chemical reactor can be written in non-dimensionalized form as

$$\begin{cases} \dot{x} = a_1 x - k_{-1} x^2 - xy - xz \\ \dot{y} = xy - a_5 y \\ \dot{z} = a_4 z - xz - k_{-5} z^2 \end{cases} \tag{2}$$

In (2), x, y and z are the mole fractions of X, Y and Z. Also, the rate constants k_1, k_3 and k_5 are incorporated in the parameters a_1, a_5 and a_4.

To simplify the notations, we express the chemical reactor system (2) as

$$\begin{cases} \dot{x} = ax - dx^2 - xy - xz \\ \dot{y} = xy - cy \\ \dot{z} = bz - xz - pz^2 \end{cases} \tag{3}$$

Huang and Yang [8] showed that the chemical reactor system (3) is *chaotic,* when the system parameters are chosen as

$$a = 30, \quad b = 16.5, \quad c = 10, \quad d = 0.415, \quad p = 0.5 \tag{4}$$

For numerical simulations, we take the initial state of the system (3) as

$$x(0) = 0.2, \quad y(0) = 0.2, \quad z(0) = 0.2 \tag{5}$$

Lyapunov exponents of the Huang-Yang chemical reactor system (3) are numerically obtained as

$$L_1 = 0.2001, \quad L_2 = 0, \quad L_3 = -10.8295 \tag{6}$$

Figure 1 shows the Lyapunov exponents of the chemical reactor system (3). Figure 2 shows the 3-D phase portrait of the chemical reactor system (3). Figures 3, 4 and 5 show the 2-D projections of the chemical reactor system (3) on (x_1, x_2), (x_2, x_3) and (x_1, x_3) planes, respectively.

The Kaplan-Yorke dimension of the chemical reactor system (3) is derived as

$$D_{KY} = 2 + \frac{L_1 + L_2}{|L_3|} = 2.2085, \tag{7}$$

which shows the complexity of the system.

The equilibrium points of the chemical reactor system (3) are obtained by solving the following system of equations:

$$\begin{cases} ax - dx^2 - xy - xz = 0 \\ xy - cy = 0 \\ bz - xz - pz^2 = 0 \end{cases} \tag{8}$$

We take the parameter values as in the chaotic case (4), i.e.

$$a = 30, \quad b = 16.5, \quad c = 10, \quad d = 0.415, \quad p = 0.5 \tag{9}$$

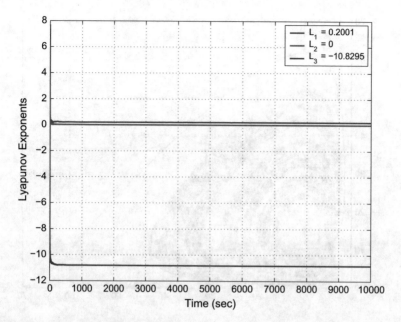

Fig. 1 Lyapunov exponents of the chemical reactor chaotic system

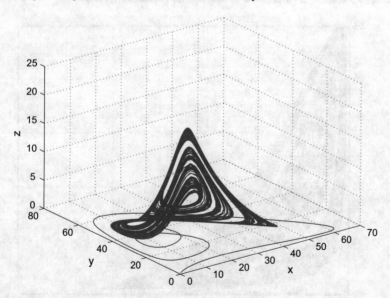

Fig. 2 3-D phase portrait of the chemical reactor chaotic system

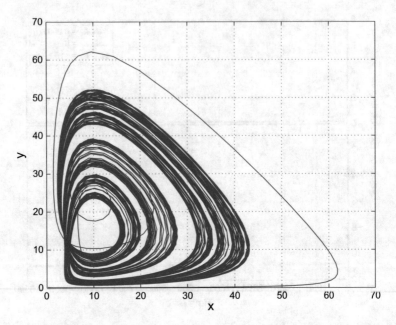

Fig. 3 2-D projection of the chemical reactor chaotic system in (x, y) plane

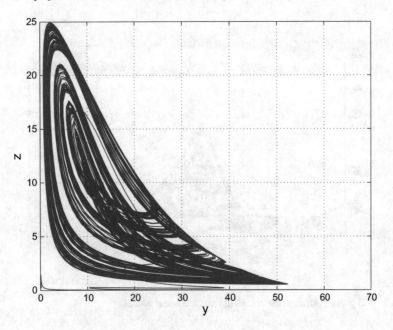

Fig. 4 2-D projection of the chemical reactor chaotic system in (y, z) plane

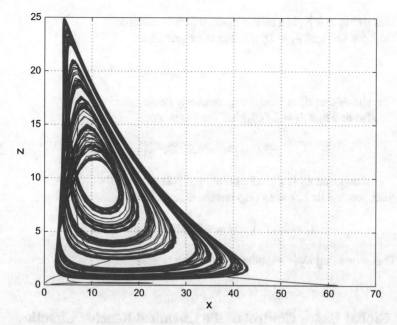

Fig. 5 2-D projection of the chemical reactor chaotic system in (x, z) plane

Solving the system (8), we obtain four equilibrium points given by

$$E_0 = \begin{bmatrix} 0 \\ 0 \\ 0 \end{bmatrix}, \quad E_1 = \begin{bmatrix} 0 \\ 0 \\ 33 \end{bmatrix}, \quad E_2 = \begin{bmatrix} 72.2892 \\ 0 \\ 0 \end{bmatrix}, \quad E_3 = \begin{bmatrix} 1.8927 \\ 0 \\ 29.2145 \end{bmatrix} \tag{10}$$

In vector notation, we can express the chemical reactor system (3) as

$$\dot{X} = f(X) \tag{11}$$

Then a simple calculation gives the Jacobian matrix of the system (3) as

$$J(X) = Df(X) = \begin{bmatrix} 30 - 0.83\,x - y - z & -x & -x \\ y & x - 10 & 0 \\ -z & 0 & 16.5 - x - z \end{bmatrix} \tag{12}$$

We find that $J_0 = Df(E_0)$ has the eigenvalues

$$\lambda_1 = -10, \quad \lambda_2 = 16.5, \quad \lambda_3 = 30 \tag{13}$$

This shows that E_0 is a saddle point, which is unstable.

Next, we find that $J_1 = Df(E_1)$ has the eigenvalues

$$\lambda_1 = -16.5, \quad \lambda_2 = -10, \quad \lambda_3 = -3 \tag{14}$$

This shows that E_1 is locally exponentially stable.

We also find that $J_2 = Df(E_2)$ has the eigenvalues

$$\lambda_1 = -30, \quad \lambda_2 = -55.7892, \quad \lambda_3 = 62.2892 \tag{15}$$

This shows that E_2 is a saddle point, which is unstable.

Next, we find that $J_3 = Df(E_3)$ has the eigenvalues

$$\lambda_1 = 2.4553, \quad \lambda_2 = -8.1073, \quad \lambda_3 = -17.8479 \tag{16}$$

This shows that E_3 is a saddle point, which is unstable.

3 Global Chaos Control of the Chemical Reactor Chaotic System

In this section, we use adaptive integral sliding mode control for the global hyperchaos control of the new memristor-based hyperchaotic system with unknown system parameters. The adaptive control mechanism helps the control design by estimating the unknown parameters [3–5, 92, 93].

The controlled chemical reactor chaotic system is described by

$$\begin{cases} \dot{x} = ax - dx^2 - xy - xz + u_x \\ \dot{y} = xy - cy + u_y \\ \dot{z} = bz - xz - pz^2 + u_z \end{cases} \tag{17}$$

where x, y, z are the states of the system and a, b, c, d, p are unknown system parameters.

The design goal is to find suitable feedback controllers u_x, u_y, u_z so as to globally stabilize the system (17) with estimates of the unknown parameters.

Based on the sliding mode control theory [25, 37, 38], the integral sliding surface of each state is defined as follows:

$$\begin{cases} s_x = \left(\frac{d}{dt} + \lambda_x\right)\left(\int_0^t x(\tau)d\tau\right) = x + \lambda_x \int_0^t x(\tau)d\tau \\ s_y = \left(\frac{d}{dt} + \lambda_y\right)\left(\int_0^t y(\tau)d\tau\right) = y + \lambda_y \int_0^t y(\tau)d\tau \\ s_z = \left(\frac{d}{dt} + \lambda_z\right)\left(\int_0^t z(\tau)d\tau\right) = z + \lambda_z \int_0^t z(\tau)d\tau \end{cases} \tag{18}$$

From Eq. (18), it follows that

$$\begin{cases} \dot{s}_x = \dot{x} + \lambda_x x \\ \dot{s}_y = \dot{y} + \lambda_y y \\ \dot{s}_z = \dot{z} + \lambda_z z \end{cases} \tag{19}$$

The Hurwitz condition is realized if $\lambda_x > 0$, $\lambda_y > 0$ and $\lambda_z > 0$.
We consider the adaptive feedback control given by

$$\begin{cases} u_x = -\hat{a}(t)x + \hat{d}(t)x^2 + xy + xz - \lambda_x x - \eta_x \operatorname{sgn}(s_x) - k_x s_x \\ u_y = -xy + \hat{c}(t)y - \lambda_y y - \eta_y \operatorname{sgn}(s_y) - k_y s_y \\ u_z = -\hat{b}(t)z + xz + \hat{p}(t)z^2 - \lambda_z z - \eta_z \operatorname{sgn}(s_z) - k_z s_z \end{cases} \tag{20}$$

where $\eta_x, \eta_y, \eta_z, k_x, k_y, k_z$ are positive constants.
Substituting (20) into (17), we obtain the closed-loop control system given by

$$\begin{cases} \dot{x} = [a - \hat{a}(t)]x - [d - \hat{d}(t)]x^2 - \lambda_x x - \eta_x \operatorname{sgn}(s_x) - k_x s_x \\ \dot{y} = -[c - \hat{c}(t)]y - \lambda_y y - \eta_y \operatorname{sgn}(s_y) - k_y s_y \\ \dot{z} = [b - \hat{b}(t)]z - [p - \hat{p}(t)]z^2 - \lambda_z z - \eta_z \operatorname{sgn}(s_z) - k_z s_z \end{cases} \tag{21}$$

We define the parameter estimation errors as

$$\begin{cases} e_a(t) = a - \hat{a}(t) \\ e_b(t) = b - \hat{b}(t) \\ e_c(t) = c - \hat{c}(t) \\ e_d(t) = d - \hat{d}(t) \\ e_p(t) = p - \hat{p}(t) \end{cases} \tag{22}$$

Using (22), we can simplify the closed-loop system (21) as

$$\begin{cases} \dot{x} = e_a x - e_d x^2 - \lambda_x x - \eta_x \operatorname{sgn}(s_x) - k_x s_x \\ \dot{y} = -e_c y - \lambda_y y - \eta_y \operatorname{sgn}(s_y) - k_y s_y \\ \dot{z} = e_b z - e_p z^2 - \lambda_z z - \eta_z \operatorname{sgn}(s_z) - k_z s_z \end{cases} \tag{23}$$

Differentiating (22) with respect to t, we get

$$\begin{cases} \dot{e}_a = -\dot{\hat{a}} \\ \dot{e}_b = -\dot{\hat{b}} \\ \dot{e}_c = -\dot{\hat{c}} \\ \dot{e}_d = -\dot{\hat{d}} \\ \dot{e}_p = -\dot{\hat{p}} \end{cases} \tag{24}$$

Next, we state and prove the main result of this section.

Theorem 1 *The chemical reactor chaotic system (17) is rendered globally asymptotically stable for all initial conditions $\mathbf{X}(0) \in \mathbf{R}^3$ by the adaptive integral sliding mode control law (20) and the parameter update law*

$$\begin{cases} \dot{\hat{a}} = s_x x \\ \dot{\hat{b}} = s_z z \\ \dot{\hat{c}} = -s_y y \\ \dot{\hat{d}} = -s_x x^2 \\ \dot{\hat{p}} = -s_z z^2 \end{cases} \tag{25}$$

where $\lambda_x, \lambda_y, \lambda_z, \eta_x, \eta_y, \eta_z, k_x, k_y, k_z$ are positive constants.

Proof We consider the quadratic Lyapunov function defined by

$$V(s_x, s_y, s_z, e_a, e_b, e_c, e_d, e_p) = \frac{1}{2}\left(s_x^2 + s_y^2 + s_z^2\right) + \frac{1}{2}\left(e_a^2 + e_b^2 + e_c^2 + e_d^2 + e_p^2\right) \tag{26}$$

Clearly, V is positive definite on $\mathbf{R}^8$.
Using (19), (23) and (24), the time-derivative of V is obtained as

$$\dot{V} = s_x\left[e_a x - e_d x^2 - k_x s_x\right] + s_y\left[-e_c y - \eta_y \operatorname{sgn}(s_y) - k_y s_y\right] - e_a \dot{\hat{a}} - e_b \dot{\hat{b}} \\ + s_z\left[e_b z - e_p z^2 - \eta_z \operatorname{sgn}(s_z) - k_z s_z\right] - e_c \dot{\hat{c}} - e_d \dot{\hat{d}} - e_p \dot{\hat{p}} \tag{27}$$

i.e.

$$\dot{V} = -\eta_x|s_x| - k_x s_x^2 - \eta_y|s_y| - k_y s_y^2 - \eta_z|s_z| - k_z s_z^2 + e_a\left[s_x x - \dot{\hat{a}}\right] \\ + e_b\left[s_z z - \dot{\hat{b}}\right] + e_c\left[-s_y y - \dot{\hat{c}}\right] + e_d\left[-s_x x^2 - \dot{\hat{d}}\right] + e_p\left[-s_z z^2 - \dot{\hat{p}}\right] \tag{28}$$

Using the parameter update law (25), we obtain

$$\dot{V} = -\eta_x|s_x| - k_x s_x^2 - \eta_y|s_y| - k_y s_y^2 - \eta_z|s_z| - k_z s_z^2 \tag{29}$$

which shows that $\dot{V}$ is negative semi-definite on $\mathbf{R}^8$.

Hence, by Barbalat's lemma [10], it is immediate that $\mathbf{X}(t)$ is globally asymptotically stable for all values of $\mathbf{X}(0) \in \mathbf{R}^3$.

This completes the proof.　　　　　　　　　　　　　　　　　　■

For numerical simulations, we take the parameter values of the new chemical reactor system (17) as in the chaotic case (4).

We take the values of the parameters as $k_x = 20$, $k_y = 20$, $k_z = 20$, $\eta_x = 0.1$, $\eta_y = 0.1$, $\eta_z = 0.1$, $\lambda_x = 50$, $\lambda_y = 50$, and $\lambda_z = 50$.

We take the estimates of the system parameters as

$$\hat{a}(0) = 12.6, \quad \hat{b}(0) = 17.3, \quad \hat{c}(0) = 15.4, \quad \hat{d}(0) = 8.4, \quad \hat{p}(0) = 6.5 \qquad (30)$$

We take the initial state of the chemical reactor chaotic system (17) as

$$x(0) = 14.9, \quad y(0) = 18.5, \quad z(0) = 23.7 \qquad (31)$$

Figure 6 shows the time-history of the controlled states x, y, z.

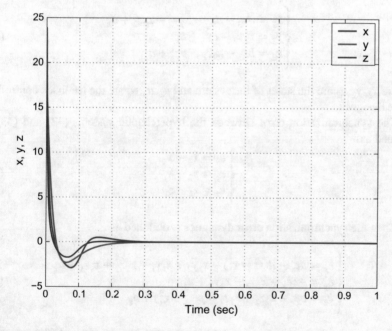

Fig. 6 Time-history of the controlled states x, y, z

4   Global Chaos Synchronization of the Chemical Reactor Chaotic Systems

In this section, we use adaptive integral sliding mode control for the global hyper-chaos synchronization of the new memristor-based hyperchaotic systems with unknown system parameters. The adaptive control mechanism helps the control design by estimating the unknown parameters [3–5, 92, 93].

As the master system, we consider the chemical reactor chaotic system given by

$$
\begin{cases}
\dot{x}_1 = ax_1 - dx_1^2 - x_1y_1 - x_1z_1 \\
\dot{y}_1 = x_1y_1 - cy_1 \\
\dot{z}_1 = bz_1 - x_1z_1 - pz_1^2
\end{cases}
\tag{32}
$$

where x_1, y_1, z_1 are the states of the system and a, b, c, d, p are unknown system parameters.

As the slave system, we consider the chemical reactor chaotic system given by

$$
\begin{cases}
\dot{x}_2 = ax_2 - dx_2^2 - x_2y_2 - x_2z_2 + u_x \\
\dot{y}_2 = x_2y_2 - cy_2 + u_y \\
\dot{z}_2 = bz_2 - x_2z_2 - pz_2^2 + u_z
\end{cases}
\tag{33}
$$

where x_2, y_2, z_2 are the states of the system and u_x, u_y, u_z are the feedback controllers to be designed.

The synchronization error between the hyperchaotic systems (32) and (33) is defined as

$$
\begin{cases}
e_x = x_2 - x_1 \\
e_y = y_2 - y_1 \\
e_z = z_2 - z_1
\end{cases}
\tag{34}
$$

Then the synchronization error dynamics is obtained as

$$
\begin{cases}
\dot{e}_x = ae_x - d(x_2^2 - x_1^2) - x_2y_2 + x_1y_1 - x_2z_2 + x_1z_1 + u_x \\
\dot{e}_y = -ce_y + x_2y_2 - x_1y_1 + u_y \\
\dot{e}_z = be_z - x_2z_2 + x_1z_1 - p(z_2^2 - z_1^2) + u_z
\end{cases}
\tag{35}
$$

Based on the sliding mode control theory [25, 37, 38], the integral sliding surface of each error variable is defined as follows:

$$\begin{cases} s_x = \left(\dfrac{d}{dt} + \lambda_x\right)\left(\displaystyle\int_0^t e_x(\tau)d\tau\right) = e_x + \lambda_x \displaystyle\int_0^t e_x(\tau)d\tau \\[3mm] s_y = \left(\dfrac{d}{dt} + \lambda_y\right)\left(\displaystyle\int_0^t e_y(\tau)d\tau\right) = e_y + \lambda_y \displaystyle\int_0^t e_y(\tau)d\tau \\[3mm] s_z = \left(\dfrac{d}{dt} + \lambda_z\right)\left(\displaystyle\int_0^t e_z(\tau)d\tau\right) = e_z + \lambda_z \displaystyle\int_0^t e_z(\tau)d\tau \end{cases} \tag{36}$$

From Eq. (36), it follows that

$$\begin{cases} \dot{s}_x = \dot{e}_x + \lambda_x e_x \\ \dot{s}_y = \dot{e}_y + \lambda_y e_y \\ \dot{s}_z = \dot{e}_z + \lambda_z e_z \end{cases} \tag{37}$$

The Hurwitz condition is realized if $\lambda_x > 0$, $\lambda_y > 0$ and $\lambda_z > 0$.
We consider the adaptive feedback control given by

$$\begin{cases} u_x = -\hat{a}(t)e_x + \hat{d}(t)(x_2^2 - x_1^2) + x_2 y_2 - x_1 y_1 + x_2 z_2 - x_1 z_1 \\ \qquad -\lambda_x e_x - \eta_x \,\mathrm{sgn}(s_x) - k_x s_x \\ u_y = \hat{c}(t)e_y - x_2 y_2 + x_1 y_1 - \lambda_y e_y - \eta_y \,\mathrm{sgn}(s_y) - k_y s_y \\ u_z = -\hat{b}(t)e_z + x_2 z_2 - x_1 z_1 + \hat{p}(t)(z_2^2 - z_1^2) - \lambda_z e_z - \eta_z \,\mathrm{sgn}(s_z) - k_z s_z \end{cases} \tag{38}$$

where $\eta_x, \eta_y, \eta_z, k_x, k_y, k_z$ are positive constants.
Substituting (38) into (35), we obtain the closed-loop error dynamics as

$$\begin{cases} \dot{e}_x = [a - \hat{a}(t)]e_x - [d - \hat{d}(t)](x_2^2 - x_1^2) - \lambda_x e_x - \eta_x \,\mathrm{sgn}(s_x) - k_x s_x \\ \dot{e}_y = -[c - \hat{c}(t)]e_y - \lambda_y e_y - \eta_y \,\mathrm{sgn}(s_y) - k_y s_y \\ \dot{e}_z = [b - \hat{b}(t)]e_z - [p - \hat{p}(t)](z_2^2 - z_1^2) - \lambda_z e_z - \eta_z \,\mathrm{sgn}(s_z) - k_z s_z \end{cases} \tag{39}$$

We define the parameter estimation errors as

$$\begin{cases} e_a(t) = a - \hat{a}(t) \\ e_b(t) = b - \hat{b}(t) \\ e_c(t) = c - \hat{c}(t) \\ e_d(t) = d - \hat{d}(t) \\ e_p(t) = p - \hat{p}(t) \end{cases} \tag{40}$$

Using (40), we can simplify the closed-loop system (39) as

$$\begin{cases} \dot{e}_x = e_a e_x - e_d(x_2^2 - x_1^2) - \lambda_x e_x - \eta_x \,\mathrm{sgn}(s_x) - k_x s_x \\ \dot{e}_y = -e_c e_y - \lambda_y e_y - \eta_y \,\mathrm{sgn}(s_y) - k_y s_y \\ \dot{e}_z = e_b e_z - e_p(z_2^2 - z_1^2) - \lambda_z e_z - \eta_z \,\mathrm{sgn}(s_z) - k_z s_z \end{cases} \tag{41}$$

Differentiating (40) with respect to t, we get

$$\begin{cases} \dot{e}_a = -\dot{\hat{a}} \\ \dot{e}_b = -\dot{\hat{b}} \\ \dot{e}_c = -\dot{\hat{c}} \\ \dot{e}_d = -\dot{\hat{d}} \\ \dot{e}_p = -\dot{\hat{p}} \end{cases} \tag{42}$$

Theorem 2 *The chemical reactor chaotic systems (32) and (33) are globally and asymptotically synchronized for all initial conditions by the adaptive integral sliding mode control law (38) and the parameter update law*

$$\begin{cases} \dot{\hat{a}} = s_x e_x \\ \dot{\hat{b}} = s_z e_z \\ \dot{\hat{c}} = -s_y e_y \\ \dot{\hat{d}} = -s_x (x_2^2 - x_1^2) \\ \dot{\hat{p}} = -s_z (z_2^2 - z_1^2) \end{cases} \tag{43}$$

where $\lambda_x, \lambda_y, \lambda_z, \eta_x, \eta_y, \eta_z, k_x, k_y, k_z$ are positive constants.

Proof We consider the quadratic Lyapunov function defined by

$$V(s_x, s_y, s_z, e_a, e_b, e_c, e_d, e_p) = \frac{1}{2}\left(s_x^2 + s_y^2 + s_z^2\right) + \frac{1}{2}\left(e_a^2 + e_b^2 + e_c^2 + e_d^2 + e_p^2\right) \tag{44}$$

Clearly, V is positive definite on $\mathbf{R}^8$.
Using (37), (41) and (42), the time-derivative of V is obtained as

$$\begin{aligned} \dot{V} = &\ s_x \left[e_a e_x - e_d(x_2^2 - x_1^2) - \eta_x \,\mathrm{sgn}(s_x) - k_x s_x\right] - e_a \dot{\hat{a}} - e_b \dot{\hat{b}} \\ &+ s_y \left[-e_c e_y - \eta_y \,\mathrm{sgn}(s_y) - k_y s_y\right] - e_c \dot{\hat{c}} - e_d \dot{\hat{d}} - e_p \dot{\hat{p}} \\ &+ s_z \left[e_b e_z - e_p(z_2^2 - z_1^2) - \eta_z \,\mathrm{sgn}(s_z) - k_z s_z\right] \end{aligned} \tag{45}$$

i.e.

$$\begin{aligned} \dot{V} = &-\eta_x |s_x| - k_x s_x^2 - \eta_y |s_y| - k_y s_y^2 - \eta_z |s_z| - k_z s_z^2 \\ &+ e_a \left[s_x e_x - \dot{\hat{a}}\right] + e_b \left[s_z e_z - \dot{\hat{b}}\right] + e_c \left[-s_y e_y - \dot{\hat{c}}\right] \\ &+ e_d \left[-s_x(x_2^2 - x_1^2) - \dot{\hat{d}}\right] + e_p \left[-s_z(z_2^2 - z_1^2) - \dot{\hat{p}}\right] \end{aligned} \tag{46}$$

Using the parameter update law (43), we obtain

$$\dot{V} = -\eta_x |s_x| - k_x s_x^2 - \eta_y |s_y| - k_y s_y^2 - \eta_z |s_z| - k_z s_z^2 \tag{47}$$

which shows that $\dot{V}$ is negative semi-definite on $\mathbf{R}^8$.

Hence, by Barbalat's lemma [10], it is immediate that $\mathbf{e}(t)$ is globally asymptotically stable for all values of $\mathbf{e}(0) \in \mathbf{R}^3$, where $\mathbf{e} = \left[e_x, e_y, e_z \right]^T$.

This completes the proof.   ∎

For numerical simulations, we take the parameter values of the chemical chaotic reactor systems (32) and (33) as in the chaotic case (4).

We take the values of the control parameters as $k_x = 20$, $k_y = 20$, $k_z = 20$, $\eta_x = 0.1$, $\eta_y = 0.1$, $\eta_z = 0.1$, $\lambda_x = 50$, $\lambda_y = 50$ and $\lambda_z = 50$.

We take the estimates of the system parameters as

$$\hat{a}(0) = 5.2, \quad \hat{b}(0) = 7.4, \quad \hat{c}(0) = 12.9, \quad \hat{d}(0) = 8.3, \quad \hat{p}(0) = 6.4 \tag{48}$$

We take the initial state of the master system (32) as

$$x_1(0) = 6.1, \quad y_1(0) = 8.4, \quad z_1(0) = 11.5 \tag{49}$$

We take the initial state of the slave system (33) as

$$y_1(0) = 15.9, \quad y_2(0) = 3.5, \quad z_2(0) = 23.1 \tag{50}$$

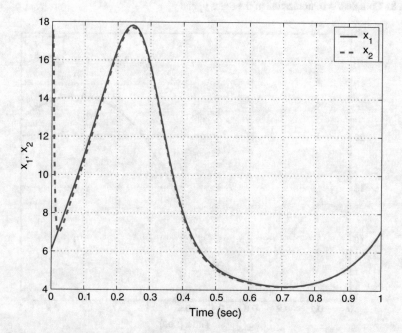

Fig. 7 Complete synchronization of the states x_1 and x_2

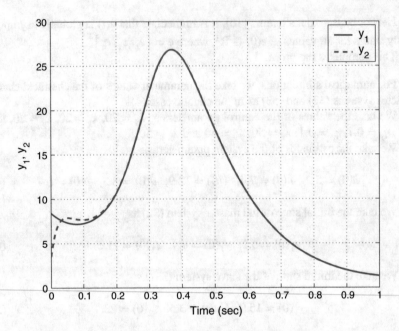

Fig. 8 Complete synchronization of the states y_1 and y_2

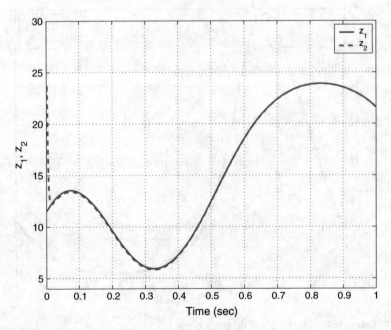

Fig. 9 Complete synchronization of the states z_1 and z_2

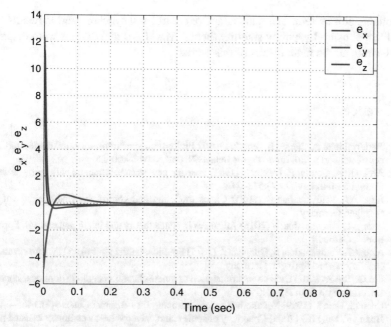

Fig. 10 Time-history of the synchronization errors e_x, e_y, e_z, e_φ

Figures 7, 8 and 9 show the complete synchronization of the chemical chaotic reactor systems (32) and (33). Figure 10 shows the time-history of the synchronization errors e_x, e_y, e_z.

5 Conclusions

In this work, we discussed the dynamics and qualitative properties of the 3-D chemical reactor chaotic obtained by Huang and Yang (2005). The phase portraits of the chemical reactor system were depicted and the qualitative properties of the chemical system were discussed in detail. The Lyapunov exponents of the chemical chaotic reactor system were obtained as $L_1 = 0.2001$, $L_2 = 0$ and $L_3 = -10.8295$. Also, the Kaplan-Yorke dimension of the chemical reactor chaotic system was derived as $D_{KY} = 2.0185$, which shows the complexity of the chemical reactor system. We showed that the chemical chaotic reactor system has three unstable equilibrium points and a stable equilibrium point. Sliding mode control is an important method used to solve various problems in control systems engineering. In robust control systems, the sliding mode control is often adopted due to its inherent advantages of easy realization, fast response and good transient performance as well as insensitivity to parameter uncertainties and disturbance. Next, using integral sliding mode control, we designed adaptive control and synchronization schemes for the chemical chaotic reactor system. The main adaptive control and synchronization results were established using Lyapunov stability theory. MATLAB simulations were displayed to illustrate all the main results of this work.

References

1. Abdurrahman A, Jiang H, Teng Z (2015) Finite-time synchronization for memristor-based neural networks with time-varying delays. Neural Netw 69:20–28
2. Arneodo A, Coullet P, Tresser C (1981) Possible new strange attractors with spiral structure. Commun Math Phys 79(4):573–576
3. Azar AT, Vaidyanathan S (2015) Chaos modeling and control systems design, vol 581. Springer, Germany
4. Azar AT, Vaidyanathan S (2016) Advances in chaos theory and intelligent control. Springer, Berlin, Germany
5. Azar AT, Vaidyanathan S, Ouannas A (2017) Fractional order control and synchronization of chaotic systems. Springer, Germany
6. Cai G, Tan Z (2007) Chaos synchronization of a new chaotic system via nonlinear control. J Uncertain Syst 1(3):235–240
7. Chen G, Ueta T (1999) Yet another chaotic attractor. Int J Bifurcat Chaos 9(7):1465–1466
8. Huang Y, Yang XS (2005) Chaoticity of some chemical attractors: a computer assisted proof. J Math Chem 38(1):107–117
9. Karthikeyan R, Sundarapandian V (2014) Hybrid chaos synchronization of four-scroll systems via active control. J Electr Eng 65(2):97–103
10. Khalil HK (2001) Nonlinear systems, 3rd edn. Prentice Hall, USA
11. Lakhekar GV, Waghmare LM, Vaidyanathan S (2016) Diving autopilot design for underwater vehicles using an adaptive neuro-fuzzy sliding mode controller. In: Vaidyanathan S, Volos C (eds) Advances and applications in nonlinear control systems. Springer, Germany, pp 477–503
12. Li D (2008) A three-scroll chaotic attractor. Phys Lett A 372(4):387–393
13. Lorenz EN (1963) Deterministic periodic flow. J Atmos Sci 20(2):130–141
14. Lü J, Chen G (2002) A new chaotic attractor coined. Int J Bifurcat Chaos 12(3):659–661
15. Moussaoui S, Boulkroune A, Vaidyanathan S (2016) Fuzzy adaptive sliding-mode control scheme for uncertain underactuated systems. In: Vaidyanathan S, Volos C (eds) Advances and applications in nonlinear control systems. Springer, Germany, pp 351–367
16. Pehlivan I, Moroz IM, Vaidyanathan S (2014) Analysis, synchronization and circuit design of a novel butterfly attractor. J Sound Vib 333(20):5077–5096
17. Pham VT, Volos CK, Vaidyanathan S, Le TP, Vu VY (2015) A memristor-based hyperchaotic system with hidden attractors: dynamics, synchronization and circuital emulating. J Eng Sci Technol Rev 8(2):205–214
18. Rasappan S, Vaidyanathan S (2013) Hybrid synchronization of n-scroll Chua circuits using adaptive backstepping control design with recursive feedback. Malays J Math Sci 73(1):73–95
19. Rasappan S, Vaidyanathan S (2014) Global chaos synchronization of WINDMI and Coullet chaotic systems using adaptive backstepping control design. Kyungpook Math J 54(1):293–320
20. Rössler OE (1976) An equation for continuous chaos. Phys Lett A 57(5):397–398
21. Sampath S, Vaidyanathan S, Volos CK, Pham VT (2015) An eight-term novel four-scroll chaotic System with cubic nonlinearity and its circuit simulation. J Eng Sci Technol Rev 8(2):1–6
22. Sarasu P, Sundarapandian V (2011) Active controller design for generalized projective synchronization of four-scroll chaotic systems. Int J Syst Signal Control Eng Appl 4(2):26–33

23. Sarasu P, Sundarapandian V (2011) The generalized projective synchronization of hyper-chaotic Lorenz and hyperchaotic Qi systems via active control. Int J Soft Comput 6(5):216–223
24. Sarasu P, Sundarapandian V (2012) Generalized projective synchronization of two-scroll systems via adaptive control. Int J Soft Comput 7(4):146–156
25. Slotine J, Li W (1991) Applied nonlinear control. Prentice-Hall, Englewood Cliffs, NJ, USA
26. Sprott JC (1994) Some simple chaotic flows. Phys Rev E 50(2):647–650
27. Sundarapandian V (2010) Output regulation of the Lorenz attractor. Far East J Math Sci 42(2):289–299
28. Sundarapandian V (2013) Adaptive control and synchronization design for the Lu-Xiao chaotic system. In: Lecture notes in, electrical engineering, vol 131, pp 319–327
29. Sundarapandian V (2013) Analysis and anti-synchronization of a novel chaotic system via active and adaptive controllers. J Eng Sci Technol Rev 6(4):45–52
30. Sundarapandian V, Karthikeyan R (2011) Anti-synchronization of hyperchaotic Lorenz and hyperchaotic Chen systems by adaptive control. Int J Syst Signal Control Eng Appl 4(2):18–25
31. Sundarapandian V, Karthikeyan R (2011) Anti-synchronization of Lü and Pan chaotic systems by adaptive nonlinear control. Eur J Sci Res 64(1):94–106
32. Sundarapandian V, Karthikeyan R (2012) Adaptive anti-synchronization of uncertain Tigan and Li systems. J Eng Appl Sci 7(1):45–52
33. Sundarapandian V, Pehlivan I (2012) Analysis, control, synchronization, and circuit design of a novel chaotic system. Math Comput Modell 55(7–8):1904–1915
34. Sundarapandian V, Sivaperumal S (2011) Sliding controller design of hybrid synchronization of four-wing chaotic systems. Int J Soft Comput 6(5):224–231
35. Suresh R, Sundarapandian V (2013) Global chaos synchronization of a family of n-scroll hyperchaotic Chua circuits using backstepping control with recursive feedback. Far East J Math Sci 7(2):219–246
36. Tigan G, Opris D (2008) Analysis of a 3D chaotic system. Chaos, Solitons Fractals 36:1315–1319
37. Utkin VI (1977) Variable structure systems with sliding modes. IEEE Trans Autom Control 22(2):212–222
38. Utkin VI (1993) Sliding mode control design principles and applications to electric drives. IEEE Trans Indus Electron 40(1):23–36
39. Vaidyanathan S (2011) Analysis and synchronization of the hyperchaotic Yujun systems via sliding mode control. In: Advances in intelligent systems and computing 176, pp 329–337
40. Vaidyanathan S (2011) Output regulation of Arneodo-Coullet chaotic system. In: Communications in computer and information, science, vol 133, pp 98–107
41. Vaidyanathan S (2011) Output regulation of the unified chaotic system. In: Communications in computer and information, science, vol 198, pp 1–9
42. Vaidyanathan S (2012) Adaptive controller and syncrhonizer design for the Qi-Chen chaotic system. In: Lecture notes of the institute for computer sciences, social-informatics and telecommunications, engineering, vol 84, pp 73–82
43. Vaidyanathan S (2012) Anti-synchronization of Sprott-L and Sprott-M chaotic systems via adaptive control. Int J Control Theor Appl 5(1):41–59
44. Vaidyanathan S (2012) Global chaos control of hyperchaotic Liu system via sliding control method. Int J Control Theor Appl 5(2):117–123
45. Vaidyanathan S (2012) Sliding mode control based global chaos control of Liu-Liu-Liu-Su chaotic system. Int J Control Theor Appl 5(1):15–20
46. Vaidyanathan S (2013) A new six-term 3-D chaotic system with an exponential nonlinearity. Far East J Math Sci 79(1):135–143
47. Vaidyanathan S (2013) Analysis and adaptive synchronization of two novel chaotic systems with hyperbolic sinusoidal and cosinusoidal nonlinearity and unknown parameters. J Eng Sci Technol Rev 6(4):53–65

48. Vaidyanathan S (2013) Analysis, control and synchronization of hyperchaotic Zhou system via adaptive control. In: Advances in intelligent systems and computing, vol 177, pp 1–10
49. Vaidyanathan S (2014) A new eight-term 3-D polynomial chaotic system with three quadratic nonlinearities. Far East J Math Sci 84(2):219–226
50. Vaidyanathan S (2014) Analysis and adaptive synchronization of eight-term 3-D polynomial chaotic systems with three quadratic nonlinearities. Eur Phys J Spec Top 223(8):1519–1529
51. Vaidyanathan S (2014) Analysis, control and synchronisation of a six-term novel chaotic system with three quadratic nonlinearities. Int J Modell Ident Control 22(1):41–53
52. Vaidyanathan S (2014) Generalized projective synchronisation of novel 3-D chaotic systems with an exponential non-linearity via active and adaptive control. Int J Modell Ident Control 22(3):207–217
53. Vaidyanathan S (2014) Global chaos synchronisation of identical Li-Wu chaotic systems via sliding mode control. Int J Modell Ident Control 22(2):170–177
54. Vaidyanathan S (2015) 3-cells Cellular neural network (CNN) attractor and its adaptive biological control. Int J PharmTech Res 8(4):632–640
55. Vaidyanathan S (2015) A 3-D novel highly chaotic system with four quadratic nonlinearities, its adaptive control and anti-synchronization with unknown parameters. J Eng Sci Technol Rev 8(2):106–115
56. Vaidyanathan S (2015) A novel chemical chaotic reactor system and its adaptive control. Int J ChemTech Res 8(7):146–158
57. Vaidyanathan S (2015) Adaptive backstepping control of enzymes-substrates system with ferroelectric behaviour in brain waves. Int J PharmTech Res 8(2):256–261
58. Vaidyanathan S (2015) Adaptive biological control of generalized Lotka-Volterra three-species biological system. Int J PharmTech Res 8(4):622–631
59. Vaidyanathan S (2015) Adaptive chaotic synchronization of enzymes-substrates system with ferroelectric behaviour in brain waves. Int J PharmTech Res 8(5):964–973
60. Vaidyanathan S (2015) Adaptive control of a chemical chaotic reactor. Int J PharmTech Res 8(3):377–382
61. Vaidyanathan S (2015) Adaptive control of the FitzHugh-Nagumo chaotic neuron model. Int J PharmTech Res 8(6):117–127
62. Vaidyanathan S (2015) Adaptive synchronization of chemical chaotic reactors. Int J ChemTech Res 8(2):612–621
63. Vaidyanathan S (2015) Adaptive synchronization of generalized Lotka-Volterra three-species biological systems. Int J PharmTech Res 8(5):928–937
64. Vaidyanathan S (2015) Adaptive synchronization of novel 3-D chemical chaotic reactor systems. Int J ChemTech Res 8(7):159–171
65. Vaidyanathan S (2015) Adaptive synchronization of the identical FitzHugh-Nagumo chaotic neuron models. Int J PharmTech Res 8(6):167–177
66. Vaidyanathan S (2015) Analysis, control and synchronization of a 3-D novel jerk chaotic system with two quadratic nonlinearities. Kyungpook Math J 55:563–586
67. Vaidyanathan S (2015) Analysis, properties and control of an eight-term 3-D chaotic system with an exponential nonlinearity. Int J Modell Ident Control 23(2):164–172
68. Vaidyanathan S (2015) Anti-synchronization of brusselator chemical reaction systems via adaptive control. Int J ChemTech Res 8(6):759–768
69. Vaidyanathan S (2015) Chaos in neurons and adaptive control of Birkhoff-Shaw strange chaotic attractor. Int J PharmTech Res 8(5):956–963
70. Vaidyanathan S (2015) Chaos in neurons and synchronization of Birkhoff-Shaw strange chaotic attractors via adaptive control. Int J PharmTech Res 8(6):1–11
71. Vaidyanathan S (2015) Coleman-Gomatam logarithmic competitive biology models and their ecological monitoring. Int J PharmTech Res 8(6):94–105
72. Vaidyanathan S (2015) Dynamics and control of brusselator chemical reaction. Int J ChemTech Res 8(6):740–749
73. Vaidyanathan S (2015) Dynamics and control of tokamak system with symmetric and magnetically confined plasma. Int J ChemTech Res 8(6):795–803

74. Vaidyanathan S (2015) Global chaos synchronization of chemical chaotic reactors via novel sliding mode control method. Int J ChemTech Res 8(7):209–221
75. Vaidyanathan S (2015) Global chaos synchronization of the forced Van der Pol chaotic oscillators via adaptive control method. Int J PharmTech Res 8(6):156–166
76. Vaidyanathan S (2015) Global chaos synchronization of the Lotka-Volterra biological systems with four competitive species via active control. Int J PharmTech Res 8(6):206–217
77. Vaidyanathan S (2015) Lotka-Volterra population biology models with negative feedback and their ecological monitoring. Int J PharmTech Res 8(5):974–981
78. Vaidyanathan S (2015) Lotka-Volterra two species competitive biology models and their ecological monitoring. Int J PharmTech Res 8(6):32–44
79. Vaidyanathan S (2015) Output regulation of the forced Van der Pol chaotic oscillator via adaptive control method. Int J PharmTech Res 8(6):106–116
80. Vaidyanathan S (2016) Global chaos regulation of a symmetric nonlinear gyro system via integral sliding mode control. Int J ChemTech Res 9(5):462–469
81. Vaidyanathan S, Azar AT (2015) Analysis, control and synchronization of a nine-term 3-D novel chaotic system. In: Azar AT, Vaidyanathan S (eds) Chaos modelling and control systems design, studies in computational intelligence, vol 581. Springer, Germany, pp 19–38
82. Vaidyanathan S, Madhavan K (2013) Analysis, adaptive control and synchronization of a seven-term novel 3-D chaotic system. Int J Control Theor Appl 6(2):121–137
83. Vaidyanathan S, Pakiriswamy S (2013) Generalized projective synchronization of six-term Sundarapandian chaotic systems by adaptive control. Int J Control Theor Appl 6(2):153–163
84. Vaidyanathan S, Pakiriswamy S (2015) A 3-D novel conservative chaotic System and its generalized projective synchronization via adaptive control. J Eng Sci Technol Rev 8(2):52–60
85. Vaidyanathan S, Rajagopal K (2011) Hybrid synchronization of hyperchaotic Wang-Chen and hyperchaotic Lorenz systems by active non-linear control. Int J Syst Signal Control Eng Appl 4(3):55–61
86. Vaidyanathan S, Rajagopal K (2012) Global chaos synchronization of hyperchaotic Pang and hyperchaotic Wang systems via adaptive control. Int J Soft Comput 7(1):28–37
87. Vaidyanathan S, Rasappan S (2011) Global chaos synchronization of hyperchaotic Bao and Xu systems by active nonlinear control. In: Communications in computer and information science vol 198, pp 10–17
88. Vaidyanathan S, Rasappan S (2014) Global chaos synchronization of n-scroll Chua circuit and Lur'e system using backstepping control design with recursive feedback. Arab J Sci Eng 39(4):3351–3364
89. Vaidyanathan S, Sampath S (2011) Global chaos synchronization of hyperchaotic Lorenz systems by sliding mode control. In: Communications in computer and information science vol 205, pp 156–164
90. Vaidyanathan S, Sampath S (2012) Anti-synchronization of four-wing chaotic systems via sliding mode control. Int J Autom Comput 9(3):274–279
91. Vaidyanathan S, Volos C (2015) Analysis and adaptive control of a novel 3-D conservative no-equilibrium chaotic system. Arch Control Sci 25(3):333–353
92. Vaidyanathan S, Volos C (2016) Advances and applications in chaotic systems. Springer, Berlin, Germany
93. Vaidyanathan S, Volos C (2016) Advances and applications in nonlinear control systems. Springer, Berlin, Germany
94. Vaidyanathan S, Volos C, Pham VT, Madhavan K, Idowu BA (2014) Adaptive backstepping control, synchronization and circuit simulation of a 3-D novel jerk chaotic system with two hyperbolic sinusoidal nonlinearities. Arch Control Sci 24(3):375–403
95. Vaidyanathan S, Idowu BA, Azar AT (2015) Backstepping controller design for the global chaos synchronization of Sprott's jerk systems. In: Studies in computational intelligence vol 581, pp 39–58
96. Vaidyanathan S, Rajagopal K, Volos CK, Kyprianidis IM, Stouboulos IN (2015) Analysis, adaptive control and synchronization of a seven-term novel 3-D chaotic system with three

quadratic nonlinearities and its digital implementation in LabVIEW. J Eng Sci Technol Rev 8(2):130–141

97. Vaidyanathan S, Sampath S, Azar AT (2015) Global chaos synchronisation of identical chaotic systems via novel sliding mode control method and its application to Zhu system. Int J Modell Ident Control 23(1):92–100

98. Vaidyanathan S, Volos CK, Kyprianidis IM, Stouboulos IN, Pham VT (2015) Analysis, adaptive control and anti-synchronization of a six-term novel jerk chaotic system with two exponential nonlinearities and its circuit simulation. J Eng Sci Technol Rev 8(2):24–36

99. Vaidyanathan S, Volos CK, Pham VT (2015) Analysis, adaptive control and adaptive synchronization of a nine-term novel 3-D chaotic system with four quadratic nonlinearities and its circuit simulation. J Eng Sci Technol Rev 8(2):181–191

100. Vaidyanathan S, Volos CK, Pham VT (2015) Global chaos control of a novel nine-term chaotic system via sliding mode control. In: Azar AT, Zhu Q (eds) Advances and applications in sliding mode control systems, studies in computational intelligence, vol 576. Springer, Germany, pp 571–590

101. Volos CK, Kyprianidis IM, Stouboulos IN, Tlelo-Cuautle E, Vaidyanathan S (2015) Memristor: a new concept in synchronization of coupled neuromorphic circuits. J Eng Sci Technol Rev 8(2):157–173

102. Wei Z, Yang Q (2010) Anti-control of Hopf bifurcation in the new chaotic system with two stable node-foci. Appl Math Comput 217(1):422–429

103. Zhou W, Xu Y, Lu H, Pan L (2008) On dynamics analysis of a new chaotic attractor. Phys Lett A 372(36):5773–5777

104. Zhu C, Liu Y, Guo Y (2010) Theoretic and numerical study of a new chaotic system. Intell Inf Manag 2:104–109

Adaptive Integral Sliding Mode Controller Design for the Control and Synchronization of a Novel Jerk Chaotic System

Sundarapandian Vaidyanathan

Abstract Chaos has important applications in physics, chemistry, biology, ecology, secure communications, cryptosystems and many scientific branches. Control and synchronization of chaotic systems are important research problems in chaos theory. Sliding mode control is an important method used to solve various problems in control systems engineering. In robust control systems, the sliding mode control is often adopted due to its inherent advantages of easy realization, fast response and good transient performance as well as insensitivity to parameter uncertainties and disturbance. In this work, we first describe the Sprott jerk chaotic system (1997), which is an important model of a jerk chaotic system with two cubic nonlinearities. Next, we derive a novel jerk chaotic system by adding a linear term to Sprott jerk chaotic system. The phase portraits of the novel jerk chaotic system are displayed and the qualitative properties of the novel jerk chaotic system are discussed. We demonstrate that the novel jerk chaotic system has a unique equilibrium point at the origin, which is a saddle-focus. The Lyapunov exponents of the novel jerk chaotic system are obtained as $L_1 = 0.2062$, $L_2 = 0$ and $L_3 = -3.8062$. The Kaplan-Yorke dimension of the novel jerk system is derived as $D_{KY} = 2.0542$, which shows the complexity of the system. Next, an adaptive integral sliding mode control scheme is proposed to globally stabilize all the trajectories of the novel jerk chaotic system. Furthermore, an adaptive integral sliding mode control scheme is proposed for the global chaos synchronization of identical novel jerk chaotic systems. The adaptive control mechanism helps the control design by estimating the unknown parameters. Numerical simulations using MATLAB are shown to illustrate all the main results derived in this work.

Keywords Chaos · Chaotic systems · Jerk systems · Chaos control · Integral sliding mode control · Adaptive control · Synchronization

S. Vaidyanathan (✉)
Research and Development Centre, Vel Tech University, Avadi, Chennai
600062, Tamil Nadu, India
e-mail: sundarcontrol@gmail.com

© Springer International Publishing AG 2017
S. Vaidyanathan and C.-H. Lien (eds.), *Applications of Sliding Mode Control in Science and Engineering*, Studies in Computational Intelligence 709,
DOI 10.1007/978-3-319-55598-0_17

393

1   Introduction

Chaos theory describes the quantitative study of unstable aperiodic dynamic behavior in deterministic nonlinear dynamical systems. For the motion of a dynamical system to be chaotic, the system variables should contain some nonlinear terms and the system must satisfy three properties: boundedness, infinite recurrence and sensitive dependence on initial conditions [3–5, 101–103].

The problem of global control of a chaotic system is to device feedback control laws so that the closed-loop system is globally asymptotically stable. Global chaos control of various chaotic systems has been investigated via various methods in the control literature [3, 4, 101, 102].

The synchronization of chaotic systems deals with the problem of synchronizing the states of two chaotic systems called as *master* and *slave* systems asymptotically with time. The design goal of the complete synchronization problem is to use the output of the master system to control of the output of the slave system so that the outputs of the two systems are synchronized asymptotically with time.

Because of the *butterfly effect* [3], which causes the exponential divergence of the trajectories of two identical chaotic systems started with nearly the same initial conditions, synchronizing two chaotic systems is seemingly a challenging research problem in the chaos literature.

The synchronization of chaotic systems was first researched by Yamada and Fujisaka [13] with subsequent work by Pecora and Carroll [22, 23]. In the last few decades, several different methods have been devised for the synchronization of chaotic and hyperchaotic systems [3–5, 101–103].

Many new chaotic systems have been discovered in the recent years such as Zhou system [120], Zhu system [121], Li system [17], Sundarapandian systems [38, 39], Vaidyanathan systems [48–53, 55, 65, 81–83, 89, 90, 93–96, 98, 100, 104–108], Pehlivan system [24], Sampath system [34], Tacha system [42], Pham systems [26, 29, 30, 32], Akgul system [2], etc.

Chaos theory has applications in several fields such as memristors [25, 27, 28, 31, 33, 109, 112, 114], fuzzy logic [6, 35, 91, 118], communication systems [10, 11, 115], cryptosystems [8, 9], electromechanical systems [12, 41], lasers [7, 18, 117], encryption [19, 20, 119], electrical circuits [1, 14, 97, 110], chemical reactions [56, 57, 59, 60, 62, 64, 66–68, 71, 73, 77, 92], oscillators [74, 75, 78, 79, 111], tokamak systems [72, 80], neurology [61, 69, 70, 76, 85, 113], ecology [58, 63, 86, 88], etc.

The adaptive control mechanism helps the control design by estimating the unknown parameters [3, 4, 101]. The sliding mode control approach is recognized as an efficient tool for designing robust controllers for linear or nonlinear control systems operating under uncertainty conditions [43, 44].

A major advantage of sliding mode control is low sensitivity to parameter variations in the plant and disturbances affecting the plant, which eliminates the necessity of exact modeling of the plant. In the sliding mode control, the control dynamics will have two sequential modes, viz. the reaching mode and the sliding mode. Basically, a sliding mode controller design consists of two parts: hyperplane design and controller design. A hyperplane is first designed via the pole-placement approach and a controller is then designed based on the sliding condition. The stability of the overall system is guaranteed by the sliding condition and by a stable hyperplane. Sliding mode control method is a popular method for the control and synchronization of chaotic systems [16, 21, 40, 45–47, 54, 84, 87, 99].

In the recent decades, there is some good interest in finding jerk chaotic systems, which are described by the third-order ordinary differential equation

$$\dddot{x} = f(x, \dot{x}, \ddot{x}) \tag{1}$$

The differential equation (1) is called a *jerk system,* because the third-order time derive in mechanical systems is called *jerk.*

By defining phase variables $x_1 = x$, $x_2 = \dot{x}$ and $x_3 = \ddot{x}$, the jerk differential equation (1) can be expressed as a 3-D system given by

$$\begin{cases} \dot{x}_1 = x_2 \\ \dot{x}_2 = x_3 \\ \dot{x}_3 = f(x_1, x_2, x_3) \end{cases} \tag{2}$$

In [37], Sprott derived a 3-D jerk chaotic system with two cubic nonlinearities. Sprott jerk system [37] is an important model of a jerk chaotic system with two cubic nonlinearities. In this work, we derive a novel jerk chaotic system by adding a linear term to the Sprott jerk chaotic system. The phase portraits of the novel jerk chaotic system are displayed and the qualitative properties of the novel jerk chaotic system are discussed. We demonstrate that the novel jerk chaotic system has a unique equilibrium point at the origin, which is a saddle-focus. The Lyapunov exponents of the novel jerk chaotic system are obtained as $L_1 = 0.2062, L_2 = 0$ and $L_3 = -3.8062$. The Kaplan-Yorke dimension of the novel jerk system is derived as $D_{KY} = 2.0542$, which shows the complexity of the system.

Next, an adaptive integral sliding mode control scheme is proposed to globally stabilize all the trajectories of the novel jerk chaotic system. Furthermore, an adaptive integral sliding mode control scheme is proposed for the global chaos synchronization of identical novel jerk chaotic systems.

This work is organized as follows. Section 2 describes the novel 3-D jerk chaotic system and its phase portraits. Section 3 details the qualitative properties of the novel jerk chaotic system. Section 4 contains new results on the adaptive integral sliding mode controller design for the global stabilization of the rod-type plasma torch chaotic system. Section 5 contains new results on the adaptive integral sliding mode controller design for the global synchronization of the novel jerk chaotic systems. Section 6 contains the conclusions of this work.

2   A Novel Jerk Chaotic System

A classical example of a cubic dissipative jerk chaotic flow was found by Sprott [37] and described by the third-order differential equation

$$\dddot{x} + a\ddot{x} - x\dot{x}^2 + x^3 = 0 \tag{3}$$

The state-space model of the Sprott differential equation (3) can be described as follows.

$$\begin{cases} \dot{x}_1 = x_2 \\ \dot{x}_2 = x_3 \\ \dot{x}_3 = -ax_3 + x_1 x_2^2 - x_1^3 \end{cases} \tag{4}$$

where x_1, x_2, x_3 are the states and a is a positive parameter.

In [37], it was observed that the jerk system (4) is *chaotic* when $a = 3.6$.

For numerical simulations, we take the initial state of the Sprott jerk chaotic system (4) as

$$x_1(0) = 0.2, \quad x_2(0) = 0.2, \quad x_3(0) = 0.2 \tag{5}$$

When $a = 3.6$, the Lyapunov exponents of the Sprott jerk system (4) are calculated by Wolf's algorithm [116] as

$$L_1 = 0.1370, \quad L_2 = 0, \quad L_3 = -3.7370 \tag{6}$$

This shows that the Sprott jerk system (4) is chaotic and dissipative. Also, the Kaplan-Yorke dimension of the Sprott jerk chaotic system (4) is determined as

$$D_{KY} = 2 + \frac{L_1 + L_2}{|L_3|} = 2.0367 \tag{7}$$

Figure 1 shows the Lyapunov exponents of the Sprott jerk chaotic system (4). Figure 2 shows the 3-D phase portrait of the Sprott jerk chaotic system (4).

In this work, we announce a new jerk chaotic system by adding a linear term bx to the Sprott differential equation (3). Thus, we obtain the third order differential equation

$$\dddot{x} + a\ddot{x} + bx - x\dot{x}^2 + x^3 = 0 \tag{8}$$

In system form, we can express the new differential equation (8) as the following jerk system.

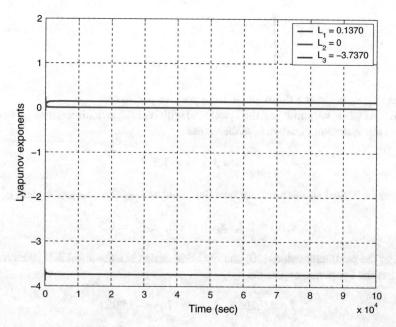

Fig. 1 Lyapunov exponents of the Sprott jerk chaotic system

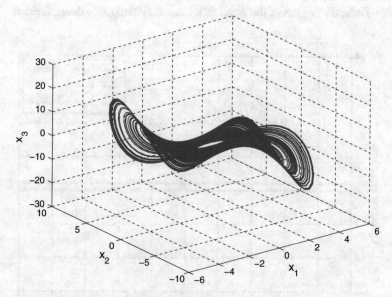

Fig. 2 3-D phase portrait of the Sprott jerk chaotic system

$$\begin{cases} \dot{x}_1 = x_2 \\ \dot{x}_2 = x_3 \\ \dot{x}_3 = -ax_3 - bx_1 + x_1 x_2^2 - x_1^3 \end{cases} \tag{9}$$

where x_1, x_2, x_3 are the states and a, b are positive parameters.

In this work, we show that the system (9) with two cubic nonlinearities is chaotic when the system parameters take the values

$$a = 3.6, \quad b = 1.3 \tag{10}$$

For numerical simulations, we take the initial state of the new jerk system (9) as

$$x_1(0) = 0.2, \quad x_2(0) = 0.2, \quad x_3(0) = 0.2 \tag{11}$$

For the parameter values (10) and the initial state (11), the novel 3-D jerk system (9) has the Lyapunov exponents

$$L_1 = 0.2062, \quad L_2 = 0, \quad L_3 = -3.8062 \tag{12}$$

This shows that the novel jerk system (9) is chaotic and dissipative. Also, the Kaplan-Yorke dimension of the novel jerk chaotic system (9) is determined as

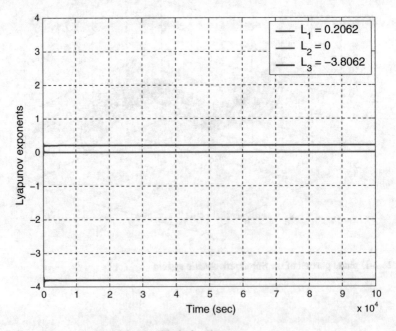

Fig. 3 Lyapunov exponents of the novel jerk chaotic system

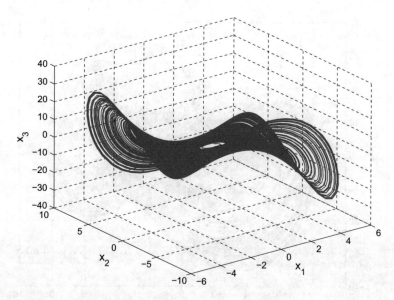

Fig. 4 3-D phase portrait of the novel jerk chaotic system

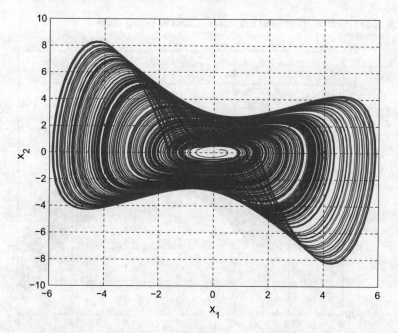

Fig. 5 2-D projection of the novel jerk chaotic system on (x_1, x_2) plane

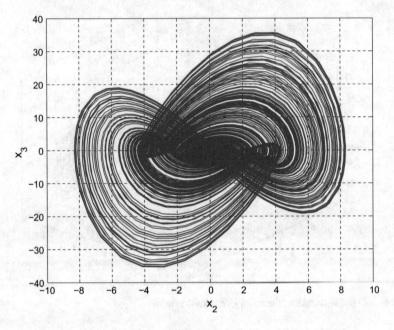

Fig. 6 2-D projection of the novel jerk chaotic system on (x_2, x_3) plane

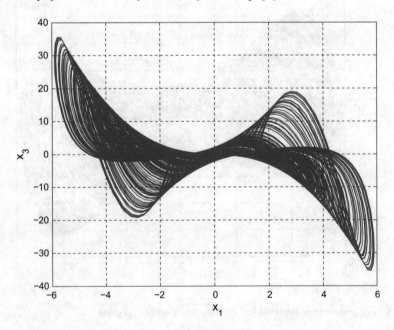

Fig. 7 2-D projection of the novel jerk chaotic system on (x_1, x_3) plane

$$D_{KY} = 2 + \frac{L_1 + L_2}{|L_3|} = 2.0542 \tag{13}$$

From (6) and (12), we note that the Maximal Lyapunov Exponent (MLE) of the new chaotic system is given by $L_1(\text{new}) = 0.2062$, which is greater than the Maximal Lyapunov Exponent (MLE) of the Sprott jerk chaotic system given by $L_1(\text{Sprott}) = 0.1370$.

From (7) and (13), we note that the Kaplan-Yorke dimension of the new chaotic system is given by $D_{KY}(\text{new}) = 2.0542$, which is greater than the Kaplan-Yorke dimension of the Sprott jerk chaotic system given by $D_{KY}(\text{Sprott}) = 2.0367$.

Figure 3 shows the Lyapunov exponents of the novel 3-D jerk chaotic system (9). Figure 4 shows the 3-D phase portrait of the novel jerk chaotic system (9). Figures 5, 6 and 7 show the 2-D projections of the novel jerk chaotic system (9) on (x_1, x_2), (x_2, x_3) and (x_1, x_3) coordinate planes, respectively.

3   Analysis of the 3-D Novel Jerk Chaotic System

In this section, we assume that the system parameters a and b are as in the chaotic case (10), i.e. $a = 3.6$ and $b = 1.3$.

3.1   Dissipativity

In vector notation, the new jerk system (9) can be expressed as

$$\dot{\mathbf{x}} = f(\mathbf{x}) = \begin{bmatrix} f_1(x_1, x_2, x_3) \\ f_2(x_1, x_2, x_3) \\ f_3(x_1, x_2, x_3) \end{bmatrix}, \tag{14}$$

where

$$\begin{cases} f_1(x_1, x_2, x_3) = x_2 \\ f_2(x_1, x_2, x_3) = x_3 \\ f_3(x_1, x_2, x_3) = -ax_3 - bx_1 + x_1 x_2^2 - x_1^3 \end{cases} \tag{15}$$

Let Ω be any region in $\mathbf{R}^3$ with a smooth boundary and also, $\Omega(t) = \Phi_t(\Omega)$, where Φ_t is the flow of f. Furthermore, let $V(t)$ denote the volume of $\Omega(t)$.

By Liouville's theorem, we know that

$$\dot{V}(t) = \int_{\Omega(t)} (\nabla \cdot f) \, dx_1 \, dx_2 \, dx_3 \tag{16}$$

The divergence of the jerk system (9) is found as:

$$\nabla \cdot f = \frac{\partial f_1}{\partial x_1} + \frac{\partial f_2}{\partial x_2} + \frac{\partial f_3}{\partial x_3} = -a < 0 \tag{17}$$

since $a = 3.6 > 0$.

Inserting the value of $\nabla \cdot f$ from (17) into (16), we get

$$\dot{V}(t) = \int_{\Omega(t)} (-a)\, dx_1\, dx_2\, dx_3 = -aV(t) \tag{18}$$

Integrating the first order linear differential equation (18), we get

$$V(t) = \exp(-at)V(0) \tag{19}$$

It is clear from Eq. (19) that $V(t) \to 0$ exponentially as $t \to \infty$. This shows that the novel jerk (9) chaotic system is dissipative. Hence, the system limit sets are ultimately confined into a specific limit set of zero volume, and the asymptotic motion of the novel jerk chaotic system (9) settles onto a strange attractor of the system.

3.2 Equilibrium Points

The equilibrium points of the novel jerk chaotic system (9) are obtained by solving the equations

$$f_1(x_1, x_2, x_3) = \qquad\qquad\qquad\qquad\qquad x_2 = \qquad 0 \tag{20a}$$
$$f_2(x_1, x_2, x_3) = \qquad\qquad\qquad\qquad\qquad x_3 = \qquad 0 \tag{20b}$$
$$f_3(x_1, x_2, x_3) = \qquad -ax_3 - bx_1 + x_1 x_2^2 - x_1^3 = \qquad 0 \tag{20c}$$

From (20a) and (20b), we get $x_2 = 0$ and $x_3 = 0$.
Substituting these values in (20c), we get

$$-bx_1 - x_1^3 = -x_1(b + x_1^2) = 0 \tag{21}$$

Since $b = 1.3 > 0$, it is immediate that $x_1 = 0$.
Thus, the novel jerk chaotic system (9) has a unique equilibrium

$$E_0 = \begin{bmatrix} 0 \\ 0 \\ 0 \end{bmatrix} \tag{22}$$

The Jacobian matrix of the new jerk system (9) at E_0 is obtained as

$$J_0 = J(E_0) = \begin{bmatrix} 0 & 1 & 0 \\ 0 & 0 & 1 \\ -b & 0 & -a \end{bmatrix} = \begin{bmatrix} 0 & 1 & 0 \\ 0 & 0 & 1 \\ -1.3 & 0 & -3.6 \end{bmatrix} \tag{23}$$

We find that the matrix J_0 has the eigenvalues

$$\lambda_1 = -3.6952, \quad \lambda_{2,3} = 0.0476 \pm 0.5912i \tag{24}$$

This shows that the equilibrium point E_0 is a saddle-focus, which is unstable.

3.3 Invariance

The new jerk chaotic system (9) is invariant under the coordinates transformation

$$(x_1, x_2, x_3) \mapsto (-x_1, -x_2, -x_3) \tag{25}$$

This shows that the new jerk chaotic system (9) has point-reflection symmetry about the origin and every non-trivial trajectory of the system (9) must have a twin trajectory.

4 Global Chaos Control of the Novel Jerk Chaotic System

In this section, we apply adaptive integral sliding mode control for the global chaos control of the novel jerk chaotic system with unknown parameters. Adaptive control mechanism helps the control design by estimating the unknown parameters [3, 4].

The controlled novel jerk chaotic system is described by

$$\begin{cases} \dot{x}_1 = x_2 + u_1 \\ \dot{x}_2 = x_3 + u_2 \\ \dot{x}_3 = -ax_3 - bx_1 + x_1 x_2^2 - x_1^3 + u_3 \end{cases} \tag{26}$$

where x_1, x_2, x_3 are the states of the system and a, b are unknown system parameters. The design goal is to find suitable feedback controllers u_1, u_2, u_3 so as to globally stabilize the system (26) with estimates of the unknown parameters.

Based on the sliding mode control theory [36, 43, 44], the integral sliding surface of each x_i $(i = 1, 2, 3)$ is defined as follows:

$$s_i = \left(\frac{d}{dt} + \lambda_i \right) \left(\int_0^t x_i(\tau) d\tau \right) = x_i + \lambda_i \int_0^t x_i(\tau) d\tau, \quad i = 1, 2, 3 \tag{27}$$

From Eq. (27), it follows that

$$\begin{cases} \dot{s}_1 = \dot{x}_1 + \lambda_1 x_1 \\ \dot{s}_2 = \dot{x}_2 + \lambda_2 x_2 \\ \dot{s}_3 = \dot{x}_3 + \lambda_3 x_3 \end{cases} \tag{28}$$

The Hurwitz condition is realized if $\lambda_i > 0$ for $i = 1, 2, 3$.
We consider the adaptive feedback control given by

$$\begin{cases} u_1 = -x_2 - \lambda_1 x_1 - \eta_1 \, \mathrm{sgn}(s_1) - k_1 s_1 \\ u_2 = -x_3 - \lambda_2 x_2 - \eta_2 \, \mathrm{sgn}(s_2) - k_2 s_2 \\ u_3 = \hat{a}(t)x_3 + \hat{b}(t)x_1 - x_1 x_2^2 + x_1^3 - \lambda_3 x_3 - \eta_3 \, \mathrm{sgn}(s_3) - k_3 s_3 \end{cases} \tag{29}$$

where $\eta_i > 0$ and $k_i > 0$ for $i = 1, 2, 3$.
Substituting (29) into (26), we obtain the closed-loop control system given by

$$\begin{cases} \dot{x}_1 = -\lambda_1 x_1 - \eta_1 \, \mathrm{sgn}(s_1) - k_1 s_1 \\ \dot{x}_2 = -\lambda_2 x_2 - \eta_2 \, \mathrm{sgn}(s_2) - k_2 s_2 \\ \dot{x}_3 = -[a - \hat{a}(t)]x_3 - [b - \hat{b}(t)]x_1 - \lambda_3 x_3 - \eta_3 \, \mathrm{sgn}(s_3) - k_3 s_3 \end{cases} \tag{30}$$

We define the parameter estimation errors as

$$\begin{cases} e_a(t) = a - \hat{a}(t) \\ e_b(t) = b - \hat{b}(t) \end{cases} \tag{31}$$

Using (31), we can simplify the closed-loop system (30) as

$$\begin{cases} \dot{x}_1 = -\lambda_1 x_1 - \eta_1 \, \mathrm{sgn}(s_1) - k_1 s_1 \\ \dot{x}_2 = -\lambda_2 x_2 - \eta_2 \, \mathrm{sgn}(s_2) - k_2 s_2 \\ \dot{x}_3 = -e_a x_3 - e_b x_1 - \lambda_3 x_3 - \eta_3 \, \mathrm{sgn}(s_3) - k_3 s_3 \end{cases} \tag{32}$$

Differentiating (31) with respect to t, we get

$$\begin{cases} \dot{e}_a = -\dot{\hat{a}} \\ \dot{e}_b = -\dot{\hat{b}} \end{cases} \tag{33}$$

Next, we state and prove the main result of this section.

Theorem 1 *The new jerk chaotic system (26) is rendered globally asymptotically stable for all initial conditions $\mathbf{x}(0) \in \mathbf{R}^3$ by the adaptive integral sliding mode control law (29) and the parameter update law*

$$\begin{cases} \dot{\hat{a}} = -s_3 x_3 \\ \dot{\hat{b}} = -s_3 x_1 \end{cases} \tag{34}$$

where λ_i, η_i, k_i are positive constants for $i = 1, 2, 3$.

Proof We consider the quadratic Lyapunov function defined by

$$V(s_1, s_2, s_3, e_a, e_b) = \frac{1}{2} \left(s_1^2 + s_2^2 + s_3^2 \right) + \frac{1}{2} \left(e_a^2 + e_b^2 \right) \tag{35}$$

Clearly, V is positive definite on $\mathbf{R}^5$.
Using (28), (32) and (33), the time-derivative of V is obtained as

$$\begin{aligned} \dot{V} = s_1(-\eta_1 \, \text{sgn}(s_1) - k_1 s_1) + s_2(-\eta_2 \, \text{sgn}(s_2) - k_1 s_2) \\ + s_3 \left[-e_a x_3 - e_b x_1 - \eta_3 \, \text{sgn}(s_3) - k_3 s_3 \right] - e_a \dot{\hat{a}} - e_b \dot{\hat{b}} \end{aligned} \tag{36}$$

i.e.

$$\begin{aligned} \dot{V} = -\eta_1 |s_1| - k_1 s_1^2 - \eta_2 |s_2| - k_2 s_2^2 - \eta_3 |s_3| - k_3 s_3^2 \\ + e_a \left(-s_3 x_3 - \dot{\hat{a}} \right) + e_b \left(-s_3 x_1 - \dot{\hat{b}} \right) \end{aligned} \tag{37}$$

Using the parameter update law (34), we obtain

$$\dot{V} = -\eta_1 |s_1| - k_1 s_1^2 - \eta_2 |s_2| - k_2 s_2^2 - \eta_3 |s_3| - k_3 s_3^2 \tag{38}$$

which shows that $\dot{V}$ is negative semi-definite on $\mathbf{R}^5$.

Hence, by Barbalat's lemma [15], it is immediate that $\mathbf{x}(t)$ is globally asymptotically stable for all values of $\mathbf{x}(0) \in \mathbf{R}^3$.

This completes the proof. ∎

For numerical simulations, we take the parameter values of the new jerk chaotic system (26) as in the chaotic case (10), i.e. $a = 3.6$ and $b = 1.3$.

We take the values of the control parameters as

$$k_i = 20, \quad \eta_i = 0.1, \quad \lambda_i = 50, \quad \text{where } i = 1, 2, 3 \tag{39}$$

We take the estimates of the system parameters as $\hat{a}(0) = 15.3$ and $\hat{b}(0) = 8.4$.
We take the initial state of the system (26) as

$$x_1(0) = 27.8, \quad x_2(0) = 16.2, \quad x_3(0) = 19.3 \tag{40}$$

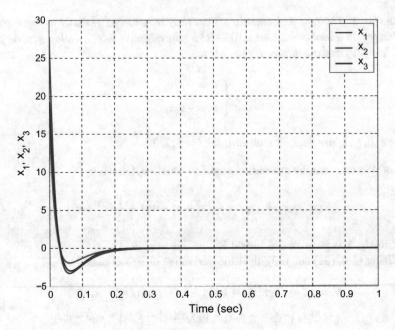

Fig. 8   Time-history of the controlled states x_1, x_2, x_3

Figure 8 shows the time-history of the controlled states x_1, x_2, x_3.

5   Global Chaos Synchronization of the Novel Jerk Chaotic Systems

In this section, we use adaptive integral sliding mode control for the global chaos synchronization of the novel jerk chaotic systems with unknown system parameters. The adaptive control mechanism helps the control design by estimating the unknown parameters [3, 4, 101].

As the master system, we consider the novel jerk chaotic system given by

$$\begin{cases} \dot{x}_1 = x_2 \\ \dot{x}_2 = x_3 \\ \dot{x}_3 = -ax_3 - bx_1 + x_1 x_2^2 - x_1^3 \end{cases} \tag{41}$$

where x_1, x_2, x_3 are the states of the system and a, b are unknown system parameters.

As the slave system, we consider the controlled rod-type plasma torch chaotic system given by

$$\begin{cases} \dot{y}_1 = y_2 + u_1 \\ \dot{y}_2 = y_3 + u_2 \\ \dot{y}_3 = -ay_3 - by_1 + y_1 y_2^2 - y_1^3 + u_3 \end{cases} \tag{42}$$

where y_1, y_2, y_3 are the states of the system and u_1, u_2, u_3 are the controllers to be designed using adaptive integral sliding mode control.

The synchronization error between the rod-type plasma torch chaotic systems (41) and (42) is defined as

$$\begin{cases} e_1 = y_1 - x_1 \\ e_2 = y_2 - x_2 \\ e_3 = y_3 - x_3 \end{cases} \tag{43}$$

Then the synchronization error dynamics is obtained as

$$\begin{cases} \dot{e}_1 = e_2 + u_1 \\ \dot{e}_2 = e_3 + u_2 \\ \dot{e}_3 = -ae_3 - be_1 + y_1 y_2^2 - x_1 x_2^2 - y_1^3 + x_1^3 + u_3 \end{cases} \tag{44}$$

Based on the sliding mode control theory [36, 43, 44], the integral sliding surface of each e_i $(i = 1, 2, 3)$ is defined as follows:

$$s_i = \left(\frac{d}{dt} + \lambda_i \right) \left(\int_0^t e_i(\tau) d\tau \right) = e_i + \lambda_i \int_0^t e_i(\tau) d\tau, \quad i = 1, 2, 3 \tag{45}$$

From Eq. (45), it follows that

$$\begin{cases} \dot{s}_1 = \dot{e}_1 + \lambda_1 e_1 \\ \dot{s}_2 = \dot{e}_2 + \lambda_2 e_2 \\ \dot{s}_3 = \dot{e}_3 + \lambda_3 e_3 \end{cases} \tag{46}$$

The Hurwitz condition is realized if $\lambda_i > 0$ for $i = 1, 2, 3$.
We consider the adaptive feedback control given by

$$\begin{cases} u_1 = -e_2 - \lambda_1 e_1 - \eta_1 \operatorname{sgn}(s_1) - k_1 s_1 \\ u_2 = -e_3 - \lambda_2 e_2 - \eta_2 \operatorname{sgn}(s_2) - k_2 s_2 \\ u_3 = \hat{a}(t)e_3 + \hat{b}(t)e_1 - y_1 y_2^2 + x_1 x_2^2 + y_1^3 - x_1^3 - \lambda_3 e_3 - \eta_3 \operatorname{sgn}(s_3) - k_3 s_3 \end{cases} \tag{47}$$

where $\eta_i > 0$ and $k_i > 0$ for $i = 1, 2, 3$.

Substituting (47) into (44), we obtain the closed-loop error dynamics as

$$
\begin{cases}
\dot{e}_1 = -\lambda_1 e_1 - \eta_1 \operatorname{sgn}(s_1) - k_1 s_1 \\
\dot{e}_2 = -\lambda_2 e_2 - \eta_2 \operatorname{sgn}(s_2) - k_2 s_2 \\
\dot{e}_3 = -[a - \hat{a}(t)]e_3 - [b - \hat{b}(t)]e_1 - \lambda_3 e_3 - \eta_3 \operatorname{sgn}(s_3) - k_3 s_3
\end{cases}
\tag{48}
$$

We define the parameter estimation errors as

$$
\begin{cases}
e_a(t) = a - \hat{a}(t) \\
e_b(t) = b - \hat{b}(t)
\end{cases}
\tag{49}
$$

Using (49), we can simplify the closed-loop system (48) as

$$
\begin{cases}
\dot{e}_1 = -\lambda_1 e_1 - \eta_1 \operatorname{sgn}(s_1) - k_1 s_1 \\
\dot{e}_2 = -\lambda_2 e_2 - \eta_2 \operatorname{sgn}(s_2) - k_2 s_2 \\
e_3 = -e_a e_3 - e_b e_1 - \lambda_3 e_3 - \eta_3 \operatorname{sgn}(s_3) - k_3 s_3
\end{cases}
\tag{50}
$$

Differentiating (49) with respect to t, we get

$$
\begin{cases}
\dot{e}_a = -\dot{\hat{a}} \\
\dot{e}_b = -\dot{\hat{b}} \\
\dot{e}_c = -\dot{\hat{c}}
\end{cases}
\tag{51}
$$

Next, we state and prove the main result of this section.

Theorem 2 *The new jerk chaotic systems (41) and (42) are globally and asymptotically synchronized for all initial conditions $\mathbf{x}(0), \mathbf{y}(0) \in \mathbf{R}^3$ by the adaptive integral sliding mode control law (47) and the parameter update law*

$$
\begin{cases}
\dot{\hat{a}} = -s_3 e_3 \\
\dot{\hat{b}} = -s_3 e_1
\end{cases}
\tag{52}
$$

where λ_i, η_i, k_i are positive constants for $i = 1, 2, 3$.

Proof We consider the quadratic Lyapunov function defined by

$$
V(s_1, s_2, s_3, e_a, e_b) = \frac{1}{2} \left(s_1^2 + s_2^2 + s_3^2 \right) + \frac{1}{2} \left(e_a^2 + e_b^2 \right)
\tag{53}
$$

Clearly, V is positive definite on $\mathbf{R}^5$.

Using (46), (50) and (51), the time-derivative of V is obtained as

$$
\begin{aligned}
\dot{V} = s_1(-\eta_1 \operatorname{sgn}(s_1) - k_1 s_1) + s_2(-\eta_2 \operatorname{sgn}(s_2) - k_1 s_2) \\
+ s_3 \left[-e_a e_3 - e_b e_1 - \eta_3 \operatorname{sgn}(s_3) - k_3 s_3 \right] - e_a \dot{\hat{a}} - e_b \dot{\hat{b}}
\end{aligned}
\tag{54}
$$

i.e.

$$
\begin{aligned}
\dot{V} = -\eta_1 |s_1| - k_1 s_1^2 - \eta_2 |s_2| - k_2 s_2^2 - \eta_3 |s_3| - k_3 s_3^2 \\
+ e_a \left(-s_3 e_3 - \dot{\hat{a}} \right) + e_b \left(-s_3 e_1 - \dot{\hat{b}} \right)
\end{aligned}
\tag{55}
$$

Using the parameter update law (52), we obtain

$$
\dot{V} = -\eta_1 |s_1| - k_1 s_1^2 - \eta_2 |s_2| - k_2 s_2^2 - \eta_3 |s_3| - k_3 s_3^2
\tag{56}
$$

which shows that $\dot{V}$ is negative semi-definite on $\mathbf{R}^5$.

Hence, by Barbalat's lemma [15], it is immediate that $\mathbf{e}(t)$ is globally asymptotically stable for all values of $\mathbf{e}(0) \in \mathbf{R}^3$.

This completes the proof. ∎

For numerical simulations, we take the parameter values of the jerk chaotic systems (41) and (42) as in the chaotic case (10), i.e. $a = 3.6$ and $b = 1.3$.

We take the values of the control parameters as

$$
k_i = 20, \quad \eta_i = 0.1, \quad \lambda_i = 50, \quad \text{where } i = 1, 2, 3
\tag{57}
$$

We take the estimates of the system parameters as $\hat{a}(0) = 14.3$ and $\hat{b}(0) = 16.7$.

We take the initial state of the master system (41) as

$$
x_1(0) = 1.5, \quad x_2(0) = 2.4, \quad x_3(0) = 3.9
\tag{58}
$$

We take the initial state of the slave system (42) as

$$
y_1(0) = 1.2, \quad y_2(0) = 4.9, \quad y_3(0) = 5.4
\tag{59}
$$

Figures 9, 10 and 11 show the complete synchronization of the novel jerk chaotic systems (41) and (42). Figure 12 shows the time-history of the synchronization errors e_1, e_2, e_3.

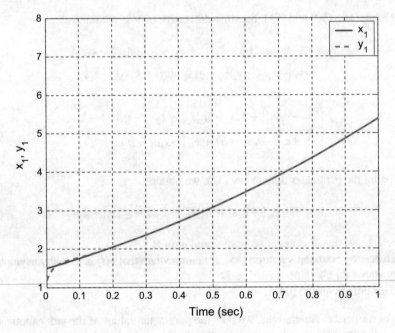

Fig. 9 Complete synchronization of the states x_1 and y_1

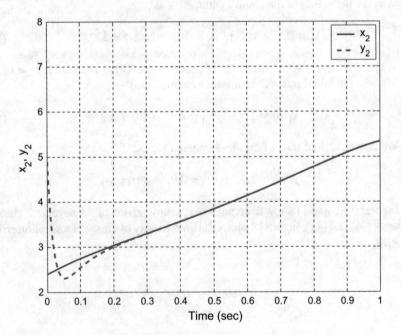

Fig. 10 Complete synchronization of the states x_2 and y_2

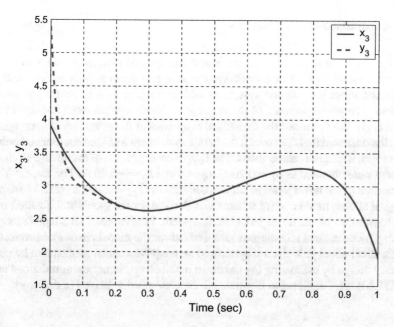

Fig. 11 Complete synchronization of the states x_3 and y_3

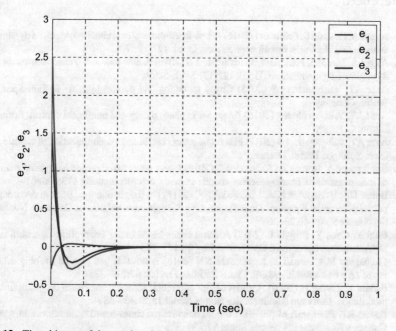

Fig. 12 Time-history of the synchronization errors e_1, e_2, e_3

6 Conclusions

In this work, we first discussed the Sprott jerk chaotic system (1997), which is an important model of a jerk chaotic system with two cubic nonlinearities. Next, we derived a novel jerk chaotic system by adding a linear term to Sprott jerk chaotic system. The phase portraits of the novel jerk chaotic system were displayed using MATLAB. We discussed the dynamical properties of the novel jerk chaotic system. We demonstrated that the novel jerk chaotic system has a unique equilibrium point at the origin, which is a saddle-focus. The Lyapunov exponents of the novel jerk chaotic system were obtained as $L_1 = 0.2062$, $L_2 = 0$ and $L_3 = -3.8062$. The Kaplan-Yorke dimension of the novel jerk system was deduced as $D_{KY} = 2.0542$. Next, an adaptive integral sliding mode control scheme was derived for the global stabilization of all the trajectories of the novel jerk chaotic system. Furthermore, an adaptive integral sliding mode control scheme was also derived for the global chaos synchronization of identical novel jerk chaotic systems. The adaptive control mechanism helps the control design by estimating the unknown parameters. Numerical simulations using MATLAB were depicted to illustrate all the main results derived in this work.

References

1. Akgul A, Hussain S, Pehlivan I (2016) A new three-dimensional chaotic system, its dynamical analysis and electronic circuit applications. Optik 127(18):7062–7071
2. Akgul A, Moroz I, Pehlivan I, Vaidyanathan S (2016) A new four-scroll chaotic attractor and its engineering applications. Optik 127(13):5491–5499
3. Azar AT, Vaidyanathan S (2015) Chaos modeling and control systems design. Springer, Berlin, Germany
4. Azar AT, Vaidyanathan S (2016) Advances in chaos theory and intelligent control. Springer, Berlin, Germany
5. Azar AT, Vaidyanathan S (2017) Fractional order control and synchronization of chaotic systems. Springer, Berlin, Germany
6. Boulkroune A, Bouzeriba A, Bouden T (2016) Fuzzy generalized projective synchronization of incommensurate fractional-order chaotic systems. Neurocomputing 173:606–614
7. Burov DA, Evstigneev NM, Magnitskii NA (2017) On the chaotic dynamics in two coupled partial differential equations for evolution of surface plasmon polaritons. Commun. Nonlinear Sci Numer Simul 46:26–36
8. Chai X, Chen Y, Broyde L (2017) A novel chaos-based image encryption algorithm using DNA sequence operations. Opti Lasers Eng 88:197–213
9. Chenaghlu MA, Jamali S, Khasmakhi NN (2016) A novel keyed parallel hashing scheme based on a new chaotic system. Chaos Solitons Fractals 87:216–225
10. Fallahi K, Leung H (2010) A chaos secure communication scheme based on multiplication modulation. Commun Nonlinear Sci Numer Simul 15(2):368–383
11. Fontes RT, Eisencraft M (2016) A digital bandlimited chaos-based communication system. Commun Nonlinear Sci Numer Simul 37:374–385
12. Fotsa RT, Woafo P (2016) Chaos in a new bistable rotating electromechanical system. Chaos Solitons Fractals 93:48–57
13. Fujisaka H, Yamada T (1983) Stability theory of synchronized motion in coupled-oscillator systems. Prog Theoret Phys 63:32–47

14. Kacar S (2016) Analog circuit and microcontroller based RNG application of a new easy realizable 4D chaotic system. Optik 127(20):9551–9561
15. Khalil HK (2002) Nonlinear systems. Prentice Hall, New York, USA
16. Lakhekar GV, Waghmare LM, Vaidyanathan S (2016) Diving autopilot design for underwater vehicles using an adaptive neuro-fuzzy sliding mode controller. In: Vaidyanathan S, Volos C (eds) Advances and applications in nonlinear control systems. Springer, Berlin, Germany, pp 477–503
17. Li D (2008) A three-scroll chaotic attractor. Phys Lett A 372(4):387–393
18. Liu H, Ren B, Zhao Q, Li N (2016) Characterizing the optical chaos in a special type of small networks of semiconductor lasers using permutation entropy. Opt Commun 359:79–84
19. Liu W, Sun K, Zhu C (2016) A fast image encryption algorithm based on chaotic map. Opt Lasers Eng 84:26–36
20. Liu X, Mei W, Du H (2016) Simultaneous image compression, fusion and encryption algorithm based on compressive sensing and chaos. Opt Commun 366:22–32
21. Moussaoui S, Boulkroune A, Vaidyanathan S (2016) Fuzzy adaptive sliding-mode control scheme for uncertain underactuated systems. In: Vaidyanathan S, Volos C (eds) Advances and applications in nonlinear control systems. Springer, Berlin, Germany, pp 351–367
22. Pecora LM, Carroll TL (1990) Synchronization in chaotic systems. Phys Rev Lett 64:821–824
23. Pecora LM, Carroll TL (1991) Synchronizing chaotic circuits. IEEE Trans Circuits Syst 38:453–456
24. Pehlivan I, Moroz IM, Vaidyanathan S (2014) Analysis, synchronization and circuit design of a novel butterfly attractor. J Sound Vib 333(20):5077–5096
25. Pham VT, Volos C, Jafari S, Wang X, Vaidyanathan S (2014) Hidden hyperchaotic attractor in a novel simple memristive neural network. Optoelectron Adv Mater Rapid Commun 8(11–12):1157–1163
26. Pham VT, Volos CK, Vaidyanathan S (2015) Multi-scroll chaotic oscillator based on a first-order delay differential equation. In: Azar AT, Vaidyanathan S (eds) Chaos modeling and control systems design, studies in computational intelligence, vol 581. Springer, Germany, pp 59–72
27. Pham VT, Volos CK, Vaidyanathan S, Le TP, Vu VY (2015) A memristor-based hyperchaotic system with hidden attractors: dynamics, synchronization and circuital emulating. J Eng Sci Technol Rev 8(2):205–214
28. Pham VT, Jafari S, Vaidyanathan S, Volos C, Wang X (2016) A novel memristive neural network with hidden attractors and its circuitry implementation. Sci China Technol Sci 59(3):358–363
29. Pham VT, Jafari S, Volos C, Giakoumis A, Vaidyanathan S, Kapitaniak T (2016) A chaotic system with equilibria located on the rounded square loop and its circuit implementation. IEEE Trans Circuits Syst II: Express Briefs 63(9):878–882
30. Pham VT, Jafari S, Volos C, Vaidyanathan S, Kapitaniak T (2016) A chaotic system with infinite equilibria located on a piecewise linear curve. Optik 127(20):9111–9117
31. Pham VT, Vaidyanathan S, Volos CK, Hoang TM, Yem VV (2016) Dynamics, synchronization and SPICE implementation of a memristive system with hidden hyperchaotic attractor. In: Azar AT, Vaidyanathan S (eds) Advances in chaos theory and intelligent control. Springer, Berlin, Germany, pp 35–52
32. Pham VT, Vaidyanathan S, Volos CK, Jafari S, Kuznetsov NV, Hoang TM (2016) A novel memristive time-delay chaotic system without equilibrium points. Eur Phys J: Spec Top 225(1):127–136
33. Pham VT, Vaidyanathan S, Volos CK, Jafari S, Wang X (2016) A chaotic hyperjerk system based on memristive device. In: Vaidyanathan S, Volos C (eds) Advances and applications in chaotic systems. Springer, Berlin, Germany, pp 39–58
34. Sampath S, Vaidyanathan S, Volos CK, Pham VT (2015) An eight-term novel four-scroll chaotic system with cubic nonlinearity and its circuit simulation. J Eng Sci Technol Rev 8(2):1–6

35. Shirkhani N, Khanesar M, Teshnehlab M (2016) Indirect model reference fuzzy control of SISO fractional order nonlinear chaotic systems. Procedia Comput Sci 102:309–316
36. Slotine J, Li W (1991) Applied nonlinear control. Prentice-Hall, Englewood Cliffs, NJ, USA
37. Sprott JC (1997) Some simple chaotic jerk functions. Am J Phys 65:537–543
38. Sundarapandian V (2013) Analysis and anti-synchronization of a novel chaotic system via active and adaptive controllers. J Eng Sci Technol Rev 6(4):45–52
39. Sundarapandian V, Pehlivan I (2012) Analysis, control, synchronization, and circuit design of a novel chaotic system. Math Comput Model 55(7–8):1904–1915
40. Sundarapandian V, Sivaperumal S (2011) Sliding controller design of hybrid synchronization of four-wing Chaotic systems. Int J Soft Comput 6(5):224–231
41. Szmit Z, Warminski J (2016) Nonlinear dynamics of electro-mechanical system composed of two pendulums and rotating hub. Procedia Eng 144:953–958
42. Tacha OI, Volos CK, Kyprianidis IM, Stouboulos IN, Vaidyanathan S, Pham VT (2016) Analysis, adaptive control and circuit simulation of a novel nonlinear finance system. Appl Math Comput 276:200–217
43. Utkin VI (1977) Variable structure systems with sliding modes. IEEE Trans Autom Control 22(2):212–222
44. Utkin VI (1993) Sliding mode control design principles and applications to electric drives. IEEE Trans Ind Electron 40(1):23–36
45. Vaidyanathan S (2011) Analysis and synchronization of the hyperchaotic Yujun systems via sliding mode control. Adv Intell Syst Comput 176:329–337
46. Vaidyanathan S (2012) Global chaos control of hyperchaotic Liu system via sliding control method. Int J Control Theory Appl 5(2):117–123
47. Vaidyanathan S (2012) Sliding mode control based global chaos control of Liu-Liu-Liu-Su chaotic system. Int J Control Theory Appl 5(1):15–20
48. Vaidyanathan S (2013) A new six-term 3-D chaotic system with an exponential nonlinearity. Far East J Math Sci 79(1):135–143
49. Vaidyanathan S (2013) Analysis and adaptive synchronization of two novel chaotic systems with hyperbolic sinusoidal and cosinusoidal nonlinearity and unknown parameters. J Eng Sci Technol Rev 6(4):53–65
50. Vaidyanathan S (2014) A new eight-term 3-D polynomial chaotic system with three quadratic nonlinearities. Far East J Math Sci 84(2):219–226
51. Vaidyanathan S (2014) Analysis and adaptive synchronization of eight-term 3-D polynomial chaotic systems with three quadratic nonlinearities. Eur Phys J: Spec Top 223(8):1519–1529
52. Vaidyanathan S (2014) Analysis, control and synchronisation of a six-term novel chaotic system with three quadratic nonlinearities. Int J Model Identif Control 22(1):41–53
53. Vaidyanathan S (2014) Generalised projective synchronisation of novel 3-D chaotic systems with an exponential non-linearity via active and adaptive control. Int J Model Identif Control 22(3):207–217
54. Vaidyanathan S (2014) Global chaos synchronisation of identical Li-Wu chaotic systems via sliding mode control. Int J Model Identif Control 22(2):170–177
55. Vaidyanathan S (2015) A 3-D novel highly chaotic system with four quadratic nonlinearities, its adaptive control and anti-synchronization with unknown parameters. J Eng Sci Technol Rev 8(2):106–115
56. Vaidyanathan S (2015) A novel chemical chaotic reactor system and its adaptive control. Int J ChemTech Res 8(7):146–158
57. Vaidyanathan S (2015) A novel chemical chaotic reactor system and its output regulation via integral sliding mode control. Int J ChemTech Res 8(11):669–683
58. Vaidyanathan S (2015) Active control design for the anti-synchronization of Lotka-Volterra biological systems with four competitive species. Int J PharmTech Res 8(7):58–70
59. Vaidyanathan S (2015) Adaptive control design for the anti-synchronization of novel 3-D chemical chaotic reactor systems. Int J ChemTech Res 8(11):654–668
60. Vaidyanathan S (2015) Adaptive control of a chemical chaotic reactor. Int J PharmTech Res 8(3):377–382

61. Vaidyanathan S (2015) Adaptive control of the FitzHugh-Nagumo chaotic neuron model. Int J PharmTech Res 8(6):117–127
62. Vaidyanathan S (2015) Adaptive synchronization of chemical chaotic reactors. Int J ChemTech Res 8(2):612–621
63. Vaidyanathan S (2015) Adaptive synchronization of generalized Lotka-Volterra three-species biological systems. Int J PharmTech Res 8(5):928–937
64. Vaidyanathan S (2015) Adaptive synchronization of novel 3-D chemical chaotic reactor systems. Int J ChemTech Res 8(7):159–171
65. Vaidyanathan S (2015) Analysis, properties and control of an eight-term 3-D chaotic system with an exponential nonlinearity. Int J Model Identif Control 23(2):164–172
66. Vaidyanathan S (2015) Anti-synchronization of brusselator chemical reaction systems via adaptive control. Int J ChemTech Res 8(6):759–768
67. Vaidyanathan S (2015) Anti-synchronization of brusselator chemical reaction systems via integral sliding mode control. Int J ChemTech Res 8(11):700–713
68. Vaidyanathan S (2015) Anti-synchronization of chemical chaotic reactors via adaptive control method. Int J ChemTech Res 8(8):73–85
69. Vaidyanathan S (2015) Anti-synchronization of the FitzHugh-Nagumo chaotic neuron models via adaptive control method. Int J PharmTech Res 8(7):71–83
70. Vaidyanathan S (2015) Chaos in neurons and synchronization of Birkhoff-Shaw strange chaotic attractors via adaptive control. Int J PharmTech Res 8(6):1–11
71. Vaidyanathan S (2015) Dynamics and control of brusselator chemical reaction. Int J ChemTech Res 8(6):740–749
72. Vaidyanathan S (2015) Dynamics and control of Tokamak system with symmetric and magnetically confined plasma. Int J ChemTech Res 8(6):795–802
73. Vaidyanathan S (2015) Global chaos synchronization of chemical chaotic reactors via novel sliding mode control method. Int J ChemTech Res 8(7):209–221
74. Vaidyanathan S (2015) Global chaos synchronization of Duffing double-well chaotic oscillators via integral sliding mode control. Int J ChemTech Res 8(11):141–151
75. Vaidyanathan S (2015) Global chaos synchronization of the forced Van der Pol chaotic oscillators via adaptive control method. Int J PharmTech Res 8(6):156–166
76. Vaidyanathan S (2015) Hybrid chaos synchronization of the FitzHugh-Nagumo chaotic neuron models via adaptive control method. Int J PharmTech Res 8(8):48–60
77. Vaidyanathan S (2015) Integral sliding mode control design for the global chaos synchronization of identical novel chemical chaotic reactor systems. Int J ChemTech Res 8(11):684–699
78. Vaidyanathan S (2015) Output regulation of the forced Van der Pol chaotic oscillator via adaptive control method. Int J PharmTech Res 8(6):106–116
79. Vaidyanathan S (2015) Sliding controller design for the global chaos synchronization of forced Van der Pol chaotic oscillators. Int J PharmTech Res 8(7):100–111
80. Vaidyanathan S (2015) Synchronization of Tokamak systems with symmetric and magnetically confined plasma via adaptive control. Int J ChemTech Res 8(6):818–827
81. Vaidyanathan S (2016) A novel 3-D conservative chaotic system with a sinusoidal nonlinearity and its adaptive control. Int J Control Theory Appl 9(1):115–132
82. Vaidyanathan S (2016) A novel 3-D jerk chaotic system with three quadratic nonlinearities and its adaptive control. Arch Control Sci 26(1):19–47
83. Vaidyanathan S (2016) A novel 3-D jerk chaotic system with two quadratic nonlinearities and its adaptive backstepping control. Int J Control Theory Appl 9(1):199–219
84. Vaidyanathan S (2016) Anti-synchronization of 3-cells cellular neural network attractors via integral sliding mode control. Int J PharmTech Res 9(1):193–205
85. Vaidyanathan S (2016) Global chaos control of the FitzHugh-Nagumo chaotic neuron model via integral sliding mode control. Int J PharmTech Res 9(4):413–425
86. Vaidyanathan S (2016) Global chaos control of the generalized Lotka-Volterra three-species system via integral sliding mode control. Int J PharmTech Res 9(4):399–412
87. Vaidyanathan S (2016) Global chaos regulation of a symmetric nonlinear gyro system via integral sliding mode control. Int J ChemTech Res 9(5):462–469

88. Vaidyanathan S (2016) Hybrid synchronization of the generalized Lotka-Volterra three-species biological systems via adaptive control. Int J PharmTech Res 9(1):179–192
89. Vaidyanathan S (2016) Mathematical analysis, adaptive Control and synchronization of a ten-term novel three-scroll chaotic system with four quadratic nonlinearities. Int J Control Theory Appl 9(1):1–20
90. Vaidyanathan S, Azar AT (2015) Analysis, control and synchronization of a nine-term 3-D novel chaotic system. In: Azar AT, Vaidyanathan S (eds) Chaos modelling and control systems design, studies in computational intelligence, vol 581. Springer, Germany, pp 19–38
91. Vaidyanathan S, Azar AT (2016) Takagi-Sugeno fuzzy logic controller for Liu-Chen four-scroll chaotic system. Int J Intell Eng Inform 4(2):135–150
92. Vaidyanathan S, Boulkroune A (2016) A novel 4-D hyperchaotic chemical reactor system and its adaptive control. In: Vaidyanathan S, Volos C (eds) Advances and applications in chaotic systems. Springer, Berlin, Germany, pp 447–469
93. Vaidyanathan S, Madhavan K (2013) Analysis, adaptive control and synchronization of a seven-term novel 3-D chaotic system. Int J Control Theory Appl 6(2):121–137
94. Vaidyanathan S, Pakiriswamy S (2015) A 3-D novel conservative chaotic system and its generalized projective synchronization via adaptive control. J Eng Sci Technol Rev 8(2):52–60
95. Vaidyanathan S, Pakiriswamy S (2016) A five-term 3-D novel conservative chaotic system and its generalized projective synchronization via adaptive control method. Int J Control Theory Appl 9(1):61–78
96. Vaidyanathan S, Pakiriswamy S (2016) Adaptive control and synchronization design of a seven-term novel chaotic system with a quartic nonlinearity. Int J Control Theory Appl 9(1):237–256
97. Vaidyanathan S, Rajagopal K (2016) Adaptive control, synchronization and LabVIEW implementation of Rucklidge chaotic system for nonlinear double convection. Int J Control Theory Appl 9(1):175–197
98. Vaidyanathan S, Rajagopal K (2016) Analysis, control, synchronization and LabVIEW implementation of a seven-term novel chaotic system. Int J Control Theory Appl 9(1):151–174
99. Vaidyanathan S, Sampath S (2011) Global chaos synchronization of hyperchaotic Lorenz systems by sliding mode control. Commun Comput Inf Sci 205:156–164
100. Vaidyanathan S, Volos C (2015) Analysis and adaptive control of a novel 3-D conservative no-equilibrium chaotic system. Arch Control Sci 25(3):333–353
101. Vaidyanathan S, Volos C (2016) Advances and applications in chaotic systems. Springer, Berlin, Germany
102. Vaidyanathan S, Volos C (2016) Advances and applications in nonlinear control systems. Springer, Berlin, Germany
103. Vaidyanathan S, Volos C (2017) Advances in memristors memristive devices and systems. Springer, Berlin, Germany
104. Vaidyanathan S, Volos C, Pham VT, Madhavan K, Idowu BA (2014) Adaptive backstepping control, synchronization and circuit simulation of a 3-D novel jerk chaotic system with two hyperbolic sinusoidal nonlinearities. Arch Control Sci 24(3):375–403
105. Vaidyanathan S, Rajagopal K, Volos CK, Kyprianidis IM, Stouboulos IN (2015) Analysis, adaptive control and synchronization of a seven-term novel 3-D chaotic system with three quadratic nonlinearities and its digital implementation in LabVIEW. J Eng Sci Technol Rev 8(2):130–141
106. Vaidyanathan S, Volos CK, Kyprianidis IM, Stouboulos IN, Pham VT (2015) Analysis, adaptive control and anti-synchronization of a six-term novel jerk chaotic system with two exponential nonlinearities and its circuit simulation. J Eng Sci Technol Rev 8(2):24–36
107. Vaidyanathan S, Volos CK, Pham VT (2015) Analysis, adaptive control and adaptive synchronization of a nine-term novel 3-D chaotic system with four quadratic nonlinearities and its circuit simulation. J Eng Sci Technol Rev 8(2):181–191
108. Vaidyanathan S, Volos CK, Pham VT (2015) Global chaos control of a novel nine-term chaotic system via sliding mode control. In: Azar AT, Zhu Q (eds) Advances and applications in sliding mode control systems, studies in computational intelligence, vol 576. Springer, Germany, pp 571–590

109. Volos CK, Kyprianidis IM, Stouboulos IN, Tlelo-Cuautle E, Vaidyanathan S (2015) Memristor: a new concept in synchronization of coupled neuromorphic circuits. J Eng Sci Technol Rev 8(2):157–173
110. Volos CK, Pham VT, Vaidyanathan S, Kyprianidis IM, Stouboulos IN (2015) Synchronization phenomena in coupled Colpitts circuits. J Eng Sci Technol Rev 8(2):142–151
111. Volos CK, Pham VT, Vaidyanathan S, Kyprianidis IM, Stouboulos IN (2016) Synchronization phenomena in coupled hyperchaotic oscillators with hidden attractors using a nonlinear open loop controller. In: Vaidyanathan S, Volos C (eds) Advances and applications in chaotic systems. Springer, Berlin, Germany, pp 1–38
112. Volos CK, Pham VT, Vaidyanathan S, Kyprianidis IM, Stouboulos IN (2016) The case of bidirectionally coupled nonlinear circuits via a memristor. In: Vaidyanathan S, Volos C (eds) Advances and applications in nonlinear control systems. Springer, Berlin, Germany, pp 317–350
113. Volos CK, Prousalis D, Kyprianidis IM, Stouboulos I, Vaidyanathan S, Pham VT (2016) Synchronization and anti-synchronization of coupled Hindmarsh-Rose neuron models. Int J Control Theory Appl 9(1):101–114
114. Volos CK, Vaidyanathan S, Pham VT, Maaita JO, Giakoumis A, Kyprianidis IM, Stouboulos IN (2016) A novel design approach of a nonlinear resistor based on a memristor emulator. In: Azar AT, Vaidyanathan S (eds) Advances in chaos theory and intelligent control. Springer, Berlin, Germany, pp 3–34
115. Wang B, Zhong SM, Dong XC (2016) On the novel chaotic secure communication scheme design. Commun Nonlinear Sci Numer Simul 39:108–117
116. Wolf A, Swift JB, Swinney HL, Vastano JA (1985) Determining Lyapunov exponents from a time series. Phys D 16:285–317
117. Wu T, Sun W, Zhang X, Zhang S (2016) Concealment of time delay signature of chaotic output in a slave semiconductor laser with chaos laser injection. Opt Commun 381:174–179
118. Xu G, Liu F, Xiu C, Sun L, Liu C (2016) Optimization of hysteretic chaotic neural network based on fuzzy sliding mode control. Neurocomputing 189:72–79
119. Xu H, Tong X, Meng X (2016) An efficient chaos pseudo-random number generator applied to video encryption. Optik 127(20):9305–9319
120. Zhou W, Xu Y, Lu H, Pan L (2008) On dynamics analysis of a new chaotic attractor. Phys Lett A 372(36):5773–5777
121. Zhu C, Liu Y, Guo Y (2010) Theoretic and numerical study of a new chaotic system. Intell Inf Manag 2:104–109

Sliding Mode Control Design for a Sensorless Sun Tracker

Ahmed Rhif and Sundarapandian Vaidyanathan

Abstract The photovoltaic sun tracker is generally used to increase the electrical energy production. The sun tracker considered in this study has two degrees of freedom (2-DOF) and characterized by the lack of sensors. In this way, the tracker will have as set point the sun position at every second during the day for a specific period of five years. After sunset, the tracker goes back to the initial position automatically, which is sunrise. Sliding mode control is an important method used to solve various problems in control systems engineering. In robust control systems, the sliding mode control is often adopted due to its inherent advantages of easy realization, fast response and good transient performance as well as insensitivity to parameter uncertainties and disturbance. In this work, we apply sliding mode control (SMC) to ensure the tracking mechanism. Also, we design a sliding mode observer to replace the velocity sensor which is affected from a lot of measurement disturbances. The control results are established using Lyapunov stability theory. Numerical simulations are shown to illustrate all the main results derived in this work. Experimental measurements show that this automatic dual-axis sun tracker increases the power production by over 40%.

Keywords Sun tracker · Solar energy · Photovoltaic panel · DC motor · Azimuth · Altitude · Sliding mode control · Sliding mode observer

A. Rhif
Laboratory of Advanced Systems, Polytechnic School of Tunisia,
University of Carthage, 1054 Carthage, Tunisia
e-mail: ahmed.rhif@gmail.com

S. Vaidyanathan (✉)
Research and Development Centre, Vel Tech University, Avadi, Chennai
600062, Tamil Nadu, India
e-mail: sundarcontrol@gmail.com

© Springer International Publishing AG 2017
S. Vaidyanathan and C.-H. Lien (eds.), *Applications of Sliding Mode Control
in Science and Engineering*, Studies in Computational Intelligence 709,
DOI 10.1007/978-3-319-55598-0_18

419

1   Introduction

The first automatic solar tracking system was presented by McFee [13]. In this work, McFee had developed an algorithm to compute the flux density distribution and the total received power in a central receiver solar power. In [23], Semma and Imamura used a microprocessor to adjust the solar collectors' positions in a photovoltaic concentrator such that they are pointed toward the sun at every moment.

Many techniques have been developed in the solar energy literature on effectively capturing the solar energy [11]. The use of the artificial solar energy is currently processed in three different processes:

1. Photothermal conversion that converts the radiation to usable heat (heat engines, water heaters, etc.)
2. Photovoltaic conversion that converts radiation into electrical current
3. Photochemical conversion that allows the store of thermal energy by the deformation of molecules.

For each of these three sectors, solar energy is received by sensors which are perpendicular to the solar radiation that will increase the amount of the collected energy collected during the day. It has been shown experimentally that if we keep the sensor perpendicular to the solar radiation, then the flat plate photovoltaic panel has an average gain of 20% compared with a horizontal sensor.

Sensors that are more sophisticated than flat plate collectors have been developed such as the concentration sensors that are useful in the thermal photovoltaic systems [45].

A sun tracker is a process which orients various payloads toward the sun such as photovoltaic panels, reflectors, lenses or other optical devices. The sun trackers of photovoltaic panels are used to minimize the incidence angle between the photovoltaic cell and the incoming light the fact that increases the amount of the produced energy by about 40%. In photovoltaic plants, it is estimated that solar trackers take places in at least 85% of new commercial installations, in 2009–2011, when production is greater than 1 MW.

There are different types of trackers, viz. (1) passive tracker, (2) chronological tracker, and (3) active tracker. They are detailed as follows.

(1) The passive trackers use a boiling point from a compressed fluid that is driven to one side to other by the solar heat which creates a gas pressure that may cause the tracker movement Semma 1980. As this process presents a bad quality of orientation precision, it turns out to be unsuitable for certain types of photovoltaic collectors. The term passive tracker is used too for photovoltaic panels that include a hologram behind stripes of photovoltaic cells. In this way, sunlight reflects on the hologram which allows the cell heat from behind, thereby increasing the modules' efficiency [13]. Moreover, the plant does not have to move while the hologram still reflects sunlight from the needed angle towards the photovoltaic cells.

Fig. 1 Two axis solar tracker

(2) The chronological tracker counteracts the rotation of the earth by turning in the opposite direction at an equal rate as the earth [45]. Actually, the rates are not exactly equal because the position of the sun changes in relation to the earth by 360° every year. A chronological tracker is a very simple and accurate solar tracker specifically used with polar mount. The process control is ensured by a motor that rotates at a very slow rate (about 15°/h).

(3) The active trackers use two motors (Fig. 1) and a gear trains to drive the tracker by a controller matched to the solar direction [22]. In fact, two axis trackers are used to orient movable mirrors of heliostats that reflect the sunlight toward the absorber of a power station central [10]. As each mirror has an individual orientation, those plants are controlled through a central computer system which also allows the system, to be shut down when necessary.

The photovoltaic sun tracker is generally used to increase the electrical energy production. The sun tracker considered in this study has two degrees of freedom (2-DOF) and characterized by the lack of sensors. In this way, the tracker will have as set point the sun position at every second during the day for a specific period of five years. After sunset, the tracker goes back to the initial position automatically, which is sunrise.

Sliding mode control is an important method used to solve various problems in control systems engineering [2–4, 42–44]. A major advantage of sliding mode control is low sensitivity to parameter variations in the plant and disturbances affecting the plant, which eliminates the necessity of exact modeling of the plant [2–4, 42–44].

In the sliding mode control, the control dynamics will have two sequential modes, viz. the reaching mode and the sliding mode. Basically, a sliding mode controller design consists of two parts: hyperplane design and controller design. A hyperplane is first designed via the pole-placement approach and a controller is then designed based on the sliding condition. The stability of the overall system is guaranteed by the sliding condition and by a stable hyperplane. Sliding mode control method is a popular method for the control and synchronization of chaotic systems [12, 14, 31, 35–41].

In this work, we apply sliding mode control (SMC) to ensure the tracking mechanism. Also, we design a sliding mode observer to replace the velocity sensor which is affected from a lot of measurement disturbances. The control results are established using Lyapunov stability theory. Numerical simulations are shown to illustrate all the main results derived in this work.

2 Dual Axis Sun Tracker (DAST)

In areas where sunlight is weak or in those areas where subsidies are not very expanded, the initial cost of solar panels can be extremely high. A solution to improve the energy production is given by the solar tracker or the sun tracker. The sun tracker allows placing the panel in relation to the best position of the sun (orthogonal to radiation, if possible). Indeed, the position of the sun varies constantly, both during the day and during different times of the year. The perfect solution is to use a solar tracker of six axes to track the exactly sun movements. This system requires a minimum of clearance to allow free movement in all directions. It is therefore incompatible with an integrated roof, except perhaps for systems with two axes.

In 1992, Agarwal presented a two axis tracking process consisted of a worm gear drives and four bar type kinematic linkages that make easy the focusing operation of the reflectors in a solar concentrator system [1]. A Dual Axis Sun Tracker (Fig. 2) has two degrees of freedom that act as axes of rotation. It has a vertical axis (primary axis) perpendicular to the ground and a horizontal one typically normal to the primary axis. The first axis is a vertical pivot shaft that allows the device to move to a compass point. The second one is a horizontal elevation pivot implanted upon the platform. To adjust for sun azimuth, azimuth motor drives the vertical axis that makes the panel rotates [10, 23]. To adjust for the sun altitude, a second motor will operate for the panel elevation acting on the horizontal axis.

Using the combinations between the two axes (azimuth and altitude), any location in the upward hemisphere could be pointed. Such systems should be operated under a computer control or a microcontroller according to the expected solar orientation. Also, it may use a tracking sensor to control motors that orient the panels toward the sun. This type of plant is also used to orient the parabolic reflectors that mount a sterling engine to produce electricity at the device. In order to control the movement of these plants, special drives are designed and rigorously tested. For that, many considerations during cloudy periods must be expected to keep the tracker out from wasting energy.

Fig. 2 The DAST system orientation

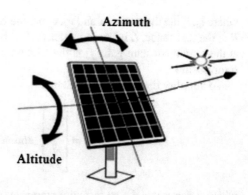

Fig. 3 The electrical model in the $(d - q)$ referential system

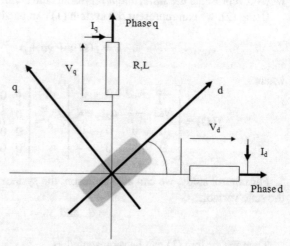

In fact after tests, a solar panel of $3\,\text{m}^2$ fixed to a surface produces about 5 kWh of electricity per day. The same installation, but equipped with a tracker, can provide up to 8 kWh per day. In conclusion, this device allows the increase of the produced energy amount by about 40% in relation to the fixed panels. So, the tracker presents a very good tool for energy production optimization.

Figure 3 describes the electrical model of the process in the $(d - q)$ referential system.

The mathematical model of the process in the $(d - q)$ referential system in Fig. 3 is described as follows.

$$\begin{cases} \frac{di_d}{dt} = \frac{1}{L}(v_d - Ri_d + NL\Omega i_q) \\ \frac{di_q}{dt} = \frac{1}{L}(v_q - Ri_q + NL\Omega i_d - K\Omega) \\ \frac{d\Omega}{dt} = \frac{1}{J}(Ki_q - f_v\Omega - C) \\ \frac{d\theta}{dt} = \Omega \end{cases} \quad (1)$$

where i_d, i_q are the currents, and v_d, v_q are the voltages in the $(d - q)$ referential. Also, R is the resistance, L is the inductance, J is the inertia, N is the number of spins, K is the torque constant gain, f_v is the friction, Ω is the angular velocity and θ is the rotor position.

To simplify the notation, we define

$$x = \begin{bmatrix} i_d \\ i_q \\ \theta \\ \Omega \end{bmatrix} \quad \text{and} \quad u = \begin{bmatrix} v_d \\ v_q \end{bmatrix} \tag{2}$$

where x represents the *state*, and u represents the *control*.

Using (2), we can represent the system (1) compactly as

$$\dot{x} = f(\Omega)x + gu + p \tag{3}$$

where

$$f(\Omega) = \begin{bmatrix} -\frac{R}{L} & N\Omega & 0 & 0 \\ N\Omega & -\frac{R}{L} & a_{23} & -K \\ 0 & 0 & 0 & 1 \\ 0 & \frac{K}{J} & 0 & -\frac{f_v}{J} \end{bmatrix}, \quad g = \begin{bmatrix} \frac{1}{L} & 0 \\ 0 & \frac{1}{L} \\ 0 & 0 \\ 0 & 0 \end{bmatrix}, \quad p = \begin{bmatrix} 0 \\ 0 \\ 0 \\ -\frac{C}{J} \end{bmatrix} \tag{4}$$

For control study, we can also represent the system (1) in another way by using the state variables

$$y_1 = \theta \quad \text{and} \quad y_2 = i_d \tag{5}$$

Then the system (1) can be represented as

$$\begin{cases} \theta = y_1 \\ \Omega = \dot{y}_1 \\ i_d = y_2 \\ i_q = \frac{1}{K}(J\ddot{y}_1 + f_v \dot{y}_1 + C) \\ v_d = L\dot{y}_2 + Ry_2 - \frac{NL}{K}\dot{y}_1(J\ddot{y}_1 + f_v\dot{y}_1 + C) \\ v_q = \frac{JL}{K}\dddot{y}_1 + \frac{1}{K}(Lf_v + RJ)\ddot{y}_1 + \left[\frac{Rf_v}{K} + K - NLy_2\right]\dot{y}_1 + \frac{R}{K}C \end{cases} \tag{6}$$

3 Sliding Mode Controller Design

Many types of controllers have been designed and implemented in the sun tracker. A comparative study of Maximum Power Point Tracking (MPPT) techniques for photovoltaic systems is given in [32].

Enslin and Snyman [7] have proposed a simplified control technique based on the positive feedback effect of the output current regarding low-power MPPT converters.

Sullivan and Powers [25] presented a MPPT technique considering a PV system used to supply electric racing cars. The experimental vehicle has six photovoltaic modules divided in nine sections; each one is connected separately to a MPP tracker and the output links are parallel-connected in order to recharge the battery [25].

Hua and Shen [9] analyzed the performance of dc-dc converters using P & O in photovoltaic systems. This choice was justified by considering the reduced number of sensors and the global complexity of the photovoltaic systems.

The appearance of the sliding mode approach occurred in Soviet Union in the 1960s with the discovery of the discontinuous control and its effect on the system dynamics. The sliding mode control approach emerged in the applications with Variable System Structure (VSS) techniques [24, 33, 34]. The sliding mode control is a popular method in control literature due its robustness against the disturbances and uncertainties in modelling [15–21].

The principle of the sliding mode control is to force the system to converge towards a selected surface called the *sliding surface* and then to evolve there in spite of uncertainties and disturbances.

The surface is defined by a set of relations between the state variables of the system. The synthesis of a control law by sliding mode includes two phases:

(1) The sliding surface is defined according to the control objectives and to the desired performances in closed loop.
(2) The synthesis of the discontinuous control is carried out in order to force the trajectories of the system state to reach the sliding surface, and then, to evolve in spite of uncertainties and parametric variations.

The tracking problem of the photovoltaic panel is treated by using sliding mode control with the nonlinear sliding surface as shown below.

$$s(t) = k_1 e(t) + k_2 \dot{e}(t) \tag{7}$$

where $e(t)$ is the system error and k_1, k_2 are positive constants.

For the second-order sliding mode control, the invariance condition is given by

$$s(x) = 0 \text{ and } \dot{s}(x) = 0 \tag{8}$$

To ensure that the state converges to the sliding surface, we have to verify the Lyapunov stability criterion [24] given as follows:

$$s\dot{s} \leq -\eta|s| \tag{9}$$

where $\eta > 0$.

The control law can be then defined as follows.

$$u = u_0 + u_1 \tag{10}$$

where u_0 is the nominal control and u_1 is the discontinuous control allowing to reject the disturbances.

To deal with the path tracking, we have to first stabilize the tracking error to the origin.

We suppose that i_{dr} and i_{qr} are the reference currents in the $(d - q)$ referential.

We also suppose that v_{dr} and v_{qr} are the reference voltages in the $(d - q)$ referential.

Let θ_r denote the reference rotor position and let Ω_r denote the reference angular velocity.

We define the tracking error as

$$\begin{cases} e_1(t) = i_d(t) - i_{dr} \\ e_2(t) = i_q(t) - i_{qr} \\ e_3(t) = \Omega(t) - \Omega_r \\ e_4(t) = \theta(t) - \theta_r \end{cases} \tag{11}$$

Differentiating the tracking errors along (1), we obtain

$$\begin{cases} \dot{e}_1 = \frac{1}{L}[v_d - v_{dr} - Re_1 + NL(e_3 e_2 + e_3 i_{qr} + e_2 \Omega_r)] \\ \dot{e}_2 = \frac{1}{L}[v_q - v_{qr} - Re_2 + NL(e_3 e_1 + e_3 i_{dr} + e_1 \Omega_r) - Ke_3] \\ \dot{e}_3 = \frac{1}{J}[Ke_2 - f_v e_3 - C_r] \\ \dot{e}_4 = e_3 \end{cases} \tag{12}$$

Now, we start to study the velocity control by a second order sliding mode control.

The sliding surface used to ensure the existence of this approach is written as follows.

$$s_\Omega = \mu e_3 + \dot{e}_3 \tag{13}$$

where $\mu > 0$ is a constant.

The derivative of Eq. (13) gives

$$\dot{s}_\Omega = \mu \dot{e}_3 + \ddot{e}_3 \tag{14}$$

i.e.

$$\dot{s}_\Omega = \frac{\mu}{J}(Ke_2 - f_v e_3 - C_r) + \frac{1}{J}(K\dot{e}_2 - f_v \dot{e}_3 - \dot{C}_r) \tag{15}$$

In the second phase, we have to deal with the position control process.

In this way, we consider the sliding surface

$$s_\theta = \mu_1 e_4 + \mu_2 \dot{e}_4 + \ddot{e}_4 \tag{16}$$

Thus, we find that

$$s_\theta = \mu_1 e_4 + \mu_2 e_3 + \frac{1}{J}(Ke_2 - f_v e_3 - C_r) \tag{17}$$

where $\mu_1, \mu_2 > 0$.

Finally, we define the sliding controls as follows.

$$u_{s1} = -U_0 \operatorname{sgn}(s_\Omega) \tag{18}$$

$$u_{s2} = -U_0 \operatorname{sgn}(s_\theta) = \dot{s}_\theta \tag{19}$$

where $U_0 > 0$.

4 The Sliding Mode Observer Design

The angular velocities of the stepper motors measurements are a hard task and could give unreliable results because of the disturbances that influence the tachometry sensor. In this way, it may be adequate to estimate the angular velocity using an observer instead of the sensor. This operation would give better results and will reduce the number of the sensors which are very costly and may have hard maintenance skills.

The observer can reconstruct the state of a system from the measurement of inputs and outputs [26–30]. The observer is used when all or part of the state vector cannot be measured. It allows the estimation of unknown parameters or variables of a system. This observer can be used to reconstruct the speed of an electric motor, for example, from the electromagnetic torque. It also allows reconstructing the flow of the machine, etc.

Sliding mode control can be used in the state observers design. These nonlinear high gain observers have the ability to bring coordinates of the estimator error dynamics to zero in finite time. Additionally, sliding mode observers (Fig. 4) have

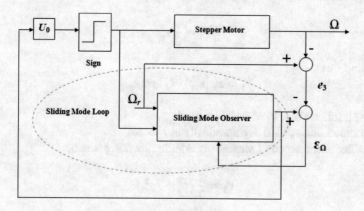

Fig. 4 Velocity control based on a sliding mode observer

an attractive measurement noise resilience which is similar to a Kalman filter. The chattering concerned with the sliding mode method can be eliminated by modifying the observer gain, without sacrificing the control robustness and precision qualities [5, 6, 8].

Let $\hat{\theta}(t)$ and $\hat{\Omega}(t)$ denote the estimates for position and angular velocity respectively.

Then the estimation errors for the position and angular velocity are defined as follows.

$$\begin{cases} \varepsilon_\theta = \theta - \hat{\theta} \\ \varepsilon_\Omega = \Omega - \hat{\Omega} \end{cases} \tag{20}$$

We note that the angular velocity Ω is the derivative of the position θ.
Thus, it follows that

$$\dot{\varepsilon}_\theta = \frac{d\theta}{dt} - \frac{d\hat{\theta}}{dt} = \Omega - \hat{\Omega} = \varepsilon_\Omega \tag{21}$$

In this case, we can estimate the system (1) as follows.

$$\begin{cases} \frac{d\hat{\Omega}}{dt} = \frac{K}{J}i_q - \frac{f_v}{J}\hat{\Omega} - \frac{C_r}{J} + \lambda_2 \,\mathrm{sgn}(\Omega - \hat{\Omega}(t)) \\ \frac{d\hat{\theta}}{dt} = \hat{\Omega}(t) + \lambda_1 \,\mathrm{sgn}(\theta - \hat{\theta}(t)) \end{cases} \tag{22}$$

where $\lambda_1, \lambda_2 > 0$.
Also, we have

$$\begin{cases} \dot{\varepsilon}_\theta = \varepsilon_\Omega - \lambda_1 \,\mathrm{sgn}(\varepsilon_\theta) \\ \dot{\varepsilon}_\Omega = \frac{d\Omega}{dt} - \frac{d\hat{\Omega}}{dt} = -\frac{f_v}{J}\varepsilon_\Omega \end{cases} \tag{23}$$

To make certain that the state converges to the sliding surface, we have to verify the Lyapunov stability criterion given by the Lyapunov quadratic function

$$V_1 = \frac{1}{2}\varepsilon_\theta^2 \tag{24}$$

We find that

$$\dot{V}_1 = \varepsilon_\theta \left[\varepsilon_\Omega - \lambda_1 \,\mathrm{sgn}(\varepsilon_\theta) \right] \tag{25}$$

Thus, it follows that $\dot{V}_1 < 0$ if $\lambda_1 > |\varepsilon_\Omega|_{max}$.
This shows that $\varepsilon_\theta \to 0$ asymptotically as $t \to \infty$.
Consider now a second Lyapunov quadratic function given by

$$V_2 = \frac{1}{2}\left(\varepsilon_\theta^2 + J\varepsilon_\Omega^2\right) \tag{26}$$

As $\varepsilon_\theta \to 0$ asymptotically, for large values of time, we have

Table 1 Experimental parameters of the stepper motor

Element	Definition	Numerical value
J	The inertia moment	$3.0145 \cdot 10^{-4}$ Kg.m^2
C	Mathematical torque	0.780 Nm
K	Torque constant	0.433 Nm/A
R	Armature resistance	3.15 Ω
L	Armature inductance	8.15 mH
f_v	The friction force	0.0172 Nms/rad

Fig. 5 Inclination angle of the azimuth motor

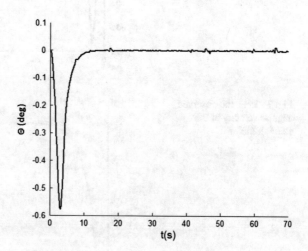

$$\dot{V}_2 = J\varepsilon_\Omega \dot{\varepsilon}_\Omega \tag{27}$$

or

$$\dot{V}_2 = -f_v \varepsilon_\Omega^2 - \varepsilon_\Omega \left[\lambda_2 \, \text{sgn}(\varepsilon_\Omega) - C_r\right] \tag{28}$$

Thus, it follows that $\dot{V}_2 < 0$ if $\lambda_2 > |C_r|_{\max}$.
This shows that $\varepsilon_\Omega \to 0$ asymptotically as $t \to \infty$.

5 Experimental Results and Discussions

Experimental results are accomplished for a second order sliding mode control of the system (2) (shown in Fig. 2) connected to a computer through a serial cable RS232.

In this experiment, we need to study the system control evolution, the real and observed angular velocities Ω_1 and Ω_2, the azimuth and the altitude inclinations θ_1 and θ_2.

Fig. 6 Inclination angle of
the altitude motor

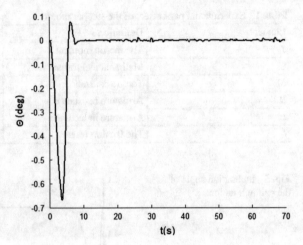

Fig. 7 Real and observed
angular velocity of the
azimuth motor

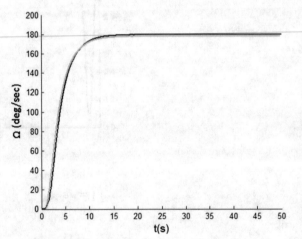

The sliding surface parameters used for this experiment are $\mu = 0.135$, $\mu_1 = 1.2$ and $\mu_2 = 0.355$.

Table 1 shows the characteristics of the stepper motor.

The experimental test is considered in a period of time which does not exceed 100 s.

Figures 5 and 6 show that using the second order sliding mode control, we can reach the steady state in a short time, viz. 10 s. Also, for the first test, we have $\theta_1 = 0.58°$, $\theta_2 = 0.68°$.

Thus, we notice that the azimuth inclination θ_1 is very close to the altitude inclination θ_2. This fact is related to the elliptic movement nature of the sun. The real and observed angular velocities shown in Figs. 7 and 8 of the two axis tracker are also very close: $\Omega_1 = 180°/s$ and $\Omega_2 = 160°/s$ (Fig. 9)

Fig. 8 Real and observed angular velocity of the altitude motor

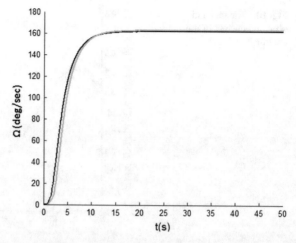

Fig. 9 System control evolutions

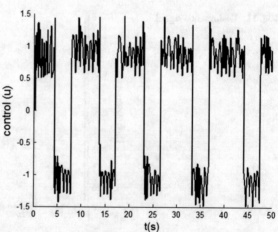

This interesting result is due to the high estimation quality of the sliding mode observer.

Moreover, the SMC shows its robustness in Figs. 4 and 5, when we inject a periodic disturbance signal (Fig. 10) where the signal period $T = 15$ s.

We see that the system deviates a little from the reference trajectory and then come back to the equilibrium position.

On the other hand, the control evolution shown in Fig. 9, with no high level and no sharp commutation frequency, can give good operating conditions for the actuators (both azimuth and altitude motors) (Fig. 11).

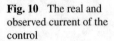

Fig. 10 The real and observed current of the control

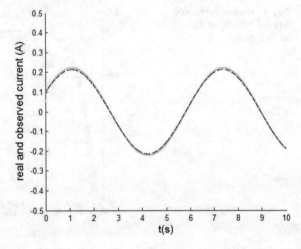

Fig. 11 Disturbance signal

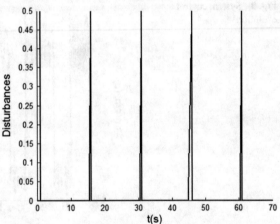

6 Conclusions

In this work, we gave a review of the literature of tracking process for the dual axis sun tracker by a sliding mode control law. In this way, different controllers have been presented. The synthesis of a control law by second order sliding mode using a nonlinear sliding surface has been treated. Then a sliding mode observer has been considerate to evaluate the angular velocities of the stepper motors. Experimental results show the effectiveness of the sliding mode control in the tracking process, its robustness and the high estimation quality of the sliding mode observer.

References

1. Agarwal AK (1992) Two axis tracking system for solar concentrators. Renew Energy 2:181–182
2. Azar AT, Vaidyanathan S (2015) Chaos modeling and control systems design. Springer, Berlin, Germany
3. Azar AT, Vaidyanathan S (2016) Advances in Chaos theory and intelligent control. Springer, Berlin, Germany
4. Azar AT, Vaidyanathan S (2017) Fractional order control and synchronization of chaotic systems. Springer, Berlin, Germany
5. Braiek EB, Rotella F (1993) Design of observers for nonlinear time variant systems. In: Proceedings of the IEEE international conference on systems, man and cybernetics. doi:10.1109/ICSMC.1993.390711
6. Braiek EB, Rotella F (1994) State observer design for analytical non-linear systems. In: Proceedings of the IEEE international conference on systems, man and cybernetics. doi:10.1109/ICSMC.1994.400164
7. Enslin JHR, Snyman DB (1992) Simplified feed-forward control of the maximum power point in PV installations. In: Proceedings of the international conference on industrial electronics, control, instrumentation, and automation, pp 548–553
8. Flota M, Alvarez-Salas R, Miranda H, Cabal-Yepez E, Romero-Troncoso RJ (2011) Nonlinear observer-based control for an active rectifier. J Sci Ind Res 70(12):1017–1025
9. Hua C, Shen C (1997) Control of DC/DC converters for solar energy system with maximum power tracking. In: Proceedings of the 23rd annual international conference on industrial electronics, control, and instrumentation, pp 827–832
10. Hua C, Shen C (1998) Comparative study of peak power tracking techniques for solar storage system. In: Proceedings of the 13th annual applied power electronics conference and exposition. doi:10.1109/APEC.1998.653972
11. Isabella O, Jäger K, Smets A, van Swaaij R, Zeman M (2016) Solar energy: the physics and engineering of photovoltaic conversion. Technologies and Systems. UIT Cambridge Ltd., Cambridge, U.K
12. Lakhekar GV, Waghmare LM, Vaidyanathan S (2016) Diving autopilot design for underwater vehicles using an adaptive neuro-fuzzy sliding mode controller. In: Vaidyanathan S, Volos C (eds) Advances and applications in nonlinear control systems. Springer, Berlin, Germany, pp 477–503
13. McFee RH (1975) Power collection reduction by mirror surface nonflatness and tracking error for a central receiver solar power system. Appl Opt 14:1493–1502
14. Moussaoui S, Boulkroune A, Vaidyanathan S (2016) Fuzzy adaptive sliding-mode control scheme for uncertain underactuated systems. In: Vaidyanathan S, Volos C (eds) Advances and applications in nonlinear control systems. Springer, Berlin, Germany, pp 351–367
15. Rhif A (2011a) A review note for position control of an autonomous underwater vehicle. IETE Tech Rev 28:486–492
16. Rhif A (2011b) Position control review for a photovoltaic system: dual axis sun tracker. IETE Tech Rev 28:479–485
17. Rhif A (2012a) A high order sliding mode control with PID sliding surface: simulation on a torpedo. Int J Inf Tech, Control Autom 2(1):1–13
18. Rhif A (2012b) Stabilizing sliding mode control design and application for a DC motor: speed control. Int J Instrum Control Syst 2(1):25–33
19. Rhif A (2014) Sliding mode-multimodel stabilising control using single and several sliding surfaces: simulation on an autonomous underwater vehicle. Int J Model, Ident Control 22(2):126–138
20. Rhif A, Kardous Z, Braiek NBH (2011) A high order sliding mode-multimodel control of nonlinear systems simulation on a submarine mobile. In: Proceedings of the eighth international multi-conference on systems, signals and devices (SSD-2011), pp 1–6

21. Rhif A, Kardous Z, Braiek NBH (2012) A high order sliding mode observer: torpedo guidance application. J Eng Tech 2(1):13–18
22. Roth P, Georgiev A, Boudinov H (2004) Design and construction of a system for sun-tracking. Renew Energy 29(3):393–402
23. Semma RP, Imamura MS (1980) Sun tracking controller for multi-kW photovoltaic concentrator system. In: Proceedings of the 3rd international photovoltaic solar energy conference, pp 27–31
24. Slotine J, Li W (1991) Applied nonlinear control. Prentice-Hall, Englewood Cliffs, NJ, USA
25. Sullivan CR, Powers MJ (1993) High-efficiency maximum power point tracker for photovoltaic arrays in a solar-powered race vehicle. In: Proceedings of the IEEE 24th annual power electronics specialist conference, pp 574–580
26. Sundarapandian V (2001a) Global observer design for nonlinear systems. Math Comput Model 35(1–2):45–54
27. Sundarapandian V (2001b) Local observer design for nonlinear systems. Math Comput Model 35(1–2):25–36
28. Sundarapandian V (2001c) Observer design for discrete-time nonlinear systems. Math Comput Model 35(1–2):37–44
29. Sundarapandian V (2002a) Nonlinear observer design for bifurcating systems. Math Comput Model 36(1–2):183–188
30. Sundarapandian V (2002b) Nonlinear observer design for discrete-time bifurcating systems. Math Comput Model 36(1–2):211–215
31. Sundarapandian V, Sivaperumal S (2011) Sliding controller design of hybrid synchronization of four-wing chaotic systems. Int J Soft Comput 6(5):224–231
32. Tofoli FL, de Castro Pereira D, de Paula WJ (2015) Comparative study of maximum power point tracking techniques for photovoltaic systems. Int J Photoenergy 2015. doi:10.1155/2015/812582
33. Utkin VI (1977) Variable structure systems with sliding modes. IEEE Trans Autom Control 22(2):212–222
34. Utkin VI (1993) Sliding mode control design principles and applications to electric drives. IEEE Trans Ind Electron 40(1):23–36
35. Vaidyanathan S (2011) Analysis and synchronization of the hyperchaotic Yujun systems via sliding mode control. Adv Intell Syst Comput 176:329–337
36. Vaidyanathan S (2012a) Global chaos control of hyperchaotic Liu system via sliding control method. Int J Control Theor Appl 5(2):117–123
37. Vaidyanathan S (2012b) Sliding mode control based global chaos control of Liu-Liu-Liu-Su chaotic system. Int J Control Theor Appl 5(1):15–20
38. Vaidyanathan S (2014) Global chaos synchronisation of identical Li-Wu chaotic systems via sliding mode control. Int J Model, Ident Control 22(2):170–177
39. Vaidyanathan S (2016a) Anti-synchronization of 3-cells cellular neural network attractors via integral sliding mode control. Int J PharmTech Res 9(1):193–205
40. Vaidyanathan S (2016b) Global chaos regulation of a symmetric nonlinear gyro system via integral sliding mode control. Int J ChemTech Res 9(5):462–469
41. Vaidyanathan S, Sampath S (2011) Global chaos synchronization of hyperchaotic Lorenz systems by sliding mode control. Commun Comput Inf Sci 205:156–164
42. Vaidyanathan S, Volos C (2016a) Advances and applications in chaotic systems. Springer, Berlin, Germany
43. Vaidyanathan S, Volos C (2016b) Advances and applications in nonlinear control systems. Springer, Berlin, Germany
44. Vaidyanathan S, Volos C (2017) Advances in memristors, memristive devices and systems, Springer, Berlin, Germany
45. Yousef HA (1999) Design and implementation of a fuzzy logic computer-controlled sun tracking system. In: Proceedings of the IEEE international symposium on industrial electronics. doi:10.1109/ISIE.1999.796768

Super-Twisting Sliding Mode Control of the Enzymes-Substrates Biological Chaotic System

Sundarapandian Vaidyanathan

Abstract Chaos has important applications in physics, chemistry, biology, ecology, secure communications, cryptosystems and many scientific branches. Control and synchronization of chaotic systems are important research problems in chaos theory. Sliding mode control is an important method used to solve various problems in control systems engineering. In robust control systems, the sliding mode control is often adopted due to its inherent advantages of easy realization, fast response and good transient performance as well as insensitivity to parameter uncertainties and disturbance. In this work, we first investigate research in the dynamic analysis and properties of the enzymes-substrates reactions system with ferroelectric behaviour in the brain waves which was studied by Enjieu Kadji et al. (Chaos Solitons Fractals 32:862–882, 2007, [10]). Next, we apply multivariable super-twisting sliding mode control to globally stabilize all the trajectories of the enzymes-substrates biological chaotic system. Furthermore, we use multivariable super-twisting sliding mode control for the global chaos synchronization of identical enzymes-substrates biological chaotic systems. Super-twisting sliding mode control is very useful for the global stabilization and synchronization of the enzymes-substrates reaction system as it achieves finite time stability for the system. Numerical simulations using MATLAB are shown to illustrate all the main results derived in this work.

Keywords Chaos · Chaotic systems · Biological systems · Chaos control · Sliding mode control · Super-twisting control

1 Introduction

Chaos theory describes the quantitative study of unstable aperiodic dynamic behavior in deterministic nonlinear dynamical systems. For the motion of a dynamical system to be chaotic, the system variables should contain some nonlinear terms and

S. Vaidyanathan (✉)
Research and Development Centre, Vel Tech University, Avadi,
Chennai 600062, Tamil Nadu, India
e-mail: sundarcontrol@gmail.com

© Springer International Publishing AG 2017
S. Vaidyanathan and C.-H. Lien (eds.), *Applications of Sliding Mode Control in Science and Engineering*, Studies in Computational Intelligence 709,
DOI 10.1007/978-3-319-55598-0_19

the system must satisfy three properties: boundedness, infinite recurrence and sensitive dependence on initial conditions [3–5, 104–106].

The problem of global control of a chaotic system is to device feedback control laws so that the closed-loop system is globally asymptotically stable. Global chaos control of various chaotic systems has been investigated via various methods in the control literature [3, 4, 104, 105].

The synchronization of chaotic systems deals with the problem of synchronizing the states of two chaotic systems called as *master* and *slave* systems asymptotically with time. The design goal of the complete synchronization problem is to use the output of the master system to control of the output of the slave system so that the outputs of the two systems are synchronized asymptotically with time.

Because of the *butterfly effect* [3], which causes the exponential divergence of the trajectories of two identical chaotic systems started with nearly the same initial conditions, synchronizing two chaotic systems is seemingly a challenging research problem in the chaos literature.

The synchronization of chaotic systems was first researched by Yamada and Fujisaka [15] with subsequent work by Pecora and Carroll [25, 26]. In the last few decades, several different methods have been devised for the synchronization of chaotic and hyperchaotic systems [3–5, 104–106].

Many new chaotic systems have been discovered in the recent years such as Zhou system [122], Zhu system [123], Li system [19], Sundarapandian systems [41, 42], Vaidyanathan systems [51–56, 58, 68, 84–86, 92, 93, 96–99, 101, 103, 107–111], Pehlivan system [27], Sampath system [38], Tacha system [45], Pham systems [29, 32, 33, 35], Akgul system [2], etc.

Chaos theory has applications in several fields such as memristors [28, 30, 31, 34, 36, 37, 112, 115, 117], fuzzy logic [6, 39, 94, 120], communication systems [11, 12, 118], cryptosystems [8, 9], electromechanical systems [13, 44], lasers [7, 20, 119], encryption [21, 22, 121], electrical circuits [1, 2, 16, 100, 113], chemical reactions [59, 60, 62, 63, 65, 67, 69–71, 74, 76, 80, 95], oscillators [77, 78, 81, 82, 114], tokamak systems [75, 83], neurology [64, 72, 73, 79, 88, 116], ecology [61, 66, 89, 91], etc.

The sliding mode control approach is recognized as an efficient tool for designing robust controllers for linear or nonlinear control systems operating under uncertainty conditions [40, 46, 47].

A major advantage of sliding mode control is low sensitivity to parameter variations in the plant and disturbances affecting the plant, which eliminates the necessity of exact modeling of the plant. In the sliding mode control, the control dynamics will have two sequential modes, viz. the reaching mode and the sliding mode. Basically, a sliding mode controller design consists of two parts: hyperplane design and controller design. A hyperplane is first designed via the pole-placement approach and a controller is then designed based on the sliding condition. The stability of the overall system is guaranteed by the sliding condition and by a stable hyperplane. Sliding mode control method is a popular method for the control and synchronization of chaotic systems [18, 23, 43, 48–50, 57, 87, 90, 102].

In this work, we first investigate research in the dynamic analysis and properties of the enzymes-substrates reactions system with ferroelectric behaviour in the brain waves which was studied by Enjieu Kadji et al. [10]. Section 2 contains the phase portraits and qualitative properties of the enzymes-substrates biological chaotic system.

Super-twisting sliding mode control [24] is very useful for the global stabilization and synchronization of the enzymes-substrates reaction systems as it achieves finite time stability for the systems.

In Sect. 3, we review the multivariable super-twisting sliding mode control [24] for nonlinear control systems.

In Sect. 4, we apply multivariable super-twisting sliding mode control [24] to globally stabilize all the trajectories of the enzymes-substrates biological chaotic system.

In Sect. 5, we use multivariable super-twisting sliding mode control for the global chaos synchronization of identical enzymes-substrates biological chaotic systems.

Numerical simulations using MATLAB are shown to illustrate all the main results derived in this work.

2   Enzymes-Substrates Biological Chaotic System

Coherent oscillations in biological systems were studied by Frohlich [14] and the following suggestions were made which are taken as a physical basis for theoretical investigation of enzymatic substrate reaction with ferroelectric behavior in brain waves model [17].

1. When metabolic energy is available, long-wavelength electric vibrations are very strongly and coherently excited in active biological systems.
2. Biological systems have metastable states with a very high electric polarization.

These long range interactions may lead to a selective transport of enzymes and hence specific chemical reactions may become possible. Enjieu Kadji et al. [10] derives enzymes-substrates reactions system with ferroelectric behavior in brain waves. Specifically, chaotic behavior was noted for the 2-D enzymes-substrates biological system.

The enzymes-substrates biological system is described by the differential equation

$$\ddot{x} - \mu(1 - x^2 + ax^4 - bx^6)\dot{x} + x = E\cos(\Omega t) \tag{1}$$

In Eq. (1), a, b are positive parameters, μ is the parameter of nonlinearity, while E and Ω are the amplitude and the frequency of the external cosinusoidal excitation, respectively.

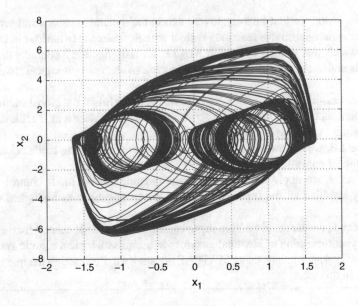

Fig. 1 2-D phase portrait of the enzymes-substrates biological system

The second-order differential equation (1) can be expressed in systems form as

$$\begin{cases} \dot{x}_1 = x_2 \\ \dot{x}_2 = \mu x_2 (1 - x_1^2 + a x_1^4 - b x_1^6) - x_1 + E \cos(\Omega t) \end{cases} \quad (2)$$

where x_1, x_2 are the states of the system.

For numerical simulations, we take the initial conditions as $x_1(0) = 0.2$ and $x_2(0) = 0.2$.

In [10], it was shown that the system (2) is *chaotic* when

$$a = 2.55, \quad b = 1.70, \quad \mu = 2.001, \quad E = 8.27, \quad \Omega = 3.465 \quad (3)$$

In this case, the Lyapunov exponents of the system (2) are obtained as $L_1 = 0.0751$ and $L_2 = -9.9450$. Since $L_1 > 0$, it follows that the biological system (2) is chaotic.

Figure 1 shows the 2-D phase portrait of the enzymes-substrates system (2).

3 Multivariable Super-Twisting Control for Nonlinear Control Systems

In this section, we shall review multivariable super-twisting control [24] for globally stabilizing nonlinear systems.

We consider a globally defined and smooth nonlinear system given by

$$\dot{x} = f(x, t) + g(x, t)u \tag{4}$$

where $x \in \mathbf{R}^n$, f is a C^1 vector field on $\mathbf{R}^n$ and $g \in C^1(\mathbf{R}^n, \mathbf{R}^n)$. We assume that $g(x, t)$ is an invertible matrix.

In (4), u is an n-dimensional control. We seek to find u so as to globally stabilize the plant (4) for all initial conditions $x(0) \in \mathbf{R}^n$.

In [24], Nagesh and Edwards derived the super-twisting control law given by

$$u = g(x, t)^{-1}[v - f(x, t)] \tag{5}$$

where v is defined via a sliding vector s by the following equations.

$$v = -k_1 \frac{s}{\|s\|^{\frac{1}{2}}} + z - k_2 s \tag{6}$$

$$\dot{s} = -k_1 \frac{s}{\|s\|^{\frac{1}{2}}} + z - k_2 s \tag{7}$$

and

$$\dot{z} = -k_3 \frac{s}{\|s\|} - k_4 s \tag{8}$$

We note that k_1, k_2, k_3, k_4 are positive gains chosen carefully so that the super-twisting control u defined by (5) renders the closed-loop control system globally asymptotically stable [24].

4 Global Chaos Control of the Enzymes-Substrates Biological System

In this section, we use super-twisting sliding mode control for the global chaos control of the enzymes-substrates biological system discussed in Sect. 2.

Thus, we consider the controlled enzymes-substrates biological system given by

$$\begin{cases} \dot{x}_1 = x_2 + u_1 \\ \dot{x}_2 = \mu x_2 (1 - x_1^2 + a x_1^4 - b x_1^6) - x_1 + E \cos(\Omega t) + u_2 \end{cases} \tag{9}$$

where u_1, u_2 are controls to be designed.

In matrix form, we can express (9) as

$$\dot{x} = f(x, t) + g(x, t)u \tag{10}$$

where

$$f(x,t) = \begin{bmatrix} x_2 \\ \mu x_2(1 - x_1^2 + a x_1^4 - b x_1^6) - x_1 + E\cos(\Omega t) \end{bmatrix} \text{ and } g(x,t) = I \quad (11)$$

As outlined in Sect. 3, a super-twisting sliding mode control for stabilizing the system (9) is given by

$$u = g(x,t)^{-1}[v - f(x,t)] \quad (12)$$

i.e.

$$\begin{cases} u_1 = -x_2 + v_1 \\ u_2 = -\mu x_2(1 - x_1^2 + a x_1^4 - b x_1^6) + x_1 - E\cos(\Omega t) + v_2 \end{cases} \quad (13)$$

where

$$\begin{cases} v_1 = -k_1 \dfrac{s_1}{\|s\|^{\frac{1}{2}}} + z_1 - k_2 s_1 \\ v_2 = -k_1 \dfrac{s_2}{\|s\|^{\frac{1}{2}}} + z_2 - k_2 s_2 \end{cases} \quad (14)$$

$$\begin{cases} \dot{s}_1 = -k_1 \dfrac{s_1}{\|s\|^{\frac{1}{2}}} + z_1 - k_2 s_1 \\ \dot{s}_2 = -k_1 \dfrac{s_2}{\|s\|^{\frac{1}{2}}} + z_2 - k_2 s_2 \end{cases} \quad (15)$$

and

$$\begin{cases} \dot{z}_1 = -k_3 \dfrac{s_1}{\|s\|} + z_1 - k_4 s_1 \\ \dot{z}_2 = -k_3 \dfrac{s_2}{\|s\|} + z_2 - k_4 s_2 \end{cases} \quad (16)$$

where k_1, k_2, k_3, k_4 are positive gains.

The super-twisting sliding mode control given by (13) globally stabilizes the enzymes-substrates biological system (9) for all initial conditions $x(0) \in \mathbf{R}^2$.

For numerical simulations, we take the parameters as in the chaotic case (3), i.e.

$$a = 2.55, \quad b = 1.70, \quad \mu = 2.001, \quad E = 8.27, \quad \Omega = 3.465 \quad (17)$$

We take the gains as $k_1 = 5$, $k_2 = 50$, $k_3 = 6$ and $k_4 = 60$.

We take the initial state of the enzymes-substrates system (9) as

$$x_1(0) = 7.2, \quad x_2(0) = 15.8 \quad (18)$$

Figure 2 shows the finite-time stability of the closed-loop control system, when the super-twisting sliding mode control law (13) is implemented.

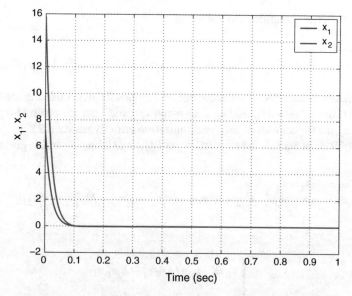

Fig. 2 Time-history of the controlled states x_1, x_2 of the enzymes-substrates system

5 Global Chaos Synchronization of the Enzymes-Substrates Biological Systems

In this section, we use super-twisting sliding mode control for the global chaos synchronization of the enzymes-substrates biological systems studied in Sect. 2.

As the master system, we consider the enzymes-substrates biological system

$$\begin{cases} \dot{x}_1 = x_2 \\ \dot{x}_2 = \mu x_2(1 - x_1^2 + ax_1^4 - bx_1^6) - x_1 + E\cos(\Omega t) \end{cases} \quad (19)$$

As the slave system, we consider the enzymes-substrates biological system

$$\begin{cases} \dot{y}_1 = y_2 + u_1 \\ \dot{y}_2 = \mu y_2(1 - y_1^2 + ay_1^4 - by_1^6) - y_1 + E\cos(\Omega t) + u_2 \end{cases} \quad (20)$$

The synchronization error between the systems (19) and (20) is defined by

$$\begin{cases} e_1 = y_1 - x_1 \\ e_2 = y_2 - x_2 \end{cases} \quad (21)$$

Then the error dynamics is obtained as

$$\begin{cases} \dot{e}_1 = e_2 + u_1 \\ \dot{e}_2 = \mu e_2 + \mu[-y_1^2 y_2 + x_1^2 x_2 + a(y_1^4 y_2 - x_1^4 x_2) - b(y_1^6 y_2 - x_1^6 x_2)] - e_1 + u_2 \end{cases} \quad (22)$$

A super-twisting sliding mode control for completely synchronizing the systems (19) and (20) is obtained by finding a super-twisting sliding mode control for globally stabilizing the trajectories of the synchronization error dynamics (22).

As outlined in Sect. 3, the required super-twisting sliding control is given by

$$\begin{cases} u_1 = -e_2 + v_1 \\ u_2 = -\mu e_2 - \mu[-y_1^2 y_2 + x_1^2 x_2 + a(y_1^4 y_2 - x_1^4 x_2) - b(y_1^6 y_2 - x_1^6 x_2)] \\ \quad + e_1 + v_2 \end{cases} \quad (23)$$

where

$$\begin{cases} v_1 = -k_1 \frac{s_1}{\|s\|^{\frac{1}{2}}} + z_1 - k_2 s_1 \\ v_2 = -k_1 \frac{s_2}{\|s\|^{\frac{1}{2}}} + z_2 - k_2 s_2 \end{cases} \quad (24)$$

$$\begin{cases} \dot{s}_1 = -k_1 \frac{s_1}{\|s\|^{\frac{1}{2}}} + z_1 - k_2 s_1 \\ \dot{s}_2 = -k_1 \frac{s_2}{\|s\|^{\frac{1}{2}}} + z_2 - k_2 s_2 \end{cases} \quad (25)$$

and

$$\begin{cases} \dot{z}_1 = -k_3 \frac{s_1}{\|s\|} + z_1 - k_4 s_1 \\ \dot{z}_2 = -k_3 \frac{s_2}{\|s\|} + z_2 - k_4 s_2 \end{cases} \quad (26)$$

where k_1, k_2, k_3, k_4 are positive gains.

For numerical simulations, we take the parameters as in the chaotic case (3), i.e.

$$a = 2.55, \quad b = 1.70, \quad \mu = 2.001, \quad E = 8.27, \quad \Omega = 3.465 \quad (27)$$

We take the gains as $k_1 = 5$, $k_2 = 50$, $k_3 = 6$ and $k_4 = 60$.
We take the initial state of the master system (19) as

$$x_1(0) = 8.5, \quad x_2(0) = 4.9 \quad (28)$$

We take the initial state of the slave system (20) as

$$y_1(0) = 2.2, \quad y_2(0) = 16.7 \quad (29)$$

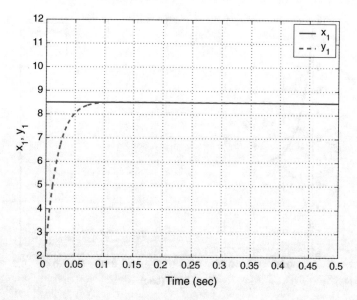

Fig. 3 Complete synchronization of the states x_1 and y_1

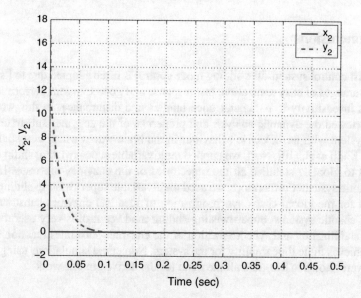

Fig. 4 Complete synchronization of the states x_2 and y_2

Figures 3 and 4 show the complete synchronization of the enzymes-substrates biological systems (19) and (20) in finite time. Figure 5 shows the finite-time convergence of the error trajectories e_1, e_2.

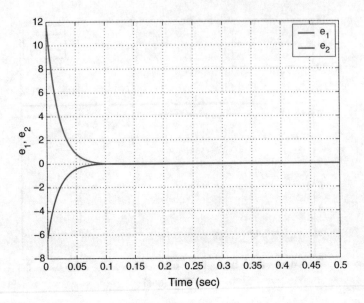

Fig. 5 Time-history of the synchronization errors e_1, e_2

6 Conclusions

In robust control systems, the sliding mode control is often adopted due to its inherent advantages of easy realization, fast response and good transient performance as well as insensitivity to parameter uncertainties and disturbance. In this work, we first reviewed the dynamic analysis and properties of the enzymes-substrates reactions system with ferroelectric behaviour in the brain waves which was studied by Enjieu Kadji et al. [10]. Next, we applied multivariable super-twisting sliding mode control to globally stabilize all the trajectories of the enzymes-substrates biological chaotic system. Furthermore, we used multivariable super-twisting sliding mode control for the global chaos synchronization of identical enzymes-substrates biological chaotic systems. Super-twisting sliding mode control is very useful for the global stabilization and synchronization of the enzymes-substrates reaction system as it achieves finite time stability for the system. Numerical simulations using MATLAB were shown to illustrate all the main results derived in this work.

References

1. Akgul A, Hussain S, Pehlivan I (2016) A new three-dimensional chaotic system, its dynamical analysis and electronic circuit applications. Optik 127(18):7062–7071
2. Akgul A, Moroz I, Pehlivan I, Vaidyanathan S (2016) A new four-scroll chaotic attractor and its engineering applications. Optik 127(13):5491–5499

3. Azar AT, Vaidyanathan S (2015) Chaos modeling and control systems design. Springer, Berlin
4. Azar AT, Vaidyanathan S (2016) Advances in chaos theory and intelligent control. Springer, Berlin
5. Azar AT, Vaidyanathan S (2017) Fractional order control and synchronization of chaotic systems. Springer, Berlin
6. Boulkroune A, Bouzeriba A, Bouden T (2016) Fuzzy generalized projective synchronization of incommensurate fractional-order chaotic systems. Neurocomputing 173:606–614
7. Burov DA, Evstigneev NM, Magnitskii NA (2017) On the chaotic dynamics in two coupled partial differential equations for evolution of surface plasmon polaritons. Commun Nonlinear Sci Numer Simul 46:26–36
8. Chai X, Chen Y, Broyde L (2017) A novel chaos-based image encryption algorithm using DNA sequence operations. Opt Lasers Eng 88:197–213
9. Chenaghlu MA, Jamali S, Khasmakhi NN (2016) A novel keyed parallel hashing scheme based on a new chaotic system. Chaos Solitons Fractals 87:216–225
10. Enjieu Kadji HG, Chabi Orou JB, Yamapi R, Woafo P (2007) Nonlinear dynamics and strange attractors in the biological system. Chaos Solitons Fractals 32:862–882
11. Fallahi K, Leung H (2010) A chaos secure communication scheme based on multiplication modulation. Commun Nonlinear Sci Numer Simul 15(2):368–383
12. Fontes RT, Eisencraft M (2016) A digital bandlimited chaos-based communication system. Commun Nonlinear Sci Numer Simul 37:374–385
13. Fotsa RT, Woafo P (2016) Chaos in a new bistable rotating electromechanical system. Chaos Solitons Fractals 93:48–57
14. Frohlich H (1968) Long range coherence and energy storage in biological systems. Int J Quantum Chem 2:641–649
15. Fujisaka H, Yamada T (1983) Stability theory of synchronized motion in coupled-oscillator systems. Progress Theoret Phys 63:32–47
16. Kacar S (2016) Analog circuit and microcontroller based RNG application of a new easy realizable 4D chaotic system. Optik 127(20):9551–9561
17. Kaiser F (1978) Coherent oscillations in biological systems, I. Bifurcation phenomena and phase transitions in an enzyme-substrate reaction with ferroelectric behavior. Z Naturforsch A 294:304–333
18. Lakhekar GV, Waghmare LM, Vaidyanathan S (2016) Diving autopilot design for underwater vehicles using an adaptive neuro-fuzzy sliding mode controller. In: Vaidyanathan S, Volos C (eds) Advances and applications in nonlinear control systems. Springer, Berlin, pp 477–503
19. Li D (2008) A three-scroll chaotic attractor. Phys Lett A 372(4):387–393
20. Liu H, Ren B, Zhao Q, Li N (2016) Characterizing the optical chaos in a special type of small networks of semiconductor lasers using permutation entropy. Opt Commun 359:79–84
21. Liu W, Sun K, Zhu C (2016) A fast image encryption algorithm based on chaotic map. Opt Lasers Eng 84:26–36
22. Liu X, Mei W, Du H (2016) Simultaneous image compression, fusion and encryption algorithm based on compressive sensing and chaos. Opt Commun 366:22–32
23. Moussaoui S, Boulkroune A, Vaidyanathan S (2016) Fuzzy adaptive sliding-mode control scheme for uncertain underactuated systems. In: Vaidyanathan S, Volos C (eds) Advances and applications in nonlinear control systems. Springer, Berlin, pp 351–367
24. Nagesh I, Edwards C (2014) A multivariable super-twisting sliding mode approach. Automatica 50:984–988
25. Pecora LM, Carroll TL (1990) Synchronization in chaotic systems. Phys Rev Lett 64:821–824
26. Pecora LM, Carroll TL (1991) Synchronizing chaotic circuits. IEEE Trans Circuits Syst 38:453–456
27. Pehlivan I, Moroz IM, Vaidyanathan S (2014) Analysis, synchronization and circuit design of a novel butterfly attractor. J Sound Vib 333(20):5077–5096
28. Pham VT, Volos C, Jafari S, Wang X, Vaidyanathan S (2014) Hidden hyperchaotic attractor in a novel simple memristive neural network. Optoelectron Adv Mater Rapid Commun 8(11–12):1157–1163

29. Pham VT, Volos CK, Vaidyanathan S (2015) Multi-scroll chaotic oscillator based on a first-order delay differential equation. In: Azar AT, Vaidyanathan S (eds) Chaos modeling and control systems design, studies in computational intelligence, vol 581. Springer, Germany, pp 59–72

30. Pham VT, Volos CK, Vaidyanathan S, Le TP, Vu VY (2015) A memristor-based hyperchaotic system with hidden attractors: dynamics, synchronization and circuital emulating. J Eng Sci Technol Rev 8(2):205–214

31. Pham VT, Jafari S, Vaidyanathan S, Volos C, Wang X (2016) A novel memristive neural network with hidden attractors and its circuitry implementation. Sci China Technol Sci 59(3):358–363

32. Pham VT, Jafari S, Volos C, Giakoumis A, Vaidyanathan S, Kapitaniak T (2016) A chaotic system with equilibria located on the rounded square loop and its circuit implementation. IEEE Trans Circuits Syst II Express Briefs 63(9):878–882

33. Pham VT, Jafari S, Volos C, Vaidyanathan S, Kapitaniak T (2016) A chaotic system with infinite equilibria located on a piecewise linear curve. Optik 127(20):9111–9117

34. Pham VT, Vaidyanathan S, Volos CK, Hoang TM, Yem VV (2016) Dynamics, synchronization and SPICE implementation of a memristive system with hidden hyperchaotic attractor. In: Azar AT, Vaidyanathan S (eds) Advances in chaos theory and intelligent control. Springer, Berlin, pp 35–52

35. Pham VT, Vaidyanathan S, Volos CK, Jafari S, Kuznetsov NV, Hoang TM (2016) A novel memristive time-delay chaotic system without equilibrium points. Eur Phys J Spec Top 225(1):127–136

36. Pham VT, Vaidyanathan S, Volos CK, Jafari S, Kuznetsov NV, Hoang TM (2016) A novel memristive time-delay chaotic system without equilibrium points. Eur Phys J Spec Top 225(1):127–136

37. Pham VT, Vaidyanathan S, Volos CK, Jafari S, Wang X (2016) A chaotic hyperjerk system based on memristive device. In: Vaidyanathan S, Volos C (eds) Advances and applications in chaotic systems. Springer, Berlin, pp 39–58

38. Sampath S, Vaidyanathan S, Volos CK, Pham VT (2015) An eight-term novel four-scroll chaotic system with cubic nonlinearity and its circuit simulation. J Eng Sci Technol Rev 8(2):1–6

39. Shirkhani N, Khanesar M, Teshnehlab M (2016) Indirect model reference fuzzy control of SISO fractional order nonlinear chaotic systems. Proc Comput Sci 102:309–316

40. Slotine J, Li W (1991) Applied nonlinear control. Prentice-Hall, Englewood Cliffs

41. Sundarapandian V (2013) Analysis and anti-synchronization of a novel chaotic system via active and adaptive controllers. J Eng Sci Technol Rev 6(4):45–52

42. Sundarapandian V, Pehlivan I (2012) Analysis, control, synchronization, and circuit design of a novel chaotic system. Math Comput Model 55(7–8):1904–1915

43. Sundarapandian V, Sivaperumal S (2011) Sliding controller design of hybrid synchronization of four-wing Chaotic systems. Int J Soft Comput 6(5):224–231

44. Szmit Z, Warminski J (2016) Nonlinear dynamics of electro-mechanical system composed of two pendulums and rotating hub. Proc Eng 144:953–958

45. Tacha OI, Volos CK, Kyprianidis IM, Stouboulos IN, Vaidyanathan S, Pham VT (2016) Analysis, adaptive control and circuit simulation of a novel nonlinear finance system. Appl Math Comput 276:200–217

46. Utkin VI (1977) Variable structure systems with sliding modes. IEEE Trans Autom Control 22(2):212–222

47. Utkin VI (1993) Sliding mode control design principles and applications to electric drives. IEEE Trans Ind Electron 40(1):23–36

48. Vaidyanathan S (2011) Analysis and synchronization of the hyperchaotic Yujun systems via sliding mode control. Adv Intell Syst Comput 176:329–337

49. Vaidyanathan S (2012) Global chaos control of hyperchaotic Liu system via sliding control method. Int J Control Theory Appl 5(2):117–123

50. Vaidyanathan S (2012) Sliding mode control based global chaos control of Liu-Liu-Liu-Su chaotic system. Int J Control Theory Appl 5(1):15–20
51. Vaidyanathan S (2013) A new six-term 3-D chaotic system with an exponential nonlinearity. Far East J Math Sci 79(1):135–143
52. Vaidyanathan S (2013) Analysis and adaptive synchronization of two novel chaotic systems with hyperbolic sinusoidal and cosinusoidal nonlinearity and unknown parameters. J Eng Sci Technol Rev 6(4):53–65
53. Vaidyanathan S (2014) A new eight-term 3-D polynomial chaotic system with three quadratic nonlinearities. Far East J Math Sci 84(2):219–226
54. Vaidyanathan S (2014) Analysis and adaptive synchronization of eight-term 3-D polynomial chaotic systems with three quadratic nonlinearities. Eur Phys J Spec Top 223(8):1519–1529
55. Vaidyanathan S (2014) Analysis, control and synchronisation of a six-term novel chaotic system with three quadratic nonlinearities. Int J Model Ident Control 22(1):41–53
56. Vaidyanathan S (2014) Generalised projective synchronisation of novel 3-D chaotic systems with an exponential non-linearity via active and adaptive control. Int J Model Ident Control 22(3):207–217
57. Vaidyanathan S (2014) Global chaos synchronisation of identical Li-Wu chaotic systems via sliding mode control. Int J Model Ident Control 22(2):170–177
58. Vaidyanathan S (2015) A 3-D novel highly chaotic system with four quadratic nonlinearities, its adaptive control and anti-synchronization with unknown parameters. J Eng Sci Technol Rev 8(2):106–115
59. Vaidyanathan S (2015) A novel chemical chaotic reactor system and its adaptive control. Int J ChemTech Res 8(7):146–158
60. Vaidyanathan S (2015) A novel chemical chaotic reactor system and its output regulation via integral sliding mode control. Int J ChemTech Res 8(11):669–683
61. Vaidyanathan S (2015) Active control design for the anti-synchronization of Lotka-Volterra biological systems with four competitive species. Int J PharmTech Res 8(7):58–70
62. Vaidyanathan S (2015) Adaptive control design for the anti-synchronization of novel 3-D chemical chaotic reactor systems. Int J ChemTech Res 8(11):654–668
63. Vaidyanathan S (2015) Adaptive control of a chemical chaotic reactor. Int J PharmTech Res 8(3):377–382
64. Vaidyanathan S (2015) Adaptive control of the FitzHugh-Nagumo chaotic neuron model. Int J PharmTech Res 8(6):117–127
65. Vaidyanathan S (2015) Adaptive synchronization of chemical chaotic reactors. Int J ChemTech Res 8(2):612–621
66. Vaidyanathan S (2015) Adaptive synchronization of generalized Lotka-Volterra three-species biological systems. Int J PharmTech Res 8(5):928–937
67. Vaidyanathan S (2015) Adaptive synchronization of novel 3-D chemical chaotic reactor systems. Int J ChemTech Res 8(7):159–171
68. Vaidyanathan S (2015) Analysis, properties and control of an eight-term 3-D chaotic system with an exponential nonlinearity. Int J Model Ident Control 23(2):164–172
69. Vaidyanathan S (2015) Anti-synchronization of brusselator chemical reaction systems via adaptive control. Int J ChemTech Res 8(6):759–768
70. Vaidyanathan S (2015) Anti-synchronization of brusselator chemical reaction systems via integral sliding mode control. Int J ChemTech Res 8(11):700–713
71. Vaidyanathan S (2015) Anti-synchronization of chemical chaotic reactors via adaptive control method. Int J ChemTech Res 8(8):73–85
72. Vaidyanathan S (2015) Anti-synchronization of the FitzHugh-Nagumo chaotic neuron models via adaptive control method. Int J PharmTech Res 8(7):71–83
73. Vaidyanathan S (2015) Chaos in neurons and synchronization of Birkhoff-Shaw strange chaotic attractors via adaptive control. Int J PharmTech Res 8(6):1–11
74. Vaidyanathan S (2015) Dynamics and control of brusselator chemical reaction. Int J ChemTech Res 8(6):740–749

75. Vaidyanathan S (2015) Dynamics and control of Tokamak system with symmetric and magnetically confined plasma. Int J ChemTech Res 8(6):795–802
76. Vaidyanathan S (2015) Global chaos synchronization of chemical chaotic reactors via novel sliding mode control method. Int J ChemTech Res 8(7):209–221
77. Vaidyanathan S (2015) Global chaos synchronization of Duffing double-well chaotic oscillators via integral sliding mode control. Int J ChemTech Res 8(11):141–151
78. Vaidyanathan S (2015) Global chaos synchronization of the forced Van der Pol chaotic oscillators via adaptive control method. Int J PharmTech Res 8(6):156–166
79. Vaidyanathan S (2015) Hybrid chaos synchronization of the FitzHugh-Nagumo chaotic neuron models via adaptive control method. Int J PharmTech Res 8(8):48–60
80. Vaidyanathan S (2015) Integral sliding mode control design for the global chaos synchronization of identical novel chemical chaotic reactor systems. Int J ChemTech Res 8(11):684–699
81. Vaidyanathan S (2015) Output regulation of the forced Van der Pol chaotic oscillator via adaptive control method. Int J PharmTech Res 8(6):106–116
82. Vaidyanathan S (2015) Sliding controller design for the global chaos synchronization of forced Van der Pol chaotic oscillators. Int J PharmTech Res 8(7):100–111
83. Vaidyanathan S (2015) Synchronization of Tokamak systems with symmetric and magnetically confined plasma via adaptive control. Int J ChemTech Res 8(6):818–827
84. Vaidyanathan S (2016) A novel 3-D conservative chaotic system with a sinusoidal nonlinearity and its adaptive control. Int J Control Theory Appl 9(1):115–132
85. Vaidyanathan S (2016) A novel 3-D jerk chaotic system with three quadratic nonlinearities and its adaptive control. Arch Control Sci 26(1):19–47
86. Vaidyanathan S (2016) A novel 3-D jerk chaotic system with two quadratic nonlinearities and its adaptive backstepping control. Int J Control Theory Appl 9(1):199–219
87. Vaidyanathan S (2016) Anti-synchronization of 3-cells cellular neural network attractors via integral sliding mode control. Int J PharmTech Res 9(1):193–205
88. Vaidyanathan S (2016) Global chaos control of the FitzHugh-Nagumo chaotic neuron model via integral sliding mode control. Int J PharmTech Res 9(4):413–425
89. Vaidyanathan S (2016) Global chaos control of the generalized Lotka-Volterra three-species system via integral sliding mode control. Int J PharmTech Res 9(4):399–412
90. Vaidyanathan S (2016) Global chaos regulation of a symmetric nonlinear gyro system via integral sliding mode control. Int J ChemTech Res 9(5):462–469
91. Vaidyanathan S (2016) Hybrid synchronization of the generalized Lotka-Volterra three-species biological systems via adaptive control. Int J PharmTech Res 9(1):179–192
92. Vaidyanathan S (2016) Mathematical analysis, adaptive control and synchronization of a ten-term novel three-scroll chaotic system with four quadratic nonlinearities. Int J Control Theory Appl 9(1):1–20
93. Vaidyanathan S, Azar AT (2015) Analysis, control and synchronization of a nine-term 3-D novel chaotic system. In: Azar AT, Vaidyanathan S (eds) Chaos modelling and control systems design, studies in computational intelligence, vol 581. Springer, Germany, pp 19–38
94. Vaidyanathan S, Azar AT (2016) Takagi-Sugeno fuzzy logic controller for Liu-Chen four-scroll chaotic system. Int J Intell Eng Inform 4(2):135–150
95. Vaidyanathan S, Boulkroune A (2016) A novel 4-D hyperchaotic chemical reactor system and its adaptive control. In: Vaidyanathan S, Volos C (eds) Advances and applications in chaotic systems. Springer, Berlin, pp 447–469
96. Vaidyanathan S, Madhavan K (2013) Analysis, adaptive control and synchronization of a seven-term novel 3-D chaotic system. Int J Control Theory Appl 6(2):121–137
97. Vaidyanathan S, Pakiriswamy S (2016) A five-term 3-D novel conservative chaotic system and its generalized projective synchronization via adaptive control method. Int J Control Theory Appl 9(1):61–78
98. Vaidyanathan S, Pakiriswamy S (2015) A 3-D novel conservative chaotic system and its generalized projective synchronization via adaptive control. J Eng Sci Technol Rev 8(2):52–60
99. Vaidyanathan S, Pakiriswamy S (2016) Adaptive control and synchronization design of a seven-term novel chaotic system with a quartic nonlinearity. Int J Control Theory Appl 9(1):237–256

100. Vaidyanathan S, Rajagopal K (2016) Adaptive control, synchronization and LabVIEW implementation of Rucklidge chaotic system for nonlinear double convection. Int J Control Theory Appl 9(1):175–197
101. Vaidyanathan S, Rajagopal K (2016) Analysis, control, synchronization and LabVIEW implementation of a seven-term novel chaotic system. Int J Control Theory Appl 9(1):151–174
102. Vaidyanathan S, Sampath S (2011) Global chaos synchronization of hyperchaotic Lorenz systems by sliding mode control. Commun Comput Inf Sci 205:156–164
103. Vaidyanathan S, Volos C (2015) Analysis and adaptive control of a novel 3-D conservative no-equilibrium chaotic system. Arch Control Sci 25(3):333–353
104. Vaidyanathan S, Volos C (2016) Advances and applications in chaotic systems. Springer, Berlin
105. Vaidyanathan S, Volos C (2016) Advances and applications in nonlinear control systems. Springer, Berlin
106. Vaidyanathan S, Volos C (2017) Advances in memristors, memristive devices and systems. Springer, Berlin
107. Vaidyanathan S, Volos C, Pham VT, Madhavan K, Idowu BA (2014) Adaptive backstepping control, synchronization and circuit simulation of a 3-D novel jerk chaotic system with two hyperbolic sinusoidal nonlinearities. Arch Control Sci 24(3):375–403
108. Vaidyanathan S, Rajagopal K, Volos CK, Kyprianidis IM, Stouboulos IN (2015) Analysis, adaptive control and synchronization of a seven-term novel 3-D chaotic system with three quadratic nonlinearities and its digital implementation in LabVIEW. J Eng Sci Technol Rev 8(2):130–141
109. Vaidyanathan S, Volos CK, Kyprianidis IM, Stouboulos IN, Pham VT (2015) Analysis, adaptive control and anti-synchronization of a six-term novel jerk chaotic system with two exponential nonlinearities and its circuit simulation. J Eng Sci Technol Rev 8(2):24–36
110. Vaidyanathan S, Volos CK, Pham VT (2015) Analysis, adaptive control and adaptive synchronization of a nine-term novel 3-D chaotic system with four quadratic nonlinearities and its circuit simulation. J Eng Sci Technol Rev 8(2):181–191
111. Vaidyanathan S, Volos CK, Pham VT (2015) Global chaos control of a novel nine-term chaotic system via sliding mode control. In: Azar AT, Zhu Q (eds) Advances and applications in sliding mode control systems, studies in computational intelligence, vol 576. Springer, Germany, pp 571–590
112. Volos CK, Kyprianidis IM, Stouboulos IN, Tlelo-Cuautle E, Vaidyanathan S (2015) Memristor: a new concept in synchronization of coupled neuromorphic circuits. J Eng Sci Technol Rev 8(2):157–173
113. Volos CK, Pham VT, Vaidyanathan S, Kyprianidis IM, Stouboulos IN (2015) Synchronization phenomena in coupled Colpitts circuits. J Eng Sci Technol Rev 8(2):142–151
114. Volos CK, Pham VT, Vaidyanathan S, Kyprianidis IM, Stouboulos IN (2016) Synchronization phenomena in coupled hyperchaotic oscillators with hidden attractors using a nonlinear open loop controller. In: Vaidyanathan S, Volos C (eds) Advances and applications in chaotic systems. Springer, Berlin, pp 1–38
115. Volos CK, Pham VT, Vaidyanathan S, Kyprianidis IM, Stouboulos IN (2016) The case of bidirectionally coupled nonlinear circuits via a memristor. In: Vaidyanathan S, Volos C (eds) Advances and applications in nonlinear control systems. Springer, Berlin, pp 317–350
116. Volos CK, Prousalis D, Kyprianidis IM, Stouboulos I, Vaidyanathan S, Pham VT (2016) Synchronization and anti-synchronization of coupled Hindmarsh-Rose neuron models. Int J Control Theory Appl 9(1):101–114
117. Volos CK, Vaidyanathan S, Pham VT, Maaita JO, Giakoumis A, Kyprianidis IM, Stouboulos IN (2016) A novel design approach of a nonlinear resistor based on a memristor emulator. In: Azar AT, Vaidyanathan S (eds) Advances in chaos theory and intelligent control. Springer, Berlin, pp 3–34
118. Wang B, Zhong SM, Dong XC (2016) On the novel chaotic secure communication scheme design. Commun Nonlinear Sci Numerical Simul 39:108–117

119. Wu T, Sun W, Zhang X, Zhang S (2016) Concealment of time delay signature of chaotic output in a slave semiconductor laser with chaos laser injection. Opt Commun 381:174–179
120. Xu G, Liu F, Xiu C, Sun L, Liu C (2016) Optimization of hysteretic chaotic neural network based on fuzzy sliding mode control. Neurocomputing 189:72–79
121. Xu H, Tong X, Meng X (2016) An efficient chaos pseudo-random number generator applied to video encryption. Optik 127(20):9305–9319
122. Zhou W, Xu Y, Lu H, Pan L (2008) On dynamics analysis of a new chaotic attractor. Phys Lett A 372(36):5773–5777
123. Zhu C, Liu Y, Guo Y (2010) Theoretic and numerical study of a new chaotic system. Intell Inf Manage 2:104–109

Super-Twisting Sliding Mode Control and Synchronization of Moore-Spiegel Thermo-Mechanical Chaotic System

Sundarapandian Vaidyanathan

Abstract Chaos has important applications in physics, chemistry, biology, ecology, secure communications, cryptosystems and many scientific branches. Control and synchronization of chaotic systems are important research problems in chaos theory. Sliding mode control is an important method used to solve various problems in control systems engineering. In robust control systems, the sliding mode control is often adopted due to its inherent advantages of easy realization, fast response and good transient performance as well as insensitivity to parameter uncertainties and disturbance. In this work, we first discuss the properties of the Moore-Spiegel thermo-mechanical chaotic system (1966). The Moore-Spiegel system is a nonlinear thermo-mechanical chaotic oscillator that describes a fluid element oscillating vertically in a temperature gradient with a linear restoring force. The Moore-Spiegel system is a classical example of a 3-D chaotic system. We show that the Moore-Spiegel system has a unique equilibrium at the origin, which is a saddle point. Next, we apply multivariable super-twisting sliding mode control to globally stabilize all the trajectories of the Moore-Spiegel chaotic system. Furthermore, we use multivariable super-twisting sliding mode control for the global chaos synchronization of the identical Moore-Spiegel chaotic systems. Super-twisting sliding mode control is very useful for the global stabilization and synchronization of the Moore-Spiegel chaotic system as it achieves finite time stability for the system. Numerical simulations using MATLAB are shown to illustrate all the main results derived in this work.

Keywords Chaos · Chaotic systems · Moore-Spiegel system · Thermo-mechanical system · Chaos control · Sliding mode control · Super-twisting control

S. Vaidyanathan (✉)
Research and Development Centre, Vel Tech University, Avadi, Chennai 600062,
Tamil Nadu, India
e-mail: sundarcontrol@gmail.com

© Springer International Publishing AG 2017
S. Vaidyanathan and C.-H. Lien (eds.), *Applications of Sliding Mode Control
in Science and Engineering*, Studies in Computational Intelligence 709,
DOI 10.1007/978-3-319-55598-0_20

451

1 Introduction

Chaos theory describes the quantitative study of unstable aperiodic dynamic behavior in deterministic nonlinear dynamical systems. For the motion of a dynamical system to be chaotic, the system variables should contain some nonlinear terms and the system must satisfy three properties: boundedness, infinite recurrence and sensitive dependence on initial conditions [4–6, 104–106].

The problem of global control of a chaotic system is to device feedback control laws so that the closed-loop system is globally asymptotically stable. Global chaos control of various chaotic systems has been investigated via various methods in the control literature [4, 5, 104, 105].

The synchronization of chaotic systems deals with the problem of synchronizing the states of two chaotic systems called as *master* and *slave* systems asymptotically with time. The design goal of the complete synchronization problem is to use the output of the master system to control of the output of the slave system so that the outputs of the two systems are synchronized asymptotically with time.

Because of the *butterfly effect* [4], which causes the exponential divergence of the trajectories of two identical chaotic systems started with nearly the same initial conditions, synchronizing two chaotic systems is seemingly a challenging research problem in the chaos literature.

The synchronization of chaotic systems was first researched by Yamada and Fujisaka [15] with subsequent work by Pecora and Carroll [25, 26]. In the last few decades, several different methods have been devised for the synchronization of chaotic and hyperchaotic systems [4–6, 104–106].

Many new chaotic systems have been discovered in the recent years such as Zhou system [122], Zhu system [123], Li system [18], Sundarapandian systems [41, 42], Vaidyanathan systems [51–56, 58, 68, 84–86, 92, 93, 96–99, 101, 103, 107–111], Pehlivan system [27], Sampath system [38], Tacha system [45], Pham systems [29, 32, 33, 35], Akgul system [2], etc.

Chaos theory has applications in several fields such as memristors [28, 30, 31, 34, 36, 37, 112, 115, 117], fuzzy logic [8, 39, 94, 120], communication systems [12, 13, 118], cryptosystems [10, 11], electromechanical systems [14, 44], lasers [9, 19, 119], encryption [20, 21, 121], electrical circuits [1, 3, 16, 100, 113], chemical reactions [59, 60, 62, 63, 65, 67, 69–71, 74, 76, 80, 95], oscillators [77, 78, 81, 82, 114], tokamak systems [75, 83], neurology [64, 72, 73, 79, 88, 116], ecology [61, 66, 89, 91], etc.

The sliding mode control approach is recognized as an efficient tool for designing robust controllers for linear or nonlinear control systems operating under uncertainty conditions [40, 46, 47].

A major advantage of sliding mode control is low sensitivity to parameter variations in the plant and disturbances affecting the plant, which eliminates the necessity of exact modeling of the plant. In the sliding mode control, the control dynamics will have two sequential modes, viz. the reaching mode and the sliding mode. Basically, a sliding mode controller design consists of two parts: hyperplane design and controller design. A hyperplane is first designed via the pole-placement approach and a controller is then designed based on the sliding condition. The stability of the overall system is guaranteed by the sliding condition and by a stable hyperplane. Sliding mode control method is a popular method for the control and synchronization of chaotic systems [17, 23, 43, 48–50, 57, 87, 90, 102].

In this work, we first discuss the properties of the Moore-Spiegel thermomechanical chaotic system [22]. The Moore-Spiegel system is a nonlinear thermomechanical chaotic oscillator that describes a fluid element oscillating vertically in a temperature gradient with a linear restoring force. The Moore-Spiegel system is a classical example of a 3-D chaotic system. We show that the Moore-Spiegel system has a unique equilibrium at the origin, which is a saddle point.

Super-twisting sliding mode control [24] is very useful for the global stabilization and synchronization of the Moore-Spiegel chaotic system as it achieves finite time stability for the systems.

In Sect. 3, we review the multivariable super-twisting sliding mode control [24] for nonlinear control systems.

In Sect. 4, we apply multivariable super-twisting sliding mode control [24] to globally stabilize all the trajectories of the Moore-Spiegel chaotic system.

In Sect. 5, we use multivariable super-twisting sliding mode control for the global chaos synchronization of identical Moore-Spiegel chaotic systems.

Numerical simulations using MATLAB are shown to illustrate all the main results derived in this work.

2 Moore-Spiegel Thermo-Mechanical Chaotic System

In [22], Moore and Spiegel derived a thermo-mechanical oscillator described by the differential equation

$$\dddot{x} + \ddot{x} + (T - R + Rx^2)\dot{x} + Tx = 0 \tag{1}$$

where R and T are constants. The physical background to the Moore-Spiegel equation (1) is in fluid mechanics. Basically, the Moore-Spiegel equation describes a small fluid element oscillating vertically in a temperature gradient with a linear restoring force. The element exchanges heat with the surrounding fluid and its buoyancy depends upon temperature. In other words, the Moore-Spiegel equation (1) is a nonlinear thermo-mechanical oscillator with displacement $x(t)$. The parameter R corresponds to the Rayleigh number and T to the tension constant that quantifies the restoring force.

To simplify the notation, we denote the constants R and T as a and b respectively.

In system form, we can express the Moore-Spiegel equation (1) as

$$\begin{cases} \dot{x}_1 = x_2 \\ \dot{x}_2 = x_3 \\ \dot{x}_3 = -bx_1 - (b-a)x_2 - x_3 - ax_1^2 x_2 \end{cases} \qquad (2)$$

We note that the Moore-Spiegel system (2) is a jerk system with a cubic nonlinearity.

In [7], it was shown that the Moore-Spiegel system (2) is chaotic when the parameter values are taken as

$$a = 100, \quad b = 27 \qquad (3)$$

For simulations, we take the initial values of the Moore-Spiegel system (2) as

$$x_1(0) = 0.2, \quad x_2(0) = 0.2, \quad x_3(0) = 0.2 \qquad (4)$$

For the parameter values (3) and the initial values (4), the Lyapunov exponents of the Moore-Spiegel system (2) are calculated as

$$L_1 = 0.1731, \quad L_2 = 0, \quad L_3 = -1.1731 \qquad (5)$$

Since $L_1 > 0$, it follows that the Moore-Spiegel system (2) is chaotic.
Since $L_1 + L_2 + L_3 = -1 < 0$, we conclude that the Moore-Spiegel system (2) is dissipative.

The Kaplan-Yorke dimension of the Moore-Spiegel system (2) is found as

$$D_{KY} = 2 + \frac{L_1 + L_2}{|L_3|} = 2.1476 \qquad (6)$$

Figure 1 shows the 3-D phase portrait of the Moore-Spiegel chaotic system (2). Figures 2, 3 and 4 show the 2-D phase portraits of the Moore-Spiegel chaotic system (2). Figure 5 shows the Lyapunov exponents of the Moore-Spiegel chaotic system (2).

The equilibrium points of the Moore-Spiegel chaotic system (2) are calculated by solving the equations

$$x_2 = 0 \qquad (7a)$$
$$x_3 = 0 \qquad (7b)$$
$$-bx_1 - (b-a)x_2 - x_3 - ax_1^2 x_2 = 0 \qquad (7c)$$

From (7a) and (7b), we find that $x_2 = 0$ and $x_3 = 0$.
Substituting $x_2 = x_3 = 0$ in (7c), we get

$$-bx_1 = 0 \text{ or } x_1 = 0 \qquad (8)$$

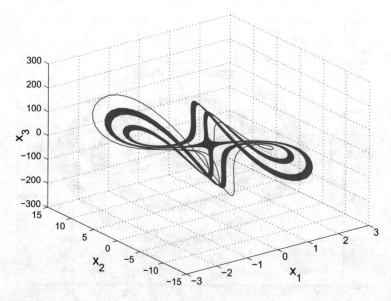

Fig. 1 3-D phase portrait of the Moore-Spiegel chaotic system

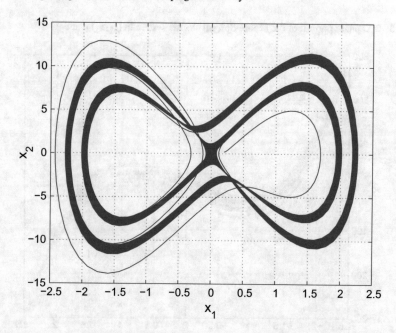

Fig. 2 2-D phase portrait of the Moore-Spiegel chaotic system in (x_1, x_2) plane

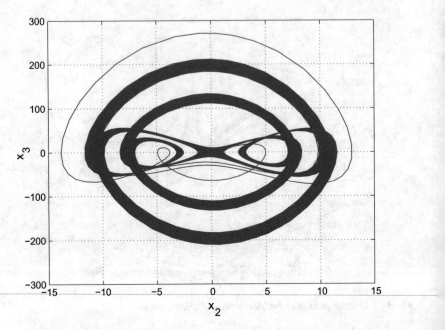

Fig. 3 2-D phase portrait of the Moore-Spiegel chaotic system in (x_2, x_3) plane

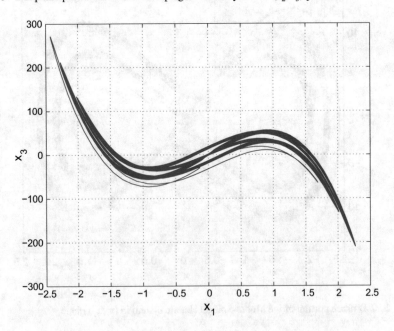

Fig. 4 2-D phase portrait of the Moore-Spiegel chaotic system in (x_1, x_3) plane

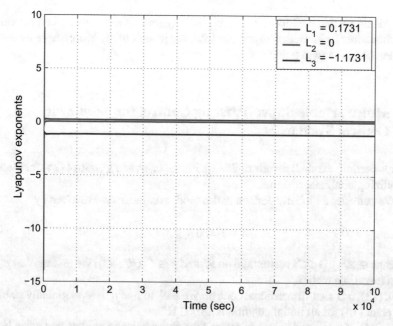

Fig. 5 Lyapunov exponents of the Moore-Spiegel chaotic system

Thus, we conclude that the Moore-Spiegel chaotic system (2) has a unique equilibrium at the origin, viz.

$$E_0 = \begin{bmatrix} 0 \\ 0 \\ 0 \end{bmatrix} \tag{9}$$

The linearization of the Moore-Spiegel chaotic system (2) at E_0 is given by

$$J_0 = J(E_0) = \begin{bmatrix} 0 & 1 & 0 \\ 0 & 0 & 1 \\ -b & -(b-a) & -1 \end{bmatrix} = \begin{bmatrix} 0 & 1 & 0 \\ 0 & 0 & 1 \\ -27 & 73 & -1 \end{bmatrix} \tag{10}$$

The eigenvalues of J_0 are numerically obtained as

$$\lambda_1 = -9.2279, \quad \lambda_2 = 0.3725, \quad \lambda_3 = 7.8554 \tag{11}$$

This shows that the equilibrium E_0 is a saddle-point and unstable.

Next, we note that the Moore-Spiegel chaotic system (2) is invariant under the coordinates transformation

$$(x_1, x_2, x_3) \mapsto (-x_1, -x_2, -x_3) \tag{12}$$

This shows that the Moore-Spiegel chaotic system (2) has point reflection symmetry about the origin. Hence, every non-trivial trajectory of the Moore-Spiegel chaotic system (2) has a twin trajectory.

3 Multivariable Super-Twisting Control for Nonlinear Control Systems

In this section, we shall review multivariable super-twisting control [24] for globally stabilizing nonlinear systems.

We consider a globally defined and smooth nonlinear system given by

$$\dot{x} = f(x, t) + g(x, t)u \tag{13}$$

where $x \in \mathbf{R}^n$, f is a C^1 vector field on $\mathbf{R}^n$ and $g \in C^1(\mathbf{R}^n, \mathbf{R}^n)$. We assume that $g(x, t)$ is an invertible matrix.

In (13), u is an n-dimensional control. We seek to find u so as to globally stabilize the plant (13) for all initial conditions $x(0) \in \mathbf{R}^n$.

In [24], Nagesh and Edwards derived the super-twisting control law given by

$$u = g(x, t)^{-1}[v - f(x, t)] \tag{14}$$

where v is defined via a sliding vector s by the following equations.

$$v = -k_1 \frac{s}{\|s\|^{\frac{1}{2}}} + z - k_2 s \tag{15}$$

$$\dot{s} = -k_1 \frac{s}{\|s\|^{\frac{1}{2}}} + z - k_2 s \tag{16}$$

and

$$\dot{z} = -k_3 \frac{s}{\|s\|} - k_4 s \tag{17}$$

We note that k_1, k_2, k_3, k_4 are positive gains chosen carefully so that the super-twisting control u defined by (14) renders the closed-loop control system globally asymptotically stable [24].

4 Global Chaos Control of the Moore-Spiegel Thermo-Mechanical Chaotic System

In this section, we use super-twisting sliding mode control for the global chaos control of the Moore-Spiegel thermo-mechanical system discussed in Sect. 2.

Thus, we consider the controlled Moore-Spiegel chaotic system given by

$$\begin{cases} \dot{x}_1 = x_2 + u_1 \\ \dot{x}_2 = x_3 + u_2 \\ \dot{x}_3 = -bx_1 - (b-a)x_2 - x_3 - ax_1^2 x_2 + u_3 \end{cases} \tag{18}$$

In matrix form, we can express (18) as

$$\dot{x} = f(x) + g(x)u \tag{19}$$

where

$$f(x) = \begin{bmatrix} x_2 \\ x_3 \\ -bx_1 - (b-a)x_2 - x_3 - ax_1^2 x_2 \end{bmatrix} \quad \text{and} \quad g(x) = I \tag{20}$$

As outlined in Sect. 3, a super-twisting sliding mode control for stabilizing the system (18) is given by

$$u = g(x)^{-1}[v - f(x)] \tag{21}$$

i.e.

$$\begin{cases} u_1 = -x_2 + v_1 \\ u_2 = -x_3 + v_2 \\ u_3 = bx_1 + (b-a)x_2 + x_3 + ax_1^2 x_2 + v_3 \end{cases} \tag{22}$$

where

$$\begin{cases} v_1 = -k_1 \dfrac{s_1}{\|s\|^{\frac{1}{2}}} + z_1 - k_2 s_1 \\[2mm] v_2 = -k_1 \dfrac{s_2}{\|s\|^{\frac{1}{2}}} + z_2 - k_2 s_2 \\[2mm] v_3 = -k_1 \dfrac{s_3}{\|s\|^{\frac{1}{2}}} + z_3 - k_2 s_3 \end{cases} \tag{23}$$

$$\begin{cases} \dot{s}_1 = -k_1 \dfrac{s_1}{\|s\|^{\frac{1}{2}}} + z_1 - k_2 s_1 \\[2mm] \dot{s}_2 = -k_1 \dfrac{s_2}{\|s\|^{\frac{1}{2}}} + z_2 - k_2 s_2 \\[2mm] \dot{s}_3 = -k_1 \dfrac{s_3}{\|s\|^{\frac{1}{2}}} + z_3 - k_2 s_3 \end{cases} \tag{24}$$

and

$$\begin{cases} \dot{z}_1 = -k_3 \frac{s_1}{\|s\|} + z_1 - k_4 s_1 \\ \dot{z}_2 = -k_3 \frac{s_2}{\|s\|} + z_2 - k_4 s_2 \\ \dot{z}_3 = -k_3 \frac{s_3}{\|s\|} + z_3 - k_4 s_3 \end{cases} \tag{25}$$

where k_1, k_2, k_3, k_4 are positive gains.

The super-twisting sliding mode control given by (22) globally stabilizes the enzymes-substrates biological system (18) for all initial conditions $x(0) \in \mathbf{R}^2$.

For numerical simulations, we take the parameters as in the chaotic case (3), i.e.

$$a = 100, \quad b = 27 \tag{26}$$

We take the gains as $k_1 = 5$, $k_2 = 50$, $k_3 = 6$ and $k_4 = 60$.
We take the initial state of the Moore-Spiegel system (18) as

$$x_1(0) = 17.2, \quad x_2(0) = 25.6, \quad x_3(0) = 12.9 \tag{27}$$

Figure 6 shows the finite-time stability of the closed-loop control system, when the super-twisting sliding mode control law (22) is implemented.

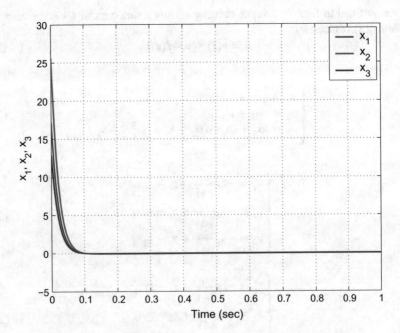

Fig. 6 Time-history of the controlled states x_1, x_2, x_3 of the Moore-Spiegel system

5 Global Chaos Synchronization of the Moore-Spiegel Thermo-Mechanical Chaotic Systems

In this section, we use super-twisting sliding mode control for the global chaos synchronization of the Moore-Spiegel thermo-mechanical chaotic systems studied in Sect. 2.

As the master system, we consider the Moore-Spiegel chaotic system

$$
\begin{cases}
\dot{x}_1 = x_2 \\
\dot{x}_2 = x_3 \\
\dot{x}_3 = -bx_1 - (b-a)x_2 - x_3 - ax_1^2 x_2
\end{cases}
\tag{28}
$$

As the slave system, we consider the Moore-Spiegel chaotic system

$$
\begin{cases}
\dot{y}_1 = y_2 + u_1 \\
\dot{y}_2 = y_3 + u_2 \\
\dot{y}_3 = -by_1 - (b-a)y_2 - y_3 - ay_1^2 y_2 + u_3
\end{cases}
\tag{29}
$$

The synchronization error between the systems (28) and (29) is defined by

$$
\begin{cases}
e_1 = y_1 - x_1 \\
e_2 = y_2 - x_2 \\
e_3 = y_3 - x_3
\end{cases}
\tag{30}
$$

Then the error dynamics is obtained as

$$
\begin{cases}
\dot{e}_1 = e_2 + u_1 \\
\dot{e}_2 = e_3 + u_2 \\
\dot{e}_3 = -be_1 - (b-a)e_2 - e_3 - a(y_1^2 y_2 - x_1^2 x_2) + u_3
\end{cases}
\tag{31}
$$

A super-twisting sliding mode control for completely synchronizing the systems (28) and (29) is obtained by finding a super-twisting sliding mode control for globally stabilizing the trajectories of the synchronization error dynamics (31).

As outlined in Sect. 3, the required super-twisting sliding control is given by

$$
\begin{cases}
u_1 = -e_2 + v_1 \\
u_2 = -e_3 + v_2 \\
u_3 = be_1 + (b-a)e_2 + e_3 + a(y_1^2 y_2 - x_1^2 x_2) + v_3
\end{cases}
\tag{32}
$$

where

$$
\begin{cases}
v_1 = -k_1 \dfrac{s_1}{\|s\|^{\frac{1}{2}}} + z_1 - k_2 s_1 \\[2mm]
v_2 = -k_1 \dfrac{s_2}{\|s\|^{\frac{1}{2}}} + z_2 - k_2 s_2 \\[2mm]
v_3 = -k_1 \dfrac{s_3}{\|s\|^{\frac{1}{2}}} + z_3 - k_2 s_3
\end{cases}
\tag{33}
$$

$$
\begin{cases}
\dot{s}_1 = -k_1 \dfrac{s_1}{\|s\|^{\frac{1}{2}}} + z_1 - k_2 s_1 \\[2mm]
\dot{s}_2 = -k_1 \dfrac{s_2}{\|s\|^{\frac{1}{2}}} + z_2 - k_2 s_2 \\[2mm]
\dot{s}_3 = -k_1 \dfrac{s_3}{\|s\|^{\frac{1}{2}}} + z_3 - k_2 s_3
\end{cases}
\tag{34}
$$

and

$$
\begin{cases}
\dot{z}_1 = -k_3 \dfrac{s_1}{\|s\|} + z_1 - k_4 s_1 \\[2mm]
\dot{z}_2 = -k_3 \dfrac{s_2}{\|s\|} + z_2 - k_4 s_2 \\[2mm]
\dot{z}_3 = -k_3 \dfrac{s_3}{\|s\|} + z_3 - k_4 s_3
\end{cases}
\tag{35}
$$

where k_1, k_2, k_3, k_4 are positive gains.

For numerical simulations, we take the parameters as in the chaotic case (3), i.e.

$$
a = 100, \quad b = 27
\tag{36}
$$

We take the gains as $k_1 = 5$, $k_2 = 50$, $k_3 = 6$ and $k_4 = 60$.
We take the initial state of the master system (28) as

$$
x_1(0) = 5.3, \quad x_2(0) = 14.9, \quad x_3(0) = 12.5
\tag{37}
$$

We take the initial state of the slave system (29) as

$$
y_1(0) = 12.4, \quad y_2(0) = 8.5, \quad y_3(0) = 21.8
\tag{38}
$$

Figures 7, 8 and 9 show the complete synchronization of the Moore-Spiegel thermo-mechanical chaotic systems (28) and (29) in finite time. Figure 10 shows the finite-time convergence of the error trajectories e_1, e_2.

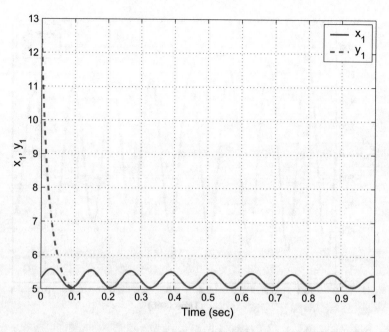

Fig. 7 Complete synchronization of the states x_1 and y_1

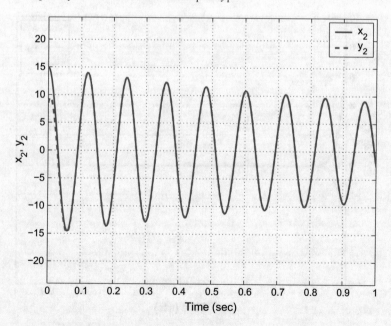

Fig. 8 Complete synchronization of the states x_2 and y_2

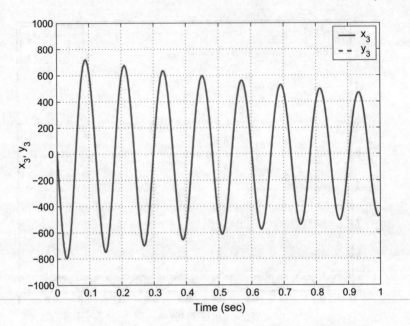

Fig. 9 Complete synchronization of the states x_3 and y_3

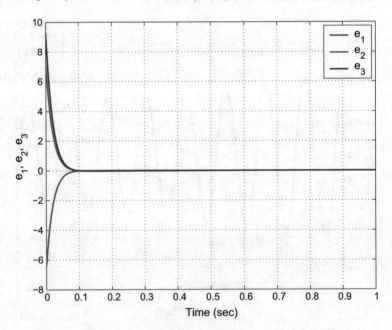

Fig. 10 Time-history of the synchronization errors e_1, e_2, e_3

6 Conclusions

Sliding mode control is an important method used to solve various problems in control systems engineering. In robust control systems, the sliding mode control is often adopted due to its inherent advantages of easy realization, fast response and good transient performance as well as insensitivity to parameter uncertainties and disturbance. In this work, we first investigated the properties of the Moore-Spiegel thermo-mechanical chaotic system (1966). The Moore-Spiegel system is a nonlinear thermo-mechanical chaotic oscillator that describes a fluid element oscillating vertically in a temperature gradient with a linear restoring force. The Moore-Spiegel system is a classical example of a 3-D chaotic system with a cubic nonlinearity. We showed that the Moore-Spiegel system has a unique, unstable, equilibrium at the origin. Next, we applied multivariable super-twisting sliding mode control to globally stabilize all the trajectories of the Moore-Spiegel chaotic system. We also applied multivariable super-twisting sliding mode control for the global chaos synchronization of the identical Moore-Spiegel chaotic systems. Super-twisting sliding mode control is very useful for the global stabilization and synchronization of the Moore-Spiegel chaotic system as it achieves finite time stability for the system. Numerical simulations using MATLAB are shown to illustrate all the main results derived in this work.

References

1. Akgul A, Hussain S, Pehlivan I (2016) A new three-dimensional chaotic system, its dynamical analysis and electronic circuit applications. Optik 127(18):7062–7071
2. Akgul A, Moroz I, Pehlivan I, Vaidyanathan S (2016) A new four-scroll chaotic attractor and its engineering applications. Optik 127(13):5491–5499
3. Akgul A, Moroz I, Pehlivan I, Vaidyanathan S (2016) A new four-scroll chaotic attractor and its engineering applications. Optik 127(13):5491–5499
4. Azar AT, Vaidyanathan S (2015) Chaos modeling and control systems design. Springer, Berlin
5. Azar AT, Vaidyanathan S (2016) Advances in chaos theory and intelligent control. Springer, Berlin
6. Azar AT, Vaidyanathan S (2017) Fractional order control and synchronization of chaotic systems. Springer, Berlin
7. Balmforth NJ, Craster RV (1997) Synchronizing Moore and Spiegel. Astrophys J 7(4):738–752
8. Boulkroune A, Bouzeriba A, Bouden T (2016) Fuzzy generalized projective synchronization of incommensurate fractional-order chaotic systems. Neurocomputing 173:606–614
9. Burov DA, Evstigneev NM, Magnitskii NA (2017) On the chaotic dynamics in two coupled partial differential equations for evolution of surface plasmon polaritons. Commun Nonlinear Sci Numer Simul 46:26–36
10. Chai X, Chen Y, Broyde L (2017) A novel chaos-based image encryption algorithm using DNA sequence operations. Opt Lasers Eng 88:197–213
11. Chenaghlu MA, Jamali S, Khasmakhi NN (2016) A novel keyed parallel hashing scheme based on a new chaotic system. Chaos, Solitons Fractals 87:216–225

12. Fallahi K, Leung H (2010) A chaos secure communication scheme based on multiplication modulation. Commun Nonlinear Sci Numer Simul 15(2):368–383
13. Fontes RT, Eisencraft M (2016) A digital bandlimited chaos-based communication system. Commun Nonlinear Sci Numer Simul 37:374–385
14. Fotsa RT, Woafo P (2016) Chaos in a new bistable rotating electromechanical system. Chaos, Solitons Fractals 93:48–57
15. Fujisaka H, Yamada T (1983) Stability theory of synchronized motion in coupled-oscillator systems. Prog Theoret Phys 63:32–47
16. Kacar S (2016) Analog circuit and microcontroller based RNG application of a new easy realizable 4D chaotic system. Optik 127(20):9551–9561
17. Lakhekar GV, Waghmare LM, Vaidyanathan S (2016) Diving autopilot design for underwater vehicles using an adaptive neuro-fuzzy sliding mode controller. In: Vaidyanathan S, Volos C (eds) Advances and applications in nonlinear control systems. Springer, Berlin, pp 477–503
18. Li D (2008) A three-scroll chaotic attractor. Phys Lett A 372(4):387–393
19. Liu H, Ren B, Zhao Q, Li N (2016) Characterizing the optical chaos in a special type of small networks of semiconductor lasers using permutation entropy. Opt Commun 359:79–84
20. Liu W, Sun K, Zhu C (2016) A fast image encryption algorithm based on chaotic map. Opt Lasers Eng 84:26–36
21. Liu X, Mei W, Du H (2016) Simultaneous image compression, fusion and encryption algorithm based on compressive sensing and chaos. Opt Commun 366:22–32
22. Moore D, Spiegel E (1966) A thermally excited nonlinear oscillator. Astrophys J 143:871–887
23. Moussaoui S, Boulkroune A, Vaidyanathan S (2016) Fuzzy adaptive sliding-mode control scheme for uncertain underactuated systems. In: Vaidyanathan S, Volos C (eds) Advances and applications in nonlinear control systems. Springer, Berlin, pp 351–367
24. Nagesh I, Edwards C (2014) A multivariable super-twisting sliding mode approach. Automatica 50:984–988
25. Pecora LM, Carroll TL (1990) Synchronization in chaotic systems. Phys Rev Lett 64:821–824
26. Pecora LM, Carroll TL (1991) Synchronizing chaotic circuits. IEEE Trans Circ Syst 38:453–456
27. Pehlivan I, Moroz IM, Vaidyanathan S (2014) Analysis, synchronization and circuit design of a novel butterfly attractor. J Sound Vibr 333(20):5077–5096
28. Pham VT, Volos C, Jafari S, Wang X, Vaidyanathan S (2014) Hidden hyperchaotic attractor in a novel simple memristive neural network. Optoelectron Adv Mater Rapid Commun 8(11–12):1157–1163
29. Pham VT, Volos CK, Vaidyanathan S (2015) Multi-scroll chaotic oscillator based on a first-order delay differential equation. In: Azar AT, Vaidyanathan S (eds) Chaos modeling and control systems design, studies in computational intelligence, vol 581. Springer, Germany, pp 59–72
30. Pham VT, Volos CK, Vaidyanathan S, Le TP, Vu VY (2015) A memristor-based hyperchaotic system with hidden attractors: dynamics, synchronization and circuital emulating. J Eng Sci Technol Rev 8(2):205–214
31. Pham VT, Jafari S, Vaidyanathan S, Volos C, Wang X (2016) A novel memristive neural network with hidden attractors and its circuitry implementation. Sci China Technol Sci 59(3):358–363
32. Pham VT, Jafari S, Volos C, Giakoumis A, Vaidyanathan S, Kapitaniak T (2016) A chaotic system with equilibria located on the rounded square loop and its circuit implementation. IEEE Trans Circ Syst II Express Briefs 63(9):878–882
33. Pham VT, Jafari S, Volos C, Vaidyanathan S, Kapitaniak T (2016) A chaotic system with infinite equilibria located on a piecewise linear curve. Optik 127(20):9111–9117
34. Pham VT, Vaidyanathan S, Volos CK, Hoang TM, Yem VV (2016) Dynamics, synchronization and SPICE implementation of a memristive system with hidden hyperchaotic attractor. In: Azar AT, Vaidyanathan S (eds) Advances in chaos theory and intelligent control. Springer, Berlin, pp 35–52

35. Pham VT, Vaidyanathan S, Volos CK, Jafari S, Kuznetsov NV, Hoang TM (2016) A novel memristive time-delay chaotic system without equilibrium points. Eur Phys J Spec Top 225(1):127–136
36. Pham VT, Vaidyanathan S, Volos CK, Jafari S, Kuznetsov NV, Hoang TM (2016) A novel memristive time-delay chaotic system without equilibrium points. Eur Phys J Spec Top 225(1):127–136
37. Pham VT, Vaidyanathan S, Volos CK, Jafari S, Wang X (2016) A chaotic hyperjerk system based on memristive device. In: Vaidyanathan S, Volos C (eds) Advances and applications in chaotic systems. Springer, Berlin, pp 39–58
38. Sampath S, Vaidyanathan S, Volos CK, Pham VT (2015) An eight-term novel four-scroll chaotic system with cubic nonlinearity and its circuit simulation. J Eng Sci Technol Rev 8(2):1–6
39. Shirkhani N, Khanesar M, Teshnehlab M (2016) Indirect model reference fuzzy control of SISO fractional order nonlinear chaotic systems. Proc Comput Sci 102:309–316
40. Slotine J, Li W (1991) Applied nonlinear control. Prentice-Hall, Englewood Cliffs
41. Sundarapandian V (2013) Analysis and anti-synchronization of a novel chaotic system via active and adaptive controllers. J Eng Sci Technol Rev 6(4):45–52
42. Sundarapandian V, Pehlivan I (2012) Analysis, control, synchronization, and circuit design of a novel chaotic system. Math Comput Model 55(7–8):1904–1915
43. Sundarapandian V, Sivaperumal S (2011) Sliding controller design of hybrid synchronization of four-wing Chaotic systems. Int J Soft Comput 6(5):224–231
44. Szmit Z, Warminski J (2016) Nonlinear dynamics of electro-mechanical system composed of two pendulums and rotating hub. Proc Eng 144:953–958
45. Tacha OI, Volos CK, Kyprianidis IM, Stouboulos IN, Vaidyanathan S, Pham VT (2016) Analysis, adaptive control and circuit simulation of a novel nonlinear finance system. Appl Math Comput 276:200–217
46. Utkin VI (1977) Variable structure systems with sliding modes. IEEE Trans Autom Control 22(2):212–222
47. Utkin VI (1993) Sliding mode control design principles and applications to electric drives. IEEE Trans Ind Electron 40(1):23–36
48. Vaidyanathan S (2011) Analysis and synchronization of the hyperchaotic Yujun systems via sliding mode control. Adv Intell Syst Comput 176:329–337
49. Vaidyanathan S (2012) Global chaos control of hyperchaotic Liu system via sliding control method. Int J Control Theory Appl 5(2):117–123
50. Vaidyanathan S (2012) Sliding mode control based global chaos control of Liu-Liu-Liu-Su chaotic system. Int J Control Theory Appl 5(1):15–20
51. Vaidyanathan S (2013) A new six-term 3-D chaotic system with an exponential nonlinearity. Far East J Math Sci 79(1):135–143
52. Vaidyanathan S (2013) Analysis and adaptive synchronization of two novel chaotic systems with hyperbolic sinusoidal and cosinusoidal nonlinearity and unknown parameters. J Eng Sci Technol Rev 6(4):53–65
53. Vaidyanathan S (2014) A new eight-term 3-D polynomial chaotic system with three quadratic nonlinearities. Far East J Math Sci 84(2):219–226
54. Vaidyanathan S (2014) Analysis and adaptive synchronization of eight-term 3-D polynomial chaotic systems with three quadratic nonlinearities. Eur Phys J Spec Top 223(8):1519–1529
55. Vaidyanathan S (2014) Analysis, control and synchronisation of a six-term novel chaotic system with three quadratic nonlinearities. Int J Model Ident Control 22(1):41–53
56. Vaidyanathan S (2014) Generalised projective synchronisation of novel 3-D chaotic systems with an exponential non-linearity via active and adaptive control. Int J Model Ident Control 22(3):207–217
57. Vaidyanathan S (2014) Global chaos synchronisation of identical Li-Wu chaotic systems via sliding mode control. Int J Model Ident Control 22(2):170–177
58. Vaidyanathan S (2015) A 3-D novel highly chaotic system with four quadratic nonlinearities, its adaptive control and anti-synchronization with unknown parameters. J Eng Sci Technol Rev 8(2):106–115

59. Vaidyanathan S (2015) A novel chemical chaotic reactor system and its adaptive control. Int J ChemTech Res 8(7):146–158
60. Vaidyanathan S (2015) A novel chemical chaotic reactor system and its output regulation via integral sliding mode control. Int J ChemTech Res 8(11):669–683
61. Vaidyanathan S (2015) Active control design for the anti-synchronization of Lotka-Volterra biological systems with four competitive species. Int J PharmTech Res 8(7):58–70
62. Vaidyanathan S (2015) Adaptive control design for the anti-synchronization of novel 3-D chemical chaotic reactor systems. Int J ChemTech Res 8(11):654–668
63. Vaidyanathan S (2015) Adaptive control of a chemical chaotic reactor. Int J PharmTech Res 8(3):377–382
64. Vaidyanathan S (2015) Adaptive control of the FitzHugh-Nagumo chaotic neuron model. Int J PharmTech Res 8(6):117–127
65. Vaidyanathan S (2015) Adaptive synchronization of chemical chaotic reactors. Int J ChemTech Res 8(2):612–621
66. Vaidyanathan S (2015) Adaptive synchronization of generalized Lotka-Volterra three-species biological systems. Int J PharmTech Res 8(5):928–937
67. Vaidyanathan S (2015) Adaptive synchronization of novel 3-D chemical chaotic reactor systems. Int J ChemTech Res 8(7):159–171
68. Vaidyanathan S (2015) Analysis, properties and control of an eight-term 3-D chaotic system with an exponential nonlinearity. Int J Model Ident Control 23(2):164–172
69. Vaidyanathan S (2015) Anti-synchronization of brusselator chemical reaction systems via adaptive control. Int J ChemTech Res 8(6):759–768
70. Vaidyanathan S (2015) Anti-synchronization of brusselator chemical reaction systems via integral sliding mode control. Int J ChemTech Res 8(11):700–713
71. Vaidyanathan S (2015) Anti-synchronization of chemical chaotic reactors via adaptive control method. Int J ChemTech Res 8(8):73–85
72. Vaidyanathan S (2015) Anti-synchronization of the FitzHugh-Nagumo chaotic neuron models via adaptive control method. Int J PharmTech Res 8(7):71–83
73. Vaidyanathan S (2015) Chaos in neurons and synchronization of Birkhoff-Shaw strange chaotic attractors via adaptive control. Int J PharmTech Res 8(6):1–11
74. Vaidyanathan S (2015) Dynamics and control of brusselator chemical reaction. Int J ChemTech Res 8(6):740–749
75. Vaidyanathan S (2015) Dynamics and control of Tokamak system with symmetric and magnetically confined plasma. Int J ChemTech Res 8(6):795–802
76. Vaidyanathan S (2015) Global chaos synchronization of chemical chaotic reactors via novel sliding mode control method. Int J ChemTech Res 8(7):209–221
77. Vaidyanathan S (2015) Global chaos synchronization of Duffing double-well chaotic oscillators via integral sliding mode control. Int J ChemTech Res 8(11):141–151
78. Vaidyanathan S (2015) Global chaos synchronization of the forced Van der Pol chaotic oscillators via adaptive control method. Int J PharmTech Res 8(6):156–166
79. Vaidyanathan S (2015) Hybrid chaos synchronization of the FitzHugh-Nagumo chaotic neuron models via adaptive control method. Int J PharmTech Res 8(8):48–60
80. Vaidyanathan S (2015) Integral sliding mode control design for the global chaos synchronization of identical novel chemical chaotic reactor systems. Int J ChemTech Res 8(11):684–699
81. Vaidyanathan S (2015) Output regulation of the forced Van der Pol chaotic oscillator via adaptive control method. Int J PharmTech Res 8(6):106–116
82. Vaidyanathan S (2015) Sliding controller design for the global chaos synchronization of forced Van der Pol chaotic oscillators. Int J PharmTech Res 8(7):100–111
83. Vaidyanathan S (2015) Synchronization of Tokamak systems with symmetric and magnetically confined plasma via adaptive control. Int J ChemTech Res 8(6):818–827
84. Vaidyanathan S (2016) A novel 3-D conservative chaotic system with a sinusoidal nonlinearity and its adaptive control. Int J Control Theory Appl 9(1):115–132
85. Vaidyanathan S (2016) A novel 3-D jerk chaotic system with three quadratic nonlinearities and its adaptive control. Arch Control Sci 26(1):19–47

86. Vaidyanathan S (2016) A novel 3-D jerk chaotic system with two quadratic nonlinearities and its adaptive backstepping control. Int J Control Theory Appl 9(1):199–219
87. Vaidyanathan S (2016) Anti-synchronization of 3-cells cellular neural network attractors via integral sliding mode control. Int J PharmTech Res 9(1):193–205
88. Vaidyanathan S (2016) Global chaos control of the FitzHugh-Nagumo chaotic neuron model via integral sliding mode control. Int J PharmTech Res 9(4):413–425
89. Vaidyanathan S (2016) Global chaos control of the generalized Lotka-Volterra three-species system via integral sliding mode control. Int J PharmTech Res 9(4):399–412
90. Vaidyanathan S (2016) Global chaos regulation of a symmetric nonlinear gyro system via integral sliding mode control. Int J ChemTech Res 9(5):462–469
91. Vaidyanathan S (2016) Hybrid synchronization of the generalized Lotka-Volterra three-species biological systems via adaptive control. Int J PharmTech Res 9(1):179–192
92. Vaidyanathan S (2016) Mathematical analysis, adaptive Control and synchronization of a ten-term novel three-scroll chaotic system with four quadratic nonlinearities. Int J Control Theory Appl 9(1):1–20
93. Vaidyanathan S, Azar AT (2015) Analysis, control and synchronization of a nine-term 3-D novel chaotic system. In: Azar AT, Vaidyanathan S (eds) Chaos modelling and control systems design, studies in computational intelligence, vol 581. Springer, Germany, pp 19–38
94. Vaidyanathan S, Azar AT (2016) Takagi-Sugeno fuzzy logic controller for Liu-Chen four-scroll chaotic system. Int J Intell Eng Inform 4(2):135–150
95. Vaidyanathan S, Boulkroune A (2016) A novel 4-D hyperchaotic chemical reactor system and its adaptive control. In: Vaidyanathan S, Volos C (eds) Advances and applications in chaotic systems. Springer, Berlin, pp 447–469
96. Vaidyanathan S, Madhavan K (2013) Analysis, adaptive control and synchronization of a seven-term novel 3-D chaotic system. Int J Control Theory Appl 6(2):121–137
97. Vaidyanathan S, Pakiriswamy S (2016) A five-term 3-D novel conservative chaotic system and its generalized projective synchronization via adaptive control method. Int J Control Theory Appl 9(1):61–78
98. Vaidyanathan S, Pakiriswamy S (2015) A 3-D novel conservative chaotic system and its generalized projective synchronization via adaptive control. J Eng Sci Technol Rev 8(2):52–60
99. Vaidyanathan S, Pakiriswamy S (2016) Adaptive control and synchronization design of a seven-term novel chaotic system with a quartic nonlinearity. Int J Control Theory Appl 9(1):237–256
100. Vaidyanathan S, Rajagopal K (2016) Adaptive control, synchronization and LabVIEW implementation of Rucklidge chaotic system for nonlinear double convection. Int J Control Theory Appl 9(1):175–197
101. Vaidyanathan S, Rajagopal K (2016) Analysis, control, synchronization and LabVIEW implementation of a seven-term novel chaotic system. Int J Control Theory Appl 9(1):151–174
102. Vaidyanathan S, Sampath S (2011) Global chaos synchronization of hyperchaotic Lorenz systems by sliding mode control. Commun Comput Inf Sci 205:156–164
103. Vaidyanathan S, Volos C (2015) Analysis and adaptive control of a novel 3-D conservative no-equilibrium chaotic system. Arch Control Sci 25(3):333–353
104. Vaidyanathan S, Volos C (2016) Advances and applications in chaotic systems. Springer, Berlin
105. Vaidyanathan S, Volos C (2016) Advances and applications in nonlinear control systems. Springer, Berlin
106. Vaidyanathan S, Volos C (2017) Advances in memristors., Memristive devices and systemsSpringer, Berlin
107. Vaidyanathan S, Volos C, Pham VT, Madhavan K, Idowu BA (2014) Adaptive backstepping control, synchronization and circuit simulation of a 3-D novel jerk chaotic system with two hyperbolic sinusoidal nonlinearities. Arch Control Sci 24(3):375–403
108. Vaidyanathan S, Rajagopal K, Volos CK, Kyprianidis IM, Stouboulos IN (2015) Analysis, adaptive control and synchronization of a seven-term novel 3-D chaotic system with three quadratic nonlinearities and its digital implementation in LabVIEW. J Eng Sci Technol Rev 8(2):130–141

109. Vaidyanathan S, Volos CK, Kyprianidis IM, Stouboulos IN, Pham VT (2015) Analysis, adaptive control and anti-synchronization of a six-term novel jerk chaotic system with two exponential nonlinearities and its circuit simulation. J Eng Sci Technol Rev 8(2):24–36

110. Vaidyanathan S, Volos CK, Pham VT (2015) Analysis, adaptive control and adaptive synchronization of a nine-term novel 3-D chaotic system with four quadratic nonlinearities and its circuit simulation. J Eng Sci Technol Rev 8(2):181–191

111. Vaidyanathan S, Volos CK, Pham VT (2015) Global chaos control of a novel nine-term chaotic system via sliding mode control. In: Azar AT, Zhu Q (eds) Advances and applications in sliding mode control systems, studies in computational intelligence, vol 576. Springer, Germany, pp 571–590

112. Volos CK, Kyprianidis IM, Stouboulos IN, Tlelo-Cuautle E, Vaidyanathan S (2015) Memristor: a new concept in synchronization of coupled neuromorphic circuits. J Eng Sci Technol Rev 8(2):157–173

113. Volos CK, Pham VT, Vaidyanathan S, Kyprianidis IM, Stouboulos IN (2015) Synchronization phenomena in coupled Colpitts circuits. J Eng Sci Technol Rev 8(2):142–151

114. Volos CK, Pham VT, Vaidyanathan S, Kyprianidis IM, Stouboulos IN (2016) Synchronization phenomena in coupled hyperchaotic oscillators with hidden attractors using a nonlinear open loop controller. In: Vaidyanathan S, Volos C (eds) Advances and applications in chaotic systems. Springer, Berlin, pp 1–38

115. Volos CK, Pham VT, Vaidyanathan S, Kyprianidis IM, Stouboulos IN (2016) The case of bidirectionally coupled nonlinear circuits via a memristor. In: Vaidyanathan S, Volos C (eds) Advances and applications in nonlinear control systems. Springer, Berlin, pp 317–350

116. Volos CK, Prousalis D, Kyprianidis IM, Stouboulos I, Vaidyanathan S, Pham VT (2016) Synchronization and anti-synchronization of coupled Hindmarsh-Rose neuron models. Int J Control Theory Appl 9(1):101–114

117. Volos CK, Vaidyanathan S, Pham VT, Maaita JO, Giakoumis A, Kyprianidis IM, Stouboulos IN (2016) A novel design approach of a nonlinear resistor based on a memristor emulator. In: Azar AT, Vaidyanathan S (eds) Advances in chaos theory and intelligent control. Springer, Berlin, pp 3–34

118. Wang B, Zhong SM, Dong XC (2016) On the novel chaotic secure communication scheme design. Commun Nonlinear Sci Numer Simul 39:108–117

119. Wu T, Sun W, Zhang X, Zhang S (2016) Concealment of time delay signature of chaotic output in a slave semiconductor laser with chaos laser injection. Opt Commun 381:174–179

120. Xu G, Liu F, Xiu C, Sun L, Liu C (2016) Optimization of hysteretic chaotic neural network based on fuzzy sliding mode control. Neurocomputing 189:72–79

121. Xu H, Tong X, Meng X (2016) An efficient chaos pseudo-random number generator applied to video encryption. Optik 127(20):9305–9319

122. Zhou W, Xu Y, Lu H, Pan L (2008) On dynamics analysis of a new chaotic attractor. Phys Lett A 372(36):5773–5777

123. Zhu C, Liu Y, Guo Y (2010) Theoretic and numerical study of a new chaotic system. Intell Inf Manage 2:104–109

Printed in the United States
By Bookmasters